AF602134

ADVANCES IN
MOLECULAR AND
CELL BIOLOGY

Volume 10 • 1994

MOLECULAR PROCESSES
OF PHOTOSYNTHESIS

ADVANCES IN MOLECULAR AND CELL BIOLOGY

MOLECULAR PROCESSES OF PHOTOSYNTHESIS

Series Editor: E. EDWARD BITTAR
Department of Physiology
University of Wisconsin
Madison, Wisconsin

Guest Editor: J. BARBER
Department of Biochemistry
Imperial College of Science, Technology, and Medicine
London, England

VOLUME 10 • 1994

Greenwich, Connecticut *London, England*

55 Old Post Road, No. 2
Greenwich, Connecticut 06836

JAI PRESS LTD.
The Courtyard
28 High Street
Hampton Hill, Middlesex TW12 1PD
England

ISBN: 1-55938-710-6

Transferred to digital printing 2005

CONTENTS

LIST OF CONTRIBUTORS

S. Anandan	Department of Biology University of California Los Angeles, California
B. Andersson	Department of Biochemistry Arrhenius Laboratories for Natural Sciences Stockholm University Stockholm, Sweden
J. Barber	Department of Biochemistry Wolfson Laboratories Imperial College of Science, Technology, and Medicine London, England
B. D. Bruce	Department of Biochemistry and The Center for Legume Research Walters Life Sciences Building University of Tennessee Knoxville, Tennessee
R. B. Buchanan	Department of Plant Biology University of California Berkeley, California
P. Chitnis	Division of Biology Kansas State University Manhattan, Kansas
W. S. Chow	CSIRO Division of Plant Industry & Cooperative Research Centre for Plant Science Canberra, Australia
R. J. Cogdell	Department of Botany The University of Glasgow Glasgow, Scotland

B. W. Dreyfuss — Department of Biology, University of California, Los Angeles, California

A. A. Gatenby — Central Research and Development, Experimental Station, E. I. du Pont de Nemours & Co., Wilmington, Delaware

A. N. Glazer — Department of Molecular and Cell Biology, University of California, Berkeley, California

S. Gomez — Department of Biology, University of California, Los Angeles, California

R. Grimm — Analytische Meßtechnik, Hewlett-Packard GmbH, Waldbronn, Germany

S. Gutteridge — Central Research and Development, Experimental Station, E. I. du Pont de Nemours & Co., Wilmington, Delaware

J. B. Jackson — School of Biochemistry, University of Birmingham, Birmingham, England

K. Keegstra — MSU-DOE, Plant Research Laboratory, Michigan State University, East Lansing, Michigan

C. Kerfeld — Department of Biology, University of California, Los Angeles, California

A. Lee — Department of Biology, University of California, Los Angeles, California

T. Lundqvist — Department of Molecular Biology, Biomedical Centrum, Swedish University of Agriculture, Uppsala, Sweden

A. G. McEwan — School of Biological Sciences, University of East Anglia, Norwich, England

D. T. Morishige — Department of Biology, University of California, Los Angeles, California

G. F. Peter — Department of Biology, University of California, Los Angeles, California

S. Preiss — Department of Biology, University of California, Los Angeles, California

V. Speth — Zellbiologie, Institut für Biologie, Freiburg, Germany

E. Takahashi — Department of Plant Biology, University of Illinois at Urbana-Champaign, Urbana, Illinois

T. Takeuchi — Department of Biology, University of California, Los Angeles, California

J. P. Thornber — Department of Biology, University of California, Los Angeles, California

P. V. Viitanen — Central Research and Development, Experimental Station, E. I. du Pont de Nemours & Co., Wilmington, Delaware

C. A. Wraight — Department of Plant Biology, University of Illinois at Urbana-Champaign, Urbana, Illinois

PREFACE

It is generally agreed that photosynthetic organisms first appeared on our planet three and a half billion years ago. If they had not done so, life would not have blossomed and diversified in the way it has and would be restricted to anaerobic chemotrophic bacteria. Instead, we live on a planet teeming with a vast and almost incomprehensible variety of microbial, plant, and animal life. More amazing is that this achievement involves the combination of the most common of all compounds: water and carbon dioxide. Sunlight is used to split water into its elemental constituents; oxygen is lost to the atmosphere as a by-product while the hydrogen is used as a reducing agent for the conversion of carbon dioxide to organic molecules. These organic molecules are the building blocks for living cells and a stored energy source for nonphotosynthetic as well as photosynthetic organisms that oxidize them by the process of respiration. Today, photosynthesis occurs on a vast scale, so much so that just in a few hours the amount of organic material produced globally is equal to the total weight of the human population of the world. In energy terms, for a given time, photosynthesis captures more than 100 times the total metabolic energy requirement of mankind (i.e., as food) and was, of course, the origin of the fossil fuels on which we are so dependent. The enormity of the process means that the carbon dioxide in the atmosphere is recycled every 300 years and oxygen every 2000 years. The molecular machines that make this possible are embedded in membrane systems of higher plants, algae, and photosynthetic bacteria, and together make up by far the greatest power source on Earth.

Research into the basic mechanisms of photosynthesis has a long and distinguished history and has consistently been at the forefront of science. The success of this research, particularly in recent years, suggests that photosynthesis may turn out to be the first complex biological system to have its structure, function, and regulation described in rigorous chemical terms at the atomic level. It is likely that such knowledge will help us to tackle perhaps the most vital problem facing mankind, namely our need for a continuous and nonpolluting source of energy. The benefit may come by providing a "blueprint" for new technologies able to carry out efficient conversion of solar energy based on the principles of biological systems, and/or creating highly efficient "energy crops" sufficiently hardy to grow in a wide range of environments. The former is likely to involve new developments in material sciences while the latter will call on the rapidly advancing techniques of genetic engineering.

The contents of this volume review some of the most important developments which are a part of the drive towards the overall goal of obtaining the complete description of the photosynthetic processes at the molecular level. The topics covered have been carefully selected and represent the wide spectrum of the subject. In producing this book I was fortunate to have the positive input of leading scientists working at the cutting edge of photosynthesis research. As a consequence the book is unique and of a very high standard. It is therefore a very valuable addition to the series *Advances in Molecular and Cell Biology* and is suitable, not only for specialists, but for all those interested in molecular cell biology in general.

The first chapter by Bertil Andersson and myself presents an overview of the structure, functional properties, and dynamics of the photosynthetic membrane of oxygenic organisms. Not only does it give an up-to-date account of knowledge in the specific areas covered but also helps to set the scene for the rest of the book. Chapters 2 and 3 by Phillip Thornber and colleagues and by Alex Glazer, respectively, give excellent accounts of the light-harvesting antenna systems. These systems service the photochemical reaction centers that are discussed briefly in Chapter 1, but in great detail in Chapter 5 by Eiji Takahashi and Colin Wraight. Their chapter focuses specificially on the reaction center of purple photosynthetic bacteria of which a great deal is known. But their discussions have implications for reaction centers from other organisms, particularly for photosystem II of higher plants, algae, and cyanobacteria. This relationship with photosystem II is dealt with in Chapter 1 where it is also pointed out that, unlike the bacterial systems, photosystem II is highly vulnerable to photoinduced damage. This damage seems to be an intrinsic and unavoidable penalty of photosystem II's ability to bring about the splitting of water with the consequential production of reactive and dangerous oxidizing species and toxic forms of oxygen. This vulnerability of the oxygenic system to damage manifests itself as the physiological stress phenomenon known as photoinhibition, a subject that is elegantly covered by Fred Chow in Chapter 4.

The remaining five chapters deal with molecular aspects of photosynthesis which do not directly involve light-induced reactions but rather deal with specific aspects of the "dark" processes. Baz Jackson and Alistair McEwan have produced a thoughtful review of the properties of membrane complexes having dehydrogenase activities and which are very important components of the metabolism of photosynthetic bacteria.

I was particularly pleased that Steve Gutteridge accepted my invitation to write a chapter on ribulose 1,5-bisphosphate (Rubisco). His joint effort with Tomas Lundqvist gives a most up-to-date review on the structure and mechanism of this key protein of the carbon fixation cycle. The activity of Rubisco and other enzymes involved in the dark reactions are regulated in such a way as to control various pathways and optimize the overall efficiency of photosynthesis and plant growth. A central control mechanism involves the ferredoxin/thioredoxin system. Who better than Bob Buchanan could write a state-of-the-art review on this important topic? He has been the main contributor to this area which is evidenced by the clarity and thoroughness of Chapter 8. Understanding how the chloroplast imports proteins synthesized in the cytoplasm and combines them with others that are encoded by plastid DNA are the subjects of Chapters 9 and 10. Tony Gatenby and colleagues review the subject area of molecular chaperones. The discovery of *chaperonins* is relatively recent and has created considerable interest to those trying to elucidate the rules that govern the folding of polypeptides to form proteins having specific functional properties. The intense interest in this subject has stemmed from work carried out with chloroplasts which use, for example, a chaperonin to assemble Rubisco. Chapter 10 complements rather nicely the chapter by Gatenby and colleagues. It deals with the topic of targeting and translocation of proteins into and within the chloroplast. The general principles presented also apply to other non-photosynthetic organelles, such as the mitochondrion.

I am most indebted to all the contributors of this book. Their chapters are excellent and they have provided, in one volume, a unique set of reviews that demonstrate the importance and level of achievement which has been attained in photosynthesis research. I must confess, however, that in the limited space available I could not include all aspects of this highly active area of research. It is a large and multidisciplinary subject that has many flavors. Nevertheless, I believe the variety of subjects covered here will give the nonspecialist not only a taste of the achievements of research into photosynthesis, but also basic observations and concepts that are applicable to other areas of molecular cell biology.

Inevitably, the success of this venture required the help of others and I am particuarly grateful to Roisin Smyth and Lyn Barber for keeping me organized.

Jim Barber
Guest Editor

COMPOSITION, ORGANIZATION, AND DYNAMICS OF THYLAKOID MEMBRANES

B. Andersson and J. Barber

Advances in Molecular and Cell Biology
Volume 10, pages 1–53.

ISBN: 1-55938-710-6

I. INTRODUCTION

The first major step in the process of photosynthesis is the transfer of reducing equivalents to NAD^+ and $NADP^+$, together with the conversion of ADP to ATP. In the case of anoxygenic photosynthetic bacteria, the reducing equivalents are supplied by oxidizing a range of compounds, including H_2S and organic acids. Oxygenic organisms, however, use H_2O for their supply of electrons and protons. In all cases, the electron/proton transfer reactions involve redox-active species which are either embedded in a membrane system or closely associated with it. In contrast, the enzymes involved in the carbon fixation process are found in the aqueous phase, whether it be the stroma of chloroplasts or the cytoplasm of prokaryotic organisms. A concept that has emerged during the past few years is that despite the morphological, physiological, and ecological differences between various types of photosynthetic organisms, at the molecular level there are striking similarities. Such similarities are particularly evident for the redox reactions where identification and sequencing of genes have allowed comparisons to be made of primary protein structures. Despite this, considerable variation on the general theme is usually encountered because photosynthetic organisms have to adapt or respond to changes in their environment. This means that they must possess a dynamic approach to their metabolism in order to survive. The ability to adapt to environmental factors becomes crucial for those organisms that are not mobile, such as higher plants, which do not have the capacity of minimizing the impact of an environmental change by seeking out more amenable surroundings.

In this chapter we review the composition, structure, and dynamics of the thylakoid membrane and outline mechanisms that exist which allow this part of the photosynthetic apparatus to cope with changing environmental conditions. Most of our discussion will refer to the thylakoid membrane of higher plants, but we will not totally ignore the other major classes of oxygenic photosynthetic organisms, notably the prokaryotic cyanobacteria and eukaryotic algae.

II. THYLAKOID MEMBRANE STRUCTURE AND FUNCTION

A. Structure

The structure of the thylakoid membrane of oxygenic organisms has been extensively studied using the techniques of electron microscopy. In transmission

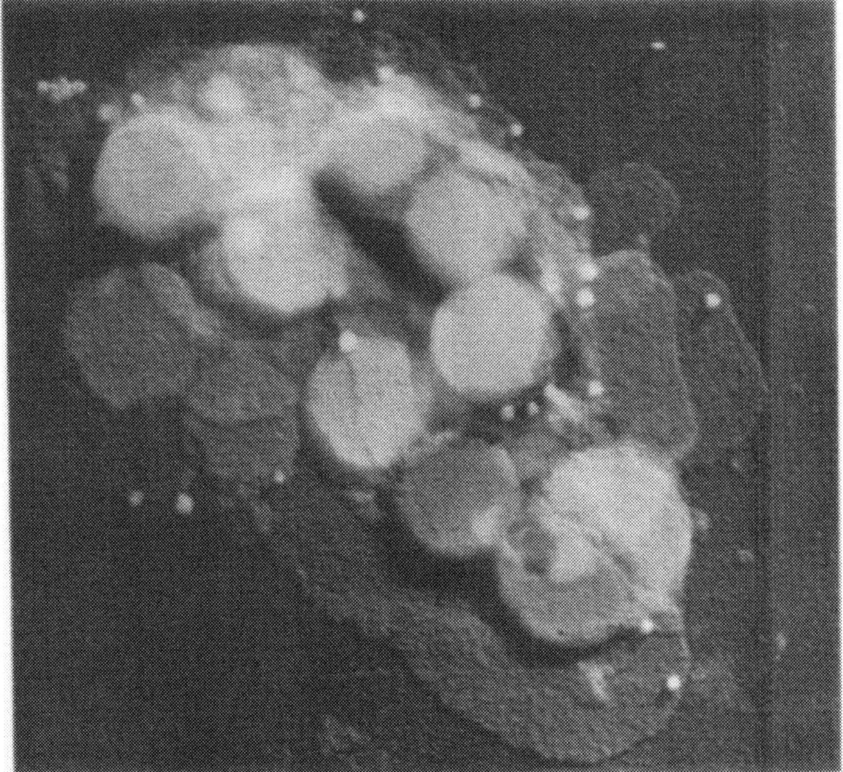

Figure 1. Electron micrograph of thylakoid membranes from a preparation of chloroplasts of *Vicia faba* isolated in 0.3 M glucose showing granal discs and stroma lamellae in face view. Dried and washed preparation shadowed with Au/Pd; × 36,000 (picture courtesy of A.D. Greenwood).

mode, the thylakoid membranes of higher plant chloroplasts are revealed as complex folded lamellae systems composed of stacked (granal) and nonstacked (stromal) regions. There are many detailed reviews that have discussed the organization of thylakoids, ranging from the early detailed book by Kirk and Tilney-Bassett (1967) to the more recent review by Staehelin (1986). The typical cross section of a transmission micrograph of higher plant chloroplasts is now well documented in the most elementary text books of plant physiology and biochemistry, but in Figure 1 we present a more unusual view of the thylakoids of *Vicia faba*. This electron micrograph indicates the disc-like nature of the granal stacks and also gives a view of the face of the stromal lamellae.

A picture that is often presented, based on electron microscopy studies using tangential and serial sectioning, shows the stromal and granal lamellae connected by a tilted network of spiral frets that follow a right-to-left helix around the surface of cylindrical grana (Paolillo, 1970; Brangeon and Mustardy, 1979). Such models, however, give a rather rigid view of a membrane system which we now know to be in a state of continuous flux during normal function. Moreover, it does not take into account variations in thylakoid organization that are found, either in a particular species grown under different conditions, or between different types of organisms. For example, eukaryotic algae often have a thylakoid membrane organization which is very different to that which is found in higher plants (Coombs and Greenwood, 1976). Nevertheless, a common feature of the thylakoid systems of algae and those of higher plants, despite their differences in organization, is that they both possess a membrane system with surfaces which are either in tight appression or exposed to the aqueous stromal phase of the chloroplast. This concept of exposed and nonexposed membrane surfaces will be dealt with later and may well be a key factor in regulation and adaptation to environmental change. However, it should be noted that this common feature of eukaryotic thylakoid membranes cannot be extrapolated to those of cyanobacteria. In the case of these oxygenic prokaryotes, the presence of phycobilisomes seems to restrict close membrane–membrane interaction, and thus the concept of differentiation into exposed and nonexposed surfaces can no longer be applied.

B. Redox Reactions

The primary role of the thylakoid membrane is to act as a matrix for the light-driven electron and proton transport systems which underly the reduction of $NADP^+$ and the conversion of ADP to ATP. Oxygenic photosynthesis involves the cooperation of two photosystems: photosystem I (PS I) and photosystem II (PS II). There is now much evidence to suggest that PS I evolved from a photosystem similar to that found in present-day photosynthetic green sulphur bacteria (Nitschke and Rutherford, 1991). Similarly, PS II has its evolutionary links with the purple photosynthetic bacteria (Barber, 1987) although it is likely that both types of reaction centers have a common origin (Nitschke and Rutherford, 1991).

When PS I absorbs a photon, the exciton is rapidly transferred via an antenna system to a special form of chlorophyll which acts as a trap (Figure 2). This trap is known as P700 and when excited readily gives up an electron to the primary electron acceptor A_o. The redox potential (E_m) of A_o^- is approximately −1.0 V while $P700^+$ is about +0.43 V. On the acceptor side, the electron is rapidly transferred through a series of secondary acceptors, denoted A_1 (E_m = −0.8 V), F_X (E_m = −0.7 V), F_B (E_m = −0.58 V), and F_A (E_m = −0.53 V), and finally to soluble ferredoxin (E_m = −0.42 V). Most evidence indicates that P700 is a dimer of chlorophyll *a* while A_o is probably a monomeric form of chlorophyll *a*. Extraction and reconstitution experiments have identified A_1 as a phylloquinone molecule while Mössbauer measurements have indicated that F_X is a [4Fe–4S] center. F_A and F_B are also likely to be iron–sulphur centers of the [4Fe–4S] type as deduced from EPR studies and

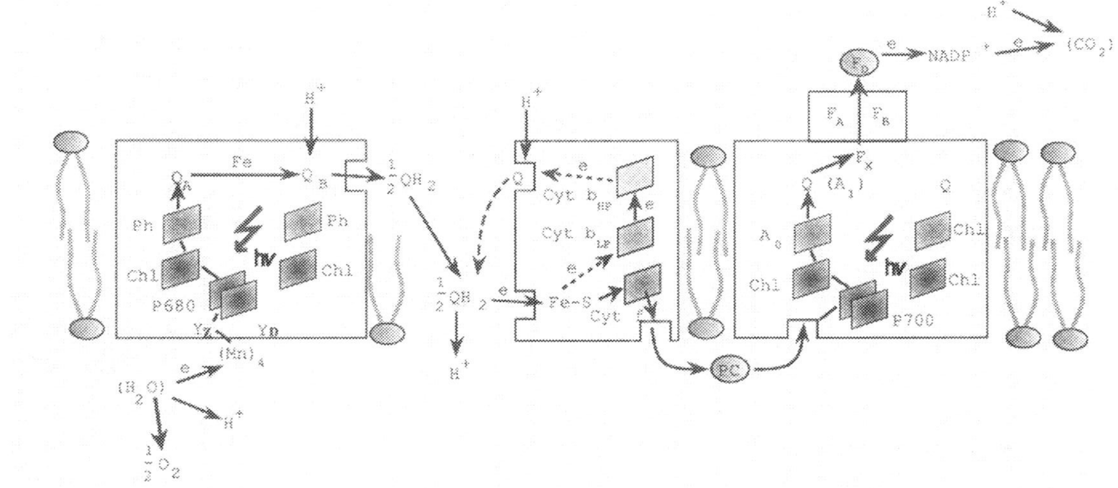

Figure 2. A schematic representation of electron and proton flow in oxygenic photosynthesis. No attempt is made to show the multipeptide nature of the complexes and the CF_o–CF_1 coupling factor that utilizes the electrochemical potential energy transthylakoid H^+ gradient has not been included. The solid arrows indicate noncyclic electron flow from H_2O to $NADP^+$ which results in the net transfer of H^+ across the membrane with a stoichiometry of 2H/e. The dotted pathway indicates the "Q-cycle" which, after initial priming, increases the H/e ratio to 3. Electron/proton transfer to CO_2 is symbolically shown but is accomplished by the reductive pentose phosphate (Calvin) cycle. The similar twofold symmetry representation of the cofactor organization in the photosystem I (PS I) and photosystem II (PS II) reaction centers is extrapolated from the general picture emerging from several approaches, but particularly from X-ray crystallography (Michel and Deisenhofer, 1986; Krauss et al., 1993). Chl = chlorophyll *a*; Ph = pheophytin; Q = quinone (plastoquinone in PS II and in membrane pool); phylloquinone in PS I (A_1). P700 and P680 are the primary electron donors of PS I and PS II, respectively; Fe is a nonheme iron of PS II; A_o is a chl which acts as a primary electron acceptor of PS I; F_X is a 4Fe–4S center bound to the PS I reaction center; F_A and F_B are also 4Fe–4S centers bound to a 9-kDa extrinsic protein (PS I-C protein); F_D is soluble ferredoxin containing a 2Fe–2S center; $(Mn)_4$ is a cluster of 4 Mn atoms involved in water splitting; Y_Z and Y_D are tyrosine residues at positions 161 on the D1 and D2 proteins, respectively; cyt b_{HP} and cyt b_{LP} are cytochromes *b*563 high potential (E_m = 0.05 V) and low potential (E_m = –0.15 V), respectively; cyt *f* = cytochrome *f*; PC = plastocyanin; e = electron.

more recently from X-ray crystallographic data (Krauss et al., 1993). This structural data is also consistent with F_X being a [4Fe–4S] center. A detailed review on the structure, function, and molecular biology of PS I has been written by Goldbeck and Bryant (1991).

The oxidized form of P700 is usually reduced by the copper-containing protein, plastocyanin ($E_m = +0.37$) although in some algae and cyanobacteria a soluble form of cytochrome *c* can act as an electron donor to $P700^+$. Thus, the overall light-driven reactions of PS I make it a plastocyanin–ferredoxin oxidoreductase. Reduced soluble ferredoxin is then used as an electron donor to $NADP^+$. This reaction is catalyzed by a flavoprotein having ferredoxin–$NADP^+$ oxidoreductase activity.

Photosystem two (PS II) also has a light harvesting antenna which transfers excitons to the photochemical trap (Figure 2). This trap is known as P680 which, like P700, also acts as a primary electron donor. However, in this case $P680^+$ is strongly oxidizing with a redox potential of around +1.1 V. The immediate electron acceptor is pheophytin ($E_m = -0.62$ V) which then passes its electron to a plastoquinone molecule known as Q_A ($E_m = -0.3$ V). The electron is then passed to the second plastoquinone acceptor Q_B ($E_m = -0.1$ V). Like $NADP^+$, Q_B is a two-electron acceptor and is not fully reduced until a second photochemical turnover occurs. In its fully reduced state, the Q_B quinone is protonated and leaves PS II as a mobile redox component able to diffuse in the lipid matrix of the membrane. On the donor side, $P680^+$ is rapidly reduced by a tyrosine residue at position 161 on the D1-protein of PS II. This oxidized tyrosine is then reduced by an electron derived from water. Since the complete oxidation of water is a four-electron process, the PS II reaction center must be excited four times in order to produce one molecule of dioxygen with the concomitant release of four protons. The ability to accumulate four oxidizing equivalents and thus coordinate the release of four electrons/four protons and one oxygen molecule from water is accomplished by a cluster of four manganese atoms which undergo sequential valency changes (Ono et al., 1992). A detailed review of PS II and the water-splitting processes has recently been published by Debus (1992).

From the above, it is clear that PS II functions as a water–plastoquinone oxidoreductase. The reduced plastoquinone is then used to re-reduce the plastocyanin oxidized by PS I, but the electron/proton exchange involves other redox-active components which together make up a protein complex known as the cytochrome b_6/f complex. Oxidized plastocyanin extracts electrons from cytochrome f ($E_m = +0.34$ V) while plastoquinol reduces a Rieske iron sulphur center ($E_m = +0.3$ V) which is also located in the cytochrome b_6/f complex. Electrons then flow from the Rieske Fe–S center to the cyt f^+. Since these reactions only involve electron exchange, the protons associated with plastoquinol are released into the aqueous phase following its oxidation. Also since the oxidation of plastoquinol is a two-electron/two-proton event, the Rieske Fe–S cyt *f*/plastocyanin redox chain must turn over twice or alternatively, the second electron is passed to a low-potential form of cytochrome b_6 ($E_m = -0.15$ V). Probably, depending on conditions such as light intensity, both pathways can operate. In the latter case, the electron has the possibility of being transferred to a second cytochrome b_6 heme with a higher redox potential (E_m = 0.05 V). According to the Q-cycle hypothesis of Mitchell (1975), this branching of the two-electron oxidation of plastoquinol allows, after the initial primary reaction of one turnover, an additional reduction of a plastoquinone molecule by the high-potential cytochrome b_6.

The above sequence of electron/proton transfer takes place vectorially across the thylakoid membrane as indicated in Figure 2. The net result is that protons are absorbed from the stromal side of the membrane and released on the lumenal side. The resulting electrochemical potential gradient of protons provides, according to the chemiosmotic hypothesis (Mitchell, 1961), the free energy to drive the conversion of ADP to ATP. This phosphorylation process is catalyzed by a complex known as the coupling factor, or the CF_1–CF_o complex. Clearly the Q-cycle around the b_6/f complex is a mechanism which increases the H^+/e^- ratio over and above that shown in Figure 2, as does ferredoxin-catalyzed cyclic electron flow (Arnon, 1991).

III. THYLAKOID MEMBRANE COMPOSITION

A. Lipids

Figure 3 shows the chemical structure of the main classes of lipids found in chloroplast thylakoids. These lipids can be divided into two types: those that are saponifiable and those that are not (see Quinn and Williams, 1983). The matrix of the membrane is made up of the saponifiable lipids known as the diacylglycerolipids. In contrast, the non-saponifiable lipids are the various forms of pigments (chlorophylls and carotenoids) and quinones. The pigments are bound within protein complexes, and the quinones (mainly plastoquinone-9) are located in the hydrophobic lipid matrix. The major class of saponifiable lipids contains the electroneutral galactolipids, monogalactosyldiglyceride (MGDG), and digalactosyldiglyceride (DGDG), which together can represent as much as 75% of the total lipid composition of the thylakoid membrane of all classes of oxygenic photosynthetic organisms. The remaining polar lipids are sulphoquinovosyldiglyceride (SQDG), phosphatidylglycerol (PG), and phosphatidylcholine (PC). Generally, the fatty acids associated with these lipid classes are rather unsaturated with linolenic acid (C18:3) dominating in particular for MGDG and DGDG. Although there is no large variation between higher plants and algae, the nature of the fatty acid and the degree of unsaturation is quite variable in cyanobacteria and depends very much on growth conditions, particularly temperature (Murata, 1989). In certain species of higher plant's (e.g., spinach) the fatty acid hexadecatrienoic acid (C16:3) occurs as a part of MGDG. Also worthy of note is that the thylakoid membrane possesses little or no sterols, and that 20% or less of the lipids (SQDG + PG) carry net negative electrical charge.

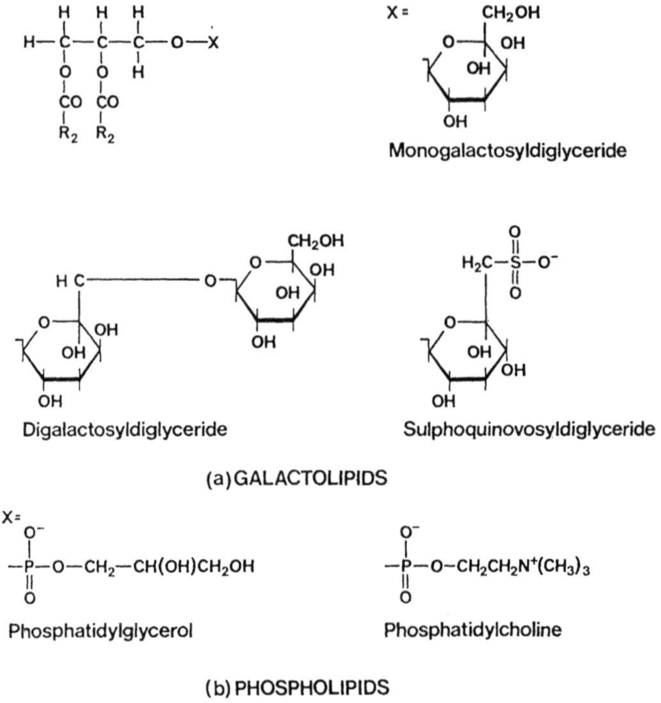

Monogalactosyldiglyceride

Digalactosyldiglyceride

Sulphoquinovosyldiglyceride

(a) GALACTOLIPIDS

Phosphatidylglycerol

Phosphatidylcholine

(b) PHOSPHOLIPIDS

Figure 3. Chemical structure of the main classes of lipids found in the thylakoid membrane of oxygenic photosynthetic organisms.

As far as regulation and adaptation to changes in growth conditions are concerned, there is evidence that PG is important. Indeed, Murata and his colleagues have shown that the sensitivity of herbaceous and woody plants to chilling is well correlated with the total levels of saturated and trans-monounsaturated molecular species of PG, such as 16:0/16:0 and 16:0/16:1 (Murata, 1983; Tasaka et al., 1990). These molecular species are thought to undergo a phase transition in the membranes of chill-sensitive plants at the chilling temperature. It is this phase change which is considered by Murata (1983) to cause chilling injury to plants. More recently Murata and colleagues (Murata, 1989; Murata and Nishida, 1989) have obtained strong evidence that the degree of unsaturation of the thylakoid lipids of cyanobacteria increases with decreasing growth temperature. Interestingly, this group has isolated, and also engineered, a mutant of the cyanobacterium, *Synechocystis* 6803, which has the gene encoding the desaturase interrupted or deleted (Wada and Murata, 1989; Wada et al., 1990). As a consequence, the mutant contrasts with the wild type in that it can grow at high, but not at low temperatures. In higher plants, no such striking change in fatty acid unsaturation levels has been observed in chill-resistent plants (Chapman et al., 1983). Chill-sensitive plants, however, have been shown to modify to some extent the unsaturation level of their thylakoid lipids in response to changes in growth temperature (Chapman et al., 1984). As mentioned above, the elegant work of Murata and colleagues (Murata, 1983; Tasaka et al., 1990) has identified changes in the saturation level of PG as being a possible key factor in the adaptation of higher plants to low-temperature growth. Despite this, Somerville and Browse (1991) have recently concluded after reviewing the literature and based on their own experiments with mutants that lipid unsaturation cannot account for naturally occurring cold-sensitivity in plants.

Chapman et al. (1983) noted that growth of chill-resistant plants at low temperatures leads to an increase in the protein to lipid ratio of thylakoid membranes. The importance of this ratio has been stressed in a recent review on the biochemistry and biophysics of thylakoid acyl lipids (Webb and Green, 1991).

Changes in lipid content, fatty acid unsaturation levels, and in the protein-to-lipid ratio would be expected to modify the overall fluidity of the membrane. The high unsaturation level of thylakoid lipids would indicate they give rise to a bilayer having a hydrophobic interior with a high degree of motion (Stubbs and Smith, 1984). Indeed, experimental evidence for this contention has been obtained from time-resolved anisotropy measurements using the fluorescence probe, 1,6-diphenyl-1,2,5-hexatriene (DPH) (Ford and Barber, 1983). Using the same techniques, Millner et al. (1984) also showed that the reconstitution of protein complexes into the bilayer of liposomes composed of thylakoid lipids decreased the motion of the acyl chains. Despite this effect and the fact that the thylakoid membrane has a relatively high protein-to-lipid ratio, the DPH probe gives an estimate for the microviscosity of pea thylakoids of 0.34 P at 25 °C (Ford and Barber, 1983). This value contrasts with 0.42 P at 35 °C for the cell membrane of human erythrocytes (which contain cholesterol, a lipid that increases the rigidity of the bilayer) and 0.82 P at 35 °C for the purple membrane of *Halobacterium halobium* (that has a very high protein-to-lipid ratio) (see Kinosita et al, 1981). Of interest is the fact that the thylakoid membrane value compares well with that of 0.29 P at 35 °C for rat liver mitochondrial membranes (Kinosita et al., 1981).

It should be noted that all the above quoted values involve measurements of the rotational diffusion of DPH and may not necessarily be indicative of viscosity factors controlling lateral movements. Moreover, the probe may selectively partition into more fluid microdomains in the membrane. Nevertheless, Shinitzki and Yuli (1982) have given convincing arguments that this technique does give an indication of fluidity levels which relate with the physiology of the membrane system under study. Indeed, the DPH fluorescence anisotropy measurements have indicated that the pea chloroplast thylakoid membrane undergoes no obvious bulk phase change over the temperature range from 55 to –20 °C (Barber et al., 1984). This absence of any phase change was observed with pea plants grown either at 7 or 17 °C. However, it was found that those grown at 7 °C had maintained a more fluid thylakoid membrane at any fixed measuring temperature as compared with the thylakoids of plants grown at 17 °C. As a consequence of this adjustment, the fluidity of the thylakoids was about the same at their respective growth temperatures, indicative of the operation of a homostatic mechanism (Barber et al., 1984). Therefore, it seems important for normal functional processes that the thylakoid membrane is maintained in a liquid crystalline state at a relatively high level of fluidity. The necessity for this fluid state has become more obvious in recent years as we have come to understand the importance of molecular trafficking along and across this membrane system. Details of these dynamic processes will be presented in later sections of this chapter.

The polar lipids of the thylakoids can be isolated and analyzed by several different procedures (Chapman and Barber, 1987). In their isolated state they can form a wide range of different structures. Most (DGDG, SQDG, PG, and PC) will form bilayers when dispersed in water or salt solutions at room temperature. In striking contrast, the naturally occurring form of MGDG (i.e., in its highly unsaturated state) does not form bilayers in aqueous dispersions but arranges into nonbilayer structures, called the hexagonal type II phase (Gounaris et al., 1983a,b). Below the phase transition temperature, when it is in its gel state, the preferred organization for DGDG is in the form of lamellar sheets (Sen et al., 1983). The gel state for naturally occurring MGDG is observed at –30 °C and below (Shipley et al., 1973). Decreasing the level of unsaturation by catalytic hydrogenation, however, dramatically raises the phase transition to well above room temperature (Gounaris et al., 1983b). Interestingly, aqueous dispersions of binary mixtures of MGDG and conventional bilayer-forming lipids can give rise to a variety of structures intermediate between bilayers and Hex-II depending on the molar ratio of MGDG to the bilayer lipid (Sen et al., 1981; Sprague and Staehelin, 1984). With total lipid extracts, the conformational states adopted depend upon conditions such as temperature, pH, electrolyte levels, and the presence of cryoprotectants. For example, the bilayer configuration of the total lipid extract is lost when the pH is lowered or the cation level raised (Gounaris et al., 1983a). Presumably, under these conditions, neutralization or screening of the electrical charges of SQDG or PG facilitates their phase separation from MGDG and thus allows the formation of nonbilayer structures due to the localization of MGDG.

Despite the fact that MGDG is nonbilayer forming *in vitro* and that it is the dominant polar lipid of all types of thylakoid membranes, there is no evidence that this membrane system, under normal conditions, consists of anything other than a bilayer structure (Gounaris et al., 1983c). Only under extreme conditions, such as heat stress (Gounaris et al., 1984a) or protein denaturation (Machold et al., 1977), have nonbilayer structures been detected in the natural membrane. Why the thylakoid membrane contains high levels of the nonbilayer-forming MGDG is unclear, although it has been suggested that its molecular shape (rather cone-like) makes it ideal for packaging large multiprotein complexes into the membrane (Gounaris and Barber, 1983; Quinn and Williams, 1985) or for facilitating sharp curvature changes such as at the ends of the granal discs (Murphy, 1983, 1986).

The question of asymmetry of lipid distribution both across and along the membrane has been addressed by several different research groups. Although no precise answers emerged from these studies, trends indicate that PG is more abundant in the outer leaflet of the bilayer, while SQDG is preferentially located on the inner half of the bilayer. It has also been reported by Unitt and Harwood (1985) that while palmitate in PG is evenly distributed linolenate, and particularly trans-Δ3-hexadecenoate, are preferentially found in the outer leaflet. These results are in accordance with the earlier work of Duvel et al. (1980). There are, however, considerable disagreements about the distribution of PC and the two electroneutral galactolipids. Radunz (1980) has argued that galactolipids were enriched at the inner surface while the opposite conclusion was deduced by Sundby and Larsson (1985) who labeled MGDG and DGDG with tritiated sodium borohydride. The work of Gounaris et al. (1984b) using monoclonal antibodies also supported the concept that MGDG was located at relatively high levels in the outer leaflet as did the chemical labeling and hydrolase studies of Rawyler and colleagues (1985, 1987).

With regard to lateral distribution, there is no evidence for large differences in lipid classes between the granal and stromal regions of thylakoid membranes of higher plants. However, there are differences in the relative levels of MGDG and DGDG (Gounaris et al., 1983d; Henry et al., 1983; Murphy and Woodrow, 1983), a trend which is discussed further in Section IVA.

B. Proteins

Four major protein complexes are present in the thylakoid membrane (Figure 2 and Table 1). The complexes that are involved in light-harvesting, electron transport, proton translocation, and enzymatic catalysis, are PS I, PS II, cytochrome b_6/f, and the ATP synthase (Anderson and Andersson, 1982). The latter two correspond structurally and functionally to related membrane complexes found in mitochondria. At present, we know of about 60 polypeptides associated with these four complexes including the various chlorophyll *a/b* proteins. Notably nearly all the corresponding genes have been sequenced (Herrmann et al., 1991). Approximately 45% of these polypeptides are encoded by the chloroplast genome while the remainder are encoded by the nuclear genome and therefore have to be imported into the chloroplast (see chapter by Bruce and Keegstra, this volume). Up to 70% of the thylakoid polypeptides of plant chloroplasts are integral membrane proteins which have one or more hydrophobic membrane-spanning regions (Table 1). The remainder are extrinsic and attached to the outer or inner thylakoid surface predominantly by electrostatic forces and/or hydrogen bonding. Another special property of the plant thylakoids is that approximately one-quarter of its membrane proteins carry chlorophyll (see chapter by Thornber et al., this volume). Apart from the protein subunits associated with the four photosynthetic complexes there is increasing experimental evidence for the presence of additional polypeptides not

Table 1. Proteins and Genes of Photosynthetic Complexes

Protein/Gene[a]	*Coded*[b]	*Mass (kDa)*[c]	*Organization*[d]	*Function*
Photosystem I				
psaA	C	82–83	intrinsic (11)	RC, binds redox groups and antenna chl *a*
psaB	C	82–83	intrinsic (11)	RC, binds redox groups and antenna chl *a*
psaC	C	9	extrinsic stromal	Binds FeS centers F_A and F_B
psaD	N	15–18	extrinsic stromal	Ferrodoxin docking?
psaE	N	8–11	extrinsic stromal	Ferrodoxin docking?
psaF	N	16–18	extrinsic lumenal	Plastocyanin docking
psaG	N	10–11	extrinsic stromal	
psaH	N	10–11	extrinsic stromal	PSI-LHCI interaction?
psaI	C	4	intrinsic (1)	
psaJ	C	5	intrinsic (1)	
psaK	N	8	intrinsic (1–2)	
psaL	N	18	intrinsic (2)	
9 kDa	N	9	extrinsic lumenal	
Plastocyanin/*pet E*	N	13	extrinsic lumenal	Electron transport
Ferrodoxin/*pet F*	N	10–11	extrinsic stromal	Electron transport
FNR/*pet H*	N	34	extrinsic stromal	$NADP^+$-reduction
Cytochrome b_6f				
Cytochrome *f*/*pet A*	C	33	intrinsic (1)	Electron transport, heme binding
Cytochrome b_6/*pet B*	C	23	intrinsic (4)	Electron transport, heme binding
Rieski protein/*pet C*	N	20	intrinsic (1)	Electron transport, binding FeS cluster
Subunit IV/*pet D*	C	16	intrinsic (3)	?
Subunit V/*pet G*	C	3,2	intrinsic (1)	?
Photosystem II				
D_1/*psbA*	C	32	intrinsic (5)	RC. Binding P680 Pheo Q_B Tyr_Z Mn
CP47/*psbB*	C	47	intrinsic (6)	Chl *a* antenna
CP43/*psbC*	C	43	intrinsic (6)	Chl *a* antenna
D_2/*psbD*	C	34	intrinsic (5)	RC. Binding 680, Q_A Tyr_D Mn?
cyt *b*-559*a*/*psbE*	C	9	intrinsic (1)	Heme binding
cyt *b*-559*b*/*psbF*	C	4	intrinsic (1)	Heme binding
Phosphoprotein/*psbH*	C	9	intrinsic (1)	
psbI	C	4,5	intrinsic (1)	RC?
psbJ	C	4,0	intrinsic (1)	
psbK	C	3,7	intrinsic (1)	
psbL	C	4,8	intrinsic (1)	
psbM	C	3,7	intrinsic (1)	

Table 1. Proteins and Genes of Photosynthetic Complexes

Protein/Gene[a]	*Coded*[b]	*Mass (kDa)*[c]	*Organization*[d]	*Function*
psbN	C	4,7	intrinsic (1)	
OEC1/*psbO*	N	33	extrinsic lumenal	Mn-stabilizing, Ca^{2+}/Cl^- sequestering
OEC2/*psbP*	N	23	extrinsic lumenal	Ca^{2+}/Cl^- sequestering
OEC3/*psbQ*	N	16	extrinsic lumenal	Ca^{2+}/Cl^- sequestering
psbR	N	10	intrinsic (1)	Docking extrinsic subunits?
psbS	N	22	intrinsic (4)	Chlorophyll binding?
7 kDa	N	7	intrinsic ?	
5 kDa	N	5	extrinsic	
3,2 kDa	N	3,2	intrinsic ?	
ATP synthase				
CF_1-α/atpA	C	60	extrinsic stromal	Regulation
CF_1-β/atpB	N	56	extrinsic stromal	Catalytic site
CF_1-γ/atpC	C	39	extrinsic stromal	Regulation
CF_1-δ/atpD	C	19	extrinsic stromal	Regulation
CF_1-e/atpE	C	14	extrinsic stromal	Inhibitor of ATPase
CF_0-I/atpF	C	15	intrinsic (1)	CF_1-binding
CF_0-II/atpG	N	13	intrinsic (1)	CF_1-binding
CF_0-III/atpH	C	8	intrinsic (2)	Proton channel
CF_0-IV/atpI	C	19	intrinsic (4)	

Notes: [a]The chlorophyll *a/b*-binding proteins are not included (see Thornber, this volume)
[b]C = chlorophyll DNA, N = nuclear DNA
[c]Apparent molecular mass (kDa)
[d]Intrinsic (number of predicted membrane spans), or extrinsic membrane proteins, lumenal or stromal side of the membrane

directly involved in the process of energy conversion but catalyzing auxiliary functions (Andersson, 1992) such as biosynthesis, protection, and regulation of the photosynthetic apparatus (see Section VI).

In dealing with the protein complement of the thylakoid membrane, one also has to consider the unexpected finding by Shinozaki et al. (1986) that the chloroplast genome carries the genes for subunits of the respiratory-chain NADH dehydrogenase of mitochondria (see chapter by Jackson and McEwan, this volume). Altogether, it is not unreasonable to assume that there could be at least 100 different polypeptides or more in the thylakoid membrane of higher plants.

Photosystem II

This is the most complicated of the four photosynthetic complexes. In plants it contains at least 20 distinct subunits (Table 1) not including five chlorophyll *a/b* proteins associated with the PS II antenna (Andersson and Styring, 1991). Twelve of these subunits are encoded by the plastid DNA. As will be described below, all PS II subunits together contain as many as 36 transmembrane helices. Despite this compositional complexity, only two proteins appear to carry all of the necessary redox components for the primary photochemistry and the water oxidation reactions. These are the D1 and D2 proteins which comprise the reaction center of PS II. The central role of these two proteins has been emphasized by their remarkable functional and structural similarities to the L- and M-subunits of the reaction center of photosynthetic purple bacteria (see chapter by Takahashi and Wraight, this volume) whose three-dimensional structure has been determined in the spectacular work of Deisenhofer et al. (1985). The D1 and D2 proteins, which are plastid-encoded by the *psbA* and *psbD* genes, respectively (see Andersson and Herrmann, 1988; Herrmann et al., 1991), can be resolved by SDS-PAGE into two diffuse bands (explaining the designation D) in the 32 to 34 kDa region (Chua and Gillham, 1977). The two proteins are homologous to each other and to the L- and M-subunits of the reaction center of the purple bacteria (Michel and Deisenhofer, 1986; Trebst, 1986; Barber, 1987). The latter gives support to the model that the D1 and D2 proteins make up a heterodimer where each of the two polypeptide chains possesses five membrane-spanning regions.

Taking the homology between the two reaction centers further using sequence alignments combined with site-directed mutagenesis, two histidines, D1-His198 and D2-His198, are suggested to coordinate the reaction center chlorophyll P680 (see Diner et al., 1991). The primary electron acceptor pheophytin is believed to be ligated to the D1 protein only via hydrogen bonding to Glu-130 (Moenne-Loccoz et al., 1989; Nabedryk et al. 1990). The primary quinone acceptor Q_A binds to the D2 protein, presumably to the peptide loop which is exposed at the stromal side of the thylakoid membrane and which connects trans-membrane helices D and E (Trebst, 1986). The secondary quinone acceptor Q_B is bound to the corresponding loop on the D1 protein.

The redox-active tyrosine which is the primary electron donor to P680 has been identified as Tyr-161 (Tyr_Z) of the D1 protein using site-directed mutagenesis (Debus et al. 1988; Metz et al. 1989). There are now several lines of evidence to suggest that the four manganese atoms of the water-splitting reaction are also associated with the reaction center heterodimer, in particular the D1 protein (see Andersson and Styring 1991; Diner et al. 1991). Detailed molecular modeling of the organization of the D1 and D2 proteins has been undertaken based on the coordinates for the purple bacterial reaction center structure (Svensson et al., 1990; Ruffle et al., 1992). Such modeling is useful but must be treated cautiously until verified by crystallographic studies or other experimental methods.

Cytochrome *b*559 is composed of two polypeptides: one subunit of 9 kDa and another of 4 kDa (Herrmann et al., 1984). Both polypeptides are encoded in the plastid DNA by genes designated *psbE* and *psbF*, respectively (Herrmann et al., 1984; Willey and Gray, 1989). Each of the two polypeptides has one membrane span with their N-termini exposed at the stromal thylakoid surface (Tae and Cramer, 1989). Structurally, the cytochrome is predicted to be a heterodimer in which the heme group is sandwiched between two histidines. The cytochrome appears to be tightly associated with the D1-D2 heterodimer as judged by its presence in the PS II reaction center preparations (Barber et al., 1987; Nanba and Satoh, 1987).

Cytochrome *b*559 can attain one high-potential form and several low-potential forms (Cramer and Whitmarsh, 1977). The function of cyt *b*559 continues to be a matter of debate, as does its stoichiometry. Many models have been proposed over the years but the concensus is now focused towards this heme acting as a protective agent against photoinhibitory damage of the PS II reaction center. Roles for the high-potential form (E_m = 400 mV) (Thompson and Brudvig, 1988) and the low-potential form (E_m = 60 to 80 mV) (Nebdal et al., 1992) have been suggested and a new scheme postulated to allow for protection against both donor and acceptor side-induced photoinhibitory damage (Barber and De Las Rivas, 1993). This new scheme is given in Figure 4 where it is assumed that there is one cytochrome per PS II reaction center which is able to undergo a reversible change in its redox potential in order to act either as an electron donor (high-potential form) to $P680^+$ or as an electron acceptor (low-potential form) from reduced pheophytin. The model presented awaits rigorous experimental support, but it does give a rational explanation for the remarkable redox shifts that this cytochrome is able to sustain and at the same time recognizes that PS II is vulnerable to photoinduced damage on both the donor and acceptor sides (Barber and Andersson, 1992).

The so-called CP43 and CP47 proteins bind chlorophyll *a* and play a role in light-harvesting and energy transfer to the reaction center (see Green, 1988). Each of the chlorophyll *a*-proteins is composed of one apo-polypeptide with apparent molecular masses of 47 and 43 kDa, respectively, and encoded by the chloroplast genes, *psbB* and *psbC* (see Andersson and Herrmann, 1988; Herrmann et al., 1991). Sequence analysis shows that the two proteins are related and each is predicted to traverse the membrane six times and expose a large hydrophilic loop at the inner thylakoid surface.

Strikingly, there are a relatively large number of quite small polypeptides (10 kDa and below) which belong to the PS II complex (Ljungberg et al., 1986b; Gray et al., 1990). Including the two subunits that are associated with cytochrome *b*559, there are 13 such small polypeptides. One of these is the product of the psbH plastid gene and is normally referred to as the 9 kDa phosphoprotein (Hird et al., 1986; Westhoff et al., 1986). The protein is composed of 72 amino acids and is predicted to have one membrane spanning region. The function is unknown but is normally presumed to be associated with the reversible phosphorylation that the protein undergoes. However, deletion of the *psbH* gene in the cyanobacterium *Synechocystis 6803* does not lead to impairment of PS II electron transport and the complex appears to be fully assembled (Mayes et al., 1993). Moreover, it is not yet clear if the *psbH* gene product is phosphorylated in cyanobacteria. Another protein of apparent molecular mass of 10 kDa is encoded by the nuclear gene *psbR* (Ljungberg

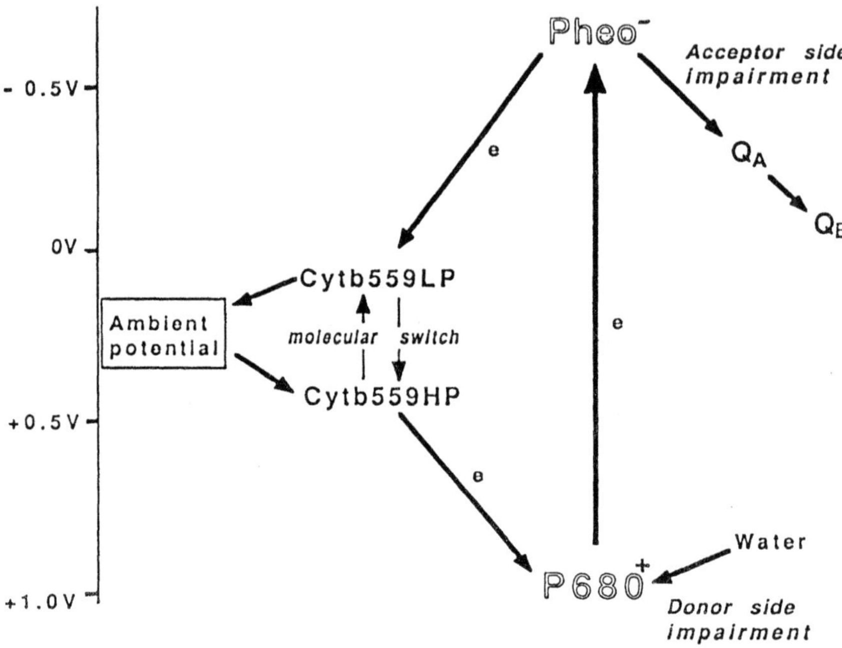

Figure 4. A scheme to emphasize the possible role of cyt *b*559 as a one electron protectant against "acceptor" or "donor" side photoinhibition. The model postulates that in order for the cyt *b*559 to act either as an electron acceptor (for acceptor side protection) or as an electron donor (for donor side protection) it shifts reversibly between its high (HP) and low (LP) potential forms, triggered by an unknown molecular switch. The reoxidation of reduced cyt $b559_{LP}$ and the reduction of oxidized cyt $b559_{HP}$ is suggested to involve electron exchange with the ambient redox system governed by factors such as the redox state of the plastoquinone pool and the presence of molecular oxygen. Model from Barber and De Las Rivas (1993).

et al., 1986a; Lautner et al., 1988). The nucleotide sequence predicts one membrane span at the extreme C-terminal end of the protein. It is suggested that the protein is located in the thylakoid lumen, anchored by its membrane-spanning C-terminal portion. It is possible that this protein has a role in the attachment of extrinsic proteins to the inner thylakoid surface. On the other hand, after inactivation of the *psbR* gene in potato plants by the antisense RNA technique (Stockhaus et al., 1990), the photosynthetic function of the transgenic plants showed a significant reduction in reoxidation of the primary quinone Q_A, a reaction which occurs towards the outer thylakoid surface.

There are six additional plastid-encoded small polypeptides which can be assigned to open reading frames in the chloroplast DNA. These have been designated *psbI, psbJ, psbK, psbL, psbM*, and *psbN* (Gray et al., 1990; Erickson and Rochaix,

1992). The functional significance of this group of one membrane-spanning polypeptides remains to be determined. The *psbI* gene product appears to be closely associated with the PS II reaction center as judged by its presence in isolated reaction center particles (Webber et al., 1989). Depletion of the *psbK* gene in *Synechocystis* 6803 did not prevent assembly of the PS II complex or impair its functioning (Ikeuchi et al., 1991; Zhang et al., 1994). On the other hand, the deletion of the *psbJ* gene from the same organism gave rise to a mutant with disassembled or unstable PSII complexes (Lind et al., 1993). Three of the small polypeptides are nuclear-encoded (Schröder et al., 1988). Two of these, with molecular masses of 7 and 3.2 kDa, show hydrophobic properties and are probably integral membrane proteins. A 5-kDa protein is water-soluble and its amino acid composition reveals a hydrophilic character, but it is not yet clear at which side of the membrane it is located and the function is not known (Ljungberg et al., 1986b).

Three extrinsic proteins, with apparent molecular masses of 33, 23, and 16 kDa, can be reversibly detached and reattached from and to the inner thylakoid surface which results in inhibition and reconstitution of water oxidation (Murata and Miyao, 1985; Andersson and Åkerlund, 1987). The three proteins are encoded by the *psbO, psbP*, and *psbQ* genes, respectively in the nuclear genome (Andersson and Herrmann, 1988). There is no evidence that the *psbP* and *psbQ* genes are present in cyanobacteria.

Despite the influence of electron transport on the donor side of PS II, there is now a general consensus that the 33-, 23-, and 16-kDa proteins have noncatalytic functions. Presently it is considered that the 23- and 16-kDa proteins facilitate binding of the Ca^{2+} and Cl^- ions which are both required as cofactors for photosynthetic water oxidation (Debus, 1992). Such a function may also be possible for the 33-kDa protein; however, the main role of this protein appears to be stabilization of the manganese cluster. An involvement of this protein in direct metal ligation is not very likely given data obtained both by biochemical and biophysical experiments (see Andersson and Styring, 1991) and by creating *psbO* deletion mutants in *Synechocystis* 6803 (Burnap and Sherman, 1991; Mayes et al., 1991).

The most recent PS II gene to be sequenced is the nuclear *psbS* gene (Kim et al., 1992; Wedel et al., 1992) which encodes a hydrophobic 22-kDa protein (Ljungberg et al., 1986a). Using the sequence information the protein is predicted to possess four membrane-spanning regions. Moreover, it shows a clear homology with chlorophyll *a/b*-binding protein which is surprising since the isolated 22-kDa protein has been found to be present in thylakoid membranes isolated from cyanobacteria (Nilsson et al., 1990). Most biochemical studies suggest that the protein has some regulatory role associated with the acceptor side of PS II and a role in assembly has been suggested, but basically its function is unknown. In contrast to the original isolation procedure (Ljungberg et al., 1986a), the 22-kDa protein has recently been obtained with bound chlorophyll molecules (Funk et al., 1994), suggesting some role for light-harvesting apparatus of PS II.

Closely attached to PS II is a large light-harvesting antenna comprised of at least 5 different, but closely related, chlorophyll *a*/*b* proteins (see chapter by Thornber et al., this volume). The major component is the so-called LHC II complex which can consist of up to three subunits in the 25- to 28-kDa region. This complex contains the main percentage of both the chlorophyll and protein mass of the thylakoid membrane. LHC II is also thought to play a central role in the short- and long-term regulation of light-harvesting, as well as influencing the organization of the thylakoid membrane system. The three-dimensional structure of this chlorophyll *a*/*b* binding protein has recently been described (Kühlbrandt et al., 1994).

Photosystem I

So far, 13 different subunits have been shown to be associated with the PS I complex (see Table 1). In contrast to PS II, most of the proteins are relatively hydrophilic and have been shown to be extrinsic membrane proteins with as many as seven being exposed at the outer or inner thylakoid surface (Goldbeck and Bryant, 1991; Andersson and Franzén, 1992). Nevertheless, there are approximately 28 hydrophilic membrane-spanning helices which can be predicted for the integral membrane proteins. Quite recently, considerable progress has been made and based upon crystals of PS I complexes obtained from the cyanobacterium *Synechococcus* structural data at 6 Å resolution is now available (Krauss et al., 1993).

As is the case with PS II, the reaction center of PS I is a heterodimer of two homologous polypeptides (Goldbeck and Bryant, 1991). However, the PS I reaction center is not analogous to that of the photosynthetic purple bacteria, but spectroscopic and biochemical evidence have shown that it is quite similar to the reaction center of green-sulphur bacteria (Nitschke and Rutherford, 1991).

The two subunits of the reaction center heterodimer of PS I are the plastid-encoded PS I-A and PS I-B polypeptides. Both polypeptides have molecular masses of 82 to 83 kDa and are predicted to form as many as 11-membrane-spanning helices each (Table 1). As indicated above, the two subunits are homologous to each other, showing approximately 45% identity. They ligate the primary electron donor P700 and the early electron acceptors A_0, A_1, and the iron–sulphur cluster F_X. In addition, and at variance with the organization of PS II, the PS I-A/PS I-B reaction center heterodimer binds about 100 antenna chlorophyll *a* molecules. PS I also contains a separate chlorophyll *a*/*b* antenna, designated LHC I (see chapter by Thornber et al., this volume). There have been several predictions and speculations about the precise arrangement of several of the ligands of the PS I reaction center, but these will not be dealt with here. A final picture will probably emerge soon from the crystallization work mentioned above.

The PS I-C protein is another functionally important subunit since it binds the iron–sulphur centers of F_A and F_B (Hayashida et al., 1987; Høj et al., 1987; Oh-oka et al., 1988). The protein is plastid-encoded and has a molecular mass of 9 kDa. It is quite hydrophilic and is not thought to possess any membrane-spanning regions.

The PS I-C protein is highly conserved between species and in all it contains nine cysteines, of which eight are expected to ligate the two iron–sulphur centers.

The PS I-D and PS I-E proteins are nuclear-encoded with molecular masses of 18 and 10 kDa, respectively (Scheller and Møller, 1990; Goldbeck and Bryant, 1991; Mann et al. 1991). They are both positively charged, hydrophilic proteins and do not contain any sequence regions that would be predicted to span the thylakoid bilayer. The majority of evidence suggests that, as in the case of PS I-C, they are extrinsic subunits located at the outer thylakoid surface (Andersson and Franzén, 1992). Both proteins are highly conserved, but their functional roles are not yet clear. The role of PS I-D has mainly been associated with the binding of soluble ferredoxin to the thylakoid membrane (Zanetti and Merati, 1987; Zilber and Malkin, 1988). However, very recent studies based upon site-directed mutagenesis combined with spectroscopic measurements *in vitro* (Rosseau et al., 1993) strongly suggest that the PS I-E protein is responsible for the binding of ferredoxin. Still, it is possible to maintain photoautotrophic growth in cyanobacteria with an inactivated *psaE* gene (Chitnis et al., 1989).

The PS I-F is yet another relatively hydrophilic protein (Steppuhn et al., 1988; Franzén et al., 1989a; Chitnis et al., 1991). It is nuclear-encoded and its molecular mass is 17 kDa. Biochemical topographical data, combined with its lumenal targeting presequence, suggest that the PS I-F protein is located at the lumenal surface of the membrane. The majority of experimental evidence suggests that this protein is required for the docking of plastocyanin to PS I (Wynn and Malkin, 1988; Hippler et al., 1989).

The PS I-G and PS I-H proteins both have molecular masses of approximately 10 kDa (Scheller and Møller 1990; Goldbeck and Bryant, 1991). They are relatively hydrophilic and are likely to be extrinsically situated at the stromal side of the thylakoid membrane. The functional role of the PS I-G and PS I-H proteins are not known although the latter has been suggested to be involved in the interaction between PS I and its LHC I antenna (Steppuhn et al. 1989).

As in the case of PS II, there are several small hydrophobic polypeptides probably all having one membrane span (see Goldbeck and Bryant,1991). These are the PS I-I, PS I-J, and PS I-K proteins. The first two proteins are chloroplast-encoded with molecular masses of 4 and 5 kDa, respectively. The PS I-K protein is nuclear encoded and somewhat larger with a molecular mass of 8.4 kDa (Franzén et al., 1989b). All three proteins appear to be tightly associated with PS I-A/PS I-B reaction center heterodimers, but their precise role remains to be established. Very recently, an additional low molecular weight nuclear-encoded polypeptide (9 kDa) has been shown to be extrinsically associated with PS I at the inner thylakoid surface (He and Malkin, 1992).

Finally, a PS I-L protein has been identified as a 22-kDa nuclear-encoded precursor protein that is probably processed to a mature protein of 18 kDa (Okkels et al. 1991). The protein is hydrophobic and predicted to have at least two membrane-spanning α-helices. To our knowledge there are no experimental results that suggest a role for this protein.

PS I is associated with a chlorophyll *a/b* antenna (LHC I) comprised of four different subunits (see Thornber et al., this volume). In contrast to PS II, the LHC I does not contribute to the majority of the pigment or protein mass of PS I.

Three additional extrinsic proteins are essential for the donor and acceptor functions of PS I although they are, in a strict sense, not part of the complex. These are the plastocyanin which is the electron donor to $P700^+$, ferredoxin, and ferredoxin-$NADP^+$ oxidoreductase which catalyzes the terminals steps of noncyclic electron transport.

Plastocyanin is a type I copper protein whose three-dimensional structure has been determined by Colman et al. (1978). It is nuclear-encoded and possesses a molecular mass of 13 kDa (Grossman et al., 1982). Plastocyanin is present in the thylakoid lumen but, as described above, its functional interaction with PS I at the inner thylakoid surface is probably mediated by the PS I-F protein.

Ferredoxin, which is also nuclear-encoded, has a molecular mass of 10.5 kDa. It is a non-heme [2Fe–2S] iron–sulphur protein which mediates electron transport from the F_A and F_B centers at the acceptor side of PS I (see Cramer et al., 1991). Ferredoxin is bound to PS I at the outer thylakoid surface probably via the PS I-E subunit through electrostatic interaction, but it is also weakly bound to the flavoprotein, ferredoxin-$NADP^+$ oxidoreductase, normally referred to as FNR. This enzyme, which is nuclear-encoded and which has a molecular mass of approximately 34 kDa, catalyzes the final reduction of $NADP^+$ during photosynthetic electron transport.

The Cytochrome b_6/f Complex

This third electron transport complex is considerably simpler than the two photosystems and its mitochondrial counterpart, the cytochrome *b/c* complex (see Cramer et al., 1990, 1991). At present there are only five subunits that are proven to belong to the cytochrome b_6/f complex (see Table 1). All of these subunits are integral membrane spanning polypeptides (altogether 10 predicted spans) and all, except one, are encoded by chloroplast genes.

Cytochrome *f* is encoded by the *petA* plastid gene (Willey et al., 1984a,b; Alt and Herrmann, 1984). The molecular mass is 33 kDa and it has one membrane-anchoring span which is located at the C-terminus of the polypeptide chain. Thus, the bulk of cytochrome *f* (250 amino acids) is located at the inner thylakoid surface. Its basic arrangement can therefore be compared to that suggested for the 10-kDa protein (*psbR*) of PS II. The lumenal portion is responsible for the heme binding. Histidine-25 may provide one ligation site for the heme, while lysine residue 145 or 222 could act as the other (Cramer et al., 1991). It has been suggested that 10 highly conserved amino acids, giving rise to a positively charged stretch between residues 58 and 154, may be of importance for the interaction with the negatively charged plastocyanin.

The apopolypeptide of cytochrome b_6 has a molecular mass of 23 kDa and is encoded by the chloroplast gene *petB* (Heinemeyer et al., 1984; Phillips and Gray, 1984). The most recent analyses of the topology of this cytochrome suggest four membrane spans with both the N- and C-termini exposed at the outer thylakoid surface (Cramer et al., 1990, 1991). The two heme groups are bound perpendicular to the plane on the membrane through four conserved histidines. It is suggested that the two hemes are located toward different sides of the membrane bridged by two of the transmembrane spans of the polypeptide. In addition there may be a plastoquinone binding site at the lumenal portion of cytochrome b_6.

The apoprotein cytochrome *b* of mitochondria and photosynthetic bacteria is considerably larger than that found in oxygenic photosynthetic organisms. Interestingly subunit IV of the cytochrome b_6/f complex (see below) is homologous to the C-terminal part of the larger respiratory cytochrome *b* protein. It therefore appears that there has been a splitting of genes in the case of the cytochrome b_6/f complex.

A 20-kDa subunit encoded by a nuclear gene (*petC*) harbors the Rieske [2Fe–2S] center (Steppuhn et al., 1987; Salter et al., 1992a). Most predictions suggest one single membrane span close to the N-terminal portion of the protein. In other words, the arrangement of the Rieske iron–sulphur protein appears to be basically similar to that of cytochrome *f*, exposing most of its mass at the lumenal side of the membrane.

Subunit IV of the cytochrome b_6/f complex is plastid encoded (*petD*) (Heinemeyer et al., 1984; Phillips and Gray, 1984) and is, as discussed above, homologous to the C-terminal part of the respiratory cytochrome *b*. It is likely that this subunit possesses three transmembrane helices. A quinone binding site has been suggested to be located at the lumenal side of this subunit.

A fifth, low molecular weight polypeptide with a molecular mass of approximately 3.2 kDa has been suggested to belong to the complex (Haley and Bogorad, 1989). It is encoded by an open reading frame in the plastid DNA which has been designated *petG*. As in the case of the low molecular mass proteins present in the two photosystems it is supposed to have one hydrophobic region that traverses the bilayer once.

As will be discussed in Section VI the cytochrome b_6/f appears to be tightly associated with and essential for the regulation of the kinase activity responsible for phosphorylation of LHC II.

The ATP Synthase (CF_1-CF_o) Complex

This complex functions as a proton translocating enzyme able to utilize the electrochemical potential gradient acting on protons across the thylakoid membrane to catalyze the synthesis of ATP (see Glaser and Norling, 1991). The complex is comprised of nine polypeptides which are arranged in two parts, CF_o which is an intrinsic component of the membrane, and CF_1 which is an extrinsic portion located at the outer thylakoid surface via the CF_o part of the complex (see Table 1). It is analogous to other H^+-ATPases such as the mitochondrial enzyme, but contains fewer polypeptide subunits.

The CF_1 contains five different subunits: α (60 kDa), β (56 kDa), γ (39 kDa), δ (19 kDa), and ε (14 kDa) (see Glaser and Norling, 1991). The α-, β- and ε-subunits are encoded by the chloroplast genes *atpA, atpB*, and *atpE*, respectively. The α- and δ-subunits in turn are encoded by the nuclear genes *atpC* and *atpD*. The stoichiometry between the subunits of the CF_1 part is α:3, β:3, γ:1, δ:1, and ε:1. The β-subunit carries the catalytic site of the enzyme, while the other subunits are thought to be essential for the organization and regulation of the complex.

The CF_o is comprised of four subunits, all being hydrophobic and integral membrane proteins. These are CF_o-I (15 kDa), CF_o-II (13 kDa), CF_o-III (8 kDa), and CF_o-IV (19 kDa) which are encoded by the genes *atpF–atpI* respectively. Apart from the *atpG* gene, which is nuclear, all of the genes of the CF_o are located in the chloroplast genome. Subunit III is thought to build the actual proton channel through the membrane which requires six copies of the polypeptide each traversing the bilayer twice. This subunit reacts covalently with dicyclohexylcarbodiimide (DCCD) which blocks proton transport through the channel. Subunit II (one membrane span) is suggested to maintain the structural organization of the channel. The roles of subunits I (one membrane span) and IV (four membrane spans) are not yet established but they could function to allow tight association of the CF_1 portion to the proton conducting channel.

C. General Comments of the Composition and Organization of the Photosynthetic Protein Complexes

All four photosynthetic complexes of the thylakoid membrane are multisubunit entities, and all contain both plastid and nuclear-encoded subunits (Table 1) (Herrmann et al., 1991). Two of the complexes, cytochrome b_6/f and ATP synthase, have their mitochondrial couterparts. Notably, the photosynthetic versions of these complexes are consistently more simple than those involved in the respiratory pathway and oxidative phosphorylation. Thus, the mitochondrial cytochrome *b/c* complexes contain 8 to 12 different subunits, as opposed to 4 to 5 in the chloroplast complex. For example, the cytochrome b_6/f complex does not contain the "core" polypeptide present in the mitochondrial complex.

The ATP synthase of the chloroplast, unlike its mitochondrial counterpart, does not have distinct additional subunits that confer oligomycin sensitivity or which function as an ATPase-inhibitor protein (see Glaser and Norling, 1991).

Nevertheless, a simple subunit composition is not a typical feature of photosynthetic complexes considering the large number of subunits within the two photosystems (see Table 1). Including its chlorophyll *a/b* light-harvesting antenna, PS II contains at least 25 different subunits and the corresponding number for PS I is at least 17. Strikingly, only a very few subunits of the photosystems are involved in the actual electron transport process. PS II, despite its complexity, only appears to

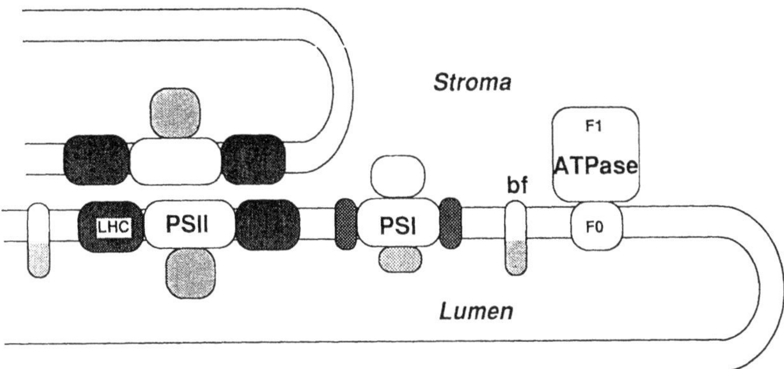

Figure 5. The thylakoid membrane and associated membrane bound protein complexes are shown in schematic form. The lateral heterogeneity of the protein complexes is indicated. The extent to which the complexes are composed of hydrophobic (*white*) or hydrophilic (*light pattern*—stromally extending, *heavy pattern*—lumenally extending) domains is indicated, and the area shown approximates to the molecular mass concerned. Ferredoxin and ferredoxin NADP reductase are included in the PS I stromal domain. Also indicated (*dark shading*) are the light harvesting complexes associated with the photosystems.

use two subunits, the D1 and D2 proteins, for the complete water–plastoquinone oxidoreductase activity. The physiological significance of this supernumerosity of the two photosystems is an essential challenge for future photosynthetic research.

Despite the considerable complexity of the two photosystems in higher plants, the number of polypeptides does not decrease significantly in more simple photosynthetic organisms. Thus, in prokaryotic cyanobacteria the number of subunits in the two photosystems is about the same as compared to their eukaryotic counterparts (see Andersson and Styring, 1991). In fact, there are additional subunits that are unique to the prokaryotic PS I and II. This is in contrast to the respiratory complexes; that is, cytochrome *c* oxidase which only has three subunits in bacteria, but as many as 14 different subunits in the human enzyme (Kadenbach et al., 1983). A similar variation in subunit numbers has also been demonstrated for the NADH–ubiquinone oxido–reductase (complex I) (Ragan, 1987).

All photosynthetic complexes, apart from the cytochrome b_6/f complex, contain extrinsic membrane proteins in addition to the hydrophobic integral ones (Figure 5). This is well established for the ATP synthase, which exposes its CF_1 portion at the outer thylakoid surface. However, experimental results which are relatively recent also point to the fact that PS I has a significant extrinsic portion exposed at the stromal side of the membrane (Andersson and Franzén, 1992). Today we know of five subunits of PS I with this location. In addition, the two extrinsic proteins, ferredoxin and FNR, are tightly associated with the extrinsic protein mass at the outer thylakoid surface, making it altogether approximately 100 kDa. In contrast, PS II has relatively few extrinsic membrane proteins and these are all thought to be located at the inner thylakoid surface (Figure 5). In fact, we know today of no extrinsic subunit of PS II exposed at the outer thylakoid surface. As will be discussed in Section IV, this organizational difference between the two photosystems may be of significance for understanding their different distribution or partitioning along the tightly stacked membrane of plant thylakoids.

IV. THYLAKOID MEMBRANE ORGANIZATION

A. Lateral Heterogeneity of the Plant Thylakoid Membrane

As mentioned in Section II.A, the thylakoid membrane shows an unusually high degree of structural complexity with a continuous network of single, nonappressed membranes interconnected with the closely paired or appressed membranes of the grana stacks. It is now well established that this structural differentiation is accompanied by a functional differentiation, and that the plant thylakoid membrane possesses a unique and quite extreme lateral heterogeneity (Figure 5) (see Anderson and Andersson, 1988; Melis, 1991). A main experimental route towards the understanding of the thylakoid membrane organization was subfractionation, which demonstrated that membrane fragments with different compositions could be easily obtained (see Andersson and Anderson, 1985). Typically, these early experiments involved detergents or mechanical fragmentation followed by centrifugation which produced a light fraction rich in PS I, plus the ATPase and a heavy fraction partially enriched in PS II. When nondetergent fractionations were combined with ultrastructural analysis it became clear that the material of the light fractions was derived from the stroma-exposed thylakoid regions, whereas the heavy fractions contained more or less intact grana stacks (Sane et al. 1970). It was some time before large membrane fragments containing virtually only PS II could be isolated. These were inside-out thylakoid membranes formed by interdisc resealing of ruptured grana and then isolated by aqueous polymer two-phase partition (Andersson et al., 1985). Consequently, this membrane fraction apart from having an everted orientation also represented the appressed region of the grana with reduced contribution from the grana membranes and margins. Another approach for isolating pure appressed thylakoid fragments was mild detergent treatment under controlled salt and pH conditions followed by centrifugation including the procedure yielding what is often referred to as the BBY preparations (Berthold et al., 1981). Consequently, by analyzing the overall composition of these fractions which represent the appressed thylakoids with that of the light fractions typical of the stromal exposed thylakoids, the properties of the two thylakoid regions could be probed.

The results from such subfractional analyses revealed a surprisingly big difference in the composition between the various subfractions based upon which the model for lateral heterogeneity of the thylakoid membrane was based (Figure 5) (Anderson and Andersson, 1982). Both the ATP synthase (CF_1-CF_o) and the PS I complexes are located only in the stroma-exposed thylakoid regions. In contrast, most of the PS II complexes and LHC II are confined to the appressed thylakoid regions. Notably, the cytochrome b_6/f complex appeared to be distributed to both thylakoid regions (Cox and Andersson, 1981; Anderson, 1982, 1992).

Simultaneously, based upon concepts in colloidal sciences, it was postulated by Barber (1980a,b) that highly charged components should be excluded through electrostatic repulsion from regions of tight membrane–membrane interaction, such as those occurring in the stacks of a granum. It was suggested that such an exclusion mechanism would apply to PS I complexes, thereby confining this photosystem to the stroma-exposed thylakoids. Thus these theoretical studies gave independent support to the subfractionation analyses.

Quite early Berzborn (1969) had shown that antibodies raised against FNR did not increase their relative binding after destacking of thylakoids in low salt. Thus, exposure of the tightly appressed thylakoid membranes to low salt did not increase the number of binding sites to these antibodies, and it was considered that this enzyme was restricted to the stroma exposed thylakoid regions. Moreover, the electron microscopy studies of Miller and Staehelin (1976) suggested an exclusive location of the ATP synthase in the nonappressed thylakoids. It should also be pointed out that it was clear from freeze-fracture electron microscopy that the appressed and nonappressed thylakoid regions possessed completely different sets of particles both with respect to shape and density (Staehelin, 1976). Despite these observations, an extreme lateral segregation of the two photosystems was not advocated from these immunological or ultrastructural studies. An important independent confirmation of the lateral heterogeneity model (Figure 5) was by the immunogold electron microscopy method and of particular importance was the study of Vallon et al. (1986) who demonstrated a total exclusion of PS I from the appressed thylakoid regions.

Since the early formation of the concept for lateral heterogenity of stacked thylakoids, several experimental observations have contributed to the development and refinement of the membrane model. In particular, the structural and functional distinction of the grana margins where the appressed and nonappressed join each other as a separate membrane domain is significant. Below we will discuss several of these later advances in some detail.

Anderson and Melis (1983) showed that the minor population of PS II centers located in the nonappressed thylakoid region were of the β-type. A more recent development of this observation is that the PS II complexes in the appressed thylakoid membranes appear to be present as dimers, while PS II in the stroma exposed thylakoid regions seems to be monomeric (Dainese et al., 1992). PS IIβ lacks the full complement of light-harvesting antenna and in particular it does not contain the 25-kDa subunit typical of the outer pool of LCH II (Larsson et al., 1987; Melis, 1991). It should also be noted that cytochrome *b*559, which is present in the stroma exposed thylakoid regions is there in its low-potential form (Cox and Andersson, 1981). Consequently PS IIβ does not contain any high-potential form of this cytochrome. The significance of this fact—which is often overlooked in the literature—with respect to the functional properties of PS IIβ remains to be established. Another difference between PS II located in the appressed and stroma-exposed regions is that the latter appear to lack the 22-kDa subunit (*psbS*) (Hundal et al., 1990).

Haehnel et al. (1990) have analyzed the lateral distribution of plastocyanin in the thylakoid lumen. Using immunogold electron microscopy, it was shown that plastocyanin was present along the whole lumenal space. However, the distribution was shown to be different in the dark and in the light. During illumination there was an increase in the plastocyanin concentration in the grana region which was accompanied by a corresponding decrease in the stroma-exposed regions. This observation clearly demonstrated lateral diffusion of plastocyanin which is significant for the question of long distance shuttling of reducing equivalents between the spatially segregated photosystems.

The question of lateral heterogeneity of lipids has also been addressed in terms of their relative abundance in appressed and stroma-exposed thylakoid regions (Gounaris et al., 1983d; Henry et al., 1983; Murphy and Woodrow, 1983). The general picture that has emerged is that compared to the protein complement there is no extreme separation of the lipid classes between the two membrane regions. One important observation, however, is that the ratio between the two major thylakoid lipids, MGDG and DGDG, is quite different. Thus the typical MGDG/DGDG ratio in the appressed thylakoid regions is 2.8, while that of the stromal thylakoids is as low as 1.2. Normally thylakoid membranes are considered to have about twice the amount of MGDG to DGDG, giving a ratio of 2. However, considering the lateral heterogeneity in the distribution of the two galactolipids, this ratio of 2 has no relevance for physiological, biochemical, or physicochemical aspects of the membrane since it is the average value for two different domains with bilayers having MGDG/DGDG ratios close to 3 and 1, respectively.

Analyses of the fatty acid content of the stromal and granal membranes did not reveal any major differences between the two membrane regions, although the appressed thylakoids were found to be slightly more unsaturated (Ford et al., 1982; Chapman et al., 1985).

Studies on the lateral distribution of pigments have shown that the stroma-exposed regions are rich in chlorophyll *a* and β-carotene (Juhler et al., 1993), consistent with the high content of PS I. The appressed membranes on the other hand are enriched in chlorophyll *b*, xanthophylls, lutein, and neoxanthin, typical for pigments associated with PS II and LHC II. Violaxanthin is evenly distributed between the two regions.

Even if the stacked thylakoid membrane of higher plant chloroplasts can be divided into the structurally distinct appressed and stroma-exposed thylakoid regions, certain subdomains can be identified. Particular emphasis has recently been paid to the grana margins as a separate functional and structural domain (Anderson, 1989; Anderson and Thomson, 1989; Albertsson et al., 1991). In fact, considering the three-dimensional organization of the thylakoid lamellae, the margins could potentially form up to 38% of the nonappressed membrane surface area depending on the extent of stacking (Anderson, 1989).

At one stage it was suggested that the cytochrome b_6/f complex may be restricted to the grana margins (Barber 1983b; Ghirardi and Melis, 1983) in an attempt to explain its presence in both the major thylakoid membrane subfractions. However, this model was not supported by immunogold electron microscopy which demonstrated a fairly even distribution of the cytochrome complex along the stacked thylakoid membrane (Allred and Staehelin, 1985; Goodchild et al., 1985). Based upon recent refined subfractionation studies, some variation in the cytochrome *f* content was found with an enrichment in the center of the appressed regions rather than in the margins (Albertsson et al., 1991).

Another view concerning the composition of the margins was taken by Murphy (1986). It was argued that the highly curved membranes in that region could not accommodate any proteins at all due to packing constraints, and that these domains were made up of cone-shaped MGDG at the inner bilayer leaflet and wedge-shaped DGDG at the outer. However, more recent and refined subfractionation studies of the thylakoid membrane, which included isolation of preparations enriched in grana margins, as well as immunogold electron microscopy, have given support to the hypothesis that indeed these border domains contain proteins and that they have distinct properties.

Webber et al. (1988) used the very mild detergent, Tween-20, and found that there was a preferential disruption along the thylakoid margins. After centrifugation, a fraction supposedly enriched in thylakoid margins was found to contain cytochrome b_6/f, PS II, and ATP synthase, but not LHC II. A subfractional approach was also taken by Andreasson et al. (1988). By developing the original two-phase partition subfractionation procedure (Andersson et al., 1985), it was possible to increase the purity of appressed and stroma-exposed membranes, as well as obtaining purified grana margins (Andreasson et al., 1988). In accordance with the subfractionation based upon solubilization with Tween, several proteins were found to be associated with the margins. However, at variance with the conclusions of Webber et al. (1988), a considerable amount of chlorophyll *a*/*b*-binding protein was also found in the margin fraction. Most significantly, evidence was presented to suggest that the PS I complexes located in the margins of the grana were different to those of the planar stroma lamellae (Svensson and Albertsson, 1989; Svensson et al., 1991). The PS I complexes of the margins were shown to have a larger antennae rich in chlorophyll *b*. Considering, the isolation of a PS I complex containing LHC II (Bassi and Simpson, 1987) it seems likely that this larger antenna is due to an extra complement of a chlorophyll *a*/*b*–protein. It therefore appears, as in the case of PS II, that PS I is heterogenous and can be divided into an α- and a β-form as suggested by Albertsson and co-workers (Svensson et al., 1991). PS Iβ

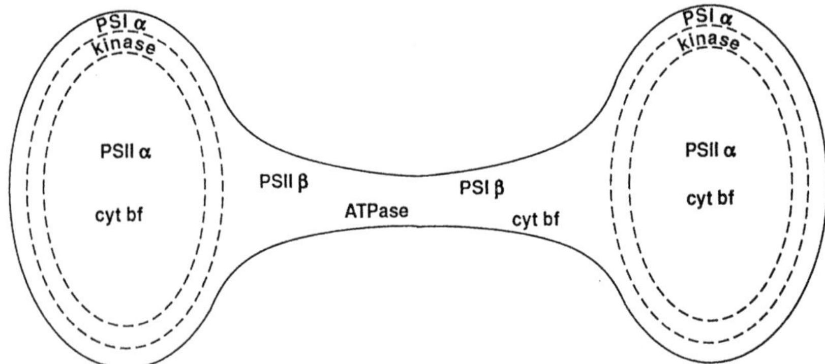

Figure 6. The distribution of protein complexes within the thylakoid membrane is shown in schematic form. Two granal stacks linked by a region of stromal lamella are shown in plan view. The distribution of the photosynthetic complexes and the subtypes thereof is indicated.

contains a small antenna and is located in the stroma lamellae regions, while PS Iα, possessing a larger antenna, is located in the grana end membranes and grana margins extending into the peripheral annulus of the actual appressed portions of the grana (Figure 6).

Yet another very interesting observation related to the grana margins has been presented by Gal et al. (1990). A 64-kDa subunit possibly associated with the kinase enzyme responsible for phosphorylation of LHC II was found by immunogold electron microscopy to be largely confined to the peripheral portion of the grana appressions close to the margins (Figure 6). Very recently, this observation has been supported by the observation of an enrichment of this subunit in grana margins purified by polymer two-phase partition (Yu et al., 1992).

The presence of a significant proportion of PS I in the margin and other stroma-exposed regions of the grana stacks, and thus in relatively close location to PS II (Figure 6), suggests that these regions of the grana stacks may be the major "reaction vessel" for linear electron transport and $NADP^+$ reduction (Anderson, 1989). PS Iβ would, in accordance with early suggestions by Sane et al. (1970), be involved in cyclic electron transport and ATP production. However, given the dynamic and continuous nature of the thylakoid membrane (see Section V) it is hard to visualize such a relatively static functional compartmentalization. Certainly, PS I complexes close to the appressed thylakoid regions (α-centers) would be the most likely to receive an electron from PS II, but that should not exclude the participation of PS Iβ centers in linear electron transport. Thus, in an instant of time there may be a functional gradient with respect to linear electron transport from the grana margins to the midportion of the stroma lamellae. The light-dependent changes in location of plastocyanin between the stromal and granal portions of the thylakoid lumen and the presence of FNR in stroma thylakoid vesicles would support this notion. Nevertheless, the possibility that the grana margin is a separate functional and organizational domain in the heterogeneous lateral organization of the thylakoid membrane is a new development that should be taken into account when dealing with functional and regulatory aspects of photosynthetic electron transport.

B. Lateral Targeting of Proteins

When analyzing the mechanism underlying the lateral distribution of the photosynthetic complexes, it is essential to realize that we are dealing with a dynamic membrane system. Location of one particular component to a restricted membrane domain may therefore not always be the case. It is useful to envisage that each component has a defined lateral index (Andersson, 1984) or partition coefficient (Albertsson, 1988) between different thylakoid regions, just as in the case of a multiphase system. Considering the stroma-exposed and appressed thylakoid regions as a two-phase system, the partition coefficient for PS I and the CF_1–CF_0 complex would be extremely high, while the partition coefficient for PS II would be rather low. Only the cytochrome b_6/f complex would have an intermediate partition coefficient, close to one. Furthermore, such a lateral partition coefficient may not be determined by the entire protein complex, but rather by one single subunit or even one particular lateral targeting sequence of such a subunit—for instance, a particular stretch of charged amino acids. Thus, if a multiprotein complex is disassembled, the individual subunits may have an entirely different lateral distribution than the complex itself. For example, if the PS II complex becomes disassembled, as is the case under strong photoinhibitory illumination *in vitro*, the various subunits show an individual lateral distribution (Hundal et al., 1990). Thus, the 22-kDa protein (psbS) remains in the appressed thylakoid regions while the chlorophyll *a*-proteins, CP47 and CP43, migrate entirely to the stroma-exposed regions. It has therefore been suggested that lateral vector or lateral targeting proteins exist which determine the relative distribution of the photosynthetic complexes between the two main thylakoid regions (Andersson and Styring, 1991). Considering that surface charge appears to play a central role in the organization of the stacked thylakoid membrane (see Section V.B), the hydrophilic extrinsic proteins located at the outer thylakoid surface appear to be likely candidates for such a lateral targeting. Notably, as discussed in Section III.C, the ATP-synthase and PS I which both have significant extrinsic portions exposed at the outer membrane surface are excluded from the tight appressions of the grana stacks (Figure 5). In contrast, the cytochrome b_6/f complex and PS II do not have any extrinsic proteins on the outside of the membrane and can be found in the appressed regions. These extrinsic proteins may also provide steric hindrance for penetration into the tight appressed membrane regions. Interestingly, it has been shown that upon removal of the extrinsic CF_1 portion from the ATPase that the integral CF_0 portion can migrate into the appressed thylakoid regions (Zakharov and Red'ko, 1988).

V. DIFFUSIONAL PROCESSES

As discussed in Section III.A, measurements with the fluorescence probe 1,6-diphenyl-1,2,5-hexatriene (DPH) indicate that the hydrophobic core of the thylakoid membrane is relatively fluid at room temperature (Ford and Barber, 1983) despite its high protein-to-lipid ratio (Chapman et al., 1983). This probe, however, measures only microviscosity as determined by its ability to rotate and "wobble" in the lipid bilayer. This measurement may, or may not, have relevance for long-range diffusion along the plane of the membrane. Nevertheless, given the lateral separation of complexes in the thylakoid membrane, particularly between granal and stromal regions, such long-range diffusion must occur (Figures 5 and 6). This diffusion occurs at two levels: redox communication between the various protein complexes, and diffusion of protein complexes themselves.

A. Diffusion of Redox Components

As far as we know, the redox-active diffusional species are plastoquinone/plastoquinol, plastocyanin, and soluble ferredoxin. Only the former occurs within the membrane itself and shuttles reducing equivalents between PS II and the cytochrome b_6/f complex. The actual quinone species is plastoquinone-9, which is a hydrophobic molecule with two methyl groups attached to its quinone ring and a side chain of nine isoprenic groups. It is located evenly throughout the lipid matrix (appressed and nonappressed regions) at a level corresponding to six or seven molecules per electron transport chain (Chapman and Barber, 1986). It seems likely that its preferred location is at the midplane of the bilayer, since the bulky isoprenoid side-chain methyl groups would not be expected to pack effectively between the acyl chains of the thylakoid lipid matrix (Millner and Barber, 1984). Given that the midplane fluidity may be very high, as indicated by DPH measurements, the lateral diffusion coefficient for plastoquinone within the thylakoid membrane could be as large as $10^{-6} cm^2s^{-1}$ (Millner and Barber, 1984). Such high values for quinone diffusion coefficients have been estimated by Lenaz and his colleagues for mitochondrial membranes (Fato et al., 1985). Using a fluorescence quenching technique and liposomes made with phosphatidylcholine, room temperature values of 10^{-7} to 3.5×10^{-7} $cm^{-2}s^{-1}$ were estimated for plastoquinone in its oxidized and reduced state, with the latter being consistently slower than the former (Blackwell et al., 1986).

Taking a value of 10^{-7} cm^2s^{-1} for plastoquinone or plastoquinol diffusion and the dimensions of a typical grana to be 250 nm, it is possible, using the 2-D diffusion equation of Einstein, $\langle x \rangle^2 = 4Dt$, to calculate the time taken for a plastoquinone molecule to diffuse from the center of a granum to its edge (x) where D is the lateral diffusion coefficient. The calculation gives approximately 0.2 ms. Since the oxidation of plastoquinol by the cytochrome b_6/f complex has a half time of about 20 ms and is the rate-limiting step of linear electron flow, we can therefore conclude that based on $D = 10^{-7}$ cm^2 s^{-1}, the lateral diffusion of plastoquinone is not rate-limiting and the redox-active step itself is reaction-limited.

However, the above conclusion becomes less certain if the presence of protein in the membrane were to lower the diffusion coefficient of plastoquinone by a factor of 10 or more either by reducing the overall fluidity of the bilayer (Millner et al., 1984) or by creating an archipelago effect . Despite this possibility Mitchell et al. (1990) favored a nondiffusion-limited mechanism after carrying out detailed simulations using a Monte Carlo approach. In contrast, Joliot and colleagues have argued that the archipelago effect is significant, and that long range plastoquinone diffusion does not occur (Lavergne and Joliot, 1991; Joliot et al., 1992).

Recent studies using a mutant of the cyanobacterium *Synechocystis* 6803, which is missing the gene which encodes its desaturase, have shown that large changes in unsaturation levels of the thylakoid lipids of this organism have little or no effect on the kinetics of noncyclic electron flow (Gombos et al., 1991). These findings indicate that there is no correlation between plastoquinol oxidation and lipid fluidity, and either favors the restricted diffusion model under all conditions or suggests that the diffusion rate of plastoquinone is sufficiently fast not to be rate-limiting, even when fatty acid unsaturation levels are low. The additional problem here is that the cyanobacterial membrane is not differentiated into stromal and granal regions so that the need for long range lateral diffusion is not obvious. The results of Murata and colleagues (Gombos et al., 1991) also showed that the inability of the desaturase mutant to grow at low temperature is not due to restricted noncyclic electron flow or to the capacity of each photosystem to work efficiently in bringing about charge separation.

Although the long-range diffusion of plastoquinone has frequently been discussed, little is known about the diffusional properties of ferredoxin and plastocyanin along the thylakoid surface. Using lipid vesicles and applying the technique of Fluorescence Recovery after Photobleaching (FRAP), Fragata and colleagues (1984) estimated a lateral diffusion coefficient of 5×10^{-8} cm s^{-1} for plastocyanin. Just how to apply this value to the thylakoid system is not clear since plastocyanin is located in the lumenal space which is clearly a highly restricted compartment (see Haehnel, 1984). Nevertheless, the experiments of Haehnel et al. (1990) mentioned in Section IV.A do demonstrate that plastocyanin can, in a light dependent manner, distribute itself within the whole of the lumenal space.

B. Diffusion of Proteins

Salt Induced Lateral Diffusion and Surface Charge Hypothesis

The first indication that the protein complexes embedded in the thylakoid membrane are able to laterally diffuse came from a study by Wang and Packer (1973). Using freeze-fracture electron microscopy they observed that changing the salt levels in the medium caused a dramatic reorganization of particles within the plane of the membrane. It had previously been shown that the same changes in electrolyte levels brought about changes in the degree of stacking of thylakoids (Izawa and Good, 1966). The existence of reversible salt-induced organizational changes was also shown by others (Ojakian and Satir, 1974) and by Staehelin (1976). From these findings and from further experimentation the concept emerged that the two salt-induced phenomena (stacking/unstacking and lateral protein diffusion) are regulated by electrostatic screening and are coupled events (Barber, 1980a, 1982a). The idea that the two events are coupled stemmed from the fact that the same salt treatments which bring about the conformational changes also induce changes in chlorophyll fluorescence, even when the PS II activity was blocked by DCMU (Barber et al., 1980). The basis of the hypothesis was that when the thylakoids were completely unstacked there was lateral intermixing of all protein complexes, and as a consequence good energy transfer from PS II, or mobile LHC II to PS I. This therefore would be the reason why the F_m level of room temperature chlorophyll fluorescence is at a minimum under these low-salt conditions. The complete unstacking and the randomization of the complexes was argued to be due to poor electrostatic screening of negative electrical charges on the exposed surfaces of the protein complexes. That is, it was suggested that coulombic repulsive forces not only keep adjacent membranes apart but also favor randomization of protein complexes along the plane of the bilayer (Barber, 1980a, 1982a,b). On adding cations (trivalents are more effective than divalents, which are more effective than monovalents, as would be expected from the theories of electrostatics) the screening of surface negative charges is improved. This in turn reduced the reduction in coulombic repulsion which allows the coming together of the membrane surface and a concomitant phase separation of protein complexes based on the differences in their surface charge densities. It was argued that the reduction in coulombic repulsion due to effective electrostatic screening allows long-range van der Waals forces to play a role in stabilizing the new conformational state. From biochemical analyses of various membrane fractions (Andersson and Anderson, 1980), it would seem that PS II complexes together with their associated LHC II have sufficiently low surface net charge densities that they are able to form tightly packed domains which facilitate strong van der Waals interactions between adjacent thylakoid surfaces, thus giving rise to the appressed regions of the grana (Barber 1980a, 1982a). On the other hand, it was reasonable to assume that the location of PS I in the nonstacked region was due to the presence of significant levels of net negative

Table 2. Lateral Migration of Protein Along Stacked Thylakoid Membranes

appressed thylakoids $\underset{b}{\overset{a}{\rightleftharpoons}}$ stroma exposed thylakoids

Physiological event	*Migrating protein(s)*	*Posttranslational modification*	*Reference**
Over excitation of PS II state I/state II transition	Outer pool of LHC II	phosphorylation[a]/dephosphorylation[b]	1, 2
State I/state II transition Reg. NADPH/ATP level	Cytochrome b_6/f complex	phosphorylation[a]/dephosphorylation[b]	3, 4
Heat stress	PS II core + inner pool of LHC II		5, 6
Photodamage and repair of PS II	PS II core	phosphorylation?[a]/palmitoylation?[b]	7, 8, 9
	PS II core + inner pool of LHC II		10
Biosynthesis of PS II	D_1 protein	palmitoylation?[b]	9
	LHC II		11

Note: *1. Allen et al. (1981). 2. Anderson and Anderson (1988). 3. Vallon et al. (1991). 4. Anderson (1992). 5. Gounaris et al. (1984a). 6. Sundby and Anderson (1985). 7. Adir et al. (1990). 8. Aro et al. (1993). 9. Matto and Edelman (1987). 10. Melis (1991). 11. Yalovski et al. (1992).

charge on its surface. Indeed, as stated in Section III.C, the PS I complex does have a number of extrinsic hydrophilic proteins bound to its outer surface (Figure 5). The location of the coupling factor complex in nonappressed regions is probably also governed by the existence of the hydrophilic and bulky CF_1 component.

The above simple picture of lateral heterogeneity for stacked thylakoids being governed by the balance of electrostatic repulsion and van der Waals attractive forces gained considerable experimental support. However, the hypothesis also has to account for the more even distribution of the cytochrome b_6/f complex (Cox and Andersson, 1981; Allred and Staehelin, 1985; Anderson, 1989, 1992) and for the lateral movements of protein complexes between appressed and nonappressed membranes (Table 2). Presumably in the case of cytochrome b_6/f complex, its surface charge properties are intermediate between PS II plus LHC II and PS I and, as such, it can partition between both membrane regions. Alternatively, the recent finding that the cytochrome b_6/f complex can be phosphorylated (Gal et al., 1992) may contribute to its heterogenous lateral distribution.

Protein Phosphorylation and Lateral Diffusion

The discovery that LHC II can be reversibly phosphorylated (Bennett, 1977, 1979a, 1991) and as a consequence migrate from the appressed to the nonappressed regions (Table 2) (Chow et al 1981; Andersson et al., 1982) is in accordance with the surface charge hypothesis (Barber, 1982 a,b). It is therefore likely that the addition of negative charges to the surface of mobile LHC II (Larsson et al., 1987) destabilizes the presence of the complex in the appressed regions so that it migrates to the more electrically charged, nonappressed regions where it can transfer absorbed light energy to PS I (Telfer et al., 1986). This occurs when PS II is over-excited relative to PS I so that a buildup of reducing equivalents in the intersystem redox carriers (plastoquinone and cytochrome b_6/f triggers the protein kinase responsible for the phosphorylation) (Allen et al., 1981; Barber, 1983a; Gal et al., 1990; Allen, 1992). The function of this seems to be either to optimize the rate of electron flow under light-limiting conditions (Williams and Allen, 1987), or to protect PS II against photodamage at high light intensities (Horton and Lee, 1985). If, on the other hand, PS I receives more light than PS II, the kinase activity is turned off and a phosphatase brings about the dephosphorylation of LHC II and, as a consequence, the complex reestablishes itself in the PS II-enriched domain in the appressed region.

The concept that the shuffling of LHC II between the appressed and nonappressed membranes can be explained in terms of electrostatic theory has been well supported by experimentation where the relative effect of LHC II phosphorylation on thylakoid membrane organization was monitored as a function of the background cation levels (Telfer et al., 1983, 1984). Whether phosphorylation of the N-terminal threonine of the LHC II polypeptide also induces intramolecular reorganization which, in turn, also contributes in some way to the trigger for its lateral diffusion (Allen, 1992) is so far unclear since no experimental data is available to support this hypothesis.

More recently, it has been demonstrated by immunogold electronmicroscopy and thylakoid membrane subfractionation (Vallon et al., 1991), that there also is regulatory lateral movement of the cytochrome b_6/f between the appressed and nonappressed thylakoid regions (Table 2). Increased levels of the cytochrome complex in the stroma thylakoids could be found in thylakoids of *Chlamydomonas* after transition from the so-called state-1 to state-2. This occurs when PS II is overreduced compared to PS I, but also when there is an increased demand for ATP (see Anderson, 1992). Thus it is thought that the lateral migration is significant for regulating noncyclic versus cyclic electron flow and hence the NADPH/ATP ratio (Vallon et al., 1991, Andersson, 1992). The lateral movement of the cytochrome b_6/f complex occurs concomitantly with LCH II migration, suggesting a dependence on protein phosphorylation. As stated above (Section V.B), it has been shown that there indeed is a phosphorylation of cytochrome b_6 (Gal et al., 1992), supporting this notion.

Migration in Connection with Protein Biosynthesis and Turnover

It is now also apparent that lateral protein trafficking occurs during biosynthesis processes (Table 2) especially in relation to the turnover of the D1 protein (Wettern, 1986; Mattoo and Edelman, 1987) and repair of the PS II complex (Adir et al., 1990; Hundal et al., 1990; Melis, 1991). This specific example is a consequence of the vulnerability of PS II to photoinhibitory damage (Barber and Andersson, 1992;

Prasil et al., 1992; Aro et al., 1993) and, as such, is a remarkable feature of the dynamics of the thylakoid membrane. Using a pulse chase procedure, it has been shown that newly synthesized D1 protein is initially inserted into the nonappressed lamellae, being the location of the chloroplast ribosomes (Wettern, 1986; Mattoo and Edelman, 1987). It is processed from a 33.5- to 32-kDa form and then diffuses laterally to the appressed regions. The trigger for this lateral migration may be the processing itself or a posttranslational palmitoylation of the D1 protein (Mattoo and Edelman, 1987). According to the "electrostatic" model, any change which makes the D1 protein less polar would facilitate its partitioning into the granal region. On the other hand, the lateral movement of the D1 protein from the appressed to the nonappressed region as a consequence of photoinhibitory damage would be aided if it were made more polar. Interestingly, like LHC II, the D1 protein undergoes phosphorylation at its N-terminus (Marder et al., 1988; Michel et al., 1988) to form a species known as $D1^*$ (Callahan et al., 1990). Recently it has been shown that $D1^*$ forms under photoinhibitory conditions (Aro et al., 1992) and it is possible that this leads to its diffusion to the nonappressed stromal lamellae where it undergoes proteolytic degradation. Whether the D1 protein migrates between the two membrane regions as a separate protein or is associated with other PS II proteins has been a matter of controversy (Adir et al., 1990; Hundal et al., 1990; Melis, 1991), although it now seems likely that *in vivo* the whole PS II complex migrates to the stromal lamellae after photoinhibitory damage where it receives a new copy of the D1 protein (Adir et al., 1990; Aro et al., 1993). It may not immediately diffuse back to the grana region, but rather exists as a "reserve" complex which is not fully functional and therefore protected against photoinhibition (Melis, 1991). In this way the photosynthetic apparatus is poised to supply a "repaired" PS II complex if and when necessary.

To date, there is no evidence for or against the concept that the shuffling of damaged and repaired complexes between granal and stromal lamellae is controlled by electrostatic forces as advocated for LHC II movement. But the discovery of $D1^*$ and its linkage with phosphorylation of the D1 protein does hint to a surface charge mechanism. To prove or disprove this requires experiments to test the effect of salts on $D1^*$ protein migration along the lines of the those conducted previously (Telfer et al., 1983, 1984). It should also be noted that lateral migration of newly synthesized LHC II from the nonappressed to the appressed regions has been shown to occur by a mechanism that is independent of its processing (Yalovsky et al., 1992) and that lateral migration of the PS II complex also occurs under heat-stress conditions (Table 2) (Gounaris et al. 1984a; Sundby and Andersson, 1985).

Clearly, the lateral diffusion of proteins requires the thylakoid membrane to be relatively fluid. Typically, proteins diffuse in biological membranes at a much slower rate than lipids or small hydrophobic molecules, such as plastoquinone. By monitoring the rate of rise in chlorophyll fluorescence due to segregation of PS I and PS II complexes induced by the addition of cations, Rubin et al. (1981) estimated a diffusion coefficient of 10^{-11} cm^2s^{-1} at 23 °C. As the temperature was lowered the rate of fluorescence rise decreased significantly with an estimated diffusion coefficient of 2×10^{-12} cm^2 s^{-1} at 10 °C. More recently, the kinetics of the migration of phosphorylated LHC II from granal to stromal membranes have been measured as a function of temperature (Carlberg et al., 1992). It was found that as the temperature was lowered from 20 to 10 °C the rate dropped by six times, and below 10 °C virtually no lateral diffusion occurred. This significant drop in mobility was not due to changes in kinase activity but was attributed to increased viscosity of the lipid matrix at lower temperatures. The data was further used to estimate diffusion coefficients using a Monte Carlo approach formulated by Mitchell et al. (1990). The values estimated were in the region of 2 to 4×10^{-12} cm^2 s^{-1} at 20 °C (Drepper et al., 1993). These low diffusion values indicate that lateral diffusion is restricted, presumably by the high protein level in the membrane, and by strong interactions between neighboring complexes and between the surfaces of adjacent membranes (Barber, 1982b; Drepper et al., 1992).

From the above work, a picture has emerged that there is a considerable amount of protein trafficking along the plane of the thylakoid membrane. Some of this movement is clearly associated with optimizing energy capture and its utilization, while others are a consequence of biosynthetic processes involving repair of photodamaged PS II centers (Table 2). For these reasons it seems that the thylakoid membrane is required to be a relatively fluid system and it is not surprising that adaptations in its composition occur in response to different growth conditions, especially temperature (see Section III.A). If such adaptations do not occur, the functional properties of thylakoids are likely to be less efficient or even inhibited. It is probably for this reason that the mutant of *Synechocystis* 6803, which has an inoperative desaturase gene (Wada et al., 1990; Gombos et al., 1991; already discussed in Section III.A of this chapter) cannot grow at low temperatures. The temperature sensitivity does not seem to be due to a reduction in the efficiency of photosynthesis but could be a consequence of increased susceptibility to photoinhibition. Since the net degree of photoinhibition is determined by a balance between the rate of damage and the rate of repair, the increased viscosity of the membrane in the mutant could inhibit the latter, an effect which will be aggrevated as the temperature is lowered.

VI. AUXILIARY FUNCTIONS OF THE THYLAKOID MEMBRANE

As discussed above, the thylakoid membrane is highly dynamic in its nature and responds to changes in the surrounding environment through both short-term and long-term acclimations. However, these changes have to be induced and regulated in a controlled manner and are therefore catalyzed by specific enzymes. Today our knowledge about these auxiliary enzymes is only in its infancy, but below we give some examples of such processes. These include: kinases and phosphatases involved in reversible thylakoid protein phosphorylation (Bennett 1991; Allen 1992); enzymes involved in other types of post-translational modification, such as acylation of proteins (Mattoo and Edelman, 1987), endogenous proteases required for processing of precursor proteins (chapter by Bruce and Keegstra, this volume), or removal of photodamaged polypeptides (Barber and Andersson, 1992; Prasil et al., 1992; chapter by Chow, this volume); fatty acid desaturases affecting membrane fluidity (Wada et al., 1989); as well as heat shock proteins (Kruse and Kloppstech, 1992) and early light-induced proteins (ELIPS) (Green et al., 1991).

Several years have passed since Bennett (1977) discovered protein phosphorylation in thylakoid membranes. Still our knowledge about the enzymes involved is not very advanced although progress with respect to the kinase responsible for the phosphorylation of LHC II has been made. A 64-kDa subunit which has been associated with the kinase is membrane-bound and can be isolated after detergent extraction (Coughland and Hind, 1986; Gal et al. 1990) and is present in substoichiometric amounts. As is common for many protein kinases, the LHC II kinase appears to be autophosphorylated (Coughland and Hind, 1987). Early on, it was suggested that the kinase is located in the appressed thylakoid regions (Cougland and Hind, 1987). However, more recent studies using electron microscopy combined with immunogold labeling (Gal et al., 1990) and subfractionation studies (Yu et al., 1992), suggest that the kinase, as revealed by the distribution of the 64-kDa subunit, has its main location at the periphery of the appressed thylakoid regions, close to the margins. The kinase has been shown to be closely associated with the cytochrome b_6/f complex, a property that appears to be essential for its activation (Gal et al., 1990; Bennett 1991; Allen, 1992). Interestingly, it has been shown that autophosphorylation of the kinase correlates with phosphorylation of the cytochrome b_6 subunit (Gal et al., 1992).

It is still not clear whether the same membrane-bound kinase is responsible for both phosphorylation of LHC II subunits and the various PS II subunits (Allen, 1992). Our understanding of the phosphatase or phosphatases responsible for dephosphorylation of the thylakoid phosphoproteins is even more limited. Very recently, however, Kieleczawa et al. (1992) isolated an alkaline phosphatase from pea thylakoids. The molecular mass of this enzyme was 51.5 kDa and it could dephosphorylate phosphorylated histones as well as phosphorylated subunits of PS II particles. As is the case with the kinase activity we still do not know if there is one or several thylakoid phosphatases.

It was originally suggested by Bennett (1979b) that the LHC II phosphatase is active in the dark and not under strict regulation. However, this would be atypical with respect to the general properties of phosphatases, which are highly regulated in various biological systems (Cohen, 1989). It therefore appears very likely that the phosphatase(s) in the thylakoid membrane is under some kind of regulatory control (Allen, 1992). Support for this notion has come from studies on D1 protein degradation and repair following photoinhibition (Aro et al., 1992, 1993). In plants the D1 protein appears to be transiently phosphorylated after being photodamaged and targeted for degradation. However, the D1 protein appears not to be degraded before the phosphate group has been removed. It is suggested that this phosphorylation is essential for avoiding premature degradation of the D1-protein before a new copy of the protein is available during the repair process, thereby ensuring stability of the remaining subunits of the PS II couples (Adir et al., 1990).

Apart from protein phosphorylation, there are very few indications of covalent posttranslational modifications of thylakoid proteins. For example, there is no evidence for the existence of any glycolsylated proteins in the thylakoid membrane. However, in an interesting paper Mattoo and Edelman (1987) showed that the D1 protein does not only undergo reversible phosphorylation but also is posttranslationally palmitoylated. This palmitoylation is light-dependent and blocked by DCMU. It has been suggested that this posttranslational modification is essential for the lateral targeting of newly synthesized D1 protein from its insertion site in the stroma exposed thylakoids to its functional site in the appressed thylakoid regions. Even though the significance of this reversible palmitoylation remains to be established both at the biochemical and physiological levels, it points to the fact that there are acyl–transferase enzymes in the thylakoid membrane. The identity and properties of such enzymes are at present entirely unknown. Yet another posttranslational modification is prenylation of proteins (Farnsworth et al., 1989) which has recently been described in plants (Swiezewska et al., 1993). Whether there are prenyl transferase activities within the chloroplast or in the thylakoid membrane remains to be established.

Another category of auxiliary enzymes in the thylakoid membrane are endogenous proteases. Such thylakoid proteases are involved in the processing of precursor proteins during biogenesis of the photosynthetic apparatus (Dalbey and von Heijne, 1992), the removal of photodamaged D1 protein (Barber and Andersson, 1992), as well as the acclimation of the light-harvesting antenna size (Lindahl and Andersson, 1992).

The biogenesis of the thylakoid membrane is a very complicated process that involves both the nuclear and plastid genomes and the transport of precursor proteins through three biomembranes (Herrmann et al., 1991; Bruce and Keegstra, this volume). With very few exceptions, newly synthesized nuclear-encoded proteins contain an amino-terminal extension which is proteolytically removed in one or two steps during import into the functional site in the chloroplast (Bruce and Keegstra, this volume; de Boer and Weisbeek, 1991). The first cleavage takes place by a stromal signal peptidase while the second cleavage requires a membrane-bound protease. The latter is analogous to signal peptidases found in the eukaryotic endoplasmic reticulum membrane, the mitochondrial inner membrane, and the prokaryotic plasma membrane (Dalbey and von Heijne, 1992). All such signal peptidases so far characterized are integral membrane proteins. It is predicted that the thylakoid signal peptidase contains two hydrophobic membrane-spanning regions with the catalytic site exposed towards the lumenal surface. These signal peptidases appear to form a new type of serine proteases, not requiring a histidine, and being possibly related to the β-lactamases. Processing activities in the thylakoid membrane have been thoroughly studied in the context of the import into the lumen of plastocyanin and the three extrinsic proteins (33, 23, and 16 kDa) associated with the water–oxidation system. The processing activity has been partially purified by solubilization with Triton X-100 (Kirwin et al., 1987). The activity is associated with the membrane, as predicted, and is exclusively located in the stroma-exposed thylakoid regions.

Much attention has also been paid to the processing of the D1 protein which involves the proteolytic cleavage of a C-terminal extension (Taylor et al., 1988; Diner et al., 1988) after its insertion into the stroma thylakoid region (Wettern, 1986). In a *Scenedesmus* LF1-mutant where the processing of the D1-protein is impaired, no ligation of manganese to the PS II complex can occur and the mutant cannot perform water oxidation (Diner et al., 1988; Taylor et al., 1988). Addition of a Triton X-100 extract from wild-type thylakoid membranes to the mutant membranes removes the C-terminal extension, and photoactivation of the mutant is permitted (Taylor et al., 1988). A membrane-bound 34-kDa protein from spinach thylakoids have been isolated and shown to be able to catalyze the correct processing of the D1 protein *in vitro* (Inagaki et al., 1989). It remains to be established whether this processing protease has any connection to the processing activity associated with the import of the lumenal proteins discussed above.

Currently, much research is focused on the proteolytic activities associated with degradation and turnover of the D1 protein during photoinhibition of photosynthesis (Barber and Andersson, 1992; Prasil et al., 1992; Aro et al., 1993; Chow, this volume). The majority of the experimental observations suggest that the protease(s) involved is of serine type and an integral part of the PS II complex (Virgin et al., 1990; Shipton and Barber, 1991; Barber and Andersson, 1992; Salter et al., 1992b). The identity of the D1 protease remains to be established. Using radioactively labeled diisopropylfluorophosphate, which covalently binds to the catalytic site of serine proteases, the chlorophyll *a*-binding protein CP43 became specifically labeled (Salter et al., 1992b). On the other hand, D1 protein proteolysis can be seen in isolated PS II reaction center particles that are devoid or largely depleted of CP43 (Shipton and Barber, 1991), and it has been suggested that D1 protein degradation is an autoproteolytic process (Virgin et al., 1990; Shipton and Barber, 1991). Possibly, the complete proteolysis of the transmembrane D1 protein requires more than one protease in a process that involves endoproteolytic cleavages at each loop exposed at the inner or outer thylakoid surfaces to be completed by exoproteolysis of intermediate fragments, which would possess one membrane span.

More recently, there is evidence for a membrane-associated protease responsible for removing excess LHC II during low-light to high-light acclimation of spinach leaves (Lindahl and Andersson, 1992).

As discussed in Sections III and V.A, the unsaturation level of thylakoid lipids varies in response to growth temperature and, at least in cyanobacteria, is essential for conferring chilling tolerance (Wada et al., 1990). The identification and characterization of desaturase enzymes should therefore provide another essential area of photosynthesis research.

Many auxiliary enzymes or components can be classified as "stress proteins". Examples of such enzymes are those involved in the so-called xanthophyll cycle which has been suggested to be essential for dissipation of excess of excitation energy (Demmig-Adams 1990). A de-epoxidase that converts violaxanthin to zeaxanthin via antheraxanthin is thought to be located at the lumenal side of the thylakoid membrane, while an epoxidase catalyzes the reverse reaction and is located at the outer thylakoid surface facing the stroma.

Another category of proteins that may play a protective role during light stress are the early light-inducible proteins (ELIPs) (Green et al., 1991). The mature ELIP has a molecular mass of 17 kDa and shows a pronounced similarity in amino acid sequence with two membrane-spanning domains of the chlorophyll *a/b*-binding proteins. It was early on suggested that ELIPs have a role in the assembly of the photosynthetic apparatus during its early stages of development (see Kruse and Kloppstech, 1992). More recently, however, several elegant studies have shown that ELIP accumulates in mature thylakoid membranes during light stress (Adamska et al., 1992 a,b, 1993). It has been suggested that ELIP could act as a chlorophyll scavenger, binding pigments released from degraded chlorophyll-containing complexes during light stress (Adamska et al., 1992a), or play a role in the repair process of PS II during photoinhibition (Adamska and Kloppstech, 1991a).

Heat stress is another threat to living organisms and studies on various forms of heat-shock proteins are a major research topic (see Kruse and Kloppstech, 1992). When it comes to heat-shock proteins associated with the thylakoid membrane, the situation is so far not entirely clear, but a heat-shock protein with a molecular mass of 22 kDa appears to be associated with the grana stacks close to PS II (Adamska and Kloppstech, 1991b; Kruse and Kloppstech, 1992). The existence of cold-shock proteins in chloroplast is another interesting possibility that should be investigated, in addition to changes in lipid properties when it comes to chilling-tolerance in plants.

Without doubt the auxiliary functions and the components thereof in the thylakoid membrane will gain more attention in the future, and many enzymes in this category have probably yet to be identified. In this respect, one should also consider the supernumerosity of the various photosynthetic complexes described in Section III.B. Several subunits which today appear to be unnecessary for the photosynthetic electron transport and the light-harvesting processes in the these complexes may well prove to possess auxiliary enzymatic functions.

ACKNOWLEDGMENTS

We thank Lynn Barber, Torill Hundal, and Hugh Salter for help in preparation of the manuscript.

ABBREVIATIONS

CF_1 - CF_o:	chloroplast ATP-synthase
CP:	chlorophyll-binding protein
DCCD:	dicyclohexylcarbodiimide
DCMU:	3-(3,4-dichlorophenyl)-1,1-dimethylurea
DGDG:	digalactosyldiglyceride
DPH:	1,6-diphenyl-1,2,5-hexatriene
ELIP:	early light-induced proteins
FNR:	ferredoxin-$NADP^+$-oxidoreductase
FRAP:	fluorescence recovery after photobleaching
Hex-II:	hexagonal type II phase
LHC:	light-harvesting complex
MGDG:	monogalactosyldiglyceride
P700:	the primary electron donor of PS I
P680:	the primary electron donor of PS II
PC:	phosphatidylcholine
PG:	phosphatidylglycerol
PS:	photosystem
Q_A:	first quinone acceptor
Q_B:	second quinone acceptor
SQDG:	sulphoquinovosyldiglyceride

REFERENCES

Adamska, I. & Kloppstech, K. (1991a). Evidence for an association of the early light-inducible protein (ELIP) of pea with photosystem II. Plant Mol. Biol. 16, 209–223.

Adamska, I. & Kloppstech, K. (1991b). Evidence for the localization of the nuclear-coded 22-kDa heat-shock protein in a subfraction of thylakoid membranes. Eur. J. Biochem. 198, 375–378.

Adamska, I., Kloppstech, K., & Ohad, I. (1992a). UV light stress induces the synthesis of the early light-inducible protein and prevents its degradation. J. Biol. Chem. 267, 24732–24737.

Adamska, I., Ohad, J., & Kloppstech, K. (1992b). Synthesis of the early light-inducible protein is controlled by blue light and related to light stress. Proc. Natl. Acad. Sci. USA 89, 2610–2613.

Adamska, I., Kloppstech, K., & Ohad, J. (1993). Early light-inducible protein in pea is stable during light stress but is degraded during recovery at low light intensity. J. Biol. Chem. 268, 5438–5444.

Adir, N., Shochat, S., & Ohad, I. (1990). Light-dependent D1 protein synthesis and translocation is regulated by reaction center II: RC II serves as an acceptor for the D1 precursor. J. Biol. Chem. 265, 12563–12568.

Albertsson, P.-Å. (1988). Analysis of the domain structure of membranes by fragmentation and separation in aqueous polymer two-phase systems. Q. Rev. Biophys. 21, 61–98.

Albertsson, P.-Å., Andreasson, E., Svensson, P., & Yu, S.-G. (1991). Localization of cytochrome f in the thylakoid membrane: Evidence for multiple domains. Biochim. Biophys. Acta 1098, 90–94.

Allen, J.F., Bennett, J., Steinback, K.E., & Arntzen, C.J. (1981). Chloroplast protein phosphorylation couples plastoquinone redox state to distribution of excitation energy between photosystems. Nature (London) 291, 1–5.

Allen, J.F. (1992). Protein phosphorylation in the regulation of photosynthesis. Biochim. Biophys. Acta 1098, 275–335.

Allred, D.R. & Staehelin, L.A. (1985). Lateral distribution of cytochrome b_6f and coupling factor ATP synthetase complexes of chloroplast thylakoid membranes. Plant Physiol. 78, 199–202.

Alt, J. & Herrmann, R.G. (1984). Nucleotide sequence of the gene for pre-apocyto-chrome *f* in the spinach plastid chromosome. Curr. Genet 8, 551–557.

Anderson, J.M. (1982). Distribution of the cytochromes of spinach chloroplasts between the appressed membranes of grana stacks and stroma-exposed thylakoid. FEBS Lett. 138, 62–66.

Anderson, J.M. & Andersson, B. (1982). The architecture of photosynthetic membranes. Lateral and transverse organization. Trends Biochem. Sci. 7, 288–292.

Anderson, J.M. & Melis, A. (1983). Localization of different photosystems in separate regions of chloroplast membranes. Proc. Natl. Acad. Sci. USA 80, 745–749.

Anderson, J.M. & Andersson, B. (1988). The dynamic photosynthetic membrane and regulation of solar energy conservation. Trends Biochem. Sci. 13, 351–355.

Anderson, J.M. (1989). The grana margins of plant thylakoid membranes. Physiol. Plant. 76, 243–248.

Anderson, J.M. & Thomson, W.W. (1989). Dynamic molecular organization of the plant thylakoid membrane. In: Photosynthesis, pp. 161–182. Alan R. Liss, New York.

Anderson, J.M. (1992). Cytochrome b_6f complex: Dynamic molecular organization, function and acclimation. Photosynth. Res. 34, 341–357.

Andersson, B. & Anderson, J.M. (1980). Lateral heterogeneity in the distribution of chlorophyll-protein complexes of the thylakoid membrane of spinach chloroplasts. Biochim. Biophys. Acta 593, 427–440.

Andersson, B., Åkerlund, H.-E., Jergil, B., & Larsson, C. (1982). Differential phosphorylation of the light-harvesting chlorophyll–protein complex in appressed and nonappressed regions of the thylakoid membrane. FEBS Lett. 149, 181–185.

Andersson, B. (1984). Isolation of inside-out thylakoid vesicles with increased photosystem II purity. Lateral index of thylakoid components. In: Advances in Photosynthesis Research (Sybesma, C., Ed.), Vol. 3, pp. 223–226. Martinus Nijhoff, The Hague.

Andersson, B. & Anderson, J.M. (1985). The chloroplast thylakoid membrane: Isolation, subfractionation and isolation of its supramolecular complexes. In: Modern Methods in Plant Analysis (Linskens, H.F. and Jackson, J.F., Eds.) Vol. I, pp. 231–258. Springer Verlag, Berlin.

Andersson, B., Sundby, C., Åkerlund, H.-E., & Albertsson, P.-Å. (1985). Inside-out thylakoid vesicles. Important tools for the characterization of the photosynthetic membrane. Physiol. Plant. 65, 322–330.

Andersson, B. & Åkerlund, H.-E. (1987). Proteins of oxygen evolving system. In: Topics in Photosynthesis (Barber, J., Ed.), Vol. 8, pp. 379–420. Elsevier, Amsterdam.

Andersson, B. & Herrmann, R.G. (1988). Structure, function and biogenesis of nuclear-encoded proteins of photosystem II. In: Plant Membranes Structure, Assembly and Function (Harwood, H.J. & Walton, T.J., Eds.), pp. 33–45. The Biochemical Society, London.

Andersson, B. & Styring, S. (1991). Photosystem II: Molecular organization, function and acclimation. In: Current Topics in Bioenergetics (Lee, C.P., Ed.), Vol. 16, pp. 1–82. Academic Press, New York.

Andersson, B. (1992). Thylakoid membrane dynamics in relation to light stress and photoinhibition. In: Trends in Photosynthesis Research (Barber, J., Guerreno, M.G., & Modrano, H., Eds.), pp.71–86. Intercept, Andover.

Andersson, B. & Franzén, L.-G. (1992). The two photosystems of oxygenic photosynthesis. In: New Comprehensive Biochemistry: Molecular Mechanisms in Bioenergetics (Ernster, L., Ed.), Vol. 23, pp. 121–143. Elsevier, Amsterdam.

Andreasson, E., Svensson, P., Weibull, C., & Albertsson. P.-Å. (1988). Separation and characterization of stroma and grana membranes: Evidence for heterogeneity in antenna size of both photosystem I and photosystem II. Biochim. Biophys. Acta 936, 339–350.

Arnon, D.I. (1991). Photosynthetic electron transport: Emergence of a concept, 1949–1959. Photosynth. Res. 29, 117–1313.
Aro, E.M., Virgin, I., & Andersson, B. (1993). Photoinhibition of photosystem II inactivation, protein damage and turnover. Biochim. Biophys. Acta 1143, 113–134.
Aro, E.M., Kettunen, R., & Tyystjärvi, E. (1992). ATP and light regulated D1 protein modification and degradation. FEBS Lett. 297, 29–33.
Barber, J. (1980a). Membrane surface charges and potentials in relation to photosynthesis. Biochim. Biophys. Acta 594, 253–308.
Barber, J. (1980b). An explanation for the relationship between salt-induced thylakoid stacking and the chlorophyll fluorescence changes associated with changes in spillover of energy from photosystem II to photosystem I. FEBS Lett. 118, 1–10.
Barber, J. (1982a). Influence of surface charges on thylakoid structure and function. Ann. Rev. Plant Physiol. 11, 261–295.
Barber, J. (1982b). The control of membrane organization by electrostatic forces. BioSci. Rep. 2, 1–13.
Barber, J. (1983a). Membrane conformational changes due to phosphorylation and the control of energy transfer in photosynthesis. Photobiochem. Photobiophys. 5, 181–190.
Barber, J. (1983b). Photosynthetic electron transport in relation to thylakoid membrane composition and organization. Plant Cell Environ. 6, 311–322.
Barber, J. (1987). Photosynthetic reaction centres: A common link. Trends Biochem Sci. 12, 321–326.
Barber, J., Chow, W.S., Schoufflaire, C., & Lannoye, R. (1980). The relationship between thylakoid stacking and salt-induced chlorophyll fluorescence changes. Biochim. Biophys. Acta. 591, 92–103.
Barber, J., Ford, R.C., Mitchell, R.A.C., & Millner, P.A. (1984). Chloroplast thylakoid membrane fluidity and its sensitivity to temperature. Planta 161, 375–380.
Barber, J., Chapman, J., & Telfer, A. (1987). Characterization of a photosystem II reaction centre isolated from the chloroplasts of *Pisum sativum*. FEBS Lett. 220, 67–73.
Barber, J. & De Las Rivas, J. (1993). Direct reduction of cytochrome b559 by photoreduced pheophytin and its possible protection against photoinhibition. Proc. Natl. Acad. Sci. USA 90, 10942–10946.
Barber, J. & Andersson, B. (1992). Too much of a good thing; light can be bad for photosynthesis. Trends Biochem. Sci. 17, 61–66.
Bassi, R. & Simpson, D. (1987). Chlorophyll–protein complexes of barley photosystem I. Eur. J. Biochem. 163, 221–230.
Bennett, J. (1977). Phosphorylation of chloroplast membrane polypeptides. Nature (London) 269, 344–346.
Bennett, J. (1979a). Chloroplast phosphoproteins. The protein kinase of thylakoid membranes is light-dependent. FEBS Lett. 103, 342–344.
Bennett, J. (1979b). Chloroplast phosphoproteins phosphorylation of polypeptides of the light-harvesting chlorophyll–protein complex. Eur. J. Biochem. 99, 133–137.
Bennett, J. (1991). Protein phosphorylation in green plant chloroplasts. Ann. Rev. Plant Physiol. 42, 281–311.
Berthold, D.A., Babcock, G.T., & Yocum, C.F. (1981). A highly resolved oxygen-evolving photosystem II preparation from spinach thylakoid membranes. FEBS Lett. 134, 231–234.
Berzborn, R.J. (1969). Untersuchungen über die Oberflächenstruktur des Thylakoidsystems der Chloroplaster mit Hilfe von Antikörpern gegen die Ferredoxin-NADP-Reduktase. Z. Naturforsch. 24b, 436–446.
Blackwell, M., Gounaris, K., & Barber, J. (1986). Evidence that pyrene excimer formation in membranes is not diffusion controlled. Biochim. Biophys. Acta 858, 221–234.
de Boer, A.D. & Weisbeek, P.J. (1991). Chloroplast protein topogenesis: Import, sorting and assembling. Biochim. Biophys. Acta 1071, 221–253.
Brangeon, J. & Mustardy, L.A. (1979). The autogenetic assembly of intra-chloroplastic lamellae viewed in three-dimensions. Biol. Cell 36, 71–80.

Burnap, R. L. & Sherman, L. A. (1991). Deletion mutagenesis in *Synechocystis* sp. PCC 6803 indicates that the Mn-stabilizing protein of photosystem II is not essential for O_2 evolution. Biochemistry 30, 440–446.

Carlberg, I., Bingsmark, S., Vennigerholz, F., Larsson, U.K., & Andersson, B. (1992). Low temperature effects on thylakoid protein phosphorylation and membrane dynamics. Biochim. Biophys. Acta 1099, 111–117.

Callahan, F.E., Ghirardi, M.L., Sopory, S.K., Mehta, M.A., Edelman, M., & Mattoo, A.K. (1990). A novel form of the 32-kDa-D1protein in the grana-localized reaction centre of photosystem II. J. Biol. Chem. 265, 15357–15360.

Chapman, D.J., De Felice, J., & Barber, J. (1983). Growth temperature effects on thylakoid membrane lipids and protein content of pea chloroplasts. Plant Physiol. 72, 225–228.

Chapman, D.J., De Felice, J., & Barber, J. (1984). Lipid, protein and plastoquinone-A content of chloroplast thylakoids: Effect of plant growth temperature. In: Advances in Photosynthesis Research (Sybesma, C., Ed.), Vol. IV, pp. 275–278. Martinus/Nijhoff Junk, The Hague.

Chapman, D.J., De Felice, J., & Barber, J. (1985). Characteristics of chloroplast thylakoid lipid composition associated with resistance to triazine herbicides. Planta 166, 280–285.

Chapman, D.J. & Barber, J. (1986). Analysis of plastoquinone-9 levels in appressed and non-appressed thylakoid membrane regions. Biochim. Biophys. Acta 850, 170–172.

Chapman, D.J., & Barber, J. (1987). Reversible inhibition of photosystem two electron transfer reactions and specific removal of the extrinsic 23-kDa polypeptide by alkaline pH. In: Progress in Photosynthesis Research (Biggins, J., Ed.), Vol. 1, pp. 669–672. Martinus Nijhoff, Dordrecht.

Chitnis, P.R., Reilly, P.A., Miedel, M.C., & Nelson, N. (1989). Structure and target mutagenesis of the gene encoding 8-kDa subunit of photosystem I from the Cyanobacterium *Synechocystis* sp. PCC 6803. J. Biol. Chem. 264, 18374–18380.

Chitnis, P.R., Purvis, D., & Nelson, N. (1991). Molecular cloning and targeted mutagenesis of the gene psaF encoding subunit III of photosystem I from the Cyanobacterium *Synechocystis* sp. PCC 6803. J. Biol. Chem. 266, 20146–20151.

Chow, W.S., Ford, R.C., & Barber, J. (1981). Possible effects of the detachment of stromal lamellae from granal stacks on salt-induced changes in spillover. Biochim. Biophys. Acta 635, 317–326.

Chua, N.H. & Gillham, N.W. (1977). The sites of synthesis of the principal thylakoid membrane polypeptides in *Chlamydomonas reinhardtii*. J. Cell. Biol. 74, 441–452.

Colman, P.M., Freeman, H.C., Guss, J.M., Murata, M., Norris, V.A., Rainshaw, J.A.M., & Venkatappa, M.P. (1978). X-ray crystal structure analysis of plastocyanin at 2.7 Å resolution. Nature (London) 272, 319–324.

Cohen, P. (1989). The structure and regulation of protein phosphatases. Ann. Rev. Biochem. 58, 453–508.

Coombs, J. & Greenwood, A.D. (1976). Compartmentation of the photosynthetic apparatus. In: Topics in Photosynthesis (Barber J., Ed.), Vol. 1, pp. 1–52. Elsevier, Amsterdam.

Coughland, S. & Hind, G. (1986). Protein kinases of the thylakoid membrane. J. Biol. Chem. 261, 14062–14068.

Coughland, S. & Hind, G. (1987). A protein kinase that phosphorylated light-harvesting complex is autophosphorylated and is associated with photosystem II. Biochemistry 26, 6515–6521.

Cox, R.P. & Andersson, B. (1981). Lateral and transvers organization of cytochromes in the chloroplast thylakoid membrane. Biochem. Biophys. Res. Commun. 103, 1336–1342.

Cramer, W.A. & Whitmarsh, J. (1977). Photosynthetic cytochromes. Ann. Rev. Plant Physiol. 28, 133–172.

Cramer, W.A., Furbacher, P.N., Szczepaniak, A., & Tae, G.-S. (1990). The chloroplast *b* cytochromes: crosslinks, topography and functions. In: Current Research in Photosynthesis (Baltscheffsky, M., Ed.), Vol. 3, pp. 221–230. Kluwer Academic, Dordrecht.

Cramer, W.A., Furbacher, P.N., Szczepaniak, A., & Tae, G.-S. (1991). Electron transport between photosystem II and photosystem I. In: Current Topics in Bioenergetics (Lee, C.P., Ed.), Vol. 16, pp. 179–222. Academic Press, New York.

Dainese, P., Santini, C., Ghiretti-Magaldi, A., Marquardt, J., Tidu, V., Mauro, S., Bergantino, E., & Bassi, R. (1992). The organization of pigment-proteins within photosystem II. In: Research in Photosynthesis (Murata, N., Ed.), Vol. II, pp. 13–20. Kluwer Academic, Dordrecht.

Dalbey, R.E. & von Heijne, G. (1992). Signal peptidases in prokaryotes and eukaryotes: A new protease family. Trends Biochem. Sci. 17, 474–478.

Debus, R.J., Barry, B.A., Sithole, I., Babcock, G.T., & McIntosh, L. (1988). Directed mutagenesis indicates that the donor to $P680^+$ in photosystem II is tyrosine-161 of the D1-polypeptide. Biochemistry 27, 9071–9074.

Debus, R.J. (1992). The manganese and calcium ions of photosynthetic oxygen evolution. Biochim. Biophys. Acta 1102, 269–352.

Deisenhofer, J., Epp, O., Miki, K., Huber, R., & Michel, H. (1985). Structure of the protein subunits in the photosynthetic reaction center of *Rhodopseudomonas viridis* at 3 Å resolution. Nature (London) 318, 618–624.

Demmig-Adams, B. (1990). Carotenoids and photoprotection in plants: A role for the xanthophyll zeaxanthin. Biochim. Biophys. Acta 1020, 1–24.

Diner, B.A., Ries, D.F., Cohen, B.N., & Metz, J.G. (1988). COOH-terminal processing of polypeptide D1 och the photosystem II reaction center of *Scenedesmus obliquus* is necessary for the assembly of the oxygen-evolving complex. J. Biol. Chem. 263, 8972–8980.

Diner, B.A., Nixon, P.J., & Farchaus, J.W. (1991). Site-directed mutagenesis of photosynthetic reaction centers. In: Current Opinions in Structural Biology (Hendrickson, W. & Klug, A., Eds.), Vol. 1, pp. 546–554. Current Biology, London.

Drepper, F., Carlberg, I., Andersson, B., & Haehnel, W. (1993). Lateral diffusion of an integral membrane protein by Monte Carlo analysis of the migration of phosphorylated light-harvesting complex II in the zhglahoid membrane. Biochemistry 32, 11915–11922.

Duvel, J.C., Dubacqs, J.P., & Tremolieres, A. (1980). In: Biogenesis and Function of Plant Lipids (Mazliak, P., Beneveniste, P., Costes, C., & Dovce, R., Eds.), pp. 91–94. Elsevier, Amsterdam.

Erickson, J.M, & Rochaix, J-D. (1992). The molecular biology of photosystem II. In: Topics in Photosynthesis (Barber, J., Ed.), Vol. 11, pp. 101–177. Elsevier, Amsterdam.

Farnsworth, C.C., Wolda, S.L., Grelb, M.H., & Glomset, J.W. (1989). Human lamin B contains a farnesylated cysteine residue. J. Biol. Chem. 264, 20422–20429.

Fato, R., Battino, M., Parenti Castelli, G., & Lenaz, G. (1985). Measurement of the lateral diffusion coefficients of ubiquinones in lipid vesicles by fluorescence quenching of 12-(9-anthroyl) stearate. FEBS Lett. 179, 238–242.

Ford, R.C., Chapman, D.J., Barber, J., Pedersen, J.Z., & Cox, R. P. (1982). Fluorescence polarization and spin label studies of the fluidity of stromal and granal chloroplast membranes. Biochim. Biophys. Acta 681, 145–151.

Ford, R.C. & Barber, J. (1983). Time dependent decay and anisotropy of fluorescence from diphenyl hexatriene embedded in the chloroplast thylakoid membrane. Biochim. Biophys. Acta 722, 341–348.

Fragata, M., Ohnishi, S., Adada, K., Ito, T., & Takahashi, M. (1984). Lateral diffusion of plastocyanin in multilamellar mixed-lipid bilayers studied by fluorescence recovery after photobleaching. Biochemistry 23, 4044–4051.

Franzén, L.-G., Frank, G., Zuber, H., & Rochaix, J.-D. (1989a). Isolation and characterization of cDNA clones encoding the 17.9 and 8.1 kDa of photosystem I from *Chlamydomonas reinhardtii*. Plant. Mol. Biol. 12, 463–474.

Franzén, L.-G., Frank, G., Zuber, H., & Rochaix, J.-D. (1989b). Isolation and characterization of cDNA clones encoding photosystem I subunits with molecular masses 11.0, 10.0, and 8.4 kDa from *Chlamydomonas reinhardtii*. Mol. Gen. Genet. 219, 137–144.

Funk, C., Schröder, W.P., Green, B.R., Renger, G., & Andersson, B. (1994). The intrinsic 22 kDa protein is a chlorophyll-binding subunit of photosystem II. FEBS Lett. 342, 261–266.

Gal, A., Herrmann, R.G., Lottspeich, F., & Ohad, I. (1992). Phosphorylation of cytochrome *b6* by the LHC II kinase associated with the cytochrome complex. FEBS Lett. 298, 33–35.

Gal, A., Hauska, G., Herrmann, R.G., & Ohad, J. (1990). Interactions between light harvesting chlorophyll-*a/b*–protein (LHC II) kinase and cytochrome *b6/f* complex. J. Biol. Chem. 265, 197423–19749.

Ghirardi, M.L. & Melis, A. (1983). Localization of photosynthetic electron transport components in mesophyll and bundle sheat chloroplasts of Zea mays. Arch. Biochem. Biophys. 224, 19–28.

Glaser, E. & Norling, B. (1991). Chloroplast and plant mitochondrial ATP synthases. In: Current Topics in Bioenergetics (Lee, C.P., Ed.), Vol. 16, pp. 223–263. Academic Press, New York.

Goldbeck, J.H. & Bryant, D.A. (1991). Photosystem I. In: Current Topics in Bioenergetics (Lee, C.P., Ed.), Vol. 16, pp. 83–177. Academic Press, New York.

Gombos, Z., Wada, H., & Murata, N. (1991). Direct evaluation of effects of fatty-acid unsaturation on the thermal properties of photosynthetic activities, as studied by mutations and transformation of *Synechocystis* PCC 6803. Plant Cell Physiol. 32, 90–164.

Goodchild, D.J., Anderson, J.M., & Andersson, B. (1985). Immunocytochemical localization of the cytochrome *b/f* complex in chloroplast thylakoid membranes. Cell. Biol. Int. Rep. 9, 715–721.

Gounaris, K. & Barber, J. (1983). Monogalactosyldiacylglycerol: The most abundant polar lipid in nature. Trends Biochem. Sci. 8, 378–381.

Gounaris, K., Sen, A., Brain, A.P.R., Quinn, P.J., & Williams, W.P. (1983a). The formation of non-bilayer structures in total polar lipid extracts of chloroplast membranes. Biochim. Biophys. Acta 728, 129–139.

Gounaris, K., Mannock, D.A., Sen, A., Brain, A.P R., Williams, W.P., & Quinn, P.J. (1983b). Polyunsaturated fatty acyl residues of galactolipids are involved in the control of bilayer/nonbilayer lipid transitions in higher plant chloroplasts. Biochim. Biophys. Acta 732, 229–242.

Gounaris, K., Brain, A.P.R., Quinn, P.J., & Williams, W.P. (1983c). Structural and functional changes associated with heat-induced phase separations of non-bilayer lipids in chloroplast thylakoid membranes. FEBS Lett. 153, 47–52.

Gounaris, K., Sundby, C., Andersson, B., & Barber, J. (1983d). Lateral heterogeneity of polar lipids in the thylakoid membranes of spinach chloroplasts. FEBS Lett. 156, 170–174.

Gounaris, K., Brain, A.P.R., Quinn, P.J., & Williams, W.P. (1984a). Structural reorganization of chloroplast thylakoid membranes in response to heat stress. Biochim. Biophys. Acta 766, 198–208.

Gounaris, K., Lambillotte, M., Barber, J., Muehlethaler, K., & Jay, F. (1984b). In: Developments in Plant Biology (Siegenthaler, P.A. & Eichenberger, W., Eds.), Vol. 9, pp. 485–488, Elsevier, Amsterdam.

Gray, J.C., Webber, A.N., Hird, S.M., Wiley, D.L., & Dyer, T.A. (1990). Genes for photosystem II polypeptides. In: Current Research in Photosynthesis (Baltscheffsky, M., Ed.), Vol. 3, pp. 461–468. Kluwer Academic, Dordrecht.

Green, B.R. (1988). The chlorophyll-protein complexes of higher plant photosynthetic membranes or Just what green band is that? Photosynth. Res. 15, 3–32.

Green, B.R., Pickersky, E., & Kloppstech, K. (1991). Chlorophyll *a/b*-binding proteins, an extended family. Trends Biochem. Sci. 16, 181–186.

Grossman, A.R., Bartlett, S.G., Schmidt, G.W., Mullet, J.E., & Chua, N.-H. (1982). Optimal conditions for posttranslational uptake of proteins by isolated chloroplasts. *In vitro* synthesis and transport of plastocyanin, ferredoxin-NADP+ oxidoreductase, and fructose-1,6-biphosphatase. J. Biol. Chem. 257, 1558–1563.

Haehnel, W. (1984). Photosynthetic electron transport in higher plants. Annu. Rev. Plant Physiol. 35, 659–693.

Haehnel, W., Mitchell, R., Ratajczak, R., Spillmann, A., & Robenck, H. (1990). Lateral diffusion of plastocyanin and plastoquinol in thylakoid membranes. In: Current Research in Photosynthesis (Baltscheffsky, M., Ed.), Vol. 2, pp. 739–746. Kluwer Academic, Dordrecht.

Haley, J. & Bogorad, L. (1989). A 4-kDa maize chloroplast polypeptide associated with the cytochrome b_6-f complex: Subunit 5, encoded by the chloroplast *petE* gene. Proc. Natl. Acad. Sci. USA 86, 1534–1538.

Hayashida, N., Matsubayashi, T., Shunosaki, K., Sugiura, M., Inoue, K., & Hiyama, T. (1987). The gene for the 9-kDa polypeptide a possible apoprotein for the iron-sulfur centers A and B of the photosystem I complex in tobacco chloroplast DNA. Curr. Genet. 12, 247–250.

He, W.Z. & Malkin, R. (1992). Specific release of a 9 kDa extrinsic polypeptide of photosystem-I from spinach chloroplasts by salt washing. FEBS Lett. 308, 298–300.

Heinemeyer, W., Alt, J., & Herrmann, R.G. (1984). Nucleotide sequence of the clustered genes for apocytochrome *b*-6 and subunit 4 of the cytochrome *b-f* complex in the spinach plastid chromosome. Curr. Genet 8, 543–550.

Henry, L.E.A., Mikkelsen, J.D., & Møller, B.L. (1983). Pigment and acyl lipid composition of photosystem I and II vesicles and of photosynthetic mutants in barley. Carlsberg Res. Comm. 48, 131–148.

Herrmann, R.G., Alt, J., Schiller, B., Widger, W.R., & Cramer, W.A. (1984). Nucleotide sequence of the gene for apocytochrome *b*-559 on the spinach plastid chromosome: Implications for the structure of the membrane protein. FEBS Lett. 176, 239–243.

Herrmann, R.G., Oelmüller, R., Bichler, J., Schneiderbauer, A., Steppuhn, J., Wedel, N., Tyagi, K., & Westhoff, P. (1991). The thylakoid membrane of higher plants: Genes, their expression and interaction. In: Plant Molecular Biology (Herrmann, R.G. & Larkins, B., Eds.), Vol. 2, pp. 411–427. Plenum Press, New York.

Hippler, M., Ratajczak, R., & Haehnel, W. (1989). Identification of the plastocyanin binding subunit of photosystem I. FEBS Lett. 230, 280–284.

Hird, S.M., Dyer, T.A., & Gray, J.C. (1986). The gene for the 10-kDa phosphoprotein of photosystem II is located in chloroplast DNA. FEBS Lett. 209, 181–186.

Høj, P.B., Svendsen, I., Scheller, H.V., & Møller, B.L. (1987). Identification of a chloroplast-encoded 9-kDa polypeptide as a 2[4Fe-4S] protein carrying centers A and B of photosystem I. J. Biol. Chem. 262, 12676–12684.

Horton, P. & Lee, P. (1985). Phosphorylation of chloroplast membrane proteins partially protects against photoinhibition. Planta 165, 37–42.

Hundal, T., Virgin, I., Styring, S., & Andersson, B. (1990). Changes in the organization of photosystem II following light-induced D1 protein degradation. Biochim. Biophys. Acta 1017, 235–241.

Ikeuchi, M., Eggers, B., Shen, G., Webber, A., Yu, J., Hirano, A., Inoue, Y., & Vermaas, W.F.J. (1991). Cloning of the *psbK* gene from *Synechocystis* sp PCC 6803 and characterization of photosystem II in mutants lacking PSII-K. J. Biol. Chem. 266, 11111–11115.

Inagaki, N., Fujita, S., & Satoh, K. (1989). Solubilization and partial purification of a thylakoidal enzyme of spinach involved in the processing of D1 protein. FEBS Lett. 246, 218–222.

Izawa, S. & Good, N.R. (1966). Effect of salt and electron transport on the conformation of isolated chloroplasts. II. Electron microscopy. Plant Physiol. 41, 544–552.

Joliot, P., Lavergne, J., & Beal, D. (1992). Plastoquinone compartmentation in chloroplasts. 1. Evidence for domains with different rates of photoreduction. Biochim. Biophys. Acta 1101, 1–12.

Juhler, R.K., Andreasson, E., Yu, S.-G., & Albertsson, P.Å. (1993). Composition of photosynthetic pigments in thylakoid membrane vesicles from spinach. Photosynth. Res. 35, 171–178.

Kadenbach, B., Ungibauer, M., Jarausch, J. Büge, U., & Kuhn-Nentwig, L. (1983). The complexity of respiratory complexes. Trends Biochem. Sci. 8, 398–400.

Kieleczawa, J., Coughlan, S.J., & Hind, G. (1992). Isolation and characterization of an alkaline phosphatase from pea thylakoids. Plant Physiol. 99, 1029–1036.

Kim, S., Sandusky, P., Bowlby, N.R., Aebersold, B., Green, B.R., Vlahakis, S., Yocum, C.F., & Pickersky, E. (1992). Characterization of a spinach psbS cDNA encoding the 22-kDa protein of photosystem II. FEBS Lett. 314, 67–71.

Kinosita, K., Kataoka, R., Kimura, Y., Gotoh, O., & Ikegami, A. (1981). Dynamic structure of biological membranes as probed by 1,6-diphenyl-1,3,5-hexatreine. A nanosecond fluorescence depolarization study. Biochemistry 20, 4270–4277.

Kirk, J.T.O. & Tilney-Bassett, R.A.E. (1967). The Plastids: Their Chemistry, Structure, Growth and Inheritance. Freeman, London.

Kirwin, P.M., Elderfield, P.D., & Robinson, C. (1987). Transport of proteins into chloroplasts. Partial purification of a thylakoidal processing peptidase involved in plastocyanin biogenesis. J. Biol. Chem., 262, 16386–16390.

Krauss, N., Hinrichs, W., Witt, I., Fromme, P., Pritzkow, W., Dauter, Z., Betzel, C., Wilson, K.S., Witt, H.T., & Sanger, W. (1993). Three-dimensional structure of system I of photosynthesis at 6 Å resolution. Nature 361 (London) 326–331.

Kruse, E.R. & Kloppstech, K. (1992). Heat-shock proteins in plants: An approach to understanding the function of plastid heat shock proteins. In: Topics in Photosynthesis (Barber J., Ed.), Vol. II, pp. 409–442. Elsevier, Amsterdam.

Kühlbrandt, W., Wang, D.N., & Fujiyoshi, Y. Atomic model of plant light-harvesting complex by electron crystallography. Nature (London) 367, 614–621.

Larsson, U.K., Sundby, C., & Andersson, B. (1987). Characterization of two different sub-populations of spinach light-harvesting chlorophyll *a/b*–protein complex (LHC II)-polypeptide composition, phosphorylation pattern, and association with photosystem II. Biochim. Biophys. Acta. 894, 56–58.

Lautner, A., Klein, R., Ljungberg, U., Bartling, D., Andersson, B., Reinke, H., Beyreyther, U., & Herrmann, R.G. (1988). Nucleotide sequence of cDNA-clones encoding the complete precursor for the 10-kDa polypeptide of photosystem II from spinach. J. Biol. Chem. 263, 10077–10081.

Lavergne, J. & Joliot, P. (1991). Restricted diffusion in photosynthetic membranes. Trends Biochem. Sci. 16, 129–134.

Lind, L.K., Shukla, V.K., Nyhus, K.J., & Pakrasi, H.B. (1993). Genetic and immunological analyses of the cyanobacterim *Synechocystis* sp. PCC 6803 show that the protein encoded by the *psbJ* gene regulates the number of photosystem II centers in thylakoid membranes. J. Biol. Chem. 268, 1575–1579.

Lindahl, M. & Andersson, B. (1992). A proteolytic activity associated with loss of outer LHCII during acclimation from low to high light. In: Research in Photosynthesis (Murata, N., Ed.), Vol. I, pp. 307–310. Kluwer Academic, Dordrecht.

Ljungberg, U., Åkerlund, H.-E., & Andersson, B. (1986a). Isolation and characterization of the 10-kDa and 22-kDa polypeptides of higher plant photosystem II. Eur. J. Biochem. 158, 477–482.

Ljungberg, U., Henrysson, T., Rochester, C.P., Åkerlund, H.-E., & Andersson, B. (1986b). The presence of low molecular weight polypeptides in spinach photosystem II core preparations. Isolation of a 5-kDa hydrophilic polypeptide. Biochim. Biophys. Acta 849, 112–120.

Machold, O., Simpson, D.J., & Høyer-Hansen, G.H. (1977). Correlation between the freeze fracture appearance and polypeptide composition of thylakoid membranes in barley. Carlsberg Res. Commun. 42, 499–516.

Marder, J., Telfer, A., & Barber, J. (1988). The D1 polypeptide subunit of the photosystem II reaction center has a phosphorylation site at its amino terminus. Biochim. Biophys. Acta 932, 362–366.

Mann, K., Schlenkrich, T., Bauer, M., & Huber, R. (1991). The amino acid sequence of three proteins of photosystem I of the cyanobacterium *Fremyella diplosiphon* (Calothrix sp PCC-7601). Biol. Chem. Hoppe-Seyler 372, 519–524.

Mattoo, A.K. & Edelman, M. (1987). Intramembrane translocation and posttranslational palmitoylation of the chloroplast 32-kDa herbicide-binding protein. Proc. Natl. Acad. Sci. USA 84, 1497–1501.

Mayes, S.R., Cook, K.M., Self, S.J., Zhang, Z., & Barber, J. (1991). Deletion of the gene encoding the PS II 33-kDa protein from *Synechocystis* PCC 6803 does not inactivate water-splitting but increases vulnerability to photoinhibition. Biochim. Biophys. Acta 1060, 1–12.

Mayes, S.R., Dubbs, J.M., Vass, I., Hideg, E., Nagy, L., & Barber, J. (1993). Further characterization of the psbH locus of *Synechocystis* sp PCC 6803: Inactivation of *psbH* impairs Q_A to Q_B electron transport in photosystem II. Biochemistry 32, 1454–1465.
Melis, A. (1991). Dynamics of photosynthetic membrane composition and function. Biochim. Biophys. Acta 1058, 87–106.
Metz, J.G., Nixon, P.J., Rögner, M., Brudvig, G.W., & Diner, B.A. (1989). Directed alteration of the D1-polypeptide of photosystem II: Evidence that tyrosine-161 is the redox component, Z, connecting the oxygen-evolving complex to the primary electron donor, P680. Biochemistry 28, 6960–6969.
Michel, H. & Deisenhofer, J. (1986). X-ray diffraction studies on a crystalline bacterial photosynthetic reaction center and conclusion on the structure photosystem II reaction centers. In: Encyclopedia of Plant Physiology (Staehelin, L.A. & Arntzen, C.J., Eds.) New Ser., Vol. 19, pp. 371–381. Springer-Verlag, Berlin.
Michel, H.P., Hunt, D.F., Shabanowitz, J., & Bennett, J. (1988). Tandem mass spectroscopy reveals that three photosystem II proteins of spinach chloroplasts contain *N*-acetyl-*O*-phosphothreonine at their NH_2 termini. J. Biol. Chem. 263, 1123–1130.
Miller, K.R. & Staehelin, L.A. (1976). Analysis of the thylakoid outer surface: Coupling factor is limited to unstacked membrane regions. J. Cell Biol. 68, 30–47.
Millner, P.A. & Barber, J. (1984). Plastoquinone as a mobile redox carrier in the photosynthetic membrane. FEBS Lett. 169, 1–6.
Millner, P.A., Mitchell, R.A.C., Chapman, D.J., & Barber, J. (1984). Fluidity properties of isolated chloroplast thylakoid lipids. Photosynth. Res. 5, 63–76.
Mitchell, P. (1961). Coupling of phosphorylation to electron and hydrogen transfer by a chemiosmotic type mechanism. Nature (London) 191, 144–148.
Mitchell, P. (1975). The proton motive Q cycle: A general formulation. FEBS Lett. 59, 137–139.
Mitchell, R.A.C., Spillmann, A., & Haehnel, W. (1990). Plastoquinone diffusion in linear photosynthetic electron transport. Biophys. J. 58, 1011–1024.
Moenne-Loccoz, P., Robert, B., & Lutz, M. (1989). A resonance raman characterization of the primary electron acceptor in photosystem II. Biochemistry 28, 3641–3645.
Murata, N. (1983). Molecular species composition of phosphatidylglycerols from chilling-sensitive and chilling-resistant plants. Plant Cell Physiol. 24, 81–86.
Murata, N. & Miyao, M. (1985). Extrinsic membrane proteins in the photosynthetic oxygen-evolving complex. Trends Biochem. Sci. 10, 122–124.
Murata, N. (1989). Low-temperature effects on cyanobacterial membranes. J. Bioenergetics Biomembranes 21, 61–75.
Murata, N. & Nishida, M. (1989). In: Chilling Injury of Horticultural Crops (Wang, C.Y., Ed.), pp. 181–189. CRC Press.
Murphy, D.J. (1983). The importance of nonpolar bilayer regions in photosynthetic membranes and their stabilization by galactolipids. FEBS Lett. 150, 19–26.
Murphy, D.J. & Woodrow, I.E. (1983). Lateral heterogeneity in the distribution of thylakoid membrane lipid and protein components and its implications for the molecular organization of photosynthetic membranes. Biochim. Biophys. Acta 725, 104–112.
Murphy, D.J. (1986). The molecular organization of the photosynthetic membranes of higher plants. Biochim. Biophys. Acta 864, 33–94.
Nabedryk, E., Andrianambinintsoa, S., Berger, G., Leonard, M., Mantele, W., & Breton, J. (1990). Characterization of bonding interactions of the intermediary electron acceptor in the reaction center of photosystem II by FTIR spectroscopy. Biochim. Biophys. Acta 1016, 49–54.
Nanba, O. & Satoh, K. (1987). Isolation of a photosystem II reaction center consisting of D1- and D-2 polypeptides and cytochrome *b*559. Proc. Natl. Acad. Sci. USA 84, 109–112.
Nedbal, J., Samson, G., & Whitmarsh, J. (1992). Redox state of a one-electron component controls the rate of photoinhibition of photosystem II. Proc. Natl. Acad. Sci. USA 89, 7929–7933.

Nilsson, F., Andersson, B., & Jansson, C. (1990). Photosystem II characteristics of a *Synechocystis* 6803 mutant lacking synthesis of the D1-polypeptide. Plant. Mol. Biol. 14, 1051–1054.

Nitschke, W. & Rutherford, A.W. (1991). Photosynthetic reaction centres: Variations on a common structural theme? Trends Biochem. Sci. 16, 241–245.

Oh-oka, H., Takahashi, Y., Kuriyama, K., Saeki, K., & Matsubara, H. (1988). The protein responsible for center A/B in spinach photosystem I: Isolation with iron-sulfur cluster(s) and complete sequence analysis. J. Biochem. 103, 962–968.

Ojakian, G.K. & Satir, P. (1974). Particle movements in chloroplast membranes: Quantitative measurements of membrane fluidity by the freeze-fracture technique. Proc. Natl. Acad. Sci. USA 21, 2052–2056.

Okkels, J.S., Scheller, H.V., Svendsen, I., & Møller, B.L. (1991). Isolation and characterization of a cDNA clone encoding an 18-kDa hydrophobic photosystem I subunit (PSI-L) from barley (*Hordeum vulgare L.*). J. Biol. Chem. 266, 6767–6773.

Ono, T-A., Noguchi, T., Inoue, Y., Kusunoki, M., Matsushita, T., & Oyanagi, H. (1992). X-ray detection of the period-four cycling of the manganese cluster in photosynthetic water oxidizing enzyme. Science 258, 1335–1337.

Paolillo, D.J. (1970). The three-dimensional arrangement of intergranal lamellae in chloroplasts. J. Cell Sci. 6, 234–255.

Phillips, A.L. & Gray, J.C. (1984). Location and nucleotide sequence of the gene for the 15.2-kDa polypeptide of the cytochrome *b-f* complex from pea chloroplasts. Mol. Gen. Genet. 194, 477–484.

Pràsil, O., Adir, N., & Ohad, I. (1992). Dynamics of photosystem II: Mechanism of photoinhibition and recovery process. In: Topics in Photosynthesis (Barber, J., Ed.), Vol. 11, pp. 295–348. Elsevier, Amsterdam.

Quinn, P.J. & Williams, W.P. (1983). The structural role of lipids and photosynthetic membranes. Biochim. Biophys. Acta 737, 223–266.

Quinn, P.J. & Williams, W.P. (1985). Environmentally induced changes in chloroplast membranes and their effects on photosynthetic function. In: Photosynthetic Mechanisms and the Environment. Topics in Photosynthesis (Barber, J. & Baker, N.R., Eds.), Vol. 6, pp. 1–47. Elsevier, Amsterdam.

Ragan, I. (1987). Structure of NADH-ubiquinone reductase (Complex I). In: Current Topics in Bioenergetics (Lee, C.P., Ed.), Vol. 15, pp. 1–36. Academic Press, New York.

Radunz, A. (1980). Binding of antibodies onto the thylakoid membrane. VI. Asymmetric distribution of lipids and proteins in the thylakoid membrane. Z. Naturforsch 35C, 1024–1031.

Rawyler, A. & Siegenthaler, P.A. (1985). Transversal localization of monogalactosyldiacylglycerol and digalactosyldiacylglycerol in spinach thylakoid membranes. Biochim. Biophys. Acta. 815, 287–298.

Rawyler, A., Unitt, M.D., Giroud, C., Davis, H., Mayor, J.-P., Harwood, J.L., & Siegenthaler, P.-A. (1987). The transmembrane distribution of galactolipids in chloroplast thylakoids is universal in a wide variety of temperate climate plants. Photosynth. Res. 11, 3–13.

Rosseau, F., Sétif, P., & Bernard, L. (1993). Evidence for the involvement of PSI-E subunit in the reduction of ferredoxin by photosystem I. EMBO J. 12, 1755–1765.

Rubin, B.T., Barber, J., Paillotin, G., Chow, W.S., & Yamamoto, Y. (1981). A diffusional analysis of the temperature sensitivity of the Mg^{2+}-induced rise of chlorophyll fluorescence from pea thylakoid membranes. Biochim. Biophys. Acta 638, 69–74.

Ruffle, S.V., Donnelly, D., Blundell, T.L., & Nugent, J.H.A. (1992). A three-dimensional model of the photosystem II reaction centre of *Picum sativum*. Photosynth. Res. 34, 287–300.

Salter, A.H., Neuman, B.J., Napier, J.A., & Gray, J.C. (1992a). Import of the precursor of the chloroplast Rieske iron–sulphur protein by pea chloroplasts. Plant Mol. Biol. 20, 569–574.

Salter, A.H., Virgin, I., Hagman, Å., & Andersson, B. (1992b). On the molecular mechanism of light-induced D1-protein degradation in photosystem II core particles. Biochemistry 31, 3990–3998.

Sane, P.V., Goodchild, D.J., & Park, R.B. (1970). Characterization of chloroplast photosystem 1 and 2 separated by a nondetergent method. Biochim. Biophys. Acta, 216, 162–178.

Scheller, H.V. & Møller, B.L. (1990). Photosystem I polypeptides. Physiol. Plant. 78, 484–494.

Schröder, W.P., Henrysson, T., & Åkerlund, H.-E. (1988). Characterization of low molecular mass proteins of photosystem II by N-terminal sequencing. FEBS Lett. 235, 289–292.

Sen, A., Williams, W. P., Brain, A.P.R., Dickins, M.J., & Quinn, P.J. (1981). Formation of inverted micelles in dispersions of mixed galactolipids. Nature (London) 293, 488–490.

Sen, A., Mannock, D.A., Collings, D.J., Quinn, P.J., & Williams, W.P. (1983). Thermotherotropic phase properties and structure of 1,2-disteroylgalactosyl-glycerols in aqueous systems. Proc. Royal Soc. London. Ser. B218, 349–364.

Shinitzki, M. & Yuli, I. (1982). Lipid fluidity at the submacroscopic level: Determination by fluorescence polarization. Chem. Phys. Lipids. 30, 261–282.

Shinozaki, K., Ohme, M., Tanaka, M., Wakasugi, T., Hayashida, N., Matsubayashi, T., Zaita, N., Chungwongse, J., Obokata, J., Yamaguchi-Shinozaki, K., Ohto, C., Torazawa, K., Meng, B.Y., Sugita, M., Kamogasshira, T., Yamada, K., Kusuda, J., Takaiwa, F., Kato, A., Tohodoh, N., Shimada, H., & Sugiura, M. (1986). The complete nucleotide sequence of tobacco chloroplast genome: Its gene organization and expression. EMBO J. 5, 2043–2049.

Shipley, G.G., Green, J.P., & Nichols, B.W. (1973). The phase behavior of monogalactosyl, digalactosyl, and sulphoquinovosyl diglycerides. Biochim. Biophys. Acta 311, 531–544.

Shipton, C.A. & Barber, J. (1991). Photoinduced degradation of the D1 polypeptide in isolated reaction centres of photosystem II: Evidence for an autoproteolytic process triggered by the oxidizing side of the photosystem. Proc. Natl. Sci. USA 88, 6691–6695.

Somerville, C. & Browse, J. (1991). Plant Lipids: Metabolism, mutants and membranes. Science 252, 80–87.

Sprague, S.G. & Staehelin, L.A. (1984). Effects of reconstitution method on the structured organization of isolated chloroplast membrane lipids. Biochim. Biophys. Acta 777, 306–322.

Staehelin, L.A. (1976). Reversible particle movements associated with unstacking and restacking of chloroplast membranes *in vitro*. J. Cell Biol. 71, 136–158.

Staehelin, L.A. (1986). Chloroplast structure and supramolecular organization of photosynthetic membranes. In: Encyclopaedia of Plant Physiology, New Series, (Staehelin, L.A. & Arntzen, C.J., Eds.), Vol. 19, pp. 1–84. Springer-Verlag, Berlin.

Steppuhn. J., Rother, C., Hermans, J., Jansen, T., Salnikow, J., Hauska, G., & Herrmann, R.G. (1987). The complex amino acid sequence of the Rieske FeS-precursor protein from spinach chloroplasts deduced from cDNA analysis. Mol. Gen. Genet. 210, 171–177.

Steppuhn, J., Hermans, J., Nechushtai, R., Ljungberg, U., Thummler, F., Lottspeich, F., & Herrmann, R.G. (1988). Nucleotide sequence of cDNA clones encoding the entire precursor polypeptides for subunits IV and V of the photosystem I reaction center from spinach. FEBS Lett. 237, 218–224.

Steppuhn, J., Hermans, J., Nechushtai, R., Herrmann, G.S., & Herrmann, R.G. (1989). Nucleotide sequences of cDNA clones encoding the entire precursor polypeptide for subunit VI and of the plastome-encoded gene for subunit VII of the photosystem I reaction center from spinach. Curr. Genet. 16, 88–108.

Stockhaus, J., Höfer, M., Renger, G., Westhoff, P., Wydrzynski, T., & Willmitzer, L. (1990). Anti-sense RNA efficiently inhibits formation of the 10-kDa polypeptide of photosystem II in transgenic potato plants: Analysis of the role of the 10-kDa protein. EMBO J. 9, 3013–3021.

Stubbs, C.D. & Smith, A.D. (1984). The modification of mammalian membrane polyunsaturated fatty acid composition in relation to membrane fluidity and function. Biochim. Biophys. Acta 779, 89–137.

Sundby, C. & Andersson, B. (1985). Temperature-induced reversible migration along the thylakoid membrane of photosystem II regulates its association with LHC-II. FEBS Lett. 191, 24–28.

Sundby, C. & Larsson, C. (1985). Transbilayer organization of the thylakoid galactolipids. Biochim. Biophys. Acta. 813, 61–67.

Svensson, B., Vass, I., Cedergren, E., & Styring, S. (1990). Structure of donor side components in photosystem II predicted by computer modeling. EMBO J. 9, 2051–2059.

Svensson, P. & Albertsson, P.-Å. (1989). Preparation of highly enriched photosystem II membrane vesicles by a nondetergent method. Photosynth. Res. 20, 249–259.

Svensson, P., Andreasson, E., & Albertsson, P.-Å. (1991). Heterogeneity among Photosystem I. Biochim. Biophys. Acta 1060, 45–50.

Swiezewska, E., Thelin, A., Dallner, G., Andersson, B., & Ernster, L. (1993). Occurrence of prenylated proteins in plant cells. Biochem. Biophys. Res. Commun. 191, 161–166.

Tae, G.-S. & Cramer, W.A. (1989). Lumen side topography of the α-subunit of the chloroplast cytochrome *b*-559. FEBS Lett. 259, 161–164.

Tasaka, Y., Nishida, I., Higashi, S., Beppu, T., & Murata, N. (1990). Fatty acid composition of phosphatidylglycerol in relation to chilling sensitivity of woody plants. Plant Cell Physiol. 31, 454–550.

Taylor, M.A., Packer, J.C.L., & Bowyer, J.R. (1988). Processing of the D1 polypeptide of the photosystem II reaction centre and photoactivation of a low fluorescence mutant (LF-1) of *Scenedesmus obliquus*. FEBS Lett. 237, 229–233.

Telfer, A., Hodges, M., & Barber, J. (1983). Analysis of chlorophyll fluorescence induction curves in the presence of DCMU as a function of magnesium concentration and NADPH activated light-harvesting chlorophyll *a/b* protein phosphorylation. Biochim. Biophys. Acta. 724, 167–175.

Telfer, A., Hodges, M., Millner, P.A., & Barber, J. (1984). The cation-dependence of the degree of protein phosphorylation-induced unstacking of pea thylakoids. Biochim. Biophys. Acta 766, 554–562.

Telfer, A., Whitelegge, J., Bottin, H., & Barber, J. (1986). Changes in the efficiency of P700 photooxidation in response to protein phosphorylation detected by flash absorption spectroscopy. J. Chem. Soc. Faraday Trans. 2, Special Issue 82, 2207–2215.

Thompson, L.K. & Brudvig, G.W. (1988). Cytochrome *b*559 may function to protect photosystem II from photoinhibition. Biochemistry 27, 6653–6658.

Trebst, A. (1986). The topology of plastoquinone and herbicide binding peptides of photosystem II in the thylakoid membrane. Z. Naturforsch. 41C, 240–245.

Unitt, M.D. & Harwood, J. (1985). Sidedness studies of thylakoid phosphatidylglycerol in higher plants. Biochem. J. 228, 707–711.

Vallon, O., Wollman, F.A., & Olive, J. (1986). Lateral distribution of the main protein complexes of the photosynthetic apparatus in *Chlamydomonas reinhardtii* and in spinach; an immunocytochemical study using intact thylakoid membranes and a PSII enriched membrane preparation. Photobiochem. Photobiophys. 12, 203–220.

Vallon, O., Bulte, L., Dainese, P., Olive, J., Bassi, R., & Wollman, F.-A. (1991). Lateral redistribution of cytochrome b_6/f complexes along thylakoid membranes upon state transitions. Proc. Natl. Acad. Sci. USA 88, 8262–8266.

Virgin, I., Ghanotakis, D.F., & Andersson, B. (1990). Light-induced D1-protein degradation in isolated photosystem II core complexes. FEBS Lett. 269, 45–48.

Wada, H. & Murata, N. (1989). *Synechocystis* PCC 6803 mutants defective in desaturation of fatty acids. Plant Cell. Physiol. 30, 971–978.

Wada, H., Gombos, Z., & Murata, N. (1990). Enhancement of chilling tolerance of a cyanobacterium by genetic manipulation of fatty acid desaturation. Nature (London) 347, 200–203.

Wang, A.Y.I. & Packer, L. (1973). Mobility of membrane particles in chloroplasts. Biochim. Biophys. Acta 305, 488–492.

Webb, M.S. & Green, B.R. (1991). Biochemical and biophysical properties of thylakoid acyl lipids. Biochim. Biophys. Acta 1060, 133–158.

Webber, A.N., Platt-Aloia, K.A., Heath, R.L., & Thomson, W.W. (1988). The marginal regions of thylakoid membranes: A partial characterization by polyoxyethylene sorbitan monolaurate (Tween-20) solubilization of spinach thylakoids. Physiol. Plant 72, 288–297.

Webber, A.N., Packman, L., Chapman, D.J., Barber, J., & Gray, J.C. (1989). A fifth chloroplast-encoded polypeptide is present in the photosystem II reaction center complex. FEBS. Lett. 242, 259–262.

Wedel, N., Klein, R., Ljungberg, U., Andersson, B., & Herrmann, R.G. (1992). The single-copy gene *psbS* codes for a phyllogenetically intriguing 22-kDa polypeptide of photosystem II. FEBS Lett. 314, 61–66.

Westhoff, P., Farchaus, J.W., & Herrmann, R.G. (1986). The gene for the M_r 10,000 phosphoprotein associated with photosystem II is part of the psbB operon of the spinach plastid chromosome. Curr. Genet. 11, 165–169.

Wettern, M. (1986). Localization of 32,000 Dalton chloroplast protein pools in thylakoid: Significance in atrazine binding. Plant Sci. 43, 173–177.

Willey, D.L., Auffret, A.D., & Gray, J.C. (1984a). Structure and topology of cytochrome *f* in pea chloroplast membranes. Cell 36, 555–562.

Willey, D.L., Howe, O.J., Auffret, A.D., Bowman, C.M., Dyer, T.A., & Gray, J.C. (1984b). Location and nucleotide sequence of the gene for cytochrome *f* in wheat chloroplast DNA. Mol. Gen. Genet. 194, 416–422.

Willey, D.L. & Gray, J.C. (1989). Two small open reading frames are cotranscribed with the pea chloroplast genes for the polypeptides of cytochrome *b*559. Curr. Genet. 15, 213–220.

Williams, W.P. & Allen, J.F. (1987). State 1/state 2 changes in higher plants and algae. Photosynth. Res. 13, 19–45.

Wynn, R.M. & Malkin, R. (1988). Interaction of plastocyanin with photosystem I: A chemical cross-linking study of the polypeptide that binds plastocyanin. Biochemistry 27, 5863–5869.

Yalovsky, S., Ne'eman, E., Schuster, G., Paulsen, H., Harel, E., & Nechushtai, R. (1992). Accumulation of a light-harvesting chlorophyll *a/b*–protein in the chloroplast grana lamellae. J. Biol. Chem. 267, 20689–20693.

Yu, S.-G., Stefansson, H., & Albertsson, P.-Å. (1992). Localization of 64-kDa LHCII-kinase in the thylakoid membrane from spinach. In: Research in Photosynthesis (Murata, N., Ed.), Vol. 1, pp. 283–286. Kluwer Academic, Dordrecht.

Zakharov, S.D. & Red'ko, T.P. (1988). Lateral distribution of C_{14}-DCCD labeled protein in the thylakoid membrane of chloroplast. Distribution of CF_0 of the ATP synthase complex. Biochimiya 53, 1549–1556.

Zanetti, G. & Merati, G. (1987). Interaction between photosystem I and ferredoxin. Identification by chemical cross-linking of the polypeptide which binds ferredoxin. Eur. J. Biochem. 169, 143–146.

Zhang, Z.-H., Mayes, S.R., Vass, I., Nagg, L., & Barber, J. (1994). Characterization of the *psb* k locus of *Synechocystis* sp PCC 68 in terms of photosystem II function. Photosyn. Res. 40, 369–378.

Zilber, A. & Malkin, R. (1988). Ferredoxin cross-links to a 22-kDa subunit of photosystem I. Plant Physiol. 88, 810–814.

ANTENNA PIGMENT–PROTEIN COMPLEXES OF HIGHER PLANTS AND PURPLE BACTERIA

J. Philip Thornber, Richard J. Cogdell, Parag Chitnis,

Daryl T. Morishige, Gary F. Peter, Stephen M. Gómez,

Shivanthi Anandan, Susanne Preiss, Beth W. Dreyfuss,

Angela Lee, Tracey Takeuchi, and Cheryl Kerfeld

Advances in Molecular and Cell Biology
Volume 10, pages 55–118.

ISBN: 1-55938-710-6

I. ORGANIZATION OF PIGMENTS AND POLYPEPTIDES IN PHOTOSYNTHETIC MEMBRANES

Most photosynthetic membrane polypeptides are constituents of one of the following multiprotein components: the photosystem(s), the cytochrome *b/c*, or the ATPase complex. Together these complexes carry out the conversion of light energy into chemical potential in an organism. Only the photosystem complex, of which there are two (photosystem I and photosystem II) in oxygenic organisms and one in purple photosynthetic bacteria, contain the chlorophyll and carotenoid molecules that harvest light and perform the primary photochemical event. Each photosystem can be envisaged as having two parts: a core component (CC) and a light-harvesting component (LHC), each of which has multiple proteins and pigments. The core component, which contains, among other things, the photochemical reaction center (RC), is required for an organism to have a stable and functional photochemical event, while the light-harvesting component functions to gather the majority of the light absorbed by the photosystem and to transfer the absorbed energy to the core complex to drive the primary photochemical event. It is now well established that most if not all of the photosynthetic pigment molecules in any organism are coordinated with, but not covalently linked to, several different proteins, and that for the large part the pigments do not occur free in the lipid bilayer, as was originally thought (for historical perspective see Thornber 1986; Gregory, 1989). Some chlorophylls in green bacteria and a small percentage of the carotenoid in plants may be exceptions to this generalization. The proteins function to orient and space precisely their associated chlorophyll and carotenoid molecules so that energy absorbed in any one of these pigment–proteins is not quenched via excimer formation but is transferred efficiently to the RC. This review focuses on the cellular biochemistry of the light-harvesting components of the higher plant photosystems and of the photosystem in purple bacteria. Other recent reviews in this area are to be found in Glazer and Melis (1987), Cogdell (1988), Green (1988), Bassi et al. (1990), Hawthornthwaite and Cogdell (1991), Thornber et al. (1991), and Zuber and Brunisholz (1991).

II. ISOLATION OF PIGMENT-CONTAINING COMPLEXES AND PIGMENT–PROTEINS

Electrophoretic, chromatographic, and centrifugal procedures have been used to fractionate surfactant extracts of photosynthetic membranes so that the multiprotein complexes in a photosystem, as well as individual caroteno-chlorophyll proteins derived from them, can be isolated (Hawthornthwaite and Cogdell, 1991; Thornber et al., 1991 for review). A long-time goal has been to find the perfect surfactant/fractionation system that will yield each pigmented entity in its native state; that is, without loss of any cofactors (particularly their carotenoid and chlorophyll molecules). Solubilization of thylakoids by one or more of many available surfactants when combined with the subsequent fractionation procedures can remove some (sometimes all) of the pigments from the polypeptide with which they are associated (see Allgood et al., 1991). This is especially true for these components in eukaryotic organisms. Thus, selection of the solubilizing surfactant(s) is critical. Of the many used in biochemical research, nonionic glycosidic surfactants (e.g., nonyl-glucoside, decyl-maltoside) come closest to the ideal for higher plant thylakoids (Camm and Green, 1980; Bassi et al., 1985; Dunahay and Staehelin, 1986; Peter and Thornber, 1990). These glycosidic surfactants have not been as extensively examined for purple bacterial membranes (however, see Ferguson et al., 1991).

A. Higher Plants

Fractionation by nondenaturing polyacrylamide gel electrophoresis (PAGE) or density gradient centrifugation in the presence of glycosidic surfactants is now a frequently used method for purifying the complexes. By using PAGE of the surfactant-solubilized green plant photosynthetic apparatus, 13 distinct caroteno-chlorophyll–proteins have been identified, many of which have now been obtained in analyzable amounts (Thornber et al., 1991). However, no single procedure yields all of the pigment–proteins in higher plant plastids in a single step. An initial fractionation to obtain either photosystem I or photosystem II as starting material is often used to simplify the number of pigment–protein complexes to be resolved in the subsequent PAGE step. Variations in the amount of pigment lost from a higher

Table 1. Characteristics of Barley Thylakoid Pigment–Protein Complexes[1]

	Apparent size (kDa):				
	Holocomplex	*Apoprotein(s)*	*CHL a/b*	*% Total CHL*	*Alternative names*
PS I	230	multiple	6.0	38	CP I*
CC I	120	58 + others	*a* only	20	CP I, CHL *a*-P1
LHC Ia	65/25	24, 22	1.4	} 18	LHCP Ia
LHC Ib	65/25	21.5, 21	2.3		LHCP Ib
LHC Ic		17			*psaF* product
LHC Id		11			
CC II-RC	80	32(D2), 30(D1), 9, 4	*a* + pheo	1	
CC IIa	55	47	*a* only	} 10	CP III, CHL *a*-P2, CP47
CC IIb	50	43	*a* only		CP IV, CHL *a*-P3, CP43
LHC IIa	35	31	2.25	4	CHL *a/b*-P1, CP29
LHC IIb	72	28, 27, 25	1.33	40	CHL *a/b*-P2, CP II
LHC IIc	30	29, 26.5	1.80	4	CP27
LHC IId	24	21	0.93	4	CP24
LHC IIe	12	11			

Note: [1]Assembled from data in Peter and Thornber (1990; 1991b) and Dreyfuss and Thornber (1994) (*Plant Phys.*, in press). See also Thornber et al. (1992).

plant complex purified by slightly different procedures and/or by the use of different surfactants, explain why over the years any one particular pigment–protein has been reported to have slightly different pigment compositions. Initially, this created uncertainty in equating data from different laboratories about pigment–proteins, and resulted in more than one name for what were identical pigment–proteins *in vivo* (Table 1).

The history behind and the rationale of the procedures for the fractionation and resolution of the higher plant photosynthetic apparatus are described in Thornber et al. (1991). In brief, the earliest evidence for the presence of multiple pigment-binding proteins in higher plant thylakoids came from studies of the electrophoresis of chloroplast lamellae solubilized with anionic surfactants such as sodium dodecyl sulfate (SDS) or sodium dodecyl benzenesulfonate. Two pigment–proteins were observed as green, protein-containing bands after electrophoresis of such extracts in polyacrylamide gel tubes (Ogawa et al., 1966; Thornber et al., 1966). The two bands were ultimately designated chlorophyll-binding proteins (CP) I and II, a terminology which neglected the fact that carotenoids were also present. Later, more gentle solubilization of higher plant thylakoids with anionic detergents, or more recently with glycosidic surfactants, as well as changes in PAGE conditions have revealed the presence of additional pigment–proteins and these are described in some detail in Section III. Table 1 summarizes the major characteristics of the higher plant pigment–proteins and correlates the names we use for them in this chapter with those of previously described components.

B. Purple Bacteria

Detergent treatment of photosynthetic membranes (chromatophores) of bacteriochlorophyll *a*-containing purple bacteria solubilizes the membrane and releases its two major components; namely, the core complex (RC-B875) and the light-harvesting antenna complexes (B800-850). Lauryl dimethylamine *N*-oxide (LDAO) is a zwitterionic detergent that has been widely used to prepare spectrally pure pigment–protein antenna complexes from a range of purple photosynthetic bacteria. It can be preferentially used throughout the purification procedure and then removed and exchanged (Hawthornthwaite and Cogdell, 1991) for one or more of a range of purer, more expensive detergents such as octyl-glucoside. This exchange is especially important for preparations of antenna complexes to be used for crystallization trials, since crystal packing is affected by the type and concentration of the detergent used to solubilize the complex, and, furthermore, not all detergents will support crystal formation.

Usually, successive cellulose ion exchange chromatography is used to isolate the light-harvesting complexes prior to further purification by molecular sieve chromatography (Hawthornthwaite and Cogdell, 1991). Alternatively, initial separation can be achieved by sucrose density gradient centrifugation to produce the B800-850 complexes for subsequent anion exchange chromatography. Two distinct bands are seen in sucrose gradients loaded with solubilized photosynthetic membranes from a range of bacteriochlorophyll *a*-containing species of purple bacteria. The upper band contains B800-850 complexes, and a lower band contains almost exclusively the RC-antenna conjugate (Hawthornthwaite and Cogdell, 1991).

III. BIOCHEMISTRY OF THE LIGHT-HARVESTING COMPONENTS

A. Purple Photosynthetic Bacteria

The absorption spectra of whole cells and membrane preparations of purple bacteria show peaks and shoulders in the NIR which correspond to their complement of different antenna complexes (Thornber, 1986). Some species of bacteria such as *R. rubrum* or *Rps. viridis* only contain a single antenna type (B880 or B1012, respectively) and show a single spectral form (Thornber et al., 1983). Others, such as *Rb. sphaeroides* contain two types [B875 (equivalent to B880 in *R. rubrum*) and B800-850 (Thornber et al., 1983)]. Still others, such as *Rps. acidophila*, contain additional antenna types such as B800-820 (Cogdell et al., 1983). This plethora of spectral forms was initially perplexing, but now a simplified picture can be drawn.

There are two main types of antenna complexes in these bacteria. The first type (e.g., B880 of *R. rubrum* or B875 of *Rb. sphaeroides*) forms the so-called core complex, and is universally present. This type of complex is intimately associated with the RC and forms a precise stoichiometric ratio with it. In some species, such as *Rps. viridis*, this core complex is reportedly seen in electron micrographs as a doughnut-like structure in which the RC is surrounded by the antenna complex. There are about 24 molecules of antenna bacteriochlorophyll present in this complex per RC (Engelhart et al., 1983; Stark et al., 1984). The other types of antenna complexes (i.e., B800-850, B800-820, etc.) form the so-called variable light-harvesting complexes. They are a diverse group, but are characterized by being present within the photosynthetic membrane in quite variable amounts with respect to the RC content. The type and amount of the variable complexes present depends upon a variety of environmental factors (see Section VIII.B) such as light intensity and the temperature at which the cells are grown (Cogdell and Scheer, 1985; Zuber, 1985; Angerhofer et al., 1986). Thus the NIR absorption spectra are quite different in *Rps. acidophila* strain 7050 cells grown at 4000 Lux compared with those grown at 200 Lux (Angerhofer et al., 1986; Ferguson et al., 1991).

All of the purple bacterial antenna complexes are built upon the same modular principle. The pigments are noncovalently bound to two (four in some organisms) low molecular weight (5–7 kDa), hydrophobic apoproteins which are soluble in 1:1 v/v $CHCl_3$:CH_3OH (Zuber, 1985). These apoproteins are the so-called α- and β-apoproteins (Cogdell et al., 1985). It should be stressed that the monomeric α plus β particle cannot be isolated in a pigmented state; only oligomeric forms can be isolated (Zuber et al., 1987). More than 20 of these apoproteins have now been sequenced (Zuber and Brunisholz, 1991) from a range of purple bacteria. Comparison of these sequences has revealed several important conserved features. The α-apoproteins contain a single conserved histidine; the β-apoproteins contain two conserved histidines. Resonance-Raman spectroscopy suggests that at least some of these histidine residues are directly liganded to the magnesium atom at the center of the bacteriochlorophyll macrocycle (Robert and Lutz, 1985). Most of the apoproteins show a tripartite character, being polar at either end and hydrophobic in the center. The central hydrophobic core is about 20 to 23 amino acids in length, and it has been suggested that it traverses the membrane in the form of a membrane-spanning α-helix (see Zuber and Brunisholz, 1991).

The native, pigmented antenna complexes are formed by aggregation of the α- and β-subunits, and association with their pigments. The intact B800-850 complex from *Rps. acidophila* strain 10050, for example, is an α-6-β-6 oligomer (Bissig et al., 1988; Papiz et al., 1989). It is worth noting that in many cases the same pigments are found in both the antenna complexes and in the core complexes, and that it is the protein that determines which function any given pigment molecule is destined to fulfill. Thus, when bacteriochlorophyll *a* is dissolved in an organic solvent in which it is monomeric, its NIR absorption band is at 772 nm (Clayton, 1963); however, when bacteriochlorophyll *a* is correctly bound in either an RC or antenna complex, this band is substantially red-shifted to 800–900 nm (Clayton, 1963; Thornber et al., 1978; Cogdell, 1986). Moreover, the single peak at 772 nm is usually transformed into several peaks and/or shoulders. This red shift arises mainly from the pigment–protein interactions within these complexes, and is now routinely used both as a way of identifying them, and as a way for judging their integrity (Cogdell and Thornber, 1979). One fascinating question is what features in the primary structures of the antenna apoproteins control where the antenna bacteriochlorophyll's Qy absorption band will be. For some years Zuber and Brunisholz in Zurich have suggested a correlation between certain conserved aromatic residue clusters around the putative bacteriochlorophyll-binding sites with the observed spectral characteristics. The differences between this particular part of the sequence (-HAAVLTTTTWLPAYYQGSA) of the B800-850 α-apoprotein of *Rb. sphaeroides*, and that (-HAAVLTATTWYAAFLQGGV) of the B800-820 α-apoprotein of *Rps. acidophilia* 7050 led Zuber and Brunisholz (1991) to suggest that the change of YY to FL is responsible for the spectral shift from 850 nm to 820 nm between these two pigment–proteins. The results of a very recent study by Fowler et al. (1992), who changed the sequence in the *Rb. sphaeroides* apoprotein by site-directed mutagenesis from YY to FY and then to FL, gave considerable support to this notion. The change from YY to FY results in the 850-nm band shifting to 830 nm, while the double mutation to FL causes a shift to 820 nm.

We now consider further details of the structure of each type of antenna complex: The B880 complex from *R. rubrum* shows a single strong absorption maximum of 880 nm. CD measurements on this NIR band suggest that it represents dimeric bacteriochlorophyll *a* (Cogdell and Scheer, 1985). This complex contains two molecules of bacteriochlorophyll *a* and one molecule of the carotenoid spirilloxanthin per α/β pair. Parkes-Loach et al. (1988) and Ghosh et al. (1988) have taken advantage of the organic solvent solubility of the B880 antenna apoproteins to investigate structure/function relationships by reconstitution techniques. These two groups have isolated and purified the individual B880 antenna apoproteins by chromatography in organic solvents. They have shown that when these apoproteins are added back to purified bacteriochlorophyll *a*, a native antenna structure is regenerated, as judged by restoration of the native absorption spectrum. These pioneering studies have opened the way for investigating the process of assembly of an antenna complex and the factors that are required to produce the different spectral forms.

The B800-850 complex from *Rb. sphaeroides* shows two strong absorption bands in the NIR, one at 800 nm and one at 850 nm. The 850-nm band is about 1.5 times as intense as the 800 nm one. CD analysis of them shows that the 850-nm band has strong dimeric character, whereas the 800 nm band is more monomeric (Cogdell and Scheer, 1985). The α- and β-B800-850 apoproteins have been sequenced (Zuber and Brunisholtz, 1991). The α-apoprotein contains 56 and the β-apoprotein 51 amino acids. These apoproteins are present in the native complexes in a 1:1 ratio (Cogdell and Thornber, 1979). There are three bacteriochlorophyll *a* molecules present per pair of apoproteins, and the bacteriochlorophyll: carotenoid ratio is 2:1. The minimum size of a complex which shows a native absorption spectrum is 80 to 100 kDa, and it is probably an α-6-β-6 oligomer (Cogdell et al., 1979; Shiozawa et al., 1982; van Grondelle et al., 1983). Interestingly, it is the carotenoid within this type of antenna complex which is responsible for the well-known electrochromic carotenoid band shift (Webster et al., 1980).

B. Light-Harvesting Complex of Photosystem I

Photosystem I (PS I) is considered to be a macromolecular association of the two pigmented multiprotein complexes: a core complex (CC I) and a light-harvesting complex I (LHC I). PS I preparations have apparent sizes of about 250 kDa (Mullet et al., 1980a,b; Bassi and Simpson, 1987; Nechushtai et al., 1987; Bruce and Malkin, 1988; Peter et al., 1988), whereas CC I preparations are generally of 110 kDa. All of the chlorophyll *b* in PS I is associated with the LHC I pigment–proteins (Anderson et al., 1983).

CC I preparations, which are obtained by removing LHC I from isolated PS I by surfactant treatment, have 60 to 110 chlorophyll *a* molecules and contain about 10 to 12 protein subunits (Bengis and Nelson, 1977; Bruce and Malkin, 1988; Zipfel and Owens, 1991). Crystals of CC I have been obtained from several cyanobacteria (Boekema et al., 1987; Ford et al., 1988; Witt et al., 1988; Rogner et al., 1990; Almog et al., 1991) but none have yet been reported from higher plants. Preliminary data indicate that CC I is a trimer *in vitro* (and maybe *in vivo* (Hladik and Sofrova, 1991)).

Development of biochemical information about LHC I came from the seminal observations of Mullet et al. (1980a,b) and of Camm and Green (1983) followed later by those of others (e.g., Haworth et al., 1983; Kuang et al., 1984; Lam et al., 1984a,b; Dunahay and Staehelin, 1985; Bassi and Simpson, 1987; Peter et al., 1988). The details of progressive increases in knowledge about LHC I have been summarized in Thornber et al. (1991). LHC I can be fractionated into at least two distinct pigment–proteins, LHC Ia and LHC Ib (see Table 1) (Lam et al., 1984b; Bassi and Simpson, 1987; Peter et al., 1988). Comparison of the subunit compositions of PS I and CC I revealed that in addition to the apoproteins of these two pigment–proteins (20–24 kDa), two other polypeptides of 17 and 11 kDa were contained in the PS I particle but not in CC I (Peter et al., 1988). No function was attributed to the 17- and 11-kDa polypeptides at that time; however, it was later determined that the bundle sheath cells of maize have three distinct pigment–proteins, LHC Ia, LHC Ib, and LHC Ic, the subunits of each being of 24, 21, and 17 kDa, respectively (Vainstein et al., 1989). Most recently, a purified LHC I holocomplex of barley was shown to have four subunits of approximately 24, 21, 17, and 11 kDa (Anandan and Thornber, 1990), confirming that these four proteins were indeed LHC I constituents. There is some evidence that all of them bind pigment cofactors (see below).

Isolation of LHC I pigment–proteins as homogeneous components is much harder to achieve than obtaining the equivalent LHC II components. Starting with the same amount of plant material, yields of individual LHC I pigment–proteins are at least an order of magnitude lower. This is due to the need to use much gentler procedures to release the individual pigment–proteins from isolated photosystem I without disrupting their pigment–protein associations. Consequently, only partial release of the LHC I components can be achieved.

A current synopsis of knowledge about these pigment–proteins is as follows:

LHC Ia. This complex has a fluorescence emission maximum of 690 nm at 77 K, and apoproteins of approximately 24 and 22 kDa (Peter et al., 1988; Dreyfuss and Thornber, in press, 1994). It binds chlorophyll *a*, chlorophyll *b*, and xanthophylls (Table 1), but neither neoxanthin nor carotenes. It has the lowest ratio of chlorophyll *a/b* of all the LHC I components (Table 1). Preliminary indications are that it, like LHC Ib and LHC IIb, occurs *in situ* in a trimeric form.

LHC Ib. There are two 21-kDa apoproteins of this the major LHC I pigment–protein. It contains chlorophylls *a* and *b* (ratio = 2.3). Lutein and violaxanthin are the only two carotenoids present. LHC Ib is responsible for the fluorescence emission maximum at 730 to 735 nm characteristically seen for photosystem I at low temperature in whole leaves. This complex can be isolated as an oligomeric, probably trimeric, pigment–protein by sucrose gradient centrifugation and Deriphat-PAGE (Preiss et al., 1993). Protease digestion of intact membranes indicated that its apoprotein(s) are deeply embedded within the thylakoid membrane (Ortiz et al., 1985).

LHC Ic. This pigment–protein complex has a single apoprotein of 17 kDa, the N-terminal sequence of which has been obtained (Anandan et al., 1989) and correlated with a published gene sequence (*psa*F of Steppuhn et al., 1988). Controversy exists whether the 17-kDa protein is that of an LHC I apoprotein or that of a nonpigmented component of CC I that functions to bind plastocyanin to the core complex (see Preiss et al., 1993 for details). In essence we believe that there are two different 17-kDa proteins but that the *psa*F gene's deduced sequence is that of the LHC Ic apoprotein while some other gene codes for another 17-kDa polypeptide, the function of which is to bind the plastocyanin subunit to CC I. Like the two LHC I pigment–proteins already mentioned, LHC Ic also binds chlorophyll *a*, maybe chlorophyll *b*, and carotenoids. Its absorption spectrum indicates that it is much more enriched in chlorophyll *a* and carotenoid than the other LHC I pigment–proteins (Preiss et al., 1993).

LHC Id. The 11-kDa component of the LHC I holocomplex has been termed LHC Id, and preliminary evidence indicates that it binds pigment cofactors (see Preiss et al., 1992).

Other LHC I components. Bassi and Simpson (1987) obtained an LHC I-680 fraction, which contained polypeptides of 21 and 24 kDa. They contended that it was equivalent to the LHC IId (CP 24) component of PS II (see Section III.C). This was subsequently refuted by Morishige et al. (1990) who used N-terminal protein sequence data to show that the isolated apoproteins of LHC IId and LHC Ia have separate and distinct sequences; the pigment–protein, LHC Ia, is responsible for the 680-nm fluorescence emission maximum in PS I. Further support for the two being separate entities came from Schwartz et al. (1991a) who compared the sequences of cDNA clones coding for various LHC I and LHC II proteins. There was little similarity between the sequence of LHC IId and those of any of the LHC I apoproteins.

The *in vivo* organization of LHC I in the thylakoid membrane has been speculated upon. Bassi and Simpson (1987) reported the isolation of two chlorophyll *a*- and *b*-containing LHC I subcomplexes from barley. These complexes had 77 K fluorescence emission maxima at 680 and 730 nm, and were termed LHC I-680 and LHC I-730, respectively. It was suggested that the two antenna complexes represent distinct PS I light-harvesting entities that are attached to CC I in different ways: LHC I-680 being peripheral to LHC I-730 which is attached directly to CC I. Our view (see Figure 3, page 96) of the organization is based on the ease with which surfactants extract pigment–proteins from PS I (Thornber et al., 1992). Our model shows some similarities to that shown in Bassi and Simpson (1987). The data suggest that since the oligomeric form of LHC Ia is the more readily extracted by glycosidic surfactants, it is more peripheral to CC I than the trimeric form of LHC Ib. If the surfactant used for extraction is Triton X-100, all LHC I proteins are readily extracted from photosystem I except the LHC Ic apoprotein, indicating that LHC Ic lies closest to and/or is the most tightly attached LHC I component to the core. Further discussion of the relative arrangement of the LHC components can be found at the end of Section III.C.

C. Light-Harvesting Complex of Photosystem II

Photosystem II (PS II) is composed of a core complex (CC II) and a light-harvesting complex (LHC II), each of which is an assembly of several pigment–proteins and, in the case of CC II, also of some polypeptides that are not associated with photosynthetic pigments. The core of photosystem II is thought to contain at least seven polypeptides: 47, 43, 34 + 32, 33, and 9 + 4 kDa derived from the CP47, CP43, and CC II RC pigment–proteins, oxygen evolution enhancer I protein, and cytochrome b_{559}, respectively. CC II, like CC I, is probably ubiquitous in oxygenic organisms. Evidence is increasing that the core complex of PS II is present *in situ* as a dimer (Peter and Thornber, 1991a). LHC II contains at least four well-defined pigment–proteins and, as far as we know, no polypeptides that are not associated with pigment.

LHC II is composed of four very similar xanthophyll–chlorophyll *a*/*b*-proteins (LHC IIa, b, c, and d) (Peter et al., 1988, 1991b), which exhibit considerable amino acid sequence homology not only to each other but also to the LHC Ia and Ib pigment–proteins of PS I (Figure 1). A fifth photosystem II pigment–protein with a 13-kDa subunit (LHC IIe) also seems to be present in LHC II (Peter and Thornber, 1991b). All of the known LHC II pigment–proteins contain xanthophylls but not carotenes. There are different proportions of lutein, neoxanthin, and violaxanthin in each LHC II component (Peter and Thornber, 1991b).

We have proposed that some of the minor LHC II's [LHC IIa, b (25-kDa-type III), c, and d] form a multiprotein complex of about 110 kDa that functions to join the major antenna component (LHC IIb) to the core complex (Peter and Thornber, 1991b; see also Peter et al., 1988). This notion and the three-dimensional structure of these LHC components is considered below after a short description of the biochemistry of each LHC II component.

LHC IIa. LHC IIa (CP29) was first observed by Machold and Meister (1979) and extensively examined by Camm and Green (1980) and later by others (Bassi et al., 1987; Peter and Thornber, 1991b; Morishige and Thornber, 1992). LHC IIa occurs in diverse species of higher plants (Green et al., 1982), in all of which it has been found to have a chlorophyll *a/b* ratios of 2.20–2.25 (Machold and Meister, 1979; Green et al., 1982; Peter and Thornber, 1991b). Its room temperature absorbance spectrum has a maximum in the red at 677 nm with an additional unusual minor peak at 645 nm which is characteristic of this pigment–protein (Machold and Meister, 1979; Camm and Green, 1980; Peter and Thornber, 1991b). LHC IIa contains the xanthophylls, lutein, and violaxanthin in approximately equal proportions, and is relatively enriched in violaxanthin in comparison with the other LHC II pigment–proteins. Smaller amounts of neoxanthin are also associated with LHC IIa (Peter and Thornber, 1991b).

The single apoprotein of LHC IIa migrates on fully denaturing SDS-PAGE with an apparent size of 30 to 31 kDa in barley and maize and of 29 kDa in spinach and pea (Camm and Green, 1980; Bassi et al., 1987; Peter and Thornber, 1991b). The reason for the difference in molecular mass between species is unknown at this time. Morishige and Thornber (1992) have isolated a cDNA for this component and confirmed its identity by comparing the deduced gene sequence for the apoprotein with sequences obtained directly on the apoprotein (see also Henrysson et al., 1989). The derived amino acid sequence of this apoprotein displays homology (38% identity and 60% similarity) to that of LHC IIb (Figure 1). An extra segment of 42 amino acids, not contained in other LHC sequences, could account for its having the largest size of all the LHC apoproteins. LHC IIa is expressed by a single or low copy number set of genes in the nucleus and is, like that of LHC IIb, light-regulated via phytochrome (Morishige and Thornber, 1992).

The chlorophyll *b*-less chlorina f2 barley mutant, initially believed to lack all LHC II subunits (e.g., Thornber and Highkin, 1974), contains detectable amounts of LHC IIa (Darr et al., 1986; White and Green, 1987b; Peter and Thornber, 1991b). Furthermore, in intermittent-light grown (IML) plants, which have elevated chlo-

```
                                                                        Transit peptide | N-terminus
LHC IaI     MAT-----QAL----------ISSSIST---SAR-AARQ-----------------IIGSRISQSVT-   RKASFVVRAASTPPVKQGANRQLWF-ASK-------
LHC IaII    MASACASSTIAAVAFSS----PSSQKNGSIVGAT----KASFLGG--KRLR-VSKFIAPVGSRSVAVSA             VAADP---DRPLWFPGS--------
LHC IbI     MASNTLMSCGIPAVC------P-SFL----SSTKSKFAAAMP------------VSVGATNSMSRFSM                      SAD-WMPG-Q-------
LHC IbII    MATVTTH--ASASIFRPCTSKP-RFLTG--SSGRLNRDLSF-----T---S-I-GSSA-KT-SSFKVEA                     KKGE-WLPGL--------
LHC IIa     MASS----VAA----------AASTFLG----TRLADPR--------PQ-N--------GRIVARFGFG  KKK---APPKKAKAPP---TTDRPLWFPGA--------
LHC IIbI    MAAS----MAL----------SSPSLVG-----K--AVK---LAPAA---SE--------VFGEGRVSM   RK-TAG-KPK---PV---SSGSPWY-GPD-RVKYL-
LHC IIbII   MAA-----SAI----------QSSAFAG-----Q-TALK---------QRDEL--VRKVGV-SDGRFSM   RR-TV--KA----VP-----QSIWY-GAD-RPKFL-
LHC IIbIII  MAS-----MAATA-SSTTVVKATP-FLG-----Q-TK--NAN-----P-LRDV------VAMGSARFTM                     SNDLWY-GPD-RVKYL-
LHC IIc     MAAAA---TSLYVS---EMLGSPVKFSG--------VARPAAPSPSS---S-------AT--FKTVALF  KKKAAAAPAKAKAAAVSPADDILAKWY-GPD-RRIFL-
LHC IId     MAAAT-SATAIVNGF------TSPFLSGG----KKSS-Q-SLLFV---NSKVGAGV-STTSRKLVVVAA      AAAPKK--------------SWIPAVKGGGNFL-

                    alpha    bravo    Hook 1                                                                     Helix I
LHC IaI     --------QSLSYSDGSLPGDFGFDPLGL-SDPEGTGG-------------------------------------------FIEPKWLAYGEVINGRFAMLGAAGAIAPEILGKA
LHC IaII    --------TPPEWLDGSLPGDFGFDPLGLGSD------------------------------------------------PESLKWNAQAELVHSRWAMLGAAGIFIPEFLTKI
LHC IbI     --------PRPSYLDGSAPGDFGFDPLGLGEV------------------------------------------------PANLERYKESELIHCRWAMLAVPGIIVPEAL---
LHC IbII    --------ASPDYLTGSLAGDNGFDPLGLAED------------------------------------------------PENLKWFVQAELVNGRWAMLGVAGMLLPEVFTKI
LHC IIa     --------QAPEYLDGTLVGDYGFDPFGLGKPAEYLQYDVDSLDQNLAQNLAGEIIGTRFEDADVKSTPFQPYAEVFGLQRFRECELIHGRWAMLATLGALTVEWLT--
LHC IIbI    GPFSG---EAPSYLTGEFPGDYGWDTAGLSAD------------------------------------------------PETFAKNRELEVIHARWAMLGALGCVFPELLARN
LHC IIbII   GPFSE---QTPSYLTGEFPGDYGWDTAGLSAD------------------------------------------------PETFAKNRELEVIHSRWAMLGALGCIFPELLSKN
LHC IIbIII  GPFSA---QTPSYLNGEFPGDYGWDTAGLSAD------------------------------------------------PEAFAKNRALEVIHGRWAMLGALGCIFPEVLEKW
LHC IIc     PEGLLDRSEIPEYLNGEVPGDYGYDPFGLSKK------------------------------------------------PEDFAKYQAYELIHARWAMLGAAGFIIPEAFNKF
LHC IId     --------D-PEWLDGSLPGDFGFDPLGLGKD------------------------------------------------PAFLKWYREAELIHGRWAMLAVLGIFVGQAWT--

                                                             Helix II                                                    charley
LHC IaI     GLIPQETALAWFQTGV---IPPAG---------TY-NYW----ADNYTLFVL-EMALMGFAEHRRF--------QDWAKPGSMGKQYFLGLEKGLGGS---GDPAYPGGPL
LHC IaII    GVLNT-PS--WYTAGEQ---------EYF---------------TDTTTLFVI-ELVLIGWAEGRRW--------ADIIKPGCVNTDPIFPNNK-LTGT-DVG---YPGGLW
LHC IbI     GLGN------WVKA-QEWAAIPGG-QA-----TYLGQPVPW-GTLPTILAI-EFLAIAFVEHQRS--------ME--KD----SE----KKK------------YPGGA-
LHC IbII    GIINV-PE--WYDAGKE---------QYF---------------ASSSTLFVI-EFILFHYVEIRRW--------QDI-KNPGSVNQDPIFKQ-YSLPKGEVG---YPGGI-
LHC IIa     GV--T-----WQDAG----KVE-LVD--G--SSYLGQPLPF-TITT-LIWI-EVLVIGYIEFQRN--------AE---------LD---PERR-----------LYPGGSY
LHC IIbI    GVK--FGEAVWFKAGSQ-IFSEGGLDYLGNPS--LVH-----AQSILAIWATQVVLMGAVEGYRV--------A-G-GPL----------------GEVVDPLYPGGS-
LHC IIbII   GVQ--FGEAVWFKAGAQ-IFSEGGLDYLGNPN--LVH-----AQSILAIWATQVVLMGLIEGYRV--------G-G-GPL----------------GEGLDPLYPGGA-
LHC IIbIII  -VKVDFKEPVWFKAGSQ-IFSDGGLDYLGNPN--LVH-----AQSILAVLGFQVVLMGLVEGFRI---------NG-LPG-V---------------GEGND-LYPGGQY
LHC IIc     GANCG-PEAVWFKTGA-LLL-DG---------NT-LNYFGKNIPINLILAVVAEVVLVGGAEYYRI--------ING-------LD---LEDK-----------LHPGGP-
LHC IId     GI-P------WFEAGA-----DPG-------------AVA--PFSFGTLLGTQLLLMGWVESKRWVDFFDPDSQD-VEWATPWSRT--AEN-FSNSTGEQG---YPGGKF
```

```
                  Hook 2                      Helix III                           Helix IV           C-terminus
LHC IaI      FNPLGFGKD--------EK-SMKELKLKEIKNGRLAMLAILGYFIQALV-TGVG--PYQNLLDHLADPVNNN---VLT--SLKFH
LHC IaII     FDPLGWGSGS------PAK--IKELRTKEIKNGRLAMLAVMGAWFQHI-YTGTG--PIDNLFAHLADPGHATI--FAA-FSPK
LHC IbI      FDPLGYSKD-------PAK--FEELKVKEIKNGRLALLAFVGFCVQQSAYPGTG--PLENLATHLADPWHNNIGDVIIPKGIFP-N
LHC IbII     FNPL----N-------FAP--TQEAKEKELANGRLAMLAFLGFVVQHNV-TGKG--PFENLLQHLSDPWHNTI--VQT----F--N
LHC IIa      FDPLGLAAD-------PEK--KETLQLAEIKHARLAMVAFLGFAVQAAA-TGKG--RLNNWATHLSDPLHTTI--FDT----FGSS
LHC IIbI     FDPLGLADD-------PEA--FAELKVKEIKNGRLAMFSMFGFFVQAIV-TGKG--PLENLADHLADPVNNNAWAFATNF---VPGK
LHC IIbII    FDPLGLADD-------PEA--FAELKVKEIKNGRLAMFSMFGFFVQAIV-TGKG--PLENLADHIADPVANNAWAFATNF---VPGK
LHC IIbIII   FDPLGLADD-------PTT--FAELKVKEIKNGRLAMFSMFGFFVQAIV-TGKG--PLENLLDHLDNPVANNAWVYATKF---VPGA
LHC IIc      FDPLGLAKD-------PDQ--AAILKVKEIKNGRLAMFSMLGFFIQAYV-TGQG--PVINLAAHLSDPFGNNLLTVIGGASERVPTL
LHC IId      FDPLSLAGTISNGVYMPDTDKLERLKLAEIKHARLAMLAMLIFYFEA----GQGKTPLGALGL
```

	Source	Gene sequence	N-terminal sequence
LHC IaI	Tomato Cab8	Pichersky *et. al.*, 1989	ND
LHC IaII	Petunia LHCI-15	Stayton *et. al.*, 1987	Ikeuchi *et. al.*, 1991
LHC IbI	Tomato Cab6B	Pichersky *et. al.*, 1987	Ikeuchi *et. al.*, 1991
LHC IbII	Arabidopsis Cab4	Zhang *et. al.*, 1991	Ikeuchi *et. al.*, 1991
LHC IIa	Barley LHCIIa-1	Morishige & Thornber, 1991	ND
LHC IIbI	Lemna ab30	Kohorn *et. al.*, 1986	Michel *et. al.*, 1991
LHC IIbII	Lemna ab19	Karlin-Neumann *et. al.*,1985	Michel *et. al.*, 1991
LHC IIbIII	Tomato Cab13	Schwartz *et. al.*, 1991	Morishige & Thornber, 1991
LHC IIc	Tomato Cab9	Pichersky & Green,1990	ND
LHC IId	Spinach CP24	Spangfort *et. al.*, 1990	Morishige *et. al.*, 1990

Figure 1. Amino acid sequence alignment obtained from translated gene sequences for the different LHC pigment–proteins (LHC Ia, LHC Ib, LHC IIa, LHC IIb, LHC IIc, LHC IId) belonging to the LHC gene family. The LHC Ic sequence has similarity to that of the PsaF gene products. The LHC Id and LHC IIe have yet to be sequenced. Where different pigment–peptide complexes have more than one type of protein involved, the sequences are labeled with an additional Roman numeral (e.g., LHC IIbIII). The N-terminal amino acid sequences of mature LHC Ia, LHC IIa, or LHC IIc have not been determined. Helix I, II, and III identify those parts of the sequence which are proposed to form the three membrane-spanning α-helices (*shaded*). Helix IV identifies a possible fourth α-helix in the lumen. Hook 1 and 2 indicate putative stromal α-helices. Alpha, bravo and charley mark the location of conserved beta-turns.

rophyll *a/b* ratios and greatly reduced amounts of LHC II antenna (Section VIII.A), the ratio of the amounts of LHC IIa to LHC IIb is greater than that found in normal light-grown plants (Darr et al., 1986). Thus, under conditions of reduced chlorophyll *b* synthesis LHC IIa is still stably integrated into the thylakoid membrane; chlorophyll molecules are thought to stabilize the pigment-binding proteins in the thylakoid membrane (Apel and Kloppstech, 1980; Bennett, 1981; Bellemare et al., 1982). LHC IIa's abundance in the mutant and in IML plants is possibly due to LHC IIa's being located closer to CC II than LHC IIb (Ghanotakis et al., 1987; Camm and Green, 1989; Peter and Thornber, 1991b; see also below). Thus, its presence would be required perhaps before LHC IIb could be added to the photosystem, and/or LHC IIa's requirement for chlorophyll *b* could be less than that of LHC IIb (see Section VII).

LHC IIb. The major chlorophyll-binding protein associated with green plant PS II is LHC IIb. Because of its abundance within the chloroplast, in which it accounts for about one-third of the total protein and binds about 45% of the chlorophyll, a large body of research has been devoted to it alone. LHC IIb has been implicated in the cation-mediated formation of grana stacks (e.g., Mullet and Arntzen, 1980; see Thornber et al., 1991 for summary). LHC IIb is also thought to be involved in the even distribution of excitation energy between photosystems. The even distribution is accomplished through phosphorylation/dephosphorylation of the LHC IIb complexes, causing a lateral movement of some of the phospho-LHC IIb from the grana stacks to the stroma lamellae; this process is correlated with the State 1–State 2 transition (Barber, 1982; Allen, 1992a,b). The N-terminal portion of LHC IIb contains a threonine or serine residue(s) at residue #3 (see Figure 1 or Appendix) which is thought to be the phosphorylation site (Mullet, 1983; Michel et al., 1991).

LHC IIb was first observed and isolated as a single, green pigment–protein, CP2 (Section II.A). Subsequently, SDS-PAGE systems resolved the LHC IIb complex into two bands. One band migrated with an apparent size of approximately 68 kDa which was thought to be a dimer (Genge et al., 1974) of the second band, which migrated at 26 kDa (Dunkley and Anderson, 1979). Under the mildest nondenaturing PAGE conditions (Peter and Thornber, 1990), LHC IIb appears to migrate solely as a trimer and often also as an even higher order oligomer in addition to the trimer. The oligomeric state of isolated LHC IIb was exhaustively analyzed by analytical centrifugation (Butler and Kuhlbrandt, 1988) and unequivocally shown to be a trimer (see also Kuhlbrandt and Wang, 1991). The oligomeric LHC IIb has an apparent size on PAGE ranging from 67 kDa (Dunkley and Anderson, 1979) to 80 kDa (Bennett et al., 1981). All normal green plant species examined to date appear to contain a trimeric form of LHC IIb.

The chlorophyll *a/b* ratio of this chlorophyll–protein complex as obtained by Deriphat-PAGE (Peter and Thornber, 1988, 1991b) is 1.33; preparations of Kuhlbrandt (1988) have a value of 1.14. Room temperature absorbance maxima are observed at 653 and 675 nm, corresponding to chlorophylls *b* and *a*, respectively. There is a distinct valley between the two peaks [cp. LHC IId (CP24)] (Peter and Thornber, 1988, 1991b). It has been calculated that a single molecule of the apoprotein would bind 15 chlorophyll molecules, corresponding to eight chlorophyll *a* and seven chlorophyll *b* molecules (Butler and Kuhlbrandt, 1988; Kuhlbrandt, 1988; see also Peter and Thornber, 1988, 1991b). It has been proposed that the chlorophyll molecules are coordinated to histidine, glutamine, and asparagine residues (Peter and Thornber, 1988). Analysis of the LHC IIb crystal structure (Section VI), however, indicates that the chlorophyll molecules are probably too far away for direct interaction of the Mg^{2+} of the porphyrin ring with the polypeptide chain (Kuhlbrandt and Wang, 1991). Therefore, the chlorophyll molecules would have to be stabilized via water molecules, or by hydrogen bonding of the porphyrin rings to the polypeptide. LHC IIb also contains xanthophylls: lutein, neoxanthin, and violaxanthin (Peter and Thornber, 1991b). A probable molar ratio is five chlorophylls per xanthophyll molecule.

On SDS-PAGE at least three apoproteins of LHC IIb are observed, migrating with apparent sizes of 28, 27–26, and 25 kDa (Peter and Thornber, 1988, 1991b). In earlier SDS-PAGE systems only two bands of 29 and 28 kDa were apparent. Addition of urea to the polyacrylamide gel matrix changes slightly the mobility of various polypeptides, among which is the 25-kDa LHC IIb apoprotein, which comigrated with the other slightly larger LHC IIb apoproteins in the absence of urea. The multiple 25–28 kDa subunits of LHC IIb do not have a simple stoichiometry, nor does it seem that there are the same number of apoproteins in all plants. Usually the largest apoprotein occurs in an order-of-magnitude greater concentration than the other two smaller ones. Although three subunits are clearly present in most plants, as many as eight can be resolved in some plants (Sigrist and Staehelin, 1992). In most species all three apoproteins have blocked N-terminal residues, most likely by an acetyl group added posttranslationally to the N-terminal arginine residue (Michel et al., 1991). The 28- and 27-kDa apoproteins of LHC IIb are not present or are greatly reduced in amount in the chlorophyll *b*-less chlorina f2 barley mutant (Section VIII). In contrast the 25-kDa LHC IIb component appears to be relatively abundant (White and Green, 1987b; Peter and Thornber, 1991b).

LHC IIb proteins are coded for in the nucleus by gene families of between 3 (Leutwiler et al., 1986) and 16 (Dunsmuir and Bedbrook, 1983; Dunsmuir et al., 1983) members in *Arabidopsis thaliana* and *Petunia hybrida*, respectively. All the apoproteins have a somewhat similar amino acid sequence (deduced from sequences of cloned genes; Green et al., 1991). Three types of LHC IIb genes occur: Type I (the most prevalent) (Stayton et al., 1986), Type II (an intron-containing gene) (Karlin-Neumann et al., 1985), and Type III (a slightly truncated gene) (Morishige and Thornber, 1990, 1991). The amino acid sequences of the three classes are very similar, but not identical, displaying at least 80% similarity (see Appendix). The expression of the LHC IIb genes and the insertion of this protein into the thylakoid membrane are described in Section VII.A.

The correlation of genes to the three different LHC IIb apoproteins is not as simple as the numbers imply. Type II genes have not yet been identified in some organisms. Type III genes have so far only been suggested to occur in barley (Morishige and Thornber, 1990), wheat (Webber and Gray, 1989), and *Arabidopsis thaliana* (Morishige and Thornber, 1991) by protein sequencing; very recently a Type III gene has been isolated from tomato (Schwartz et al., 1991b). It is not entirely clear whether the differences in apparent size of the LHC IIb apoproteins are due to expression of different genes of the LHC IIb family and/or to different posttranslational modifications. There is some evidence which suggests some LHC IIb apoproteins are palmitylated (Mattoo and Edelman, 1987), but it is unknown how this posttranslational modification might affect the rate of migration of the apoproteins, or even what percentage of them are acylated. In contrast, another study found no evidence of covalent attachment of fatty acids to the LHC IIb apoprotein (Jansson et al., 1990). Comparison of the LHC IIb apoproteins using monoclonal antibodies do indicate differences among the three subunits (Darr et al., 1986; Peter and Thornber, 1991b) but without exact knowledge of the epitopes reacting with these antibodies, unequivocal conclusions cannot be reached. It is now more apparent that the differences in size of the LHC IIb apoproteins are most probably due to expression of different LHC IIb gene family members. Sigrist and Staehelin (1992) have identified up to eight different LHC IIb apoproteins using antibodies produced against Type I or II specific synthetic peptides. In agreement with the greater number of Type I genes isolated so far, Type I apoproteins were more numerous and of greater size than the Type II proteins in each of the plant species investigated. Direct protein sequencing has shown that the LHC IIb oligomer contains at least the Type I and II gene products in Scots pine (Jansson et al., 1990). Similarly, in *A. thaliana* at least two different LHC IIb proteins derived from Type I and III genes have been identified by protein sequencing (Morishige and Thornber, 1991).

LHC IIc. Originally it was surmised that since reelectrophoresis of LHC IIb oligomer fractions on a second nondenaturing gel had yielded a monomeric form of LHC IIb (Camm and Green, 1980; Larsson and Andersson, 1985), the green chlorophyll–protein band found in the LHC IIb monomer region was solely that of LHC IIb. However, in the mildest nondenaturing gels, which keep all the LHC IIb in its trimeric or higher oligomeric state, a different pigment–protein (LHC IIc or CP27) has been observed migrating in the region of the LHC IIb monomer (Bassi et al., 1987; Dunahay et al., 1987; Peter and Thornber, 1991b). LHC IIc has different spectral and subunit characteristics than either LHC IIa or LHC IIb (Peter and Thornber, 1991b), thus substantiating that it is a distinct LHC II pigment–protein. LHC IIc is the most difficult of the LHC II components to obtain pure because of its overlap on Deriphat-PAGE with LHC IIa and d (Peter and Thornber, 1988, 1991b). Its apoprotein(s) has been reported to be a single polypeptide of 26 kDa in spinach (Dunahay et al., 1987), of 27 kDa in pea, and 29 and 26.5 kDa in barley

(Peter and Thornber, 1988, 1991b). Bassi and co-workers (1987) reported that a similar nonphosphorylatable pigment–protein in *Zea mays* contains two polypeptides of 28.5 and 29 kDa. A recently cloned gene (Pichersky et al., 1991), purportedly that of CP29 (LHC IIa), is indeed that of the larger LHC IIc component (Morishige and Thornber, 1992). The LHC IIc apoprotein is not phosphorylatable under *in vitro* conditions (Dunahay et al., 1987). It has a chlorophyll *a/b* ratio of 1.8–2.0 (Bassi et al., 1987; Peter and Thornber, 1991b) and is relatively enriched in the xanthophyll lutein, with neoxanthin and small amounts of violaxanthin also present (Peter and Thornber, 1991b). Pigmented LHC IIc appears to be present in normal, perhaps even elevated, amounts in the chlorina f2 chlorophyll *b*-less barley mutant (Peter and Thornber, 1991b).

LHC IId. LHC IId (CP24) was first isolated from octyl-glucoside solubilized PS II fractions of spinach, separated on mildly denaturing polyacrylamide gel electrophoresis (Dunahay and Staehelin, 1986). Generally this complex is reported to contain one 21 kDa subunit, but in some instances as many as four subunits have been reported (Dunahay and Staehelin, 1986). A red absorption maximum of 668 nm has been recorded (Dunahay and Staehelin, 1986; Bassi et al., 1987); however, Peter and Thornber (1991b) found their preparation had a considerably longer wavelength maximum at 674 nm. Approximately 3 to 4% of the total chlorophyll in the chloroplast is bound to this pigment–protein (chlorophyll *a/b* = 0.8) (Table 1) (Dunahay and Staehelin, 1986; Peter and Thornber, 1991b). When compared to both LHC IIa and LHC IIb, LHC IId is reported to have a much higher chlorophyll/protein molar ratio (4.5, 7.8, 10–11, respectively) (Dunahay and Staehelin, 1986); however, the value determined for LHC IIb is far removed from that of Kuhlbrandt (1988). Lutein comprises the largest portion of the xanthophyll content of LHC IId, with relatively small amounts of neoxanthin and violaxanthin also present. In particular, its neoxanthin content is much lower than in the other LHC II components (Peter and Thornber, 1991b). This complex is apparently not phosphorylated under *in vitro* conditions (Bassi et al., 1987; Dunahay and Staehelin, 1987; Peter and Thornber, 1991b).

Morishige and Thornber (1990) obtained the sequence of the N-terminal 66 residues of LHC IId, and used it to confirm that the complete sequence deduced from a putative CP24 gene by Spangfort et al. (1990) was indeed that of this pigment–protein. Its lack of correspondence to a 21-kDa pigment–protein in photosystem I, LHC Ia (LHC I-680) (Bassi et al., 1987) has been commented on in Section III.B. It is not clear if LHC IId is present in the chlorina f2 chlorophyll *b*-less barley mutant, because in this mutant the apoprotein can be confused with that of LHC Ib (see Peter and Thornber, 1991b). Its function as part of the "connector" unit that links CC II and the peripheral LHC IIb pigment–proteins (Dunahay and Staehelin, 1986; Peter and Thornber, 1991b) is discussed below.

Other LHC II's. Recently, a novel pigmented complex was isolated from spinach that contained a single apoprotein of 14 kDa (Irrgang et al., 1990). The room temperature red absorption maximum for this complex is 670 nm. Antibodies produced against the 14 kDa apoprotein react with other chlorophyll-binding proteins, such as CP47, CP43, LHC II, and CP24, indicating epitope similarities and therefore a relatedness between this newly discovered LHC and other chlorophyll-binding proteins. A similar LHC II component has been reported in barley and has been termed LHC IIe (Peter and Thornber, 1988, 1991b). LHC IIe has a chlorophyll *a/b* ratio of 1.4 and an apoprotein of 12 to 13 kDa. Relative to the other LHC II's, LHC IIe is highly enriched in xanthophylls. Because of its high xanthophyll content, LHC IIe has been hypothesized to serve in a photoprotective role for PS II or as a carotenoid carrier protein (Peter and Thornber, 1991b). Whether LHC IIe is equivalent to the pigmented complex described by Irrgang et al. (1990) is not clear.

D. Organization of the LHC II Pigment–Proteins within the Photosystems

Although the pigment–proteins that constitute LHC II are reasonably well defined, there is as yet no unequivocal model for their relative positioning within the higher plant photosystem. Our present understanding of the arrangement comes from correlating ultrastructural, photochemical, and biochemical information with changes that occur either during assembly and disassembly of LHC II or in mutants deficient in one or more LHC II pigment–proteins. It is still uncertain which LHC II subunits contact which CC II subunits or whether LHC II is organized into one large or several smaller complexes that each interact with CC II.

Biochemical approaches dominate the current view of LHC II organization, because only these can distinguish between the very similar LHC II subunits present in thylakoids (see below). Thus, the most successful approaches so far have been to isolate PS II subcomplexes that contain different complements of LHC II components (e.g., Ikeuchi et al., 1985; Bassi et al., 1987, 1992; Ghanotakis et al., 1987; Peter and Thornber, 1991a,b).

Morrissey et al. (1989) investigated the organization of LHC II subunits with respect to CC II by correlating increases in the number of LHC II chlorophylls per P680 with increases in the abundance of LHC II apoproteins during development of the y9y9 mutant of soybean. Isolated PS II holocomplexes contain 250 chlorophyll molecules per unit (Melis and Anderson, 1983; Ghirardi et al., 1986), of which CC II is thought to contain 35 and LHC II the other 215 chlorophyll molecules (Morrissey et al., 1989). Assuming that each LHC II subunit binds 15 chlorophyll molecules, they deduced that LHC IIa was closest to CC II because its relative abundance with respect to the CC II proteins changed the least as the number of LHC II chlorophylls/P680 increased. They also observed that the relative abundance of the largest LHC IIb subunit increased the most during greening, perhaps indicating a more peripheral location for it within the photosystem. It should be noted that the changes in the relative abundance of LHC II polypeptides detected by their procedures were hard to correlate with all known LHC II subunits because of comigration of some of the similar sized LHC II subunits (e.g., LHC IIc's and the 27-kDa LHC IIb subunit) in their system. In addition, it is still equivocal whether each LHC II subunit binds 15 chlorophyll molecules.

Isolation of PS II subcomplexes that contain discrete complements of LHC II subunits has led to the conclusion that LHC IIa and LHC IIc are the most tightly associated with CC II and must bind to some CC II subunit. Ikeuchi et al. (1985) used octyl-glucoside extraction of PS II holocomplexes followed by gradient centrifugation to obtain a product which had the CC II components plus only the LHC II apoprotein of 29 kDa (LHC IIa). However, the equivalent preparation from pea leaves also contained small amounts of LHC IIb subunits and a 27-kDa polypeptide, probably one of the LHC IIc apoproteins (Peter and Thornber, 1987). Ghanotakis et al. (1987) similarly used octyl-glucoside extraction, but in contrast to Ikeuchi et al. (1985) they obtained their CC II complex by differential salt precipitation. Their product contained the CC II subunits plus an LHC II subunit of 28 kDa. Camm and Green (1989) substantiated the 28-kDa subunit to be the LHC IIa (CP 29) apoprotein; they also found the preparation had a 26-kDa polypeptide (probably an apoprotein of LHC IIc). Other analyses of material prepared in a similar way from barley showed that it contained small amounts of LHC IIb, and the LHC IIa, LHC IIc, and LHC IIe pigment–proteins (Barbato et al., 1989; Peter and Thornber, 1991b).

Peter and Thornber (1991a) studied how the LHC II subunits are arranged with respect to CC II by using decyl-maltoside treatment of PS II holocomplexes to isolate three different PS II subcomplexes. These subcomplexes contained all of the CC II subunits, and some or all of the following: LHC IIa, LHC IIb, and LHC IIc. The strength of association with the core was LHC IIa > LHC IIc > the 28-kDa LHC IIb subunit > the 27-kDa LHC IIb subunit > the 25-kDa LHC IIb subunit. The use of longer chain glycosidic surfactants yielded an additional subcomplex which contained LHC IIa, LHC IIc, and the 28-kDa LHC IIb subunit, and all of the CC II subunits except that of CP43. LHC IIa and/or LHC IIc must then bind directly to CP47 and models which depict LHC II associated *only* with CP43 must be incorrect. Peter and Thornber (1991b) also studied how the LHC II subunits are arranged with respect to each other. A large LHC IIb subcomplex (probably composed of three LHC IIb trimers) which contained only the 28- and 27-kDa subunits was described. Another multimeric subcomplex composed of LHC IIa, LHC IId, and a trimeric LHC IIb unit having two copies of the 28-kDa subunit and one of the 25-kDa subunit was isolated. This indicated that LHC IIb subunits are organized into different types of LHC IIb subcomplexes. It was proposed that the LHC II subcomplex containing the minor LHC II components functions as a connector to enable energy to flow from the large LHC IIb subcomplex to CC II. (Such an organization is depicted in Figure 3, page 96).

Bassi's group (Bassi et al., 1987; Bassi and Dainese, 1992) has also proposed a model depicting the *in vivo* arrangement of the LHC II components within PS II. Although there are differences between it and that in Figure 3, both contain a common theme: The minor LHC II pigment–proteins are arranged closer to the PS II core components than the abundant LHC IIb complexes. Bassi and Dainese (1989, 1992) also isolated a multiprotein complex of essentially identical composition as the putative connector described above (Peter and Thornber, 1991b). Furthermore, upon phosphorylation of this multiprotein complex, they found that the LHC IIb material is released, implying that LHC IId lies between LHC IIa and LHC IIb with LHC IIc separately linked to the CC II. In contrast to the Peter and Thornber (1991b) model (see Figure 3, page 96), Bassi and co-workers (1987, 1992) have evidence that CP43 is required for binding of the LHC II subunits to the photosystem II RC core. They have also observed a population of LHC IIb tightly bound to CC II in the absence of any other LHC II members. Obviously, further studies are needed to resolve some of the discrepancies that have arisen.

IV. GENES AND THE PRIMARY STRUCTURES OF THE HIGHER PLANT LHC PIGMENT–PROTEINS

Biochemical studies have been used to investigate the interrelationship of the LHC apoproteins. Monoclonal and polyclonal antibodies were used to show that epitopes exist which are common to both LHC I and LHC II polypeptides (Darr et al., 1986; Evans and Anderson, 1986; Bassi et al., 1987; White and Green, 1987a; Hoyer-Hansen et al., 1988). Such cross-reactivity, however, need only indicate relatedness at small domains in the apoproteins. A further difficulty is that clean preparations of individual LHC subunits to use as antigens are hard to obtain because many of the LHC I and II apoproteins are of a very similar size. This can and has confused interpretations. Moreover, a lack of cross-reactivity between different LHC's (e.g., that of the Williams and Ellis's (1986) LHC I monoclonal antibodies with LHC II apoproteins) does not preclude the fact that the various apoproteins might still be related at other regions to which no antibodies were produced. Analysis from immunological data therefore, while useful in establishing possible similarities at certain small regions of polypeptides, cannot give exact pictures regarding relatedness between polypeptides. Unequivocal interrelationships have had to await comparisons of the primary structures of the different LHC polypeptides deduced from their gene sequences. Such information has conclusively demonstrated that the LHC I and LHC II apoproteins are indeed related to each other (see, for example, Figure 1 and Green et al., 1991).

Molecular genetic approaches have been applied to isolate and sequence many of the genes encoding photosystem subunits. With respect to the pigment–proteins, all of those apoproteins coded in the chloroplast genome (i.e., those of the core complexes) have known sequences in some organism, and a considerable number are known for the antenna apoproteins that are coded in plant nuclear DNA or bacterial DNA (Zuber and Brunisholz, 1991). Correlation of the amino acid sequence deduced from a particular gene to an actual protein component of the photosynthetic membrane is far from simple for the LHC's of green plants, since there are many of them and they are of very similar sizes and sequences (Figure 1). Uptake and assembly of an *in vitro* transcribed and translated nuclear gene into isolated intact plastids can be used (see Section VII.A), although a less equivocal procedure is to determine part of the amino acid sequence of the suspected protein product, in the same plant species, and examine whether it is contained in the sequence derived from the gene.

LHC apoproteins are encoded by a family of nuclear genes. LHC Ia, LHC Ib, LHC IIa, LHC IIb, LHC IIc, and LHC IId are all members of this multigene family (Green et al., 1991). LHC Ic does not appear to be a member (Section III.B), and whether LHC Id and LHC IIe are is unknown. The family also includes the green algal LHC IIb's (Imbault et al., 1988; Long et al., 1989; LaRoche et al., 1990; Larouche et al., 1991; Houlne and Schantz, 1987), the chlorophyll *a/c*-binding proteins of diatoms (Grossman et al., 1990), and the early light-inducible proteins (ELIPs) (Grimm et al., 1989). The progenitor(s) of these genes may have been originally located in the chloroplast genome, or in the photosynthetic bacterial genome, and transferred to the plant nuclear genome in a mechanism similar to that proposed for the nuclear genes *tuf*A (Baldauf and Palmer, 1990) and *cox*II (Nugent and Palmer, 1991), originally of the chloroplast and mitochondrion, respectively. The genes encoding LHC polypeptides are termed *cab* genes, although an alternative nomenclature has been proposed recently which uses the terms *lha* for genes encoding LHC I apoproteins and *lhb* for genes encoding LHC II apoproteins (Jansson and Gustafsson, 1991). The genes have typical eukaryotic 3′ and 5′ flanking regions including TATA boxes, CAAT boxes, and polyadenylation sites. The divergence found in the flanking sequences of members of the *cab* genes could be responsible for the differential regulation of their expression (Section VIII.A). Each of the genes fully characterized, so far, code for a precursor polypeptide consisting of a transit peptide and a mature peptide (Section VII). For example, LHC IIb's have a transit peptide of 30 to 42 amino acids with a prevalence of basic amino acids, and a mature peptide of about 222 to 233 amino acids.

The deduced sequences of the LHC apoproteins (Figure 1) are highly conserved within a class, but are less conserved when compared to the apoproteins belonging to other LHC classes. For example, the LHC Ia protein of tomato is over 90% identical to the LHC Ia protein from petunia (Pichersky et al., 1988), while its overall sequence similarity to that of LHC Ib's apoprotein is 35 to 45%, which is only marginally greater than its similarity to LHC II polypeptides (Pichersky et al., 1989).

Sixty-one LHC IIb amino acid, translated genomic DNA, and translated cDNA sequences have been aligned in an effort to identify which regions of the proteins are conserved among species (Appendix), and compared with the alignments to other LHC classes in Figure 1. Regions which are highly conserved are likely to be important for the correct folding of the LHC proteins in the thylakoid membrane and/or are involved in the binding and orienting of the chlorophylls in the membrane. On the other hand, nonconserved regions are likely to be involved in the regulation and formation of the trimeric forms of the LHC, or the assembly of the trimers into larger oligomeric structures. There are two highly conserved regions in all members of the multigene family. These regions are the result of a probable duplication event in the ancestral LHC-like gene. The two duplicated regions are likely to constitute (see Section V) two of the putative membrane-spanning α-helices (Helix I and Helix III in Figure 1), and the two putative α-helices on the stromal surface of the membrane (Hook 1 and Hook 2 in Figure 1). The alignments also show three putative β-turns (Figure 1) that are highly conserved. The C-terminal end of a third putative membrane-spanning helix (Helix II in Figure 1) has an absolutely conserved arginine residue among all the family members. The major differences between the various LHC apoprotein sequences (Figure 1) are in the residues connecting the three membrane-spanning helices (see Section V). The charged residues and hydrophobic portions of these loops may determine whether a protein will be in the grana stacks (LHC II) or in the intergrana membranes (LHC I) (Barber, 1982). They may also regulate the assembly of the higher order oligomeric structures.

V. STRUCTURE AND FOLDING OF THE LHC PIGMENT–PROTEINS IN THE MEMBRANE

Studies have focused on the three-dimensional structure of the LHC IIb component. Early data indicated that all LHC IIb polypeptides are oriented in the same manner in the thylakoid membrane, with the N-terminal portion of the polypeptide exposed to the chloroplast stroma and the C-terminal portion in the thylakoid lumen (Steinback et al., 1979; Carter and Staehelin, 1980, Andersson et al., 1982; Karlin-Neumann et al., 1985; Buergi et al., 1987). Computer analyses of the deduced amino acid sequences of their polypeptides have allowed hypotheses to be made of the folding with respect to the lipid bilayer (Karlin-Neumann et al., 1985; Buergi et al., 1987). Hydrophobic, membrane-spanning α-helices are the most likely way for the LHC pigment–proteins to be associated with the lipid bilayer (essentially a stretch of 20–23 hydrophobic residues is needed to cross the thylakoid membrane directly). For LHC IIb, three such helices of 23 residues were first suggested (Karlin-Neumann et al., 1985), but LHC IIb appears now to have longer membrane-spanning helices (28 residues) due to their not lying exactly normal to the membrane bilayer (Kuhlbrandt and Wang, 1991). Circular dichroism analyses have shown that about 44% of the LHC IIb polypeptide is in an α-helical conformation (Nabedryk et al., 1984).

Crystals of higher plant antenna pigment–proteins, suitable for high resolution X-ray analysis of the structure, have not yet been reported; however, electron diffraction studies on two-dimensional arrays of LHC IIb are yielding exciting insights into the LHC structure (Kuhlbrandt and Wang, 1991). Earlier studies on such arrays showed that the LHC IIb molecule was a 65 Å elongated molecule, with its long axis lying perpendicular to the lipid bilayer. It asymmetrically spanned the bilayer with about 20 Å exposed on the stromal side of the membrane and about 7 Å on the lumenal side (Kuhlbrandt, 1984; Li, 1985; Lyon and Unwin, 1988). Kuhlbrandt (1988) provided images to 3.7 Å resolution which suggested that there were indeed three membrane-spanning alpha-helices and that LHC IIb was organized into trimeric complexes (Section III.C).

Recently, images at 6 Å resolution have been obtained which show that each LHC IIb polypeptide is oriented in an identical fashion in the trimer (Kuhlbrandt and Wang, 1991; cf. Kuhlbrandt, 1984; Li, 1985; Lyon and Unwin, 1988). The possible locations for the 15 chlorophyll molecules is depicted; the location of the xanthophylls could not be determined. The Kuhlbrandt and Wang (1991) images also show that the first and third membrane-spanning α-helices are oriented 25° and 31° from the membrane normal, respectively, are bilaterally symmetric in the upper portion of the membrane, and cross each other in the membrane some 7 Å from the stromal surface of the membrane. The third membrane-spanning α-helix (Helix II) is oriented 11° from the membrane normal.

An updated prediction of the folding of the LHC IIb polypeptide chain is shown in Figure 2. It is based on the sequence of the mature polypeptide deduced from the *Lemna gibba ab30* gene (Kohorn et al., 1986), and has been constructed using the information from the 6-Å electron diffraction images and the polypeptide alignments of members of the LHC multigene family (Thornber et al., 1992). This model is consistent with the dimensions of the electron diffraction images and the percent of helical structure measured by circular dichroism. The model has 14 charged amino acids (seven acidic and seven basic residues) in the membrane; such a situation is energetically unfavorable. However, it is not possible, due to the increased length of the helices, to fold the protein into any structure consistent with the 6-Å images without having more charged residues in the membrane than there are in the Karlin-Neumann et al. (1985) model. The bilateral symmetry seen in the images is probably due to the putative helices I and III that resulted from the duplication event in the ancestral gene. It is suggested that two of the positively charged amino acids and two of the negatively charged amino acids (one each from Helix I and Helix III) form salt bridges between the two helices to stabilize their interaction. The majority of negative charges located in Helix I and Helix III and those in Hook 1 and Hook 2 could form a pocket of negative charge at the surface of the membrane that is exposed to the aqueous stroma. This would provide a possible molecular mechanism for the involvement of the LHC IIbs in thylakoid membrane stacking in which the positively charged N-terminal region of one LHC IIb molecule on one membrane surface would fit into the negatively charged pocket

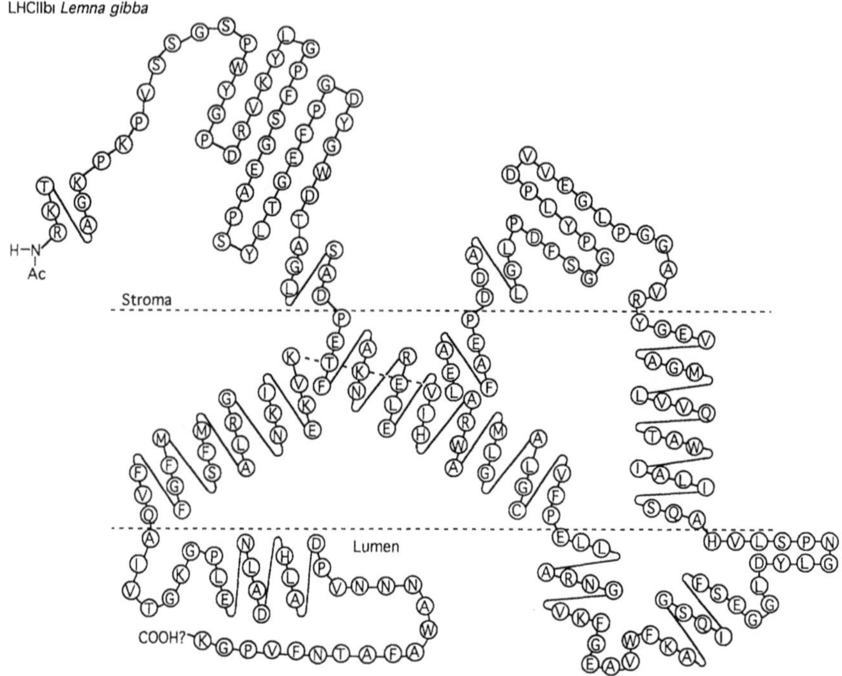

Figure 2. Proposed folding of one of the LHC polypeptides with respect to the lipid bilayer. The folding is an extension of that proposed earlier (Karlin-Neumann et al, 1985, Kohorn et al, 1986) and makes use of the initial structures obtained by electron diffraction studies of LHC IIb (Kuhlbrandt and Wang 1991). The portions protruding from the lipid bilayer probably lie along the membrane's surface, but for ease of depiction they are shown as if they extended far into the stroma. Helices I and III are shown crossing each other a few angstroms below the stromal surface of the lipid bilayer. Note the preponderance of charged residues in this region. The conserved beta-turns (including the 'hooks'), and other potential alpha-helical portions are also depicted. The exact C-terminal residue has not been unequivocally determined; the possibility exists that there is C-terminal processing.

of another LHC IIb molecule on the opposing granal surface. The remaining charged residues on Helix I and Helix III seem to be oriented in the general direction of the symmetric chlorophylls (probably chlorophyll *b*) as found in the 6-Å electron diffraction images (Kuhlbrandt and Wang, 1991).

The evidence that the magnesium ion in the chlorophyll molecules is coordinated to histidines in the polypeptide chain in the purple bacterial reaction center and antenna components is convincing (e.g., Diesenhofer et al., 1985). While this may also be the case for the plant's core complexes, it cannot be so for the LHCs because they have too few (three to five) histidine residues to accommodate 15 chlorophyll molecules. The proposal that glutamine and asparagine residues may also be involved in this coordination has been made (Wechsler et al., 1985; Peter and Thornber, 1988). However, there are only two glutamine, one asparagine, and one histidine residue in the membrane-spanning helices (see Figure 2). The other putative coordinating residues are in the loops and are possibly exposed to the lumen and stroma media. It has been proposed that the majority of the chlorophylls are coordinated to other amino acid side chains via water molecules, or are held in place by hydrogen-bonding to the porphyrin ring (Kuhlbrandt and Wang, 1991).

VI. CAROTENOIDS IN PHOTOSYNTHETIC ORGANISMS

Carotenes and xanthophylls account for 20% of the photosynthetic pigments in most oxygenic organisms, and for an even greater percentage in purple photosynthetic bacteria and brown algae. Almost invariably they occur in association with chlorophyll in the pigment–protein complexes. Carotenoids have two roles in photosynthesis:

1. Carotenoids act as antenna pigments. Light energy absorbed by carotenoids in photosynthetic membranes is used to drive the primary photochemical reactions after singlet–singlet energy transfer to the chlorophylls. Experiments with model compounds (Moore et al., 1990) have shown that this process, which can occur with 100% efficiency [e.g., in the case of the peridinin–chlorophyll *a* complex from dinoflagellates (Song et al., 1976)], only occurs when the donor carotenoid is in van der Waals contact with the acceptor chlorophylls. The antenna apoproteins are responsible for holding these pigments close enough for this singlet–singlet energy transfer to occur (cf. Nechushtai et al., 1988).
2. Carotenoids protect the photosynthetic apparatus from photodestruction. The light-harvesting role of carotenoids is a bonus for most photosynthetic organisms but is not really an essential function. On the other hand, without carotenoids to act as photoprotective agents there would not be any photosynthesis on Earth. If chlorophyll is excited by light, one of the possible products is triplet chlorophyll. If this state lasts long enough to collide with molecular oxygen then singlet oxygen will be formed. Singlet oxygen is lethal. It can oxidize chlorophylls, lipids, proteins, and indeed nucleic acids. When cells are exposed to singlet oxygen they are rapidly killed. Carotenoids prevent this harmful series of reactions. Their main protective mechanisms is by rapidly quenching the triplet chlorophyll before it has time to encounter oxygen. In addition, carotenoids are also effective direct scavengers of singlet oxygen. There are no wild-type photosynthetic organisms which lack carotenoids.

Additionally, higher plants and certain classes of green algae have a xanthophyll cycle (Krinsky, 1966), in which the deepoxidation of violaxanthin to zeaxanthin is thought to dissipate excess excitation energy, particularly in PS II (cf. Demmig et al., 1987). In leaves exposed to high light, a rapidly relaxing component of fluorescence quenching is believed to be indicative of a radiationless dissipation of excess excitation energy and has been correlated with the appearance of zeaxanthin. Plants in which the conversion of violaxanthin to zeaxanthin has been blocked with dithiothreitol, no longer exhibit this rapid relaxation of fluorescence quenching. Zeaxanthin has been proposed to quench chlorophyll excitation by interaction with singlet chlorophyll. However, the exact mechanism of excess energy dissipation by zeaxanthin and its interaction with other components of the photosynthetic apparatus is still unknown.

In higher plants, there is relatively little information on the detailed carotenoid composition of the different antenna pigment–proteins. It is therefore as yet unclear whether some or any of the complexes show specificity with regard to which types of carotenoids they will accept. Nevertheless, as described above, it would appear that the different LHCs obtained from plants grown under the same condition do have slightly different contents, but how their carotenoid content is affected by different growth conditions and stage of development is virtually uninvestigated. Such information is particularly important for understanding the function of those carotenoids involved in the xanthophyll cycle.

VII. CELL BIOLOGY AND ASSEMBLY OF HIGHER PLANT LHCs

The assembly of LHC IIb and other similar thylakoid protein complexes involves many intricate steps as well as an interplay between cytoplasmic and chloroplastic products: translocation of the precursor polypeptide (pLHCP) across chloroplast envelopes; its processing to its mature form; its insertion into thylakoid membranes; the binding of chlorophyll and carotenoid molecules to it; and, its association with other LHCP molecules to form an oligomeric and functional LHC IIb (Schmidt and Mishkind, 1988; Keegstra et al., 1989). An exact description of each step involved in the import, processing, and assembly of thylakoid proteins is still unknown. The assembly process is further complicated because the precursors are water-soluble whereas the processed products are transmembrane proteins with three highly hydrophobic regions. Furthermore, the site of membrane translocation (envelope) and membrane integration (thylakoids) are separated by an aqueous stroma. Two plausible pathways can be envisioned to accomplish this complex process. In the "soluble intermediate" pathway, the precursor protein could pass through the envelope and subsequently travel to the thylakoids in a soluble form, possibly brought about by a folding of the precursor that is quite different from that of the mature polypeptide. After passage through the stroma it is integrated into or transported across the thylakoid bilayer after further modification of its folding. Results of several studies involving reconstitution of individual steps during the assembly of LHC IIb are consistent with this pathway (Cline et al., 1985, 1989; Chitnis et al., 1986, 1987; Cline, 1986, 1988; Fulsom and Cline, 1988; Reed et al., 1990; Payan and Cline, 1991). In the "transport vesicles" pathway, the imported thylakoid protein could insert into the inner envelope membrane, then travel to the thylakoid in vesicles that bud from the inner envelope membrane and subsequently fuse with the thylakoids. Electron microscopic studies on differentiating thylakoid membranes in the greening plastids of the y-1 mutant of *Chlamydomonas reinhardtii* lend some support to this hypothesis (Hoober et al., 1991).

Several approaches have been used to dissect the different steps in the assembly of LHC IIb. *In vitro* uptake of labeled polypeptides obtained by *in vitro* translation of total poly A^+-RNA by isolated intact plastids is the most popular one (Chua and Schmidt, 1979; Grossman et al., 1980; Schmidt et al., 1981; Mullet and Chua, 1983; Cline et al., 1985; Wasmann et al., 1986). Less equivocal data are obtained if only one particular precursor polypeptide is added to the plastids. Such a polypeptide can be made by using SP6, T3, or T7 promoters in front of the gene of interest to synthesize the specific mRNA *in vitro* (Chitnis et al., 1986; Kohorn et al., 1986), which is then translated in a protein-synthesizing system. Another way to study assembly is to separate and analyze the different steps *in vitro* and later reconstitute the entire assembly process. Some of the many steps in the assembly of LHC IIb have been successfully repeated *in vitro*. For example, pLHCP can be inserted into isolated thylakoids (Cline, 1986; Chitnis et al., 1987) and be processed to the mature size by a processing peptidase in the soluble fraction of chloroplasts (Lamppa and Abad, 1987).

A. Translocation of pLHCP

The import of pLHCP into the plastids is posttranslational (Schmidt et al., 1981), energy-dependent (Grossman et al., 1980), requires cytosolic factors to make pLHCP import-competent (Waagemann et al., 1990), and is probably mediated through specific receptors in chloroplast envelopes (Cline et al., 1985). After the uptake of the *in vitro* synthesized pLHCP(s) by the isolated plastids, labeled mature polypeptide(s) can be detected in the pigmented LHC IIb in the thylakoids. *In vitro*-import experiments using *in vitro*-mutated pLHCP of *L. gibba* have revealed the importance of the amino acid charge distribution in the membrane-spanning helices (Section V) of the polypeptide for the stability of the newly imported LHCP in the thylakoids and for its assembly into LHC IIb (Kohorn and Tobin, 1987). The precursor form of LHCP can be seen in the thylakoids and also in LHC IIb under certain conditions; for example, when either barley or maize plastids (Chitnis et al., 1986), but not when *L. gibba* etiochloroplasts (Kohorn et al., 1986) import *in vitro* synthesized *L. gibba* pLHCP, the mature as well as the precursor polypeptides are observed as integral thylakoid proteins. Both of these polypeptides migrate specifically with the LHC IIb band in partially denaturing PAGE. The occurrence of both forms is also observed when *in vitro* synthesized barley pLHCP is imported into barley plastids isolated from etiolated plants which have been illuminated for less than 24 hr; when plastids from plants greened for 24 hr are used, only LHCP is seen in the thylakoids (Chitnis et al., 1988). Thus the presence of pLHCP in thylakoids is not due solely to the heterologous nature of the system used, but is largely due to the influence of the developmental stage of plastids. Pulse-chase experiments showed that the pLHCP integrated into thylakoids of intact plastids can be processed to LHCP. Therefore, it was suggested that integration of pLHCP into thylakoids precedes its processing to LHCP and the leader peptide contains information for targeting the protein not only to a specific organelle (chloroplast) but also to a specific subcompartment (thylakoids) within that organelle (Chitnis and Thornber, 1988). However, it was later shown that when the transit peptide of the precursor of the small subunit of Rubisco, a stromal protein, was fused to LHCP, the chimeric protein was successfully imported and integrated into thylakoid membranes (Lamppa, 1988). Thus, the mature protein contains sufficient information for proper intraorganeller targeting of LHCP.

B. Processing of pLHCP

Chloroplasts, like mitochondria, appear to contain a general soluble processing enzyme that recognizes the large diversity of imported precursors which initially enter and traverse the stroma (Robinson and Ellis, 1984; Abad et al., 1989). The processing of pLHCP has been shown to occur in a complex way in the stroma. Import of single gene products into plastids from *L. gibba* (Kohorn et al., 1986), wheat (Lamppa and Abad, 1987), pea (Cline, 1988; Kohorn and Yakir, 1990), tomato (Pichersky et al., 1987), tobacco (Chaumont et al., 1990), and corn (Dietz and Bogorad, 1987) also yields mature proteins of multiple electrophoretic mobilities. When imported into wheat or pea chloroplasts *in vitro*, wheat pLHCP is processed at two sites giving rise to peptides of approximately 25 and 26 kDa (Lamppa and Abad, 1987). Cleavage usually occurs preferentially at the primary site between residues #37 and 38 of pLHCP producing the 26-kDa form. However, only the 25-kDa polypeptide is produced in an organelle-free assay enriched for the chloroplast soluble-processing enzyme (Abad et al., 1989). There are distinct determinants for the cleavage of pLHCP. A mutant of pLHCP with four amino acids inserted between the transit peptide and mature protein results in loss of cleavage at the secondary site in the organelle-free reaction, without accumulation of the 26-kDa protein. On the other hand, the same mutant precursor gave rise to the 26-kDa peptide upon import (Clark et al., 1989). There is a requirement of a basic residue in the carboxy portion of the transit peptide of pLHCP for the primary processing event to occur (Clark and Lamppa, 1991). In contrast, removal of basic residues from the carboxyl terminus of the transit peptides of the precursors of the small subunit of Rubisco and of Rubisco activase, two stromal proteins, by site-specific mutagenesis has essentially no effect on precursor maturation during import.

The soluble enzyme that processes pLHCP has been partially purified and characterized. This enzyme is inhibited by millimolar concentrations of EDTA and 1,10-phenanthroline. The processing activity is insensitive to iodoacetate. Optimal processing occurs between pH 8 and 9 at 26 °C. Its apparent molecular mass is estimated to be 240 kDa from the gel filtration chromatography. These properties are similar to the processing enzyme previously shown to cleave precursors of the small subunit of Rubisco and plastocyanin. The partially purified processing enzyme that cleaves pLHCP at the secondary site also cleaves several other chloroplast precursors (Abad et al., 1991). Thus, pLHCP is processed at the secondary site by the general processing protease in the stroma. The enzyme that cleaves pLHCP at the secondary processing site is not yet identified.

C. Membrane Integration of LHCP

When radioactive pLHCP synthesized *in vitro* is incubated with isolated thylakoids, it associates with the membranes. Its insertion into the membranes absolutely requires both Mg-ATP and a stromal protein factor (Cline, 1986; Chitnis et al., 1987). Earlier studies on the integration of LHCP were performed using pLHCP. Later it was shown that the transit sequence is not required for the correct integration of LHCP into membranes and its assembly into LHC IIb (Viitanen et al., 1988). The stromal integration factor physically associates with LHCP and forms a 120-kDa intermediate. This association prevents aggregation of LHCP and maintains LHCP's competence for thylakoid insertion. The intermediate still requires addition of stroma for integration, thereby suggesting the involvement of at least two stromal factors in the integration of LHCP (Payan and Cline, 1991). Recently, it has been shown that chloroplast HSP70 can functionally substitute for stroma, enabling pLHCP to integrate into membranes. The chloroplast HSP70 retards folding of pLHCP and thus keeps it in a form competent to integrate into membranes (Yalofsky and Nechushtai, personal communication). Previously it was thought that pigments, especially chlorophyll *b*, are also required for the membrane insertion of LHCP and/or its stability in the membranes. When LHCP is expressed in *E. coli* as a fusion protein with a presequence of a periplasmic protein, it is found stably but less efficiently integrated in the inner membranes of *E. coli* (Kohorn and Auchincloss, 1991). Thus, it has been proposed that pigments are required for LHCP to integrate into a membrane. However, the question remains whether the protein inserted into *E. coli* membranes has a similar conformation to that in thylakoid membranes.

The informational requirements for the integration of LHCP into membranes have been deciphered by mutational analysis of LHCP integration. LHCP integration into thylakoids does not require the leader peptide (Lamppa, 1988; Viittanen et al., 1988). Analyses of deletion mutants of LHCP have suggested that the third membrane-spanning region of LHCP is necessary for thylakoid targeting but is not sufficient for insertion (Cline et al., 1989; Kohorn and Tobin, 1989). A detailed deletion mutagenesis of LHCP study failed to identify any short portion of the transit peptide that is required for stable integration of LHCP into thylakoids. Efficient insertion has been found to rely upon the integrity of all but the amino-terminus and a short lumenal loop of LHCP (Auchincloss et al., 1992).

Thylakoids in the chloroplasts of higher plants show structural and functional differentiation in the form of stacked and unstacked membranes (Murphy, 1986). Under normal light conditions photosystem II and its light-harvesting components are present in the stacked thylakoids. During *in vitro* experiments radiolabeled LHCP or pLHCP integrates mainly into the unstacked thylakoids and then migrates to the stacked region (Kohorn and Yakir, 1990; Yalovsky et al., 1990). Migration of the newly inserted LHCP is unaffected by light intensity or absence of the amino-terminal threonine that is thought to be phosphorylated (Section III.C) (Kohorn and Yakir, 1990).

D. Binding of Pigments to LHCP

The assembly of purified LHC IIb with the PS II core complex in membranes of intermittent light-grown plants that lack LHC IIb was demonstrated in some earlier studies; however, it occurred with low efficiency (Day et al., 1984; Darr et al., 1986). Reconstitution experiments using purified LHCP and chlorophyll molecules showed that chlorophyll molecules can bind to the apoprotein(s) *in vitro*, and that xanthophylls play a crucial role in permitting this binding and in the formation of the LHC IIb holocomplex (Plumley and Schmidt, 1987). A similar procedure was used by Paulsen et al. (1990) to reconstitute LHC IIb from LHCP overexpressed in *E. coli*. These experiments revealed that the association of pigments to the apoproteins is a highly specific process and the pigment composition of complexes formed *in vitro* cannot be modulated by the variations in the ratio of pigments present during the reconstitution. In contrast, using other chlorophyll derivatives significantly changes the stability and stoichiometry of the resulting reconstituted complexes (Paulsen et al., 1991). Deletion of some segments of LHCP, including portions of the amino terminal and carboxyl terminal hydrophilic domains, does not impair the ability of LHCP to form stable complexes that have a complete set of pigments present in the resulting LHC IIb (Paulsen et al., 1991). The hydrophobic domains and the hydrophobic sequences immediately after the carboxy-proximal putative membrane-spanning region are essential for the formation of stable LHC IIb. None of the mutant LHCPs tested forms incomplete or intermediate complexes, thus indicating a highly synergistic stabilization of LHCP–pigment complexes.

VIII. BIOGENESIS OF ANTENNA PIGMENT–PROTEINS

A. In Higher Plants

Regulation of Gene Expression for Plant LHCs

Since LHC IIb polypeptides are among the most abundant plant proteins, their genes were among the first plant genes to be cloned and the regulation of their expression studied. Many techniques have been used to probe different steps in the expression of these genes: One of the earliest ways was to immunoprecipitate translation products of polyadenylated mRNAs (e.g., Tobin, 1978, 1981). This procedure gives levels of translatable mRNAs for this protein in a tissue. Northern blots are used to measure the level of a particular message in the total RNA population (e.g., Stiekema et al., 1983). To determine the rate of transcription, *in vitro* nuclear run-off transcription has been used by many laboratories (e.g., Silverthorne and Tobin, 1984). Using these and other biochemical techniques, the influence of several intrinsic (e.g., developmental cues) and extrinsic (e.g., light) factors on the expression of *cab* genes have been studied (for details see Tobin and Silverthorne, 1985; Chitnis and Thornber, 1988; Thompson and White, 1991).

The RNA levels encoding apoproteins of both LHC I and LHC II are regulated by phytochrome (Silverthorne and Tobin, 1984; Thompson and White, 1991; Anandan et al., 1993) as well as by blue and UV-blue receptors (Marrs and Kaufman, 1989; Oelmuller et al., 1989; Wehmeyer et al., 1990). Accumulation of *cab* mRNA is mediated partly by a very low fluence response, as well as by the low fluence response. Morishige et al. (1992) and Anandan et al. (1993) have used cDNA clones from barley for the two most prevalent LHC Ib apoproteins and for LHC IIa and LHC IIb apoproteins to show that, in addition to their expression being regulated by phytochrome, their mRNAs are present in relatively low amounts in dark-grown seedlings and accumulate at different rates with increased exposure to constant illumination. The regulation of *cab* genes has been reported to occur at many different levels of gene expression. For example: phytochrome affects transcription of these genes (Silverthorne and Tobin, 1984); intermittent red light has an influence on the translation of LHCP mRNA (Slovin and Tobin, 1982); and changes in light intensity affect posttranslational modification (phosphorylation) (Bennett, 1979; Bennett et al., 1981). The expression of *cab* genes is regulated in both a quantitative as well as a qualitative manner. For example, the *cab* genes are expressed in an organ-specific manner in tobacco; however, individual members of the gene family have different patterns of expression in the various organs (Simpson et al., 1985, 1986a). The 5′ region of the genes contain information for the organ specific and light-regulated expression of *cab* genes. (Simpson et al., 1986a,b; Nagy et al., 1986a,b; Castresana et al., 1988; Gidoni et al., 1989).

Biogenesis of LHC I and LHC II during Higher Plant Plastid Development

Chloroplast development involves intricate biochemical and morphological changes (Leech, 1984; Klein et al., 1986; Briggs et al., 1987). Studies on the synthesis and assembly of photosynthetic complexes during plastid development require a homogeneous population of plastids. Traditionally, developmentally equivalent plastids have been obtained either by isolation of a subpopulation of plastids, or by arrest of all plastids at an identical developmental stage. Biogenesis studies have focused mainly on the correlation of *cab* mRNA levels with accumulation of the LHC IIb apoproteins within the thylakoids, or on the assembly of the apoproteins into pigmented multiprotein complexes.

In monocotyledonous plants, a gradient of plastid development occurs along the leaf length with the more immature plastids found in the basal regions (Leech, 1984). Defined sections along the length of the leaf can yield relatively homogeneous populations of plastids of the same developmental stage (Baker et al., 1984). Low levels of *cab* mRNA can be detected in the basal sections of maize and wheat leaves with maximum levels in the lower to middle section and diminished levels in the more mature tip sections. The LHC IIb message accumulation clearly precedes the accumulation of the apoprotein (Viro and Kloppstech, 1980; Lamppa et al., 1985; Martineau and Taylor, 1985). LHC IIb apoproteins have been shown to be present in the basal 1 cm of young wheat leaves, and their relative quantity to that of other proteins increases greatly in segments toward the tip of the blade (Viro and Kloppstech, 1980; Lamppa et al., 1985; Martineau and Taylor, 1985; Brendenkamp and Baker, 1988). In contrast, the LHC I apoproteins of 20 to 24 kDa are not detected until midway up the leaf blades and considerably later than the *psa* A/B gene products of CC I appear, demonstrating a lack of coordination with either the accumulation of CC I or of LHC IIb (Brendenkamp and Baker, 1988). The appearance of the leaf's 77-K fluorescence emission at 735 nm, characteristic of the presence of at least one LHC I component within PS I (Mullet et al., 1980a,b; see Section III.B), is observed only in the mature tip segments, indicating that LHC I becomes a major component of photosystem I only after establishment of multiple photochemically functional CC I units (Brendenkamp and Baker, 1988).

Growth of angiosperms in the dark arrests the plastid as an etiochloroplast which will complete development only upon exposure to light. Homogeneous preparations of plastids at different stages of development can be obtained by harvesting plants after different periods of greening. It is, however, difficult to separate the effects of light and plastid development from each other during this greening process. Different plant species show varying responses during this light-triggered plastid development (Burkey, 1987; Mathis and Burkey, 1987). In pea, a lack of coordination between *cab* mRNA and apoprotein accumulation has been observed (Bennett et al., 1984; Mathis and Burkey, 1987) with detectable levels of the mRNA being present in etiolated plants while LHC IIb apoproteins are absent. The accumulation of both *cab* mRNA and apoprotein appears to be more closely coordinated in barley and soybean (Hoyer-Hansen and Simpson, 1977; Apel and Kloppstech, 1978; Hiller et al., 1978a,b; Mathis and Burkey, 1987).

While etiochloroplasts of different species vary in their response to light, a strong correlation exists between LHC IIb apoprotein accumulation and chlorophyll accumulation, seemingly independent of the levels of *cab* mRNA (Apel and Kloppstech, 1980; Mathis and Burkey, 1989). Increasing the rate of chlorophyll synthesis by treatment with benzyladenine or decreasing the rate by treatment with levulinic acid during greening of cucumber cotyledons demonstrated a similar correlation between the accumulation of LHC IIb apoproteins and the synthesis of chlorophyll (Shimada et al., 1990). In both treatments a limited amount of the LHC IIb apoprotein accumulated in the absence of chlorophyll. After the end of the lag phase of chlorophyll *a* synthesis, further accumulation of LHC IIb apoproteins and initiation of chlorophyll *b* synthesis required the supply of chlorophyll *a*. Equivalent information on the synthesis of the minor LHC II apoproteins and LHC I apoproteins is virtually nonexistent during the greening of etiochloroplasts due to the relatively recent identification of their apoproteins and cloning of their corresponding genes.

Angiosperm plastid development can also be arrested at a stage known as the protochloroplast by growth of etiolated plants in intermittent light (IML). The agranal protochloroplasts synthesize chlorophyll *a* selectively and are essentially devoid of chlorophyll *b* (Argyroudi-Akoyunoglou and Akoyunoglou, 1970; Argyroudi-Akoyunoglou et al., 1971a,b; Armond et al., 1976). The amount of chlorophyll *b* formed is dependent upon the intensity and duration of the light flash and the period of the cycle (Argyroudi-Akoyunoglou and Akoyunoglou, 1970). Functional core complexes of PS I and PS II are, however, present in the protochloroplasts (Armond et al., 1976; Hiller et al., 1978a,b; Argyroudi-Akoyunoglou and Akoyunoglou, 1979a,b). In spite of the presence of substantial amounts of *cab* mRNA in IML-grown plants (Cuming and Bennett, 1981; Slovin and Tobin, 1982; Viro and Kloppstech, 1982), there is very little accumulation of either LHC IIb apoproteins (Viro and Kloppstech, 1982; Day et al., 1984; White and Green, 1988; Jaing et al., 1992) or the 20- to 24-kDa LHC I apoproteins (Mullet et al., 1980a,b; Ryrie and Young, 1984; White and Green, 1988). The apoprotein of LHC IIa (CP29) is detected in barley IML thylakoids (White and Green, 1988), as well as a 25-kDa polypeptide that may represent the type III *cab* gene product of LHC IIb, or an apoprotein of either LHC IIc (CP27) or LHC IId (CP24) (White and Green, 1988). The apoproteins of all light-harvesting complexes accumulate rapidly upon continuous illumination as the protochloroplast continues development to a mature chloroplast (Argyroudi-Akoyunoglou et al., 1971; Armond et al., 1976; Mullet et al., 1980a,b; Day et al., 1984; Ryrie and Young, 1984; Jaing et al., 1992).

Treatment with specific inhibitors of chloroplastic translation during growth in IML causes an accumulation of chlorophyll *b*, and the apoproteins of LHC IIb and other LHC II pigment–protein complexes in maize seedlings (Sarvari et al., 1989) and cucumber cotyledons (Shimada et al., 1990). These results have been explained in terms of the apoproteins having variable affinities for chlorophyll, with the apoproteins of the chlorophyll *a*-binding proteins of CC I and CC II possessing the greatest affinity. Thus, when chlorophyll synthesis is limited, such as under IML conditions, RCs would be preferentially formed. Therefore, inhibition of RC synthesis would allow some chlorophyll *a* to be incorporated into LHC IIb apoproteins either directly or after conversion to chlorophyll *b*.

The accumulation of pigment–protein complexes during light-triggered etiochloroplast and protochloroplast development has been studied primarily using mild SDS-PAGE systems to fractionate them. Studies on the developing thylakoids are complicated by difficulties in thoroughly solubilizing and clearly resolving the individual pigmented complexes while minimizing the loss of pigment from them. The large quantity of free pigment often observed has come to be thought of as characteristic of immature thylakoid membranes (Burkey, 1986). Initial investigations on pigment–protein complex formation showed the appearance and steady accumulation of CP II (i.e., LHC IIb) during exposure of seedlings to continuous illumination (Argyroudi-Akoyunoglou et al., 1971; Argyroudi-Akoyunoglou and Akoyunoglou, 1973; Hiller et al., 1973, 1978; Tanaka and Tsuji, 1985). The timing and rate of accumulation of LHC IIb shows species differences but in all cases coincides with the rapid increase of chlorophyll *b* synthesis (Hiller et al., 1978; Kalosakas et al., 1981; Tanaka and Tsuji, 1985; Burkey, 1986, 1987; Jaing et al., 1992). The increased accumulation of LHC IIb is accompanied by a reduction in the free pigment observed during the electrophoresis, suggesting further stabilization of the complexes after their initial assembly as pigment–proteins. With the recent refinements of electrophoretic procedures for separation of intact pigmented multiprotein complexes, it has been more readily observed that the biogenesis of oligomeric LHC IIb in bean and pea occurs via the initial accumulation of pigmented monomeric components prior to their incorporation into a ~75-kDa trimeric-pigmented complex (Argyroudi-Akoyunoglou and Akoyunoglou, 1979a,b; Kalosakas et al., 1981; Jaing et al., 1992; Dreyfuss and Thornber, in press, 1994).

The biogenesis of LHC I constituents in thylakoids of IML-grown pea seedlings has been observed as their addition to CC I to yield an increasing amount of a larger complete PS I unit during greening. There is a corresponding appearance of the 735-nm 77-K fluorescence emission maximum characteristic of the presence of the peripheral antennae within PS I (Jaing et al., 1992). The assembly of multiprotein complexes during the greening of IML-grown pea seedlings shows that monomeric LHC IIb accumulates and forms trimers prior to the detection of a complete PS I unit, further suggesting that biogenesis of LHC I is delayed in comparison to LHC IIb (Jaing et al., 1992; Dreyfuss and Thornber, in press, 1994). A minor chlorophyll–protein with characteristics similar to LHC IIa has been detected in IML-grown maize seedling thylakoids even prior to continuous illumination (Sarvari and Gigler, 1984).

Genetic mutations that delay or halt plastid development at early stages can be used to study the effects of development on different steps in LHC II and LHC I synthesis and assembly. In chlorophyll-deficient mutants of maize, plastids are arrested prior to mature chloroplast formation (Mascia and Robertson, 1978), while carotenoid-deficient mutants contain plastids that are arrested at a rudimentary stage of development (Bachmann et al., 1973). Studies with such mutants reveal that events at early stages of plastid development, such as synthesis of pigments, influence accumulation of *cab* mRNA (Harpster et al., 1984; Taylor et al., 1986).

Treatment with herbicides that block carotenoid synthesis and growth under extremely low light has enabled the creation of seedlings devoid of carotenoids. Plastidogenesis and the accumulation of chlorophylls *a* and *b* is normal (Oelmuller, 1989). The accumulation of the apoproteins of CC II and LHC IIb is dramatically reduced and PS II activity is absent (Markgraf and Oelmuller, 1991). However, functional PS I complexes are present with a full complement of LHC I apoproteins. This indicates that accumulation and association of LHC I within a PS I complex does not require carotenoid biosynthesis.

Summary. The accumulation of the light-harvesting apoproteins is not primarily governed by the levels of *cab* mRNA but by posttranslational stabilization, in which pigment availability is believed to play a necessary role. The formation of different pigment–protein complexes may also be regulated by the different affinities of the apoproteins for chlorophyll *a*. A substoichiometric constitution of pigments may be sufficient for stabilization of apoproteins. The requirements for accumulation of LHC I appear to differ from those for LHC IIb accumulation, and most likely differ among all individual light-harvesting pigment–protein complexes. Assembly of pigmented multiprotein complexes during chloroplast development appears to involve an organization of relatively simple complexes into further organized supramolecular structures.

During chloroplast development not only is the synthesis of pLHCP and chlorophyll-triggered, but so also is the machinery for the import and assembly of pLHCP. Immature plastids from interior leaves of lettuce have been found to be more efficient at importing pLHCP *in vitro* than mature chloroplasts from pea (Schmidt et al., 1981). Similarly when barley plastids of different developmental stages are used to import *L. gibba* pLHCP *in vitro*, the relative amount of precursor and processed forms observed in the thylakoids changes significantly (Chitnis et al., 1986, 1987). So at least one of the steps involved in the assembly of LHC IIb is dependent on plastid development. The insertion of pLHCP into thylakoid membranes also depends on the stage of plastid development for both the appearance of the stromal factor and the thylakoid membrane's receptivity for insertion (Chitnis et al., 1987). The synthesis or activity of the processing enzyme for pLHCP could also be under the control of light.

Mutants Lacking LHCs

A number of mutants of higher plants in which one or more components of the LHCs are missing have been obtained and characterized. So far in all of them, lack of LHC is due to the absence of one of the pigments associated with the complex. Most lack chlorophyll *b* (see Sommerville, 1986) while some lack carotenoids (Mayfield and Taylor, 1984). The most commonly used and best characterized mutant, chlorina f2 of barley, lacks chlorophyll *b* but shows normal PS I and II activities (Boardman and Highkin, 1966). Similar mutants have been reported of *A. thaliana* (Hirono and Redei, 1963), *Zea mays* (Miles et al., 1979), *Mellilotus alba* (Markwell et al., 1985), *Triticum aestivum* (Allen et al., 1988), *Pisum sativum* (Schwarz and Kloppstech, 1982), and *Chlamydomonas reinhardtii* (Michel et al., 1983). Another class of mutants, the virescent mutants, have delayed greening due to retardation of chlorophyll *b* biosynthesis which is often affected by light and temperature (Alberte et al., 1974; Kyle and Zalik, 1982a,b; Allen et al., 1988; Droppa et al., 1988).

The precise biochemical defects in these chlorophyll *b*-less mutants are not known largely because the pathway of chlorophyll *b* biosynthesis is uncertain (Castelfranco and Beale, 1983). The chlorophyll *b*-less mutants in general show the absence and/or reduced amounts of one or more LHCs, particularly those of LHC IIb (Thornber and Highkin, 1974; Burke et al., 1979; Markwell et al., 1985; White and Green, 1988; Knoetzel and Simpson, 1991; Peter and Thornber, 1991a,b). The levels of mRNAs encoding these apoproteins remain unaffected (Knoetzel and Simpson, 1991; Murray and Kohorn, 1991). In contrast, the carotenoid-deficient mutant of maize contains greatly reduced amounts of mRNAs for apoproteins of LHCs (Mayfield and Taylor, 1984). The plastids obtained from chlorophyll *b*-less mutants are able to import *in vitro* synthesized pLHCP and then integrate it into thylakoids (Bellemare et al., 1982; Chitnis et al., 1988). It has been proposed that the absence of some components of the LHCs in these mutants is not due to mutations in the nuclear genes encoding these apoproteins but rather to the instability of these LHCs in the absence of chlorophyll *b*. Turnover, in the light, of the 20 to 24 kDa-apoproteins of LHC I and of the apoproteins of LHC IIb from three mutants of rice with varying degrees of chlorophyll *b* deficiency exhibited an increase in turnover rate parallel with the extent of chlorophyll *b* deficiency (Terao and Katoh, 1989). The LHC I apoprotein levels were less affected than those of LHC IIb by the degree of chlorophyll *b* deficiency (cf. Greene et al., 1988).

One theory proposed to explain the mechanisms of thylakoid stacking and cation-induced changes in the distribution of excitation energy between PS I and PS II (Section III.C) invokes LHC IIb as the mediator of these effects (Barber, 1982; Staehelin and Arntzen, 1983). The chlorina f2 mutant of barley (Bassi et al., 1985) and ch-1 mutant of *A. thaliana* (Murray and Kohorn, 1991) have greatly reduced amounts of chlorophyll *b*, and hence an absence of the major LHC IIb pigment–protein, yet these mutants are reported to show stacking of thylakoid membranes.

B. In the Purple Bacteria

Regulation of Expression of puf and puc Genes

The genes encoding the antenna apoproteins have been cloned and sequenced from several different bacteria (e.g., Youvan et al., 1984; Youvan and Ismail, 1985). The genes for the "core" antenna apoproteins (*puf*A and *puf*B gene products) are located in the same operon as the RC's L and M genes. Genes *puf*A and *puf*B are located in an approximate 50 kb region of chromosomal DNA which also contains genes for pigment biosynthesis (Bauer et al., 1991). The genes for the variable or peripheral antenna apoproteins (*puc*A and *puc*B) are located outside this main photosynthetic gene cluster. In both cases, however, the structural genes for the α- and β-apoproteins are located next to each other and are transcribed on the same mRNA (Zhu and Hearst, 1986). Those species such as *Rps. palustris* and *Rps. acidophila* that contain more than one type of peripheral antenna complex, have a family of genes which code for the multiple α/β types found in these different complexes (Tadros and Waterkamp, 1989; Cogdell et al., 1990). In *Rps. palustris*, for example, four gene pairs have been cloned and sequenced.

In *Rb. capsulatus*, it has been demonstrated that the genes downstream from the structural genes are essential for assembly of the B800-850 complex (Tichy et al., 1989). Environmental conditions affect expression of the genes from the light-harvesting proteins (see Section VII.B). The purple bacteria normally grow under anaerobic conditions. Oxygen drastically represses transcription of *puf* and *puc* operons of purple bacteria, thus inhibiting synthesis of both pigments as well as apoproteins (Biel and Marrs, 1983; Zhu and Hearst, 1986; Bauer and Marrs, 1988; Bauer et al., 1988; Sganga and Bauer, 1992; see Yildiz et al., 1991 for an exception). Comparison of the promoters of the *puf* and *puc* operons, which are similarly regulated by oxygen, shows some homology; the consensus sequence, however, does not show any similarity to the promoters of the sigma-70 regulated genes of *E. coli* (Bauer et al., 1991). Therefore, a different sigma factor seems to be involved in the oxygen regulated expression of these genes. The regulatory protein PufQ, a product of the *puf* operon itself, is involved as a mediator of the effect of oxygen of the transcription of *puf* and *puc* operons (Bauer and Marrs, 1988). Sganga and Bauer (1992) have very recently provided the first description of a regulatory factor, RegA, which is responsible for promoting high-level anaerobic expression of the light-harvesting and RC structural genes. The expression of genes encoding apoproteins of B800-850 light-harvesting complex of *Rb. sphaeroides* and *Rb. capsulatus* is also regulated at posttranscriptional level (Zucconi and Beatty, 1988; Lee et al., 1989). The differential stability of different segments of mRNA encoding these proteins is the key mechanism in allowing expression of the genes in the polycistronic *puf* operon to various levels (Belasco et al., 1985). The differential degradation rates in this operon are in turn determined by the combined actions of multiple hairpin loop structures and sites of rate-limiting endonucleolytic cleavage (Klug and Cohen, 1990). In those species such as *Rps. palustris* and *Rps. acidophila*, where there are multiple types of variable antenna complexes, there is at present no data on how, at the molecular level, the expression of the different members of the gene family are regulated. But, this promises to be a fascinating story.

Environmental Effects on the Cellular Content of Antenna Complexes in Purple Bacteria

The purple bacterial photosynthetic unit is not, in most cases, a fixed structure; rather it is plastic. The number of units per cell, their size, and the type of variable antenna complex present are all strongly influenced by such factors as the light intensity or the temperature at which the cells were grown.

The Effect of Light-Intensity. Three types of responses to light intensity have been described so far (Aagard and Sistrom, 1972; Drews and Oelze, 1981; Cogdell et al., 1983). In those species that contain only the RC plus the core antenna complex (e.g., *R. rubrum* or *Rps. viridis*), the size of the photosynthetic unit cannot be changed (Aagard and Sistrom, 1972; Thornber et al., 1983). These species respond to growth at lower than normal light intensities by synthesizing, *de novo*, more photosynthetic units. The cells become more pigmented, with more extensive intracytoplasmic membranes; however, the bacteriochlorophyll:RC ratio remains constant at 25–30:1 (Thornber et al., 1983; van Grondelle et al., 1983).

Species that contain the core and the variable light-harvesting antennae (e.g., *Rb. sphaeroides* or *Rb. capsulatus*) show more complexity in their response to growth at different light intensities. When such cells are subjected to a downshift in light intensity, their cells become more pigmented and have a more extensive intracytoplasmic membrane system. Furthermore, the size of the newly synthesized photosynthetic units is also expanded (Aagaard and Sistrom, 1972; Drews and Oezle, 1981); they can contain as many as 250 to 350 bacteriochlorophyll *a* molecules per RC. This increase is accomplished by a large, relative increase in the synthesis of the variable B800-850-antenna complexes (Aagaard and Sistrom, 1972; Drews and Oezle, 1981). These species not only make more photosynthetic units per cell at lower light intensities, but these units are also larger. Zucconi and Beatty (1988) examined the mRNA levels for the B800-850 *Rb. capsulatus* complex in cells grown at different light intensities. The levels are greater in the cells grown under high light intensity, and these authors concluded that the relative amounts of B800-850 complexes are controlled by translational and/or posttranslational mechanisms (see also Lee et al., 1989).

A third type of response is shown by species such as *Rps. acidophila* and *Rps. palustris* (Cogdell et al., 1983; Evans, 1989). These species not only regulate the size of their photosynthetic units in response to changes in light intensity but they can also alter the type of variable antenna complex that is synthesized. In *Rps. acidophila*, strain 7050, for example, at intermediate light intensities a *Rb. sphaeroides*-like B800-850 complex is synthesized, whereas at much lower light intensities a B800-820-complex is made while the synthesis of the B800-850 complex is turned off (Cogdell et al., 1983). Species that have such a capability are able to grow at extremely low light intensities, and it is tempting to assume that these extra variable antenna complexes must work more efficiently; however, there is as yet no firm evidence on this point. In *Rb. capsulatus* and *Rb. sphaeroides* the genes for all of the light-harvesting apoproteins have been cloned and sequenced (see Kiley and Kaplan, 1988 for review).

The Effect of Temperature. In *Rps. acidophila*, strain 7750, and *Chr. vinosum*, strain D, the temperature at which the cells are grown controls the type of variable antenna complex that is synthesized (Hayashi and Morita, 1980; Evans, 1989). If *Rps. acidophila*, strain 7750, is grown at constant light intensity but either at 22 °C or at 30 °C, then different types of antenna complexes are preferentially synthesized. At 22 °C the B800-820 complex is predominantly synthesized while at 30 °C, only the B800-850 complex is made. Interestingly, the B800-820 complex in *Rps. acidophila*, strain 7750, grown at the lower temperature is homologous to the B800-820 complex that is induced when *Rps. acidophila*, strain 7050, is grown at low light intensities. When cells of *Chr. vinosum* are grown at 28 to 30 °C the main variable antenna complexes synthesized are the B800-820 complex and a type II B800-850 complex. If, however, the cells are grown at 38 to 40 °C a *Rb. sphaeroides*-like type I B800-850 complex is preferentially synthesized (Hayashi and Morita, 1980; Evans, 1989; see also Thornber, 1970). In each case these different complexes have different apoproteins and are therefore quite separate molecular species. How these effects are mediated at the level of gene expression is being elucidated (MacKenzie, Kaiser, and Cogdell, to be published). Other environmental factors, such as the nature of the carbon source supplied for the growth medium, also affects the types of variable antenna complexes synthesized, but these types of reactions have not yet received any systematic study.

IX. CONCLUDING REMARKS

The photosynthetic pigment–proteins form a very interesting class of proteins, having an unusually large number of prosthetic groups per unit amount of protein (about one pigment molecule for every 3.5 kDa of protein). They are also interesting functionally. In order to understand the precise molecular details of their function in light-harvesting a detailed biochemical and structural description of each pigment–protein is essential. To achieve this, it is necessary to isolate each pigment–protein in its *native* state. Methods are now available to do this, and hence the field is well situated to complete our knowledge of these components. An unequivocal designation of the primary structure to each antenna apoprotein is close to completion. But we are a long way from having answers to the following: How is each apoprotein folded with respect to the lipid bilayer? How are pigments associated with them? How is the pigment–protein assembled? And, how does each pigment–protein interact with the others and with other polypeptides to yield the multiprotein structures which form the photosystem? Some answers will come when their three-dimensional structures are known. In the interim, more protein chemistry and cell biology studies are needed. For example, more rigorous analysis of each pigment–protein is still needed to quantitate unequivocally the number of pigment molecules associated with each apoprotein. Increased attention is being paid to the regulation of the synthesis and assembly of the pigment–proteins, but little is known yet about the mechanism(s) and sequence(s) of attachment of the photosynthetic pigments to their apoproteins, or how pigment synthesis within the chloroplast is coordinated to gene expression of the apoproteins. Furthermore, it has yet to be determined whether posttranslational modifications of the proteins play an important role in the assembly/disassembly of the photosystems. All these topics need to be described before we can truly say we understand light-harvesting in photosynthesis.

Although common features are readily apparent between the RC complexes of the various groups of photosynthetic organisms, they are far less apparent between their antenna complexes. This is perhaps to be expected since the organisms carry out the same energy conversion process but are required to be diverse in their light-capturing abilities so that they can occupy a specific niche in the biosphere. Thus, the smallest unit of photosynthetic RCs is composed of two homologous but not identical proteins in all organisms: the L and M subunits of purple bacteria; the D1 and D2 proteins of photosystem II in green plants; and, the two slightly different 68-kDa subunits of CC I in photosystem I. High-resolution crystallography has revealed that the site of the primary photochemical event in the purple bacterial RC is between the L-and M-subunits (Diesenhofer et al., 1985). The P680-driven reaction in photosystem II will probably be similarly located as will the primary event driven by P700 in photosystem I (Golbeck and Bryant, 1990). Whereas the closest one can come to general features for the antenna complexes is that they all contain pigments having four pyrrole rings and protein, but even then the latter is disputed for some of the pigments in the green bacteria. If, in an equivalent manner to the D1 and D2 higher plant RC proteins, the genes for the higher plant LHC proteins are derived from those of the purple bacterial light-harvesting apoproteins, it might be expected that the amino acid sequences of the larger LHC apoproteins in plants would show a repeating motif that bore some relationship to the purple bacterial sequences. Comparison of the data bases for their proteins has not revealed any such features. There is, however, a single repeat of about 44 amino acids in the plant LHCs which is putatively correlated with two of the three membrane-spanning α-helices. But, the repeated sequence shows no obvious homology to the sequences of the purple bacterial antenna apoproteins. If the green plant LHC's have been derived from ancestral photosynthetic bacteria, these ancestors must either have been green bacteria or as yet undiscovered purple bacteria.

So far, the only antenna complex whose structure is known to high resolution is the water-soluble bacteriochlorophyll *a*-protein from *P. aesturarii* (Fenna et al., 1974; Tronrud et al., 1986). Interestingly, this complex is a *trimer* of 150 kDa. Each monomer consists of 366 amino acids and binds seven bacteriochlorophyll *a* molecules. The whole structure of the folded polypeptide, which is largely in the form of beta-sheets, is like a string bag. Five of the bacteriochlorophyll molecules are complexed to histidines, one to a carbonyl oxygen and one to water. Since this is a water-soluble complex, the bacteriochlorophyll molecules are located within the "bag" where they are protected from the solvent. The major feature of the structure is that the bacteriochlorophyll molecules, though seemingly randomly organized, are all precisely separated so that they are close enough for efficient singlet–singlet energy transfer, but not so close that concentration quenching occurs. Although there is little primary structure homology between the higher plant LHC sequences and the bacteriochlophyll *a*-protein, it is possible that the organization of the chlorophyll molecules in the two is similar. It will be interesting when the Kuhlbrandt group's structure of LHC IIb has been refined to high resolution to compare it with that of the bacteriochlorophyll *a*-protein's.

A common feature of most photosynthetic antenna systems is that they show the phenomenon of "wavelength programming" which ensures that energy transfer within the antenna is funneled energetically downhill towards the RC (Cogdell, 1988). Thus, in the purple bacteria, the shortest wavelength-absorbing antenna, the B800-820 and B800-850 complexes, are arranged furthest away from the RC, with the B880 complexes next and then the RC itself. It now appears that this basic arrangement is still further reinforced by the presence of minor, longer wavelength-absorbing forms in each type of antenna complex which may function as energy transfer links between the different antenna types (van Grondelle and Sandstrom, 1988). This phenomenon is most clearly seen in the case of the pigments in the phycobilisomes (Glazer, next chapter). It is, however, nowhere nearly so obvious in chlorophyll *a*- and *b*-containing organisms. The spectral resolution between the various chlorophyll *a* forms in the different types of antenna complex in these organisms is very small. Such a situation is perplexing for generalizing the importance of wavelength programming for energy transfer between individual antenna pigment–proteins in an organism.

We use a diagram to summarize this review. Figure 3 depicts our view of the relative arrangement and number of copies of each pigment–protein in each photosystem. It will obviously require updating as more data become available, but in the interim summarizes what is a quite complex situation.

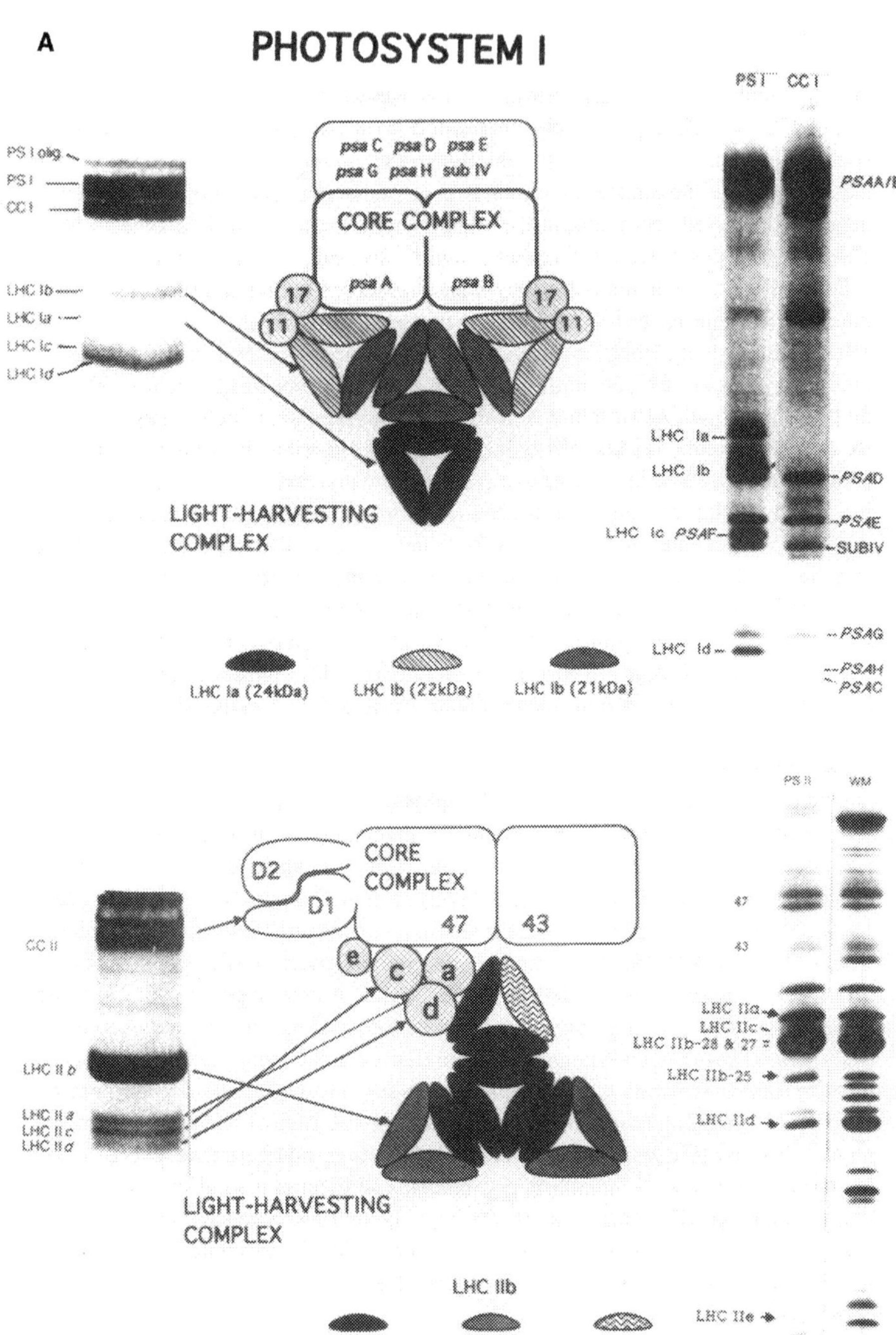
A
PHOTOSYSTEM I
PS I olig
PS I
CC I
LHC Ib
LHC Ia
LHC Ic
LHC Id
psa C psa D psa E
psa G psa H sub IV
CORE COMPLEX
psa A
psa B
17
11
LIGHT-HARVESTING
COMPLEX
PS I CC I
PSAA/B
LHC Ia
LHC Ib
PSAD
PSAE
LHC Ic PSAF
SUBIV
PSAG
LHC Id
PSAH
PSAC
LHC Ia (24kDa)
LHC Ib (22kDa)
LHC Ib (21kDa)
CC II
LHC II b
LHC II a
LHC II c
LHC II d
D2
D1
CORE
COMPLEX
47
43
e
c
a
d
LIGHT-HARVESTING
COMPLEX
PS II
WM
LHC IIa
LHC IIc
LHC IIb-28 & 27
LHC IIb-25
LHC IId
LHC IIb
LHC IIe
28kDa
27kDa
25kDa

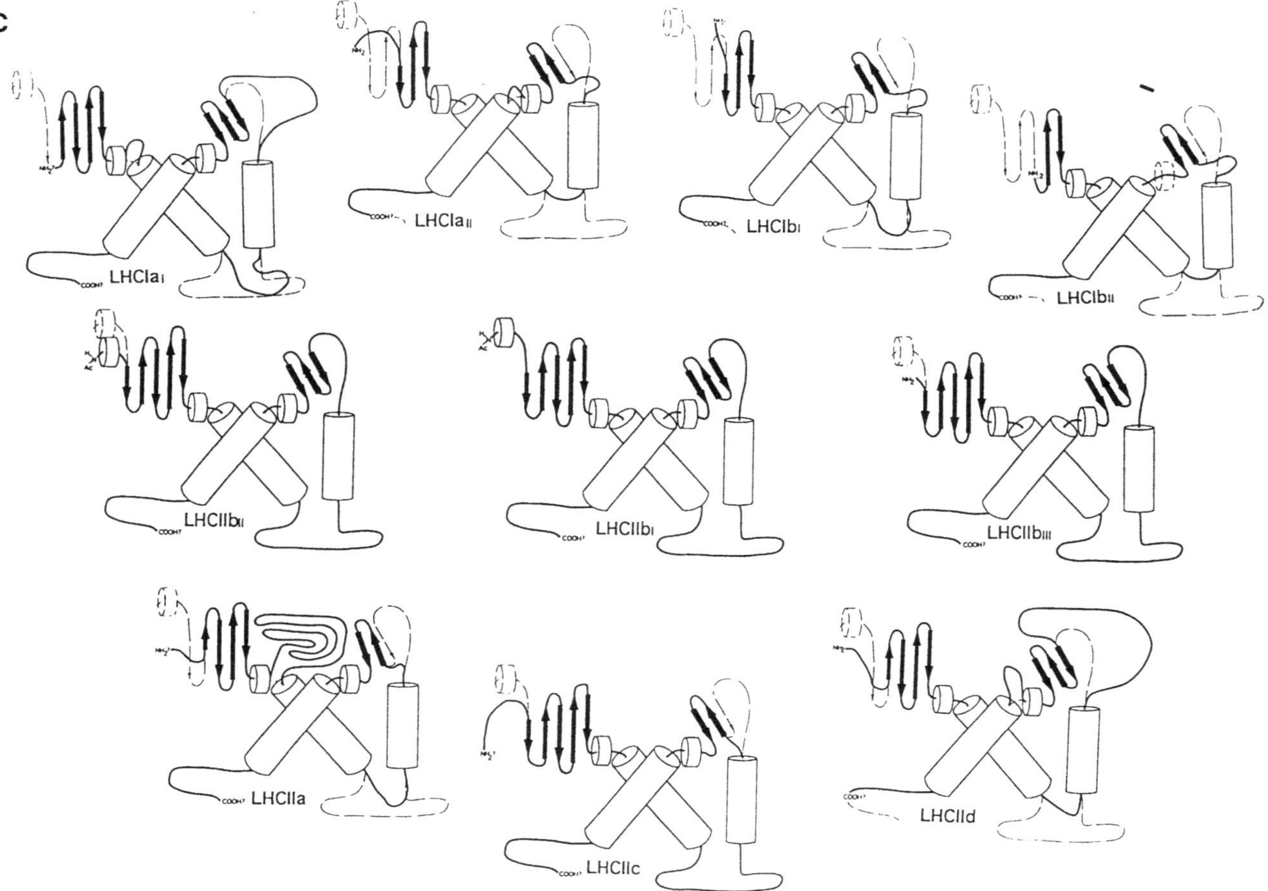
C
LHCIa I
LHCIa II
LHCIb I
LHCIb II
LHCIIb II
LHCIIb I
LHCIIb III
LHCIIa
LHCIIc
LHCIId

Figure 3. Parts **A** and **B**: Models for the organization of the barley LHC pigment-proteins in (**A**) photosystem I, and (**B**) photosystem II. On the left of each model is a photograph of a polyacrylamide gel showing the electrophoretic separation of the pigment-proteins under non-denaturing conditions. On the right is a photograph of a gel electrophoretogram run under fully denaturing conditions to show the protein subunits in each isolated photosystem. The LHC apoproteins and the CC I subunits have been labeled. The subunits in each photosystem can be compared with those in the core complex I (Part **A**), or with those in whole thylakoids (Part **B**). In the former case most of those polypeptides present in Photosystem I and absent from CC I are those belonging to LHCI. Arrows are used to connect the pigment-proteins in the left-hand gel with their proposed location in the model. A key is given below the model to identify different types within an LHC class when that class appears as a trimer in the model. Part (**C**) Schematic diagram depicting the difference in the postulated folding of the various LHC polypeptides belonging to the gene family. The proposed models for the organization of photosystems I and II (Parts **A** and **B**) are used to orient the reader to the different LHC classes. Each folded polypeptide is compared with that of Type I-LHC IIb (cf. Fig 2). Dashed regions along the polypeptide indicate where there is an absence of part of the LHC IIb sequence in that LHC apoprotein, while outlined regions indicate additional parts in that LHC sequence.

APPENDIX

The following pages are an assembly of translated GENOMIC SEQUENCES, translated cDNA gene sequences, and some *direct protein sequences of N-terminal region* of all the higher plant LHC IIb sequences known to us. Sequences are assembled by the three types of sequences in this LHC class. Type II can be subdivided as shown.

Key

GENOMIC SEQUENCE	Æ N-acetylarginine	– inserted gap
cDNA sequence	¥ O-phosphothreonine	• identity to *Lemna gibba* ab30
AMINO ACID SEQUENCE	$ O-phosphoserine	incomplete sequence

```
Type I                                      Transit Peptide              |Mature Peptide
LEMNA ab30             MAAS----MALSSPSLVGKAVKLAPAA-SEVF--------GEGRVSM RKTAGKPKPV-SSGSPWYGPDRVKYLGPFSG-EA
BARLEY LHCIIb          ...AT--.....STFA.....NLSSS-...Q---------.DA.... ....ATK.VG---..................-.S
WHEAT ab-1.6           ...TT--.....S.FA.....NL.S--.ALI---------.DA..N. ....A.A.Q.-..S.....S...L....L..-.P
RICE Cab1R             ...AT--......VMARA.----.STS.AL.---------..A.IT. ....A....AA........A...L....L..R.P
RICE Cab2R             ...AT--......A.A...----A.TK.AL.----------....IT. ..S.A....AA........A...L....L..R.P
rice Cab 2120          ...AT--......VMARA.----.STS.AL.---------.A.IT.. ....A....AA........A...L....L..-.P
MAIZE Cab1 (cab-m7)    ....T---..I..TAMA.------TPIKVGS.----------....-T. ...V....-.AA...................-.P
maize "LHCP"(cab-m5) ..S.T--.....TAFA....N-V.SS---S.---------..A..T. ....A.A..AAA...........L....L..-.P
maize cab-m9           ..S.T--.....TAFA....N-V.----SL.---------..A..T. ....A.A..AA............L....L..-.P
ARABIDOPSIS cab1       ....T--......AFA......S...-..FL---------.S...T. ...V-A.PKG-P.......S...........-.S
ARABIDOPSIS cab2,cab3....T--......AFA....N.S...-..FL---------.S...T. ...V-A.PKG-P.......S...........-.S
radish Cab-124
TOBACCO Cab-E          ....T---T.....-FA......S.SS-...T---------.N.K.T. ....N.A...-....................-.S
TOBACCO Cab-C          ....T--.....S.FA......S.SS-..IT---------.N.K.T. ...LS.A...-..S......N..........-.S
petunia Cab-91R        ...AT--........FA.....FS.SS-..IT---------.N.KAT. ...VT.A...-....................-..
petunia Cab-13         ...AT--.....SPFA....N-V.SS-..IT--------RN.K.T. ...VT.A...-....................-..
petunia Cab-25         ...AT--..I..SPFA....N-V.SS-.QIT---------.N.KAT. ...VT.A...-....................-..
petunia Cab-22L        ...AT--.....STFA..V...S.SS-..IT---------.N.KAT. ....T.A...-....................-..
petunia Cab-22R        ...TT--.....SPFA......SSSS-..IT---------.N.K.T. ...VT.A...-..S.................-..
```

```
petunia Cab-3
petunia Cab-146
petunia Cab-4
petunia Cab-102
TOMATO Cab-1A          ...AA---.......FA.Q....S.S.-..NS--------.N..IT. ..AV--A.SA-P.S...
TOMATO Cab-1B          ...AT---.......FA.Q....S.S.-..IS--------.N..IT. ..AV--A.SA-P.S..................-.S
TOMATO Cab-1C          ...AT---.......FA.Q....S.S.-..IS--------.N..IT. ..AV--A.SA-P.S...
TOMATO Cab-1D
TOMATO Cab-3A          ....T---.....STFA..T.....SS-..IT--------.N..IT. ....A....A-......
TOMATO Cab-3B          ....T---.....STFA......S.SS-..IS--------.N..IS. ....A....A-......
TOMATO Cab-3C          ..T.T---.....STFA......S.SS-..IT--------.N...T. ....T.A..A-.....................-.S
PEA ab-80,66           ....SSSS......T.A..QL..N.S--.QEL--------.AA.FT. ..S.TTK.VA-.....................-.S
pea ab-96                                                                  TTK.VA-..S...H..............-.S
PEA Cab-8              ....S---......T.T..P.ETSANPS.QEL--------.GA.FT. ..S.TTK.VA-.....................-.S
SOYBEAN Cab2           ....T---.....SS.A.Q.M....S--TPEL--------.V..... ....--..T.-........................
SOYBEAN Cab3           ...ASS--........A......G.S--APE---------V..... ...VT.--Q.-.....................-.P
SOYBEAN Cab4           ....T---.....SS.A.Q.I....S--TPQL--------.V..... ....S.--T.-T....................-.P
SOYBEAN Cab5           ....T---.....SS.A.Q.I....S--TPQL--------.V..... ....S.--T.-.....................-.P
SOYBEAN ab2.3                                AS.S--TPQL--------.V..... ....S.--T.-.....................-.P
cucumber G-9                        .T.A.N....S.N.-P.IQ--------.NAKFT. ....--S.S.-.....................-.P
spinach LHCIIb         ..S.T---...A.A.A......G.T.-..II--------....IT. .......T.Q-.....................-.S
SPINACH T2                                                             Æ.¥.....T
SPINACH T1b                                                            Æ.¥.....N
SPINACH T1a                                                            Æ.$.....N
APPLE AB10             .YIPR-----..VSICL.PLLIYKI.KP.KLNPISSTLI-VTAGFT. L.H.--A.SKV..STCDR---...........-.W
pine Cab-II/1A         ..SCG----IG.RCAFA.AQLSSVKPQNNQLLGVGGA-H-..A.LT. ..AT..KSVAASID.........L........-.P
pine Cab-II/1B         ..SCG---IG.-CAFA.GQISSLKPHTNQLLGVGAGVH-..ARLT. ..ATT.KLTASAST.........L........-.P
Type II
LEMNA ab19             ....-----.IQ.SAFA.QTA--LK--QRDELVRKVGV--SD..F.. .R.V-.AV.----Q.I...A..P.F.....E-QT
rice Cab 2123          ....------.HQTTSFLGTA-----PRRDDLVRRVGD--SG..IT. .R.V-.SA.----Q.I......P.......E-QT
silene AB-1            ..T.-----TIQQSAFA.QTL--LKP--QNELVRKVGG--NG..... .R.I-.SA.----E.I......P.F.....E-QT
PETUNIA Cab-1B         ..T.-----.IQQSAFA.QTA--LKS--QNELVRKIGSF-.G..AT. .R.I-.SA.----Q.I...E..P.F.....E-QT
tomato Cab-4           ..TC-----.IQQSAF..QAV--GKS--QNEFIRKVGNF-....IT. .R.V-..--.-.-Q.I...E..P.F.....E-QT
tomato Cab-5                                                SF-EG..VT. .R.V-..--.-.-Q.I...E..P.F.....E-QT
PEA Cab-215            ..T.-----.IQQSAFT..TG--LR--QGNE.IRKRGNF-.QA.FT. .R.V-.SA.----E.I......P.......E-QI
SPINACH P2                                                             ÆR¥V-.SA.----Q
pine LHC2176           ..TAS----.IQ.S..A.QTL--LRP-QQNELVKKVGT--AQA.IT. .R.V-RSA.----E.I......P.......E-GT
pine Cab-11/2A
Type IIa
POLYSTICHUM Cab-f3     ..TST---A..-TSTFS.QQ---LKP--VNELSRKVGA--..A..Q. MAPK-.....---...I......PLT......-SP
PHYSCOMITRELLA AB1     ...AT--A.HST--A.A.QSL--VKP--VNELSRKVLGN-V.A..T. .R.VS.SA---G.DTI...A..P.F......-.T
Type III
BARLEY 25kD                                                                       GND-......I........A-QT
WHEAT 24kD                                                                        GND-L....?.........A-QT
TOMATO Cab13           ..-.----..ATAS.TTVVKATPFLGQTKNANPLRDVVAM.SA.FT.            .ND-L.............A-QT
```

```
Type I
PSYLTGEFPGDYGWDTAGLSADPETFAKNRELEVIHARWAMLGALGCVFPELLARNGVKFGE-AV-WFKAGSQIFSEGGLDYLGNPSLVHAQSILAIWATQV
....................................G.........................-..-...K..........Q..................C..
....................................C.........................-.G-..........D................L.....C..
....................................S.........................-..-......................I..........V..
....................................S......G..................-..-......................I..........V..
....................................L..S......................-..-......................I..........V..
....................................S.........................-..-......................I..........C..
....................................C.........................-..-......................I..........C..
....................................S.........................-..-......................I..........C..
...........................R..................................-..-..........D.........................
...........................R..................................-..-..........D.........................
                                                                               .......................
....................................C.........................-..-.................................C..
....................................C.........................-..-...G......Q......................C..
.................E..................C.........................-..-.................................C..
.......................A......K.....C..T......................-..-.........K.......................C..
....................................C.........................-..-.................................C..
....................................C....................A....-..-.................................C..
.........S..........................C....................I....-..-.....A...........................C..
                                                                  .IP.V.......E.........I.......V..C..
                                                                             .E....................C..
                                                                       .......E.................S..C..

                                                                                      .............C..
....................................C.........................-..-..........Q......................C..
                                                                                      .............C..
                                                                                      .............C..
                                                                                      .............C..
                                                                                      .............C..
....................................C.........................-..-..........Q......................C..
..........................S.........S................S........-..-....................................
....................................S................S....L...-..-....................................
..........................S.........S................S........-..-....................................
........H.--------..................S.........................-..-......................I.............
....................................S................S........-..-......................I.............
....................................S................S........-..-......................I.............
-I..................................S.........................-..-......................I.............
....................................S.........................-.C-...................Q..I.............
....................................S................S........-..-.................................C..
....................................C.........................-..-.................................C..


.....................Y..............S.C..SA....I.....SVM.QG...-..-.....A..........................T.K.
....................................S......................R..-..-.....A............S.Q.I..........C..
........................N...........C......................R..-..-.....A...........................C..
```

Type II

```
...................................S.........I.....SK...Q..-..-.....A..............N..............
..........................R.....L..S.........I.....SK......-..-.....A..............N...........V..
...................................C................K......-..-.........Q..........N...........C..
..........................R........C.............I..K...T..-..-....................N.I.........A..
..........................R........C.............I.SK......-..-..........Q.........N..............
..........................R........C.............I.SK...T..-..-..........Q.........N.I.........S..
..........................R........S.........T.....EK......-..-.........A..........N.I............

................AV.................C................K..L...-..-.....A..........A...N.I.........C..
                                                ...--......-..-.....A..............N.I.........C..
```

Type IIa

```
....S..............................S..........T.....K......-..-.........A......................C..
....N...A..........S......R..................LT.....KS.....-..-.....A..........................C..
```

Type III

```
.K..N................I.H..R..A.I..KG.
..
....N..................AF....A.....G.........I...V.EK-W..VD.KEPV.........D.........N........VLGF..
```

Type I

```
VLMGAVEGYRVAG-GPLGEVVDPLYPGGS-FDPLGLADDPEAFAELKVKEIKNGRLAMFSMFGFFVQAIVTGKGPLENLADHLADPVNNNAWAFATNF-VPGK (1
.............-...............-....A...............................................................-.... (2
..........I..-.....I.........-.......ER.Q....................................GD...I.........LI....-.... (3
..........I..-..............A-............C......K...........................................Y....-.... (4
..........I..-..............A-...............................................................Y....-.... (4
..........I..-..............A-......................K........................................Y....-.... (5
..........I..-...............-....................L...............................I..........Y....-...N (6
..........I..-...............-.............G......L.K.....L.......................I..........Y....-.... (7
.....I.......-..............T-..............DV....L..........................................Y....-.... (8
I............N.....AE.L......-.......T............L........................I......................-.... (9
I............N.....AE.L......-.......T............L........................I......................-.... (9
I............D.....AE.L......-....A..T............................................................-.... (10
.............E-..............-.......E..............................L.......................SY....-.... (11
.............E-..............-.......E.......................................................Y....-.... (11
.............-...............-..............................................................SY....-.... (12
.............-......I........-................E...................................................-.... (12
.............-......I........-....................................................................-.... (12
.............-....V..........-.......E............T..............I.................V........SY....-..R. (12
.............-......I........-.......E......................................................SY....-.... (12
.............-......I...C....-.............................................F................SY....-.... (13
.............-......I........-.......E......................................................SY....-.... (13
.............-......I........-..............................................................SY....-.... (13
                                        .................A........................................-.... (13
..........I..-...............-.......E............................................................-.... (14
..........I..-...............-.......E............................................................-.... (14
..........I..-...............-.......E............................................................-.... (14
..........I..-...............-.......E......................................................SY....-.... (14
..........I..-...............-.......E............................................I...............-.... (14
..........I..-...............-.......E............................................I...............-.... (14
..........I..-...............-....................................................................-.... (14
I.........I..-...............-....................L.........................................SY....-.... (15
I.........I..-...............-.......EV...........L.........................................SY....-.... (16
I.........I..-...............-....................L................................S........SY....-.... (17
I.........I..-......T..I.....-....................L..........................................Y..KL-C... (18
I.........I..-......T..I.....-............L.......L..........................................Y....-.... (18
I.........I..-......T..I.....-....................L..........................................Y....-.... (19
I.........I..-......T..I.....-....................L..........................................Y....-.... (19
I.........I..-......T..I.....-....A...............L..........................................Y....-.... (20
..........I..-......T..I.....-....................L..........................................Y....-.... (21
I.........I..-...............-...............................................................N....-.... (22
                                                                                                        (23
                                                                                                        (23
                                                                                                        (23
I.........I.R-......T......-.-..S....E.T..........L...................SR.DR........GWT.....LSNV...-...N (24
I....I.......-......T..I....N-..........D..................................I.................Y....-.... (25
I.........I..-......T..I.....-.......E.....................................I.................Y....-.... (25
```

Type II

```
....LI.....G.-.....GL.......A-....................................................I....A..........-.... (26
....F......G.-.....GL.KV....A-..........DT........L...................R....I...F..VT...A....SY....-.... (5
...........G.-.....GL.Q.....A-.......E....                                                              (27
....F......G.-.....GL.KI....A-.......E.....................................I...Y..V....A..........-.... (28
....F......G.-.....GL.KI....A-.............................................I...S..IN...A.....Y....-.... (29
....F......G.-.....GL.KI....A-.............................................I...S..I....A.....Y....-.... (29
....F......G.-.....GL.......A-..........DS........L........................IQ..Y..V....A..........-...Q (30
                                                                                                        (23
....LI.....G.-.T...GL...L...A-............C................................I...Y.......A.....Y....-.... (31
....LI.....G.-.....GL......DA-............K................................I...Y.......A.....Y....-.... (25
```

Type IIa

```
I............-......E..I.....-....................L........................I...S......AV.....Y....-T... (32
.............-......T..I.....-..........DT...............................K.....N.......A.....Y.PTSP..TR (33
```

Type III

```
                                                                                                        (34
                                                                                                        (35
....L...F.IN.LPGV..GN.-.....QY..........TT.....................................L...DN..A....VY..K.-...A (36
```

1)Kohorn et al.,1986; 2)Chitnis et al.,1988; 3)Lamppa et al.,1985; 4)Luan and Borograd,1989; 5)Matsuoka, 1990; 6)Sullivan et al.,1989; 7)Matsuoka et al.,1987; 8)Viret et al.,1990; 9)Leutwiler et al.,1986; 10)Fourcroy et al.,1990; 11)Castresana et al.,1987; 12)Dunsmuir,1985; 13)Dunsmuir et al.,1983; 14)Pichersky et al.,1985; 15)Cashmore,1984; 16)Coruzzi et al.,1983; 17)Alexander et al.,1991; 18)Walling et al.,1988; 19)Demmin et al.,1989; 20)Hsaio et al.,1988; 21)Greenland et al.,1987; 22)Mason,1989; 23)Michel et al.,1991; 24)Chen et al., 1990; 25)Jansson and Gustafsson,1990; 26)Karlin-Neumann et al.,1985; 27)Smeekens et al.,1986; 28)Stayton et al.,1986; 29)Pichersky et al.,1987; 30)Falconet et al.,1991; 31)Yamamoto et al.,1988; 32)Pichersky et al., 1990; 33)Long et al.,1989; 34)Morishige and Thornber, 1990; 35)Webber and Gray,1989; 36)Schwartz et al.,1991

ACKNOWLEDGMENTS

The authors' research described in this chapter has been supported by grants from the NSF, USDA, and SERC.

AUTHORS' NOTE

This review has not been updated since its completion in April 1992.

ABBREVIATIONS

CP: chlorophyll-protein
RC: reaction center
CC: core complex
LHC: light-harvesting complex
PAGE: polyacrylamide gel electrophoresis
SDS: sodium dodecyl sulfate

REFERENCES

Aagaard, J. & Sistrom, W.R. (1972). Control of synthesis of reaction center bacteriochlorophyll in photosynthetic bacteria. Photochem. Photobiol. 15, 209–225.

Abad, M.S., Clark, S.E., & Lamppa, G. (1989). Properties of a chloroplast enzyme that cleaves the chlorophyll *a/b* binding protein precursor. Plant Physiol. 90, 117–124.

Abad, M.S., Oblong, J.E., & Lamppa, G.K. (1991). Soluble chloroplast enzyme cleaves preLHCP made in *E. coli* to a mature form lacking a basic N-terminal domain. Plant Physiol. 96, 1220–1227.

Alberte, R.S., Hesketh, J.D., Hofstra, G., Thornber, J.P., Naylor, A.W., Barnard, R.L., Brim, C., & Endrezzi, J. (1974). Composition and photosynthetic activity of the photosynthetic apparatus in temperature-sensitive mutants of higher plants. Proc. Natl. Acad. Sci. USA 71, 2412–2418.

Alexander, L., Falconet, D., Fristensky, B.W., White, M.J., Watson, J.C., Roe, B.A., & Thompson, W.F. (1991). Nucleotide sequence of Cab-8, a new type 1 gene encoding a chlorophyll *a/b* binding protein of LHC 2 in Pisum. Plant Mol. Biol. 17, 523–526.

Allen, J.F. (1992a). How does protein phosphorylation regulate photosynthesis?. Trends Biochem. Sci. 17, 12–17.

Allen, J.F. (1992b). Protein phosphorylation in regulation of photosynthesis. Biochim. Biophys. Acta 1098, 275–335.

Allen, K.D., Duysen, M.E., & Staehelin, L.A. (1988). Biogenesis of thylakoid membranes is controlled by light intensity on the conditional chlorophyll *b*-deficient CD3 mutant of wheat. J. Cell Biol. 107, 907–919.

Allgood, L., Curtright, R.D., & Markwell, J.P. (1991). Solubilization of photosynthetic pigment–protein complexes. Photochem. Photobiol. 54, 459–463.

Almog, O., Shoham, G., Michaeli, D., & Nechushtai, R. (1991). Monomeric and trimeric forms of photosystem I reaction center of *Mastigocladus laminosus*: Crystallization and preliminary characterization. Proc. Natl. Acad. Sci. USA 88, 5312–5316.

Anandan, S. & Thornber, J.P. (1990). Isolation of the LHC I complex of barley containing multiple pigment–proteins. Current Research in Photosynthesis (Baltscheffsky, M., Ed.), Vol. 2, pp. 285–288. Kluwer Academic, Dordrecht.

Anandan, S., Morishige, D.T., & Thornber, J.P. (1993). Light-induced biogenesis of light-harvesting complex I (LHC I) during chloroplast development in barley (*Hordeum vulgare*). Plant Physiol. 101, 227–236.

Anandan, S., Vainstein, A., & Thornber, J.P. (1989). Correlation of some published amino acid sequences for photosystem I polypeptides to a 17-kDa LHC I pigment–protein and to subunits 3 and 4 of the core complex. FEBS Lett. 256, 150–154.

Anderson, J.M., Brown, J.S., Lam, E., & Malkin, R. (1983). Chlorophyll *b*: An integral component of photosystem I of higher plant chloroplasts. Photochem. Photobiol. 38, 205–210.

Andersson, B., Anderson, J.M., & Ryrie, I.J. (1982). Transbilayer organization of the chlorophyll–proteins of spinach thylakoids. Eur. J. Biochem. 123, 465–472.

Angerhofer, A., Cogdell, R.J., & Hipkins, M.F. (1986). A spectral characterization of the light-harvesting pigment–protein complexes from *Rhodopseudomonas-acidophila*. Biochim. Biophys. Acta 848, 333–341.

Apel, K. & Kloppstech, K. (1978). Plastid membranes of barley (*Hordeum vulgare*) light-induced appearance of messenger-RNA coding for apoprotein of light-harvesting chlorophyll *a/b*–protein . Eur. J. Biochem. 85, 581–588.

Apel, K. & Kloppstech, K. (1980). The effect of light on the biosynthesis of the light-harvesting chlorophyll *a/b* protein. Evidence for the requirement of chlorophyll a for the stabilization of the apoprotein. Planta 150, 426–430.

Argyroudi-Akoyunoglou, J.H. & Akoyunoglou, G. (1970). Photoinduced changes in the chlorophyll *a* to chlorophyll *b* ratio in young bean plants. Plant Physiol. 46, 247–249.

Argyroudi-Akoyunoglou, J.H. & Akoyunoglou, G. (1973). On the formation of photosynthetic membranes in bean plants. Photochem. Photobiol. 18, 219–228.

Argyroudi-Akoyunoglou, J.H. & Akoyunoglou, G. (1979a). The chlorophyll-protein complexes of the thylakoids in greening plastids of *Phaseolus vulgaris*. FEBS Lett. 104, 78–84.

Argyroudi-Akoyunoglou, J.H. & Akoyunoglou, G. (1979b). On the formation of photosynthetic membranes in bean plants. Photochem. Photobiol. 18, 219–228.

Argyroudi-Akoyunoglou, J.H., Feleki, Z., & Akoyunoglou, G. (1971). Formation of 2 chlorophyll–protein complexes during greening of etiolated bean leaves. Biochem. Biophys. Res. Commun. 45, 606–613.

Argyroudi-Akoyunoglou, J.H., Feleki, Z., & Akoyunoglou, G. (1971). Photo-induced formation of two different chlorophyll–protein complexes at early stages of greening. Proc. Int. Congr. Photosyn. Res. 2nd. 3, 2417–2426.

Armond, P.A., Arntzen, C.J., Briantais, J.M., & Vernotte, C. (1976). Differentiation of chloroplast lamellae. Light-harvesting efficiency and grana development. Arch. Biochem. Biophys. 175, 54–63.

Auchincloss, A.H., Alexander, A., & Kohorn, B.D. (1992). Requirement for three membrane-spanning alpha-helices in the posttranslational insertion of a thylakoid membrane protein. J. Biol. Chem. 267, 10439–10446.

Bachmann, M.D., Robertson, D.S., Bowen, C.C., & Anderson, I.C. (1973). Chloroplast ultrastructure in pigment-deficient mutants of Zea mays under reduced light. J. Ultrastruc. Res. 45, 384–406.

Baker, N.R., Webber, A.N., Bradbury, M., Markwell, J.P., Baker, M.G., & Thornber, J.P. (1984). Development of photochemical competence during growth of the wheat leaf. UCLA Symp. Mol. Cell. Biol. New Ser. 14, 237–255.

Baldauf, S.L. & Palmer, J.D. (1990). Evolutionary transfer of the chloroplast tufA gene to the nucleus. Nature (London) 344, 262–265.

Barbato, R., Rigoni, F., Giardi, M.T., & Giacometti, G.M. (1989). The minor antenna complexes of an oxygen evolving photosystem II preparation: Purification and stoichiometry. FEBS Letts. 251, 147–154.

Barber, J. (1982). Influence of surface charges on thylakoid structure and function. Annu. Rev. Plant Physiol. 33, 261–295.

Bassi, R. & Simpson, D. (1987). Chlorophyll–protein complexes of barley photosystem I. Eur. J. Biochem. 163, 221–230.

Bassi, R. & Dainese, P. (1989). The role of light-harvesting complex II and of the minor chlorophyll *a/b* antenna system. Current Research in Photosynthesis (Baltscheffsky, M., Ed.), Vol. 2, pp. 209–216. Kluwer Academic, Dordrecht.

Bassi, R. & Dainese, P. (1992). A supramolecular light-harvesting complex from chloroplast photosystem II membranes. Eur. J. Biochem. 204, 317–326.

Bassi, R., Hinz, U., & Barbato, R. (1985). The role of the light-harvesting complex and photosystem II in thylakoid stacking in the chlorina-f2 barley mutant. Carlsberg Res. Commun. 50, 347–367.

Bassi, R., Hoeyer-Hansen, G., Barbato, R., Giacometti, G.M., & Simpson, D.J. (1987). Chlorophyll–proteins of the photosystem II antenna system. J. Biol. Chem. 262, 13333–13341.

Bassi, R., Rigoni, F., & Giacometti, G.M. (1990). Chlorophyll-binding proteins with antenna function in higher plants and green algae. Photochem. Photobiol. 52, 1187–1206.

Bauer, C.E. & Marrs, B.L. (1988). *Rhodobacter capsulatus puf* operon encodes a regulatory protein (PufQ) for bacteriochlorophyll synthesis. Proc. Natl. Acad. Sci. USA 85, 7074–7078.

Bauer, C.E., Young, D.A., & Marrs, B.L. (1988). Analysis of the *Rhodobacter capsulata puf* operon. Localization of the oxygen-regulated promoter region and the identification of an additional *puf*-encoded gene. J. Biol. Chem. 263, 4820–4827.

Bauer, C.E., Buggy, J.J., Yang, Z., & Marrs, B.L. (1991). The superoperonal organization of genes for pigment biosynthesis and reaction center proteins is a conserved feature in *Rhodobacter capsulatus*: Analysis of overlapping *bchB* and *pucA* transcripts. Mol. Gen. Genet. 228, 433–444.

Belasco, J.G., Beatty, J.T., Adams, C.W., Von Gabain, A., & Cohen, S.N. (1985). Differential expression of photosynthetic genes in *R. capsulata* results from segmental differences in stability within polycistronic transcript. Cell 40, 171–181.

Bellemare, G., Bartlett, S.G., & Chua, N. (1982). Biosynthesis of chlorophyll a/b-binding polypeptides in wild type and the chlorina f2 mutant of barley. J. Biol. Chem. 257, 7762–7767.

Bengis, C. & Nelson, N. (1977). Subunit structure of chloroplasts photosystem I reaction center. J. Biol. Chem. 252, 4564–4569.

Bennett, J. (1979). Chloroplast phosphoproteins. The protein kinase of thylakoid membranes is light-dependent. FEBS Lett. 103, 342–344.

Bennett, J. (1981). Biosynthesis of the light-harvesting chlorophyll *a/b* protein. Polypeptide turnover in darkness. Eur. J. Biochem. 118, 61–70.

Bennett, J., Jenkins, G.I., & Hartley, M.R. (1984). Differential regulation of the accumulation of the light-harvesting chlorophyll *a/b* complex and ribulose bisphosphate carboxylase/oxygenase in greening pea leaves. J. Cell Biochem. 25, 1–13.

Bennett, J., Markwell, J.P., Skrdla, M.P., & Thornber, J.P. (1981). Higher plant chlorophyll *a/b*-protein complexes: Studies on the phosphorylated apoproteins. FEBS Lett. 131, 325–330.

Biel, A.J. & Marrs, B.L. (1983). Transcriptional regulation of several genes for bacteriochlorophyll biosynthesis in *Rhodopseudomonas capsulata* in response to oxygen. J. Bacteriol. 156, 686–694.

Bissig, I., Brunisholz, R.A., Suter, F., Cogdell, R.J., & Zuber, H. (1988). The complete amino acid sequences of the B800-850 antenna polypeptides from Rhodopseudomonas acidophila strain 7750. Z. Naturforsch. 43, 77–87.

Boardman, N.K. & Highkin, H.R. (1966). Studies on a barley mutant lacking chlorophyll *b* I. Photochemical activity of isolated chloroplasts. Biochim. Biophys. Acta 126, 189–199.

Boekema, E.J., Dekker, J.P., Van Heel, M.G., Roegner, M., Saenger, W., Witt, I., & Witt, H.T. (1987). Evidence for a trimeric organization of the photosystem I complex from the thermophilic cyanobacterium *Synechococcus* sp. FEBS Lett. 217, 283–286.

Bredenkamp, G.J. & Baker, N.R. (1988). The changing contribution of LHC I to photosystem I activity during chloroplast biogenesis in wheat. Biochim. Biophys. Acta 934, 14–21.

Briggs, W.R., Mosinger, E., Batschauer, A., Apel, K., & Schafer, E. (1987). Molecular events in photoregulated greening in barley leaves. UCLA Symp. Mol. Cell Biol. (New Ser.) 44, 413–423.

Bruce, B.D. & Malkin, R. (1988). Subunit stoichiometry of the chloroplast photosystem I complex. J. Biol. Chem. 263, 7302–7308.

Buergi, R., Suter, F., & Zuber, H. (1987). Arrangement of the light-harvesting chlorophyll *a/b* protein complex in the thylakoid membrane. Biochim. Biophys. Acta 890, 346–351.

Buetow, D.E., Chen, H.Q., Erdos, G., & Yi, L.S.H. (1988). Regulation and expression of the multigene family coding light-harvesting chlorophyll *a/b*-bonding proteins of photosystem II. Photosyn. Res. 18, 61–97.

Burke, J.J., Steinback, K.E., & Arntzen, C.J. (1979). Analysis of the light- harvesting pigment–protein complex of wild type and a chlorophyll-*b*-less mutant of barley. Plant Physiol. 63, 237–243.

Burkey, K.O. (1986). Chlorophyll–protein complex composition and photochemical activity in developing chloroplasts from greening barley seedlings. Photosynth. Res. 10, 37–49.

Burkey, K.O. (1987). Chlorophyll–protein complex composition during chloroplast development: a species comparison. Photosynth. Res. 11, 211–224.

Butler, P.J.G. & Kuhlbrandt, W. (1988). Determination of the aggregate size in detergent solution of the light-harvesting chlorophyll *a/b*–protein complex from chloroplast membranes. Proc. Natl. Acad. Sci. USA 85, 3797–3801.

Camm, E.L. & Green, B.R. (1980). Fractionation of thylakoid membranes with the nonionic detergent octyl-D-glucopyranoside. Plant Physiol. 66, 428–432.

Camm, E.L. & Green, B.R. (1983). Isolation of PS II reaction center and its relationship to the minor chlorophyll–protein complexes. J. Cell. Biochem. 23, 171–179.

Camm, E.L. & Green, B.R. (1989). The chlorophyll *ab* complex, CP29, is associated with the photosystem II reaction center core. Biochim. Biophys. Acta 974, 180–184.

Carter, D.P. & Staehelin, L.A. (1980). Proteolysis of chloroplast thylakoid membranes. I. Selective degradation of thylakoid pigment-protein complexes at the outer membrane surface. Arch. Biochem. Biophys. 200, 364–373.

Cashmore, A.R. (1984). Structure and expression of a pea nuclear gene encoding a chlorophyll *a/b*-binding polypeptide. Proc. Natl. Acad. Sci. USA 81, 2960–2964.

Castelfranco, P.A. & Beale, S.I. (1983). Chlorophyll biosynthesis: Recent advances and areas of current interest. Annu. Rev. Plant Physiol. 34, 241–278.

Castresana, C., Staneloni, R., Malik, V.S., & Cashmore, A.R. (1987). Molecular characterization of two clusters of genes encoding the type I CAB polypeptides of PSII in Nicotiana plumbaginifolia. Plant Mol. Biol. 10, 117–126.

Castresana, C., Garcia-Luque, I., Alonso, E., Malik, V.S., & Cashmore, A.R. (1988). Both positive and negative regulatory elements mediate expression of a photoregulated CAB gene from *Nicotiana plumbaginifolia*. EMBO J. 7, 1929–1936.

Chaumont, F., O'Riordan, V., & Boutry, M. (1990). Protein transport into mitochondria is conserved between plant and yeast species. J. Biol. Chem. 265, 16856–16862.

Chen, H., Korban, S.S., & Buetow, D.E. (1990). Nucleotide sequence of an apple nuclear gene encoding a light-harvesting chlorophyll *a/b* binding polypeptide of photosystem II. Nucleic Acids Res. 18, 679.

Chitnis, P.R. & Thornber, J.P. (1988). The major light-harvesting complex of photosystem II: Aspects of its cell and molecular biology. Photosyn. Res. 16, 41–64.

Chitnis, P.R., Harel, E., Kohorn, B.D., Tobin, E.M., & Thornber, J.P. (1986). Assembly of the precursor and processed light-harvesting chlorophyll *a/b* protein of *Lemna* into the light-harvesting complex II of barley etiochloroplasts. J. Cell Biol. 102, 982–988.
Chitnis, P.R., Morishige, D.T., Nechushtai, R., & Thornber, J.P. (1988). Assembly of the barley light-harvesting chlorophyll *a/b*-proteins in barley etiochloroplasts involves processing of the precursor on thylakoids. Plant Mol. Biol. 11, 95–107.
Chitnis, P.R., Nechushtai, R., & Thornber, J.P. (1987). Insertion of the precursor of the light-harvesting chlorophyll *a/b*-protein into the thylakoids requires the presence of a developmentally regulated stromal factor. Plant Mol. Biol. 10, 3–11.
Chua, N-H. & Schmidt, G.W. (1979). Transport of proteins into mitochondria and chloroplasts. J. Cell Biol. 81, 461–483.
Clark, S.E., Abad, M.S., & Lamppa, G.K. (1989). Mutations at the transit peptide-mature protein junction separate two cleavage events during chloroplast import of the chlorophyll *a/b*-binding protein. J. Biol. Chem. 264, 17544–17550.
Clark, S.E. & Lamppa, G.K. (1991). Determinants for cleavage of the chlorophyll *a/b* binding protein precursor: A requirement for a basic residue that is not universal for chloroplast imported proteins. J. Cell Biol. 114, 681–688.
Clayton, R.K. (1963). Absorption spectra of photosynthetic bacteria and their chlorophylls. In: Bacterial Photosynthesis (Gest, H., San Pietro, A., & Vernon, L.P., Eds.), pp. 495–512. Antioch Press, Yellow Springs, Ohio.
Cline, K. (1986). Import of proteins into chloroplasts. Membrane integration of a thylakoid precursor protein reconstituted in chloroplast lysates. J. Biol. Chem. 261, 14804–14810.
Cline, K. (1988). Light-harvesting chlorophyll *a/b*-protein. Membrane insertion, proteolytic processing, assembly into LHC II, and localization to the appressed membranes occurs in chloroplast lysates. Plant Physiol. 86, 1120–1126.
Cline, K., Fulsom, D.R., & Viitanen, P.V. (1989). An imported thylakoid protein accumulates in the stroma when insertion into thylakoids is inhibited. J. Biol. Chem. 264, 14225–14232.
Cline, K., Wernerwas, M., Lubben, T.H., & Keegstra, K. (1985). Precursors to two nuclear-encoded chloroplast proteins bind to the outer envelope membrane before being imported into chloroplasts. J. Biol. Chem. 260, 3691–3696.
Cogdell, R.J. (1986). Light-harvesting complexes in purple photosynthetic bacteria. Encl. Plant Physiol. 19, 252–284.
Cogdell, R.J. (1988). The biochemistry of light-harvesting complexes. In: Photosynthetic Light-Harvesting Systems (Scheer, H. & Schneider, S., Eds.), pp. 1–10. Walter de Gruyter, Berlin.
Cogdell, R.J. & Scheer, H. (1985). Circular-dichroism of light-harvesting complexes from purple photosynthetic bacteria. Photochem. Photobiol. 42, 669–678.
Cogdell, R.J. & Thornber, J.P. (1979). The preparation and characterization of different types of light-harvesting pigment–protein complexes from some purple bacteria. CIBA Found. Symp. 61, 61–79.
Cogdell, R.J., Durant, I., Valentine, G., Lindsay, J.G., & Schmidt, K. (1983). The isolation and partial characterization of the light-harvesting pigment–protein complement of *Rhodopseudomonas acidophila*. Biochem. Biophys. Acta 722, 427–435.
Cogdell, R.J., Hawthornthwaite, A.M., Ferguson, L.A., Evans, M.B., Li, M., Gardiner, A., Mackenzie, R.C., Thornber, J.P., Brunisholz, R.A., Zuber, H., Van Grondelle, R., & Van Mourik, F. (1990). The structure and function of some unusual variable antenna complexes. In: The molecular Biology of membrane-bound complexes in phototrophic bacteria (Drews, G., & Dawes, E.A., Eds.), pp. 211–217. Plenum Press, New York.
Cogdell, R.J., Lindsay, J.G., MacDonald, W., & Reid, G.P. (1979). The subunit structure of the B800-850 light-harvesting pigment–protein complex from *Rhodopseudomonas sphaeroides* strain 2.4.1. Biochem. Soc. Trans. 7, 184–187.

Cogdell, R.J., Zuber, H., Thornber, J.P., Drews, G., Gingras, G., Niederman, R.A., Parson, W.W., & Feher, G. (1985). Recommendations for the naming of photochemical reaction centers and light-harvesting pigment–protein complexes from purple photosynthetic bacteria. Biochim. Biophys. Acta 806, 185–186.

Coruzzi, G., Broglie, R., Cashmore, A.R., & Chua, N-H. (1983). Nucleotide sequence of two pea cDNA clones encoding the small subunit of ribulose 1, 5-bisphosphate carboxylase and the major chlorophyll *a/b*-binding thylakoid polypeptide. J. Biol. Chem. 258, 1399–1402.

Cuming, A.C. & Bennett, J. (1981). Biosynthesis of the light-harvesting chlorophyll *a/b* protein. Control of messenger RNA activity by light. Eur. J. Biochem. 118, 71–80.

Darr, S.C., Somerville, S.C., & Arntzen, C.J. (1986). Monoclonal antibodies to the light-harvesting chlorophyll *a/b* protein complex of photosystem II. J. Cell Biol. 103, 733–740.

Day, D.A., Ryrie, I.J., & Fuad, N. (1984). Investigations of the role of the main light-harvesting chlorophyll–protein complex in thylakoid membranes. Reconstitution of depleted membranes from intermittent-light-grown plants with the isolated complex. J. Cell Biol. 98, 163–172.

Deisenhofer, J., Michel, H., & Huber, R. (1985). The structural basis of photosynthetic light reactions in bacteria. Trends Biochem. Sci. 10, 243–248.

Demmig, B., Winter, K., Kruger, A., & Czygan, F-C. (1987). Photoinhibition and zeaxanthin formation in intact leaves: A possible role of the xanthophyll cycle in the dissipation of excess light energy. Plant Physiol. 84, 218–224.

Demmin, D.S., Stockinger, E.J., Chang, Y.C., & Walling, L.L. (1989). Phylogenetic relationships between the chlorophyll *a/b* binding protein (CAB) multigene family: An intra- and interspecies study. J.Mol.Evol. 29, 266–279.

Dietz, K.J. & Bogorad, L. (1987). Plastid development in *Pisum sativum* during greening II. Posttranslational uptake by plastids as an indicator system to monitor changes in translatable mRNA for nuclear encoded plastid polypeptides. Plant Physiol. 85, 816–822.

Drews, G. & Oelze, J. (1981). Organization and differentiation of membranes of photrophic bacteria. Adv. Microb. Physiol. 22, 1–92.

Droppa, M., Ghirardi, M.L., Horvath, G., & Melis, A. (1988). Chlorophyll *b* deficiency in soybean mutants. II. Thylakoid membrane development and differentiation. Biochim. Biophys. Acta 932, 138–145.

Dunahay, T.G., Schuster, G., & Staehelin, L.A. (1987). Phosphorylation of spinach chlorophyll-protein complexes. CPII, but not CP29, CP27, or CP24, is phosphorylated *in vitro*. FEBS Lett. 215, 25–30.

Dunahay, T.G. & Staehelin, L.A. (1985). Isolation of photosystem I complexes from octyl glucoside/sodium dodecyl sulfate solubilized spinach thylakoids: Characterization and reconstitution into liposomes. Plant Physiol. 78, 606–613.

Dunahay, T.G. & Staehelin, L.A. (1986). Isolation and characterization of a new minor chlorophyll *a/b*-protein complex (CP24) from spinach. Plant Physiol. 80, 429–434.

Dunahay, T.G. & Staehelin, L.A. (1987). Immunolocalization of the Chl *a/b*-light-harvesting complex and CP29 under conditions favoring phosphorylation and dephosphorylation of thylakoid membranes (state 1–state 2 transitions). In: Current Research in Photosynthesis (Baltscheffsky, M., Ed.), Vol. 2, pp. 701–704. Kluwer Academic, Dordrecht.

Dunkley, P.R. & Anderson, J.M. (1979). The light-harvesting chlorophyll *a/b*-protein complex from barley thylakoid membranes. Polypeptide composition and characterization of an oligomer. Biochim. Biophys. Acta 545, 175–187.

Dunsmuir, P. (1985). The petunia chlorophyll *a/b*-binding protein genes: a comparison of Cab genes from different gene families. Nucleic Acids Res. 13, 2503–2518.

Dunsmuir, P. & Bedbrook, J. (1983). Chlorophyll *a/b* binding proteins and the small subunit of ribulose bisphosphate carboxylase are encoded by multiple genes in petunia. NATO Adv. Sci. Inst. Ser. Ser. A 63–221.

Dunsmuir, P., Smith, S.M., & Bedbrook, J. (1983). The major chlorophyll *a/b* binding protein of petunia is composed of several polypeptides encoded by a number of distinct nuclear genes. J. Mol. Appl. Genet. 2, 285–200.

Engelhardt, H., Baumeister, W., & Saxton, W.O. (1983). Electron microscopy of photosynthetic membranes containing bacteriochlorophyll *b*. Arch. Microbiol. 135, 169–175.

Evans, M.B. (1989). The structure and function of the light-harvesting complexes of purple photosynthetic bacteria. PhD Thesis, Glasgow University.

Evans, P.K. & Anderson, J.M. (1986). The chlorophyll *a/b*-proteins of PS I and PS II are immunologically related. FEBS Lett. 199, 227–233.

Falconet, D., White, M.J., Fristensky, B.W., Dobres, M.S., & Thompson, W.F. (1991). Nucleotide sequence of *cab-215*, a Type 2 gene encoding a photosystem II chlorophyll *a/b*-binding protein in *Pisum*. Plant Mol. Biol. 17, 135–139.

Ferguson, L., Halloran, E., Hawthornthwaite, A.M., Cogdell, R.J., Kerfeld, C., Peter, G.F., & Thornber, J.P. (1991). The use of non-denaturing deriphat-gel electrophoresis to fractionate pigment–protein complexes of purple bacteria. Photosyn. Res. 30, 139–143.

Fenna, R.E., Matthews, B.W., Olson, J.M., & Shaw, E.K. (1974). Structure of a bacteriochlorophyll–protein from the green photosynthetic bacterium *Chlorobium limicola*: Crystallographic evidence for a trimer. J. Mol. Biol. 84, 231–240.

Ford, R.C., Pauptit, R.A., & Holzenberg, A. (1988). Structural studies on improved crystals of the photosystem I reaction center from *Phormidium laminosum*. FEBS Lett. 238, 385–389.

Fourcroy, P., Guidet, F., & Klein-ende, D. (1990). The chlorophyll *a/b* binding protein cDNA of *Raphanus sativus*: partial sequence light dependent expression and DNA polymorphism in the *Cruciferae*. Plant Physiol. Biochem. 28, 509–515.

Fowler, G.J.S., Visschers, R.W., Grief, G.G., van Grondelle, R., & Hunter, C.N. (1992). Genetically modified photosynthetic antenna complexes with blueshifted absorbance bands. Nature (London) 355, 848–850.

Fulson, D.R. & Cline, K. (1988). A soluble protein factor is required *in vitro* for membrane insertion of the thylakoid precursor protein, LHCP. Plant Physiol. 88, 1146–1153.

Genge, S., Pilger, D., & Hiller, R.G. (1974). The relationship between chlorophyll *b* and pigment–protein complex II. Biochim. Biophys. Acta 347, 22–30.

Ghanotakis, D.F., Demetriou, D.M., & Yocum, C.F. (1987). Isolation and characterization of an oxygen-evolving photosystem II reaction center core preparation and a 28-kDa Chl-a-binding protein. Biochim. Biophys. Acta 891, 15–22.

Ghirardi, M.L., McCauley, S.W., & Melis, A. (1986). Antennae size. Biochim. Biophys. Acta 851, 331–339.

Ghosh, R., Hauser, H., & Bachofen, R. (1988). Reversible dissociation of the B873 light-harvesting complex from *Rhodospirillum rubrum* G9+. Biochemistry 27, 1004–1014.

Glazer, A.N. & Melis, A. (1987). Photochemical reaction centers: Structure organization and function. Annu. Rev. Plant Physiol. 38, 11–45.

Golbeck, J.H. & Bryant, D.A. (1990). Photosystem 1. Curr. Top. Bioenerg. 16, 83–177.

Green, B.R. (1988). The chlorophyll–protein complexes of higher plant photosynthetic membranes, or just what green band is that? Photosynth. Res. 15, 3–32.

Green, B.R., Camm, E.L., & VanHouten, J. (1982). The chlorophyll–protein complexes of *Acetabularia*. A novel chlorophyll *a/b* complex which forms oligomers. Biochim. Biophys. Acta 681, 248–255.

Green, B.R., Pichersky, E., & Kloppstech, K. (1991). Chlorophyll *a/b*-binding proteins: An extended family. Trends Biochem. Sci. 16, 181–186.

Greene, B.A., Allred, D.R., Morishige, D.T., & Staehelin, L.A. (1988). Hierarchical response of light-harvesting chlorophyll–proteins in a light-sensitive chlorophyll *b* deficient mutant of maize. Plant Physiol. 87, 357–364.

Greenland, A.J., Thomas, M.V., & Walden, R.M. (1987). Expression of two nuclear genes encoding chloroplast proteins during early development of cucumber seedlings. Planta 170, 99–110.

Gregory, R.P.F. (1989). Biochemistry of Photosynthesis, 3rd ed. Wiley & Sons, New York.

Grimm, B., Kruse, E., & Kloppstech, K. (1989). Transiently expressed early light-inducible thylakoid proteins share transmembrane domains with light-harvesting chlorophyll-binding proteins. Plant Mol. Biol. 13, 583–593.

Grossman, A.R., Bartlett, S.G., & Chua, N-H. (1980). Energy-dependent uptake of cytoplasmically synthesized polypeptides by chloroplasts. Nature (London) 285, 625–628.

Grossman, A.R., Manodori, A., & Snyder, D. (1990). Light-harvesting proteins of diatoms: Their relationship to the chlorophyll *a/b*-binding proteins of higher plants and their mode of transport into plastids. Mol. Gen. Genet. 224, 91–100.

Harpster, M.H., Mayfield, S.P., & Taylor, W.C. (1984). The effect of pigment-deficient mutants on the accumulation of photosynthetic proteins in maize. Plant Mol. Biol. 3, 59–79.

Haworth, P., Watson, J.L., & Arntzen, C.J. (1983). The detection, isolation, and characterization of a light-harvesting complex which is specifically associated with photosystem I. Biochim. Biophys. Acta 724, 151–158.

Hawthornthwaite, A.M. & Cogdell, R.J. (1991). Bacteriochlorophyll-binding proteins. In: Chlorophylls (Scheer, H., Ed.), pp. 493–528. CRC Press, Boca Raton.

Hayashi, H. & Morita, S. (1980). Near-infrared absorption spectra of light-harvesting bacteriochlorophyll protein complexes from *Chromatium vinosum*. J. Biochem. 88, 1251–1258.

Henrysson, T., Schroder, W.P., Spangfort, M., & Akerlund, H-E. (1989). Isolation and characterization of chlorophyll *a/b*-protein complex CP29 from spinach. Biochim. Biophys. Acta (B), 977, 301–308.

Hiller, R.G., Pilger, D., & Genge, S. (1973). Photosystem II activity and pigment–protein complexes in flashed bean leaves. Plant Sci. Lett. 1, 81–88.

Hiller, R.G., Pilger, T.B.G., & Genge, S. (1978a). Formation of chlorophyll–protein complexes during greening of etiolated barley leaves. Dev. Plant Biol. 2, 215–220.

Hiller, R.G., Pilger, T.B.S., & Genge, S. (1978b). Photosystem II activity and pigment–protein complexes in flashed bean leaves. Plant Sci. Lett. 1, 81–88.

Hirono, Y. & Redei, G.P. (1963). Multiple allelic control of chlorophyll b level in *Arabidopsis thaliana*. Nature (London) 197, 1324–1325.

Hladik, J. & Sofrova, D. (1991). Does the trimeric form of the photosystem I reaction center of cyanobacteria *in vivo* exist? Photosynth. Res. 29, 171–175.

Hoeyer-Hansen, G. & Simpson, D.J. (1977). Changes in the polypeptide composition of internal membranes of barley plastids during greening. Carlsberg Res. Commun. 42, 379–389.

Hoeyer-Hansen, G., Bassi, R., Hoenberg, L.S., & Simpson, D.J. (1988). Immunological characterization of chlorophyll *a/b*-binding proteins of barley thylakoids. Planta 173, 12–21.

Hoober, J.K., Boyd, C.O., & Paavola, L.G. (1991). Origin of thylakoid membranes in *Chlamydomonas reinhardtii y-1* at 38 °C. Plant Physiol. 96, 1321–1328.

Houlne, G. & Schantz, R. (1987). Molecular analysis of the transcripts encoding the light-harvesting chlorophyll *a/b*–protein in Euglena gracilis: unusual size of the mRNA. Curr. Genet. 12, 611–616.

Hsaio, K.C., Erdos, G., & Buetow, D.E. (1988). Cloning and nucleotide sequencing of a soybean encoding a light-harvesting chlorophyll *a/b*-binding protein of photosystem II. Plant Mol. Biol. 10, 473–474.

Ikeuchi, M., Yuasa, M., & Inoue, Y. (1985). Simple and discrete isolation of an oxygen-evolving PS II reaction center complex retaining manganese and the extrinsic 33-kDa protein. FEBS Lett. 185, 316–322.

Ikeuchi, M., Hirano, A., & Inoue, Y. (1991). Correspondence of apoproteins of light-harvesting chlorophyll *a/b* complexes associated with photosystem I to *cab* genes: Evidence for a novel type 4 apoprotein. Plant Cell Physiol. 32, 103–112.

Imbault, P., Wittemer, C., Johanningmeier, U., Jacobs, J.D., & Howell, S.H. (1988). Structure of the *Chlamydomonas reinhardtii cabII-1* gene encoding a chlorophyll-*a/b*-binding protein. Gene 73, 397–407.

Irrgang, K-D., Bechtel, C., Vater, J., & Renger, G. (1990). A new chl *a/b* binding protein in photosystem II from spinach with a M_r of 14 kDa. In: Current Research in Photosynthesis (Baltscheffsky, M., Ed.), Vol. 2, pp. 747–753. Kluwer Academic, Dordrecht.

Jaing, J.T., Welty, B.A., Morishige, D.T., & Thornber, J.P. (1992). Assembly of the photosystem I multiprotein complex and the oligomeric form of the major light-harvesting chlorophyll *a/b*–protein in pea seedlings grown in flashed light followed by continuous illumination. In: Regulation of Chloroplast Biogenesis (Argyroudi-Akoyunoglou, J.H., Ed.), pp. 282–294. Plenum Press, New York.

Jansson, S. & Gustafsson, P. (1990). Type 1 and Type 2 genes for the chlorophyll *a/b*-binding protein in the gymnosperm *Pinus sylvestris* (Scots pine): cDNA cloning and sequence analysis. Plant Mol. Biol. 14, 287–296.

Jansson, S. & Gustafsson, P. (1991). Evolutionary conservation of the chlorophyll *a/b*-binding proteins: cDNAs encoding Type I, II and III LHC-I polypeptides from the gymnosperm Scots pine. Mol. Gen. Genet. 229, 67–76.

Jansson, S., Selstrom, E., & Gustafsson, P. (1990). The rapidly phosphorylated 25-kDa polypeptide of the light-harvesting complex of photosystem II is encoded by a type 2 *cab-II* gene. Biochim. Biophys. Acta 1019, 110–114.

Kalosakas, K., Argyroudi-Akoyunoglou, J.H., & Akoyunoglou, G. (1981). The formation of the pigment-protein complexes in thylakoids of *Phaseolus vulgaris* during chloroplast development. Prog. Photosynth. Res. 5, 569–580.

Karlin-Neumann, G.A., Kohorn, B.D., Thornber, J.P., & Tobin, E.M. (1985). A chlorophyll *a/b*–protein encoded by a gene containing an intron with characteristics of a transposable element. J. Mol. Appl. Genet. 3, 45–61.

Keegstra, K., Olsen, L.J., & Theg, S.M. (1989). Chloroplastic precursors and their transport across the envelope membranes. Ann. Rev. Plant Physiol. Plant Mol. Biol. 40, 471–501.

Kiley, P.J. & Kaplan, S. (1988). Molecular genetics of photosynthetic membrane biosynthesis in *Rhodobacter sphaeroides*. Microbiol. Revs. 52, 50–69.

Klein, R.R., Gamble, P.E., & Mullet, J.E. (1986). Regulation of transcription and translation during chloroplast biogenesis. Curr. Topics in Plant Biochem. Physiol. 5, 74–87.

Klug, G. & Cohen, S.N. (1990). Combined actions of multiple hairpin loop structure and sites of rate-limiting endonucleolytic cleavage determines differential degradation rates of individual segments within polycistronic *puf* mRNA. J. Bacteriol. 171, 3391–3405.

Knoetzel, J. & Simpson, D. (1991). Expression and organization of antenna proteins in the light- and temperature-sensitive barley mutant chlorina-104. Planta 185, 111–123.

Kohorn, B.D. & Auchincloss, A.H. (1991). Integration of a chlorophyll-binding protein into *Escherichia coli* membranes in the absence of chlorophyll. J. Biol. Chem. 266, 12048–12052.

Kohorn, B.D. & Tobin, E.M. (1987). Amino acid charge distribution influences the assembly of apoprotein into light-harvesting complex II. J. Biol. Chem. 262, 12897–12899.

Kohorn, B.D. & Tobin, E.M. (1989). A hydrophobic, carboxy-proximal region of a light-harvesting chlorophyll *a/b*–protein is necessary for stable integration into thylakoid membranes. Plant Cell 1, 159–166.

Kohorn, B.D. & Yakir, D. (1990). Movement of newly imported light-harvesting chlorophyll-binding protein from unstacked to stacked thylakoid membranes is not affected by light treatment or absence of amino-terminal threonines. J. Biol. Chem. 265, 2118–2123.

Kohorn, B.D., Harel, E., Chitnis, P.R., Thornber, J.P., & Tobin, E.M. (1986). Functional and mutational analysis of the light-harvesting chlorophyll *a/b*–protein of thylakoid membranes. J. Cell Biol. 102, 972–981.

Krinsky, N.I. (1966). In: Biochemistry of Chloroplasts (Goodwin, T.W., Ed.), Vol. 1, pp. 423–430.

Kuang, T.Y., Argyroudi-Akoyunoglou, J.H., Nakatani, H.Y., Watson, J., & Arntzen, C.J. (1984). The origin of the long-wavelength fluorescence emission band (77 K) from photosystem I. Arch. Biochem. Biophys. 235, 618–627.

Kuhlbrandt, W. (1984). Three-dimensional structure of the light-harvesting chlorophyll *a/b*-protein complex. Nature (London) 307, 478–484.

Kuhlbrandt, W. (1988). Structure of the light-harvesting chlorophyll *a/b*–protein complex from photosynthetic membranes. In: Photosynthetic Light-Harvesting Systems (Scheer, H. & Schneider, S., Eds.), pp. 211–216. Walter de Gruyter, Berlin.

Kuhlbrandt, W. & Wang, D.N. (1991). Three-dimensional structure of plant light-harvesting complex determined by electron crystallography. Nature (London) 350, 130–134.

Kyle, D.J. & Zalik, S. (1982a). Development of photochemical activity in relation to pigment and membrane protein accumulation in chloroplasts of barley and its virescens mutant. Plant Physiol. 69, 1392–1400.

Kyle, D.J. & Zalik, S. (1982b). Photosystem II activity, plastoquinone A levels, and fluorescence characterization of a virescens mutant of barley. Plant Physiol. 70, 1026–1031.

Lam, E., Ortiz, W., & Malkin, R. (1984a). Chlorophyll *a/b* proteins of photosystem I. FEBS Lett. 168, 10–14.

Lam, E., Ortiz, W., Mayfield, S., & Malkin, R. (1984b). Isolation and characterization of a light-harvesting chlorophyll *a/b*–protein complex associated with photosystem I. Plant Physiol. 74, 650–655.

Lamppa, G.K. (1988). The chlorophyll *a/b*-binding protein inserts into the thylakoid independent of its cognate transit peptide. J. Biol. Chem. 263, 14996–14999.

Lamppa, G.K. & Abad, M.S. (1987). Processing of a wheat light-harvesting chlorophyll *a/b*–protein precursor by a soluble enzyme from higher plant chloroplasts. J. Cell Biol. 105, 2641–2648.

Lamppa, G.K., Morelli, G., & Chua, N-H. (1985). Structure and developmental regulation of a wheat gene encoding the major chlorophyll *a/b*-binding polypeptide. Mol. Cell. Biol. 5, 1370–1378.

LaRoche, J., Bennett, J., & Falkowski, P.G. (1990). Characterization of a cDNA encoding for the 28.5-kDa LHCII apoprotein from the marine unicellular chlorophyte, *Dunaliella tertiolecta*. Gene 95, 165–171.

Larouche, L., Tremblay, C., Simard, C., & Bellemare, G. (1991). Characterization of a cDNA encoding a PS2-associated chlorophyll *a/b*-binding protein (CAB) from a *Chlamydomonas moewusii* fitting into neither type-1 nor type-2. Curr. Genet. 19, 285–288.

Larsson, U.K. & Andersson, B. (1985). Different degrees of phosphorylation and lateral mobility of two polypeptides belonging to the light-harvesting complex of photosystem II. Biochim. Biophys. Acta 809, 396–402.

Lee, J.K., Kiley, P.J., & Kaplan, S. (1989). Posttranscriptional control of puc operon expression of B800-850 light-harvesting complex formation in *Rhodobacter sphaeroides*. J. Bacteriol. 171, 3391–3405.

Leech, R.M. (1984). Chloroplast development in angiosperms: Current knowledge and future prospects. Chloroplast Biogenesis (Baker, N.R. & Barber, J. Eds.), pp. 1–21. Elsevier, Amsterdam.

Leutwiler, L.S., Meyerowitz, E.M., & Tobin, E.M. (1986). Structure and expression of three light-harvesting chlorophyll *a/b*-binding protein genes in *Arabidopsis thaliana*. Nucleic Acids Res. 14, 4051–4064.

Li, J. (1985). Light-harvesting chlorophyll *a/b*–protein: Three-dimensional structure of a reconstituted membrane lattice in negative stain. Proc. Natl. Acad. Sci. USA 82, 386–390.

Long, Z., Wang, S.Y., & Nelson, N. (1989). Cloning and nucleotide sequence analysis of genes coding for the major chlorophyll-binding protein of the moss Physcomitrella patens and the halotolerant alga *Dunaliella salina*. Gene 76, 299–312.

Luan, S. & Bogorad, L. (1989). Nucleotide sequence of two genes encoding the light harvesting chlorophyll *a/b*-binding protein of rice. Nucleic Acids Res. 17, 2357–2358.

Lyon, M.K. & Unwin, P.N.T. (1988). Two-dimensional structure of the light-harvesting chlorophyll *a/b* complex by cryoelectron microscopy. J. Cell Biol. 106, 1515–1523.

Machold, O. & Meister, A. (1979). Resolution of the light-harvesting chlorophyll *a/b*–protein of *Vicia faba* chloroplasts into two different chlorophyll–protein complexes. Biochim. Biophys. Acta 546, 472–480.

Markgraf, T. & Oelmuller, R. (1991). Evidence that carotenoids are required for the accumulation of a functional photosystem II, but not photosystem I in the cotyledons of mustard seedlings. Planta 185, 97–104.

Markwell, J.P., Webber, A.N., & Lake, B. (1985). Mutants of sweetclover (*Mililotus alba*) lacking chlorophyll *b*. Studies on pigment–protein complexes and thylakoid protein phosphorylation. Plant Physiol. 77, 948–951.

Marrs, K.A. & Kaufman, L.S. (1989). Blue light regulation of transcription for nuclear genes in pea. Proc. Natl. Acad. Sci. USA 86, 4492–4495.

Martineau, B. & Taylor, W.C. (1985). Photosynthetic gene expression and cellular differentiation in developing maize leaves. Plant Physiol. 78, 399–404.

Mascia, P.N. & Robertson, D.S. (1978). Studies in chloroplast development in four mutants defective in chlorophyll biosynthesis. Planta 143, 207–211.

Mason, J.G. (1989). Nucleotide sequence of a cDNA encoding the light-harvesting chlorophyll *a/b*-binding protein from spinach. Nucleic Acids Res. 17, 5387.

Mathis, J.N. & Burkey, K.O. (1987). Regulation of light-harvesting chlorophyll–protein biosynthesis in greening seedlings. A species comparison. Plant Physiol. 85, 971–977.

Mathis, J.N. & Burkey, K.O. (1989). Light intensity regulates the accumulation of the major light-harvesting chlorophyll–protein in greening seedlings. Plant Physiol. 90, 560–566.

Matsuoka, M. (1990). Classification and characterization of cDNA that encodes the light-harvesting chlorophyll *a/b*-binding protein of photosystem II from rice. Plant Cell Physiol. 31, 519–526.

Matsuoka, M., Kano-Murakami, Y., & Yamamoto, N. (1987). Nucleotide sequence of cDNA encoding the light-harvesting chlorophyll *a/b*-binding protein from maize. Nucleic Acids Res. 15, 6302.

Mattoo, A.K. & Edelman, M. (1987). Intramembrane translocation and posttranslational palmitoylation of the chloroplast 32-kDa herbicide-binding protein. Proc. Natl. Acad. Sci. USA 84, 1497–1501.

Mayfield, S.P. & Taylor, W.C. (1984). Carotenoid-deficient maize seedlings fail to accumulate light-harvesting chlorophyll *a/b*-binding protein (LHCP) mRNA. Eur. J. Biochem. 144, 79–84.

Melis, A. & Anderson, J.M. (1983). Structural and functional organization of the photosystems in spinach chloroplasts. Antenna size, relative electron-transport capacity, and chlorophyll composition. Biochim. Biophys. Acta 724, 473–484.

Michel, H., Tellenbach, M., & Boschetti, A. (1983). A chlorophyll *b*-less mutant of *Chlamydomonas reinhardtii* lacking in the light-harvesting chlorophyll *a/b*–protein complex but not in its apoproteins. Biochim. Biophys. Acta 725, 417–424.

Michel, H., Hunt, D.F., Shabanowitz, J., & Bennett, J. (1988). Tandem mass spectrometry reveals that three photosystem II proteins of spinach chloroplasts contain *N*-acetyl-*O*-phosphothreonine at their amino-termini. J. Biol. Chem. 263, 1123–1130.

Michel, H., Griffin, P.R., Shabanowitz, J., Hunt, D.F., & Bennett, J. (1991). Tandem mass spectrometry identifies sites of three posttranslational modifications of spinach light-harvesting chlorophyll–protein II. J. Biol. Chem. 266, 17584–17591.

Miles, C.D., Markwell, J.P., & Thornber, J.P. (1979). Effect of nuclear mutation in maize on photosynthetic activity and content of chlorophyll–protein complexes. Plant Physiol. 64, 690–694.

Moore, T.A., Gust, D., & Moore, A.L. (1990). The function of carotenoid pigments in photosynthesis and the possible role in the evolution of higher plants. In: Carotenoids: Chemistry and Biology (Krinsky, N.I., Ed.), pp. 223–228. Plenum Press, New York.

Morishige, D.T., Anandan, S., Jaing, J.T., & Thornber, J.P. (1990). Amino-terminal sequence of the 21-kDa apoprotein of a minor light-harvesting pigment–protein complex of the photosystem II antenna (LHC IId/CP24). FEBS Lett. 264, 239–242.

Morishige, D.T. & Thornber, J.P. (1990). The major light-harvesting chlorophyll *a/b*–protein (LHC IIb): The smallest subunit is a novel cab gene product. In: Current Research in Photosynthesis (Baltscheffsky, M., Ed.), Vol. 2, pp. 261–264. Kluwer Academic, Dordrecht.

Morishige, D.T. & Thornber, J.P. (1991). Correlation of apoproteins with the genes of the major chlorophyll *a/b*-binding protein of photosystem II in *Arabidopsis thaliana*: Confirmation for the presence of a third member of the LHC IIb gene family. FEBS Lett. 293, 183–187.

Morishige, D.T. & Thornber, J.P. (1992). Identification and analysis of a barley cDNA clone encoding the 31-kilodalton LHC IIa (CP29) apoprotein of the light-harvesting antenna complex of photosystem II. Plant Physiol. 98, 238–245.

Morrissey, P.J., Glick, R.E., & Melis, A. (1989). Supramolecular assembly and function of subunits associated with the chlorophyll *a-b* light-harvesting complex II (LHC II) in soybean chloroplasts. Plant Cell Physiol. 30, 335–344.

Mullet, J.E. (1983). The amino acid sequence of the polypeptide segment which regulates membrane adhesion (grana stacking) in chloroplasts. J. Biol. Chem. 258, 9941–9948.

Mullet, J.E. & Arntzen, C.J. (1980). Simulation of grana stacking in a model membrane system. Mediation by a purified light-harvesting pigment-protein complex from chloroplasts. Biochim. Biophys. Acta 589, 100–117.

Mullet, J.E., Burke, J.J., & Arntzen, C.J. (1980a). Chlorophyll proteins of photosystem I. Plant Physiol. 65, 814–822.

Mullet, J.E., Burke, J.J., & Arntzen, C.J. (1980b). A developmental study of photosystem I peripheral chlorophyll proteins. Plant Physiol. 65, 823–827.

Mullet, J.E. & Chua, N-H. (1983). *In vitro* reconstitution of synthesis, uptake, and assembly of cytoplasmically synthesized chloroplast proteins. Methods Enzymol. 97, 502–509.

Murphy, D.J. (1986). The molecular organization of the photosynthetic membranes of higher plants. Biochim. Biophys. Acta 864, 33–94.

Murray, D.L. & Kohorn, B.D. (1991). Chloroplasts of *Arabidopsis thaliana* homozygous for the ch-1 locus lack chlorophyll *b*, lack stable LHCPII and have stacked thylakoids. Plant Mol. Biol. 16, 71–79.

Nabedryk, E., Andrianambinintsoa, S., & Breton, J. (1984). Transmembrane orientation of alpha-helixes in the thylakoid membrane and in the light-harvesting complex. A polarized infrared spectroscopy study. Biochim. Biophys. Acta 765, 380–387.

Nagy, F., Fluhr, R., Kuhlemeier, C., Kay, S., Boutry, M., Green, P., Poulsen, C., & Chua, N.H. (1986a). *cis*-Acting elements for selective expression of two photosynthetic genes in transgenic plants. Philos. Trans. R. Soc. London, B, 314–493.

Nagy, F., Kay, S.A., Boutry, M., Hsu, M.Y., & Chua, N.H. (1986b). Phytochrome-controlled expression of a wheat *cab* gene in transgenic tobacco seedlings. EMBO J. 5, 1119–1124.

Nechushtai, R., Peterson, C.C., Peter, G.F., & Thornber, J.P. (1987). Purification and characterization of a light-harvesting chlorophyll *a/b*–protein of photosystem I of *Lemna gibba*. Eur. J. Biochem. 164, 345–350.

Nechushtai, R., Thornber, J.P., Patterson, L.K., Fessenden, R.W., & Levanon, H. (1988). Photosensitization of triplet carotenoid in photosynthetic light-harvesting complex of photosystem II. J. Phys. Chem. 92, 1165–1168.

Nugent, J.M. & Palmer, J.D. (1991). RNA-mediated transfer of the gene *cox II* from the mitochondrion to the nucleus during flowering plant evolution. Cell 66, 473–481.

Oelmuller, R. (1989). Photooxidative destruction of chloroplasts and its effect on nuclear gene expression and extraplastidic enzyme levels. Photochem. Photobiol. 49, 229–239.

Oelmuller, R., Kendrick, R.E., & Briggs, W.R. (1989). Blue-light mediated accumulation of nuclear-encoded transcripts coding for proteins of the thylakoid membrane is absent in the phytochrome-deficient aurea mutant of tomato. Plant Mol. Biol. 13, 223–232.

Ogawa, T., Obata, F., & Shibata, K. (1966). Two pigment–proteins in spinach chloroplasts. Biochim. Biophys. Acta 112, 223–234.

Ortiz, W., Lam, E., Chollar, S., Munt, D., & Malkin, R. (1985). Topography of the protein complexes of the chloroplast thylakoid membrane. Studies of photosystem I using a chemical probe and proteolytic digestion. Plant Physiol. 77, 389–397.

Papiz, M.Z., Hawthornwaite, A.M., Cogdell, R.J., Woolley, K.J., Wightman, P.A., Ferguson, L.A., & Lindsay, J.G. (1989). Crystallization and characterization of two crystal forms of the B800-850 light-harvesting complex from *Rhodopseudomonas acidophila* strain 10050. J. Mol. Biol. 209, 833–835.

Parkes-Loach, P.S., Sprinkle, J.R., & Loach, P.A. (1988). Reconstitution of the B873 light-harvesting complex of *Rhodospirillum rubrum* from the separately isolated α-polypeptide and β-polypeptide and bacteriochlorophyll-*a*. Biochem. 27, 2718–2727.

Paulsen, H., Hobe, S., & Eisen, C. (1991). Reconstitution of LHCP-pigment complexes with mutant LHCP and chlorophyll analogs. In: Regulation of Chloroplast Biogenesis (Argyroudi-Akoyunoglou, J., Ed.), pp. 343–348. Plenum Press, New York.

Paulsen, H., Rumler, U., & Rudiger, W. (1990). Reconstitution of pigment-containing complexes from light-harvesting chlorophyll *a/b*-binding protein overexpressed in *Escherichia coli*. Planta 181, 204–211.

Payan, L.A. & Cline, K. (1991). A stromal protein factor maintains the solubility and insertion competence of an imported thylakoid membrane protein. J. Cell Biol. 112, 603–613.

Peter, G.F. & Thornber, J.P. (1987). The antenna components of photosystem II with emphasis on the major pigment–protein LHC IIb. Curr. Res. Photosyn. (Biggins, J., Ed.), 2, 101–104.

Peter, G.F. & Thornber, J.P. (1988). The antenna components of photosystem II with emphasis on the major pigment–protein LHC IIb. In: Photosynthetic Light-Harvesting Systems (Scheer, H. & Schneider, S., Eds.), pp. 175–186. Walter de Gruyter, Berlin.

Peter, G.F. & Thornber, J.P. (1990). Electrophoretic procedures for fractionation of photosystem I and II pigment–proteins of higher plants and for determination of their subunit composition. Methods Plant Biochem. 5, 194–212.

Peter, G.F. & Thornber, J.P. (1991a). Biochemical evidence that the higher plant photosystem II core complex is organized as a dimer. Plant Cell Physiol. 32, 1237–1250.

Peter, G.F. & Thornber, J.P. (1991b). Biochemical composition and organization of higher plant photosystem II light-harvesting pigment–proteins. J. Biol. Chem. 266, 16745–16754.

Peter, G.F., Machold, O., & Thornber, J.P. (1988). Identification and isolation of photosystem I and photosystem II pigment–proteins from higher plants. In: Plant Membranes: Structure, Assembly, and Function (Harwood, J.L. & Walton, T.J., Eds.), pp. 17–31. The Biochemical Society London.

Pichersky, E., Bernatzky, R., Tanksley, S.D., Breidenbach, R.B., Kausch, A.P., & Cashmore, A.R. (1985). Molecular characterization and genetic mapping of two clusters of genes encoding chlorophyll *a/b*-binding proteins in *Lycopersicon esulentum* (tomato). Gene 40, 247–258.

Pichersky, E., Hoffman, N.E., Malik, V.S., Bernatzky, R., Tanksley, S.D., Szabo, L., & Cashmore, A.R. (1987a). The tomato *cab-4* and *cab-5* genes encode a second type of CAB polypeptides localized in photosystem II. Plant Mol. Biol. 9, 109–120.

Pichersky, E., Hoffman, N.E., Bernatzky, R., Piechulla, B., Tanksley, S.D., & Cashmore, A.R. (1987b). Molecular characterization and genetic mapping of DNA sequences encoding the type I chlorophyll *a/b*-binding polypeptide of photosystem I in *Lycopersicon esculentum* (tomato). Plant Mol. Biol. 9, 205–216.

Pichersky, E., Tanksley, S.D., Piechulla, B., Stayton, M.M., & Dunsmuir, P. (1988). Nucleotide sequence and chromosomal location of *cab-7*, the tomato gene encoding the type I chlorophyll *a/b*-binding polypeptide of photosystem I. Plant Mol. Biol. 11, 69–71.

Pichersky, E., Brock, T.G., Nguyen, D., Hoffman, H.E., Piechulla, B., Tanksley, S.D., & Green, B.R. (1989). A new member of the *cab* gene family: structure, expression and chromosomal location of *cab-8*, the tomato gene encoding the Type III chlorophyll *a/b*-binding polypeptide of photosystem I. Plant Mol. Biol. 12, 257–270.

Pichersky, E. & Green, B.R. (1990). The extended family of chlorophyll *a/b*-binding proteins of PS I and PS II. In: Current Research in Photosynthesis (Baltscheffsky, M., Ed.), Vol. 2, pp. 553–557. Kluwer Academic, Dordrecht.

Pichersky, E., Solis, D., & Solis, P. (1990). Defective chlorophyll *a/b*-binding protein genes in the genome of homosporous fern. Proc. Natl Acad. Sci. USA 87, 195–199.

Pichersky, E., Subramaniam, R., White, M.J., Reid, J., Aebersold, R., & Green, B.R. (1991). Chlorophyll *a/b*-binding (CAB) polypeptides of CP29, the internal chlorophyll *a/b* complex of PS II: Characterization of the tomato gene encoding the 26-kDa (type 1) polypeptide and evidence for a second CP29 polypeptide. Mol. Gen. Genet. 227, 277–284.

Plumley, F.G. & Schmidt, G.W. (1987). Reconstitution of chlorophyll *a/b* light- harvesting complexes: Xanthophyll-dependent assembly and energy transfer. Proc. Natl. Acad. Sci. USA 84, 146–150.

Preiss, S., Peter, G.F., Morishige, D.T., & Thornber, J.P. (1993). The multiple pigment–proteins of the photosystem I antenna. Photochem. Photobiol. 57, 152–157.

Reed, J.E., Cline, K., Stephens, L.C., Bacot, K.O., & Viitanen, P.V. (1990). Early events in the import/assembly pathway of an integral thylakoid protein. Eur. J. Biochem. 194, 33–42.

Robert, B. & Lutz, M. (1985). Structures of antenna complexes of several *Rhodospirillales* from their resonance Raman-spectra. Biochim. Biophys. Acta 807, 10–23.

Robinson, C. & Ellis, R.J. (1984). Transport of proteins into chloroplasts: Partial purification of a chloroplast protease involved in the processing of imported precursor polypeptides. Eur. J. Biochem. 142, 337–342.

Rogner, M., Muhlenhoff, U., Boekema, E.J. & Witt, H.T. (1990). Mono- di-, and trimeric-PS I reaction-center complexes isolated from the thermophilic cyanobacterium *Synechococcus* sp. Size shape and activity. Biochim. Biophys. Acta 1015, 415–424.

Ryrie, I.J. & Young, S. (1984). The polypeptides of photosystem I in developing thylakoids in intermittent light-grown barley. Adv. Photosynth. Res. 4, 677–680.

Sarvari, E. & Gigler, G. (1984). Partial characterization of a minor chlorophyll–protein found in primary thylakoids of intermittently illuminated maize. Photosynth. Res. 5, 159–167.

Sarvari, E., Nyitrai, P., & Keresztes, A. (1989). Relative accumulation of LHCP II in mesophyll plastids of intermittently illuminated maize seedlings under lincomycin treatment. Biochem. Physiol. Pflanzen 184, 37–47.

Schmidt, G.W. & Mishkind, M.L. (1986). The transport of proteins into chloroplasts. Ann. Rev. Biochem. 55, 879–912.

Schmidt, G.W., Bartlett, S.G., Grossman, A.R., Cashmore, A.R., & Chua, N-H. (1981). Biosynthetic pathways of two polypeptide subunits of the light-harvesting chlorophyll *a/b*-protein complex. J. Cell Biol. 91, 468–478.

Schwartz, E., Shen, D., Aebersold, R., McGrath, J.M., Pichersky, E., & Green, B.R. (1991a). Nucleotide sequence and chromosomal location of *cab 11* and *cab 12*, the genes for the fourth polypeptide of the photosystem I light-harvesting antenna (LHCI). FEBS Lett. 280, 229–234.

Schwartz, E., Stasys, R., Aebersold, R., McGrath, J.M., Green, B.R., & Pichersky, E. (1991b). Sequence of a tomato gene encoding a third type of LHC II chlorophyll *a/b*-binding polypeptide. Plant Mol. Biol. 17, 923–925.

Schwarz, H.P. & Kloppstech, K. (1982). Effects of nuclear gene mutations on the structure and function of plastids in pea. The light-harvesting chlorophyll *a/b*–protein. Planta 155, 116–123.

Sganga, M.W. & Bauer, C.E. (1992). Regulatory factors controlling photosynthetic reaction center and light-harvesting gene expression in *Rhodobacter capsulatus*. Cell 68, 945–954.

Shimada, Y., Tanaka, A., Tanaka, Y., Takabe, A., Takabe, Y., & Tsuji, H. (1990). Formation of chlorophyll–protein complexes during greening. Distribution of newly synthesized chlorophyll among apoproteins. Plant Cell Physiol. 31, 639–647.

Shiozawa, J.A., Welte, W., Hodapp, N., & Drews, G. (1982). Studies on the size and composition of the isolated light-harvesting B800-850 pigment–protein complex of *Rhodopseudomonas capsulata*. Arch. Biochem. Biophys. 213, 473–485.

Sigrist, M. & Staehelin, L.A. (1992). Identification of type 1 and type 2 light-harvesting chlorophyll *a/b*-binding proteins using monospecific antibodies. Biochim. Biophys. Acta 1098, 191–200.

Silverthorne, J. & Tobin, E.M. (1984). Demonstration of transcriptional regulation of specific genes by phytochrome action. Proc. Natl. Acad. Sci. USA 81, 1112–1116.

Simpson, J., Timko, M.P., Cashmore, A.R., Schell, J., Van Montagu, M., & Herrea-Estrella, L. (1985). Light-inducible and tissue-specific expression of a chimeric gene under the control of the 5′-flanking sequence of a pea chlorophyll *a/b*-binding protein gene. EMBO J. 4, 2723–2729.

Simpson, J., Schell, J., Van Montagu, M., & Herrea-Estrella, L. (1986b). Light-inducible and tissue-specific pea lhcp gene expression involves and upstream element combining enhancer- and silencer-like properties. Nature (London) 323, 551–554.

Simpson, J., Van Montagu, M., & Herrera-Estrella, L. (1986a). Photosynthesis-associated gene families: Differences in response to tissue-specific and environmental factors. Science 233, 34–38.

Slovin, J.P. & Tobin, E.M. (1982). Synthesis and turnover of the light-harvesting chlorophyll *a/b*-protein in *Lemna gibba* grown with intermittent red light: Possible translational control. Planta 154, 465–472.

Smeekens, S., Van Oosten, J., De Groot, M., & Weisbeek, P. (1986). Silene cDNA clones for a divergent chlorophyll *a/b*-binding protein and a small subunit of ribulose bisphosphate carboxylase. Plant Mol. Biol. 7, 433–440.

Sommerville, C.R. (1986). Analysis of photosynthesis with mutants of higher plants and algae. Annu. Rev. Plant Physiol. Plant Mol. Biol. 37, 467–507.

Song, P-S., Koka, P., Prezelin, B., & Haxo, F.T. (1976). Molecular topology of the photosynthetic light-harvesting pigment complex, peridinin-chlorophyll *a*–protein from marine dinoflagellates. Biochemistry. 15, 4422–4427.

Spangfort, M., Larsson, U.K., Ljungberg, U., Ryberg, M., & Andersson, B. (1990). The 20-kDa apopolypeptide of the chlorophyll *a/b*–protein complex CP24: Characterization and complete primary amino acid sequence. Current Research in Photosynthesis (Baltscheffsky, M., Ed.), Vol. 2, pp. 253–256. Kluwer Academic, Dordrecht.

Staehelin, L.A. & Arntzen, C.J. (1983). Regulation of chloroplast membrane function: Protein phosphorylation changes the spatial organization of membrane components. J. Cell Biol. 97, 1327–1337.

Stark, W., Kuhlbrandt, W., Wildhaber, I., & Muhlethaler, K. (1984). The structure of the photoreceptor unit of *Rhodopseudomonas viridis*. EMBO J. 3, 777–786.

Stayton, M.M., Black, M., Bedbrook, J., & Dunsmuir, P. (1986). A novel chlorophyll *a/b*-binding (*cab*) protein gene from petunia which encodes the lower molecular weight CAB precursor protein. Nucleic Acids Res. 14, 9781–9796.

Stayton, M.M., Brosio, P., & Dunsmuir, P. (1987). Characterization of a full length petunia cDNA encoding a polypeptide of the light-harvesting complex associated with photosystem I. Plant Mol. Biol. 10, 127–137.

Steinback, K.E., Burke, J.J., & Arntzen, C.J. (1979). Evidence for the role of surface-exposed segments of the light-harvesting complex in cation-mediated control of chloroplast structure and function. Arch. Biochem. Biophys. 195, 546–557.

Stiekema, W.J., Wimpee, C.F., Silverthorne, J., & Tobin, E.M. (1983). Phytochrome control of the expression of two nuclear genes encoding chloroplast proteins in *Lemna gibba L.* Plant Physiol 72, 717–724.

Steppuhn, J., Hermans, J., Nechushtai, R., Ljungberg, U., Thummler, F., Lottspeclh, F., & Herrmann, R.G. (1988). Nucleotide sequence of cDNA clones encoding the entire precursor polypeptide for subunits IV and V of the photosystem I reaction center from spinach. FEBS Lett. 237, 108–112.

Sullivan, T.D., Christensen, A.H., & Quail, P.H. (1989). Isolation and characterization of a maize chlorophyll *a/b*-binding protein gene that produces high levels of mRNA in the dark. Mol. Gen. Genet. 215, 441–446.

Tadros, M.H. & Waterkamp, K. (1989). Multiple copies of the coding regions for the light-harvesting B800-850 alpha- and beta-apoproteins are present in *Rps. palustris* genome. EMBO J. 5, 1303–1308.

Tanaka, A. & Tsuji, H. (1985). Appearance of chlorophyll–protein complexes in greening barley seedlings. Plant Cell Physiol. 26, 893–902.

Taylor, W.C., Burgess, D.G., & Mayfield, S.P. (1986). The use of carotenoid deficiencies to study nuclear–chloroplast regulatory interactions. Curr. Topics in Plant Biochem. Physiol. 5, 117–127.

Terao, T. & Katoh, S. (1989). Synthesis and breakdown of the apoproteins of light-harvesting chlorophyll *a/b* proteins in chlorophyll *b*-deficient mutants of rice. Plant Cell Physiol. 30, 571–580.

Thompson, W.F. & White, M.J. (1991). Physiological and molecular studies of light-regulated nuclear genes in higher plants. Annu. Rev. Plant Physiol. Plant Mol. Biol. 42, 423–466.

Thornber, J.P. (1970). Photochemical reactions of purple bacteria as revealed by studies of three spectrally different caroteno-bacteriochlorophyll–protein complexes isolated form *Chromatium*, strain D. Biochemistry 9, 2688–2698.

Thornber, J.P. (1986). Biochemical characterization and structure of pigment-proteins of photosynthetic organisms. Encl. Plant Physiol. New Series 19, 98–142.

Thornber, J.P. & Highkin, H.R. (1974). Composition of the photosynthetic apparatus of normal barley leaves and a mutant lacking chlorophyll *b*. Eur. J. Biochem. 41, 109–116.

Thornber, J.P., Cogdell, R.J., Pierson, B.K., & Seftor, R.E.B. (1983). Pigment–protein complexes of purple photosynthetic bacteria: An overview. J. Cell Biochem. 23, 159–169.

Thornber, J.P., Smith, C.A., & Bailey, J.L. (1966). Partial characterization of two chlorophyll–protein complexes isolated from spinach-beet chloroplasts. Biochem. J. 100, 14–15.

Thornber, J.P., Trosper, T.L., & Strouse, C.E. (1978). Bacteriochlorophyll *in vivo*: Relationship of spectral forms to specific membrane components. In: Photosynthetic Bacteria (Clayton, R.K. & Sistrom, W.R., Eds.), pp. 133–172. Plenum Press, New York.

Thornber, J.P., Morishige, D.T., Anandan, S., & Peter, G.F. (1991). Chlorophyll-carotenoid proteins of higher plant thylakoids. In: Chlorophylls (Scheer, H., Ed.), pp. 550–585. CRC Press, Boca Raton.

Thornber, J.P., Peter, G.F., Morishige, D.T., Gomez, S., Anandan, S., Welty, B.A., Lee, A., Takeuchi, T., & Preiss, S. (1992). Photosystem I and II pigment–protein complexes. Biochem. Soc Trans., in press.

Tichy, H.V., Oberle, B., Stiehle, H., Schiltz, E., & Drews, G. (1989). Genes downstream from pucB and pucA are essential for formation of the B800-850 complex of *Rhodobacter capsulata*. J. Bacteriol. 171, 4914–4921.

Tobin, E.M. (1978). Light regulation of specific mRNA species in *Lemna gibba* L. G-3. Proc. Natl. Acad. Sci. USA 75, 4749–4753.

Tobin, E.M. (1981). Phytochrome-mediated regulation of messenger RNAs for the small subunit of ribulose 1,5-bisphosphate carboxylase and the light-harvesting chlorophyll *a/b*–protein in *Lemna gibba*. Plant Molec. Biol. 1, 35–51.

Tobin, E.M. & Silverthorne, J. (1985). Light regulation of gene-expression in higher plants. Ann. Rev. Plant Physiol. 36, 569–593.

Tronrud, D.E., Schmid, M.F., & Matthews, B.W. (1986). Structure and X-ray amino acid sequence of a bacteriochlorophyll *a*–protein from *P. aestuarii* refined at 1.9 Å resolution. J. Mol. Biol. 188, 443–454.

Vainstein, A., Peterson, C.C., & Thornber, J.P. (1989). Light-harvesting pigment-proteins of photosystem I in maize: subunit composition and biogenesis. J. Biol. Chem. 264, 4058–4063.

Van Grondelle, R. & Sundstrom, S. (1988). Excitation energy transfer in photosynthesis. In: Photosynthetic Light-Harvesting Systems (Scheer, H. & Schneider, S., Eds.), pp. 403–438. Walter de Gruyter, Berlin.

Van Grondelle, R., Hunter, C.N., Bakker, J.G.C., & Kramer, H.J.M. (1983). Size and structure of antenna complexes of photosynthetic bacteria as studied by singlet–singlet quenching of the bacteriochlorophyll fluorescence yield. Biochim. Biophys. Acta 723, 30–36.

Viitanen, P.V., Doran, E.R., & Dunsmuir, P. (1988). What is the role of the transit peptide in thylakoid integration of the light-harvesting chlorophyll *a/b*–protein? J. Biol. Chem. 263, 15000–15007.

Viret, J-F., Schanz, M-L., & Schanz, R. (1990). Nucleotide sequence of a maize cDNA coding for a light-harvesting chlorophyll *a/b*-binding protein of photosystem II. Nucleic Acids Res. 18, 7179–.

Viro, M. & Kloppstech, K. (1980). Differential expression of the genes for ribulose-1,5-bisphosphate carboxylase and light-harvesting chlorophyll *a/b*–protein in the developing barley leaf. Planta 150, 41–45.

Viro, M. & Kloppstech, K. (1982). Expression of genes for plastid membrane proteins in barley under intermittent light conditions. Planta 154, 18–23.

Waegemann, K., Paulsen, H., & Soll, J. (1990). Translocation of proteins into isolated chloroplasts requires cytosolic factors to obtain import competence. FEBS Lett. 261, 89–92.

Walling, L.L., Chang, Y.C., Demmin, D.S., & Holzer, F.M. (1988). Isolation, characterization and evolutionary relatedness of three members of the soybean multigene family encoding the chlorophyll *a/b*-binding proteins. Nucleic Acids Res. 16, 10477–10492.

Wasmann, C.C., Reiss, B., Bartlett, S.G., & Bohnert, H.J. (1986). The importance of the transit peptide and the transported protein-for-protein import into chloroplasts. Mol. G. Genet. 205, 446–453.

Webber, A.N. & Gray, J.C. (1989). Detection of calcium binding by photosystem II polypeptides immobilized onto nitrocellulose membrane. FEBS Lett. 249, 79–82.

Webster, G.D., Cogdell, R.J., & Lindsay, J.G. (1980). Identification of the carotenoid present in the B800-850 antenna complex from *Rhodopseudomonas capsulata* as that which responds electrochromically to transmembrane electric fields. Biochim. Biophys. Acta 591(2), 321–330.

Wechsler, T., Brunisholz, R., Suter, F., Fuller, R.C., & Zuber, H. (1985). The complete amino acid sequence of a bacteriochlorophyll a binding polypeptide isolated from the cytoplasmic membrane of the green photosynthetic bacterium *Chloroflexus aurantiacus*. FEBS Lett. 191, 34–38.

Wehmeyer, B., Cashmore, A.R., & Schafer, E. (1990). Photocontrol of the expression of the genes encoding chlorophyll *a/b*-binding proteins and small subunit of ribulose bisphosphate carboxylase in etiolated seedings of *Lycopersicon esculentun* L and *Nicotiana tabacum* l. Plant Physiol. 93, 990–997.

White, M.J. & Green, B.R. (1987a). Antibodies to the photosystem I chlorophyll a + b antenna cross-react with polypeptides of CP29 and LHCII. Eur. J. Biochem. 163, 545–551.

White, M.J. & Green, B.R. (1987b). Polypeptides belonging to each of the three major chlorophyll a + b–protein complexes are present in a chlorophyll b-less barley mutant. Eur. J. Biochem. 165, 531–535.

White, M.J. & Green, B.R. (1988). Intermittent-light chloroplasts are not developmentally equivalent to chlorina F2 chloroplasts in barley. Photosynth. Res. 15, 195–203.

Williams, R.S. & Ellis, R.J. (1986). Immunological studies on the light-harvesting polypeptides of photosystems I and II. FEBS Lett. 203, 295–300.

Witt, I., Witt, H.T., Di Fiore, D., Roegner, M., Hinrichs, W., Saenger, W., Granzin, J., Betzel, C., & Dauter, Z. (1988). X-ray characterization of single crystals of the reaction center I of water-splitting photosynthesis. Ber. Bunsen-Ges. Phys. Chem. 92, 1503–1506.

Yalovsky, S., Schuster, G., & Nechushtai, R. (1990). The apoprotein precursor of the major light-harvesting complex of photosystem II (LHC2b) is inserted primarily into stromal lamellae and subsequently migrates to the grana. Plant Mol. Biol. 14, 753–764.

Yamamato, N., Matsuoka, M., Kano-Murakami, Y., & Tanaka, Y. (1988). Nucleotide sequence of a full length cDNA clone of light-harvesting chlorophyll *a/b*- binding protein gene from green dark-brown pine (*Pinus thunbergii*) seedling. Nucleic Acid Res. 16, 11830.

Yildiz, F.H., Gest, H., & Bauer, C.E. (1991). Attenuated effect of oxygen on photopigment synthesis in *Rhodospirillum centenum*. J. Bacteriol. 173, 5502–5506.

Youvan, D.C., Bylina, E.J., Alberti, M., Begushi, H., & Hearst, J.E. (1984). Nucleotide and deduced polypeptide sequences of the photosynthetic reaction centre, B870 antenna and flanking polypeptides from *Rb. capsulatus*. Cell 37, 949–957.

Youvan, D.C. & Ismail, S. (1985). Light-harvesting II (B800-850-complex) structural genes from *Rb. capsulatus*. Proc. Natl. Acad. Sci. USA 82, 58–62.

Zhang, H., Hanley, S., & Goodman, H.M. (1991). Isolation, characterization, and chromosomal location of a new *cab* gene from *Arabidopsis thaliana*. Plant Physiol. 96, 1387–1388.

Zhu, Y.S. & Hearst, J.E. (1986). Regulation of expression of genes for light-harvesting antenna proteins LH-1 and LH-II; reaction center polypeptides RC-L, RC-M, and RC-H; and enzymes of bacteriochlorophyll and carotenoid biosynthesis in *Rhodobacter capsulata* by light and oxygen. Proc. Natl. Acad. Sci. USA 83, 7613–7617.

Zipfel, W. & Owens, T.G. (1991). Calculation of absolute photosystem I absorption cross sections from P700 photooxidation kinetics. Photosynth. Res. 29, 23–35.

Zuber, H. (1985). Structure and function of light-harvesting complexes and their polypeptides. Photochem. Photobiol. 42, 821–844.

Zuber, H. & Brunisholz, R.A. (1991). Structure and function of antenna polypeptides and chlorophyll–protein complexes: Principles and variability. In: Chlorophylls (Scheer, H., Ed.), pp. 627–703. CRC Press, Boca Raton.

Zuber, H., Brunisholz, R., & Sidler, W. (1987) Structure and function of light-harvesting pigment–protein complexes. In: Photosynthesis (Amesz, J., Ed.), pp. 233–271. Elsevier, Amsterdam.

ADAPTIVE VARIATIONS IN PHYCOBILISOME STRUCTURE

Alexander N. Glazer

Advances in Molecular and Cell Biology
Volume 10, pages 119–149.

ISBN: 1-55938-710-6

ABSTRACT

Phycobilisomes are macromolecular light-harvesting antenna complexes attached to the cytoplasmic face of the thylakoid membranes in cyanobacterial cells and red algal chloroplasts. These complexes, of about 6000 to 8000 kDa, are made up entirely of proteins and contain 500 to 700 linear tetrapyrroles (bilins) responsible for the absorption of visible light. Phycobiliproteins are the major components of the phycobilisome and make up about 85% of the mass of the complex. Within the phycobiliproteins, the bilins are covalently attached to cysteinyl residues through thioether linkages. Four different isomeric bilins—phycocyanobilin, phycobiliviolin, phycoerythrobilin, and phycourobilin—have been found in cyanobacterial and red algal phycobiliproteins. Bilins are formed from biliverdin IXα by two sequential ferredoxin-mediated two-electron reduction steps followed by isomerization reactions. Attachment of the bilin to the apo-α subunit of C-phycocyanin has been shown to involve the participation of the products of two genes (*cpcE* and *cpcF*) that are a part of the phycocyanin operon. Phycobilisomes have two morphologically distinct domains: a core with allophycocyanin as the main component, and rods with phycocyanin and (sometimes) phycoerythrin or phycoerythrocyanin as the major components. The assembly of the phycobilisomes is governed by the interaction of the phycobiliproteins with a family of polypeptides that determine site-specific assembly, the linker polypeptides. Special structural features of the phycobilisome minimize random walk of excitation energy among the bilins and assure directional energy transfer towards the terminal energy acceptors in the phycobilisome. The phycobilisomes of open ocean marine cyanobacteria and red algae are rich in phycobiliproteins which absorb strongly in the green region of the spectrum. Adaptive adjustments to light intensity, light quality, and nutrient limitation are achieved in part through changes in size or composition of the rod substructures of the phycobilisome.

I. INTRODUCTION

The cyanobacteria possess all of the hallmarks of cellular organization typical of prokaryotes (Stanier and Cohen-Bazire, 1977). However, these organisms perform oxygen-evolving photosynthesis characteristic of plants and employ chlorophyll *a* rather than a bacteriochlorophyll as a photosynthetic pigment. Because of the prominence of these features, the cyanobacteria were claimed by the botanists as blue-green algae and classified (with some reservations) among the algae until the late 1970s (Stanier et al., 1978).

To the student of plant photosynthesis the cyanobacteria present a fascinating mixture of the familiar and the novel. These photoautotrophs employ photosystem II and I reaction center complexes and photosynthetic electron carriers homologous to those of higher plants, but also possess an elaborate, distinctive light-harvesting complex, the phycobilisome, present among the eukaryotes only in the red algae. Much molecular evidence points to the fact that a cyanobacterium-like ancestor gave rise to the chloroplasts of higher plants (Bryant, 1987, 1992). At some point in evolution, the descendants of the ancestral organism became committed to different antenna complexes. Along one branch, chlorophyll *b* became an important antenna chromophore; the bilin chromophores of the phycobiliproteins lay along another branch. The alternative antenna complexes lead to chloroplasts with differences in the arrangement of the thylakoids. Higher order organization of the photosynthetic apparatus in chlorophyll *b*-containing organisms involves reversible stacking of the thylakoids dependent on the integral light-harvesting chlorophyll *a/b* complexes within the photosynthetic membranes. In cyanobacteria, the attachment of the phycobilisomes to the cytoplasmic face of the thylakoid membranes precludes stacking of the photosynthetic lamellae (Gantt, 1980).

Phycobilisomes absorb light over a wide range of wavelengths and transfer the excitation energy by radiationless processes to the reaction centers. A great deal is known about the phycobilisomes of cyanobacteria (for recent reviews, see Zuber, 1987; Glazer, 1989; Grossman, 1990; Bryant, 1991; Tandeau de Marsac, 1991), while much less about those of red algae (Mörschel and Rhiel, 1987; Gantt, 1990). Phycobilisomes consist exclusively of proteins; phycobiliproteins make up about 85% of the phycobilisome. Under nutrient replete conditions and low light intensity, phycobiliproteins may represent as much as 40% of the protein of a cyanobacterial cell.

The cyanobacteria are a very successful group of organisms. They are prominent members of the picoplankton community of the oceans and thrive in soil, rock, and freshwater environments. The different ways in which the cyanobacteria modify the properties of the phycobilisome, and in which they utilize it as a store of nutrients, provide a glimpse of the kinds of molecular adaptations that have ensured the survival of this group of organisms in a wide range of environments. This review focuses narrowly on the structure, function, and biosynthesis of the phycobilisome, with special attention to molecular adaptations in this light-harvesting complex seen in response to changes in light intensity, light quality, and nutrient limitation.

II. PHYCOBILIPROTEINS: BUILDING BLOCKS OF PHYCOBILISOMES

Phycobiliproteins are a family of homologous proteins which carry covalently attached bilin (linear tetrapyrrole) prosthetic groups. These proteins are multimers of an $\alpha\beta$ heterodimer, where α and β are polypeptides of 160 to 180 residues in length (Glazer, 1981).

The phycobiliproteins can be conveniently grouped into three classes on the basis of amino acid sequence and number of bilins attachment sites (Figure 1). Allophycocyanins carry a single bilin on each subunit. Phycocyanins carry three bilins, one on the α-subunit and two on the β-subunit, but are very diverse with respect to bilin composition. Four different bilins occur in various combinations on one or another

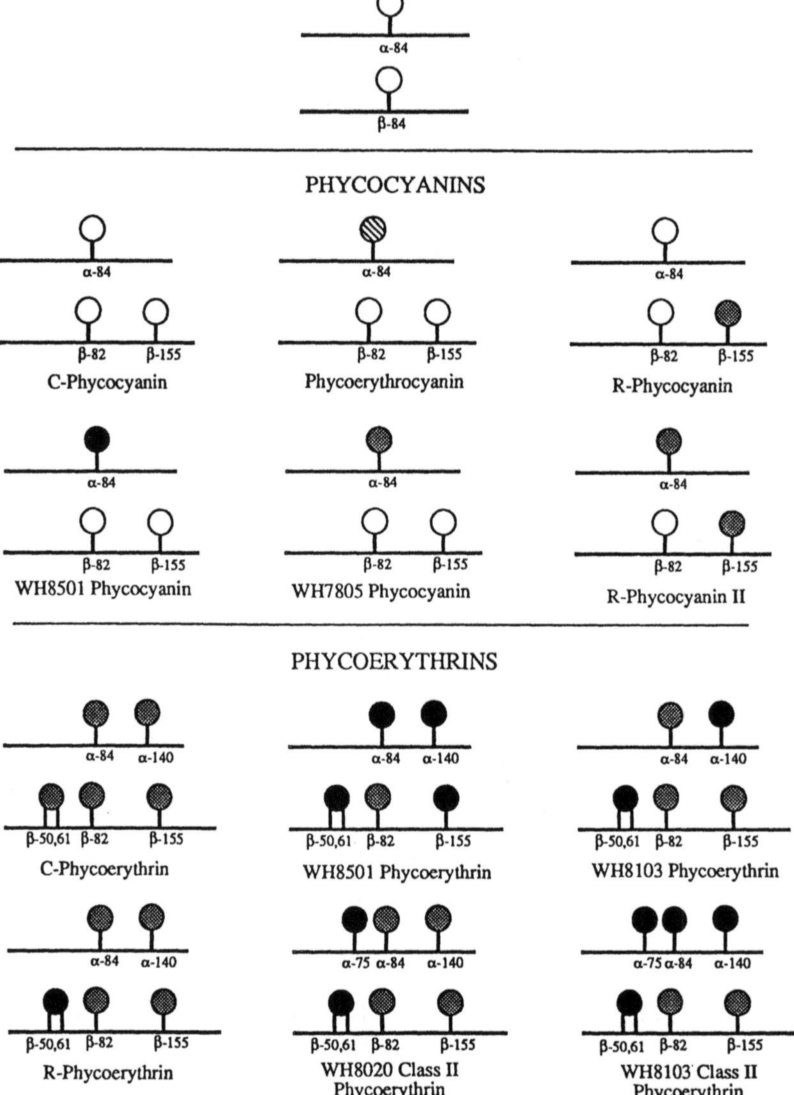

Figure 1. Classification of phycobiliproteins. Bilin location and composition in allophycocyanin, phycocyanins, and phycoerythrins is illustrated. Designations such as α-84, β-82, etc., indicate the subunit (α or β) and the position of the cysteine residue(s) in the amino acid sequence at which a bilin is attached through a thioether linkage. The following symbols are used for bilin prosthetic groups (for structures, see Figure 2): *open circle*, phycocyanobilin (PCB); *cross-hatched circle*, phycoerythrobilin (PEB); *filled-in circle*, phycourobilin (PUB); and *striped circle*, phycobiliviolin (cryptoviolin; PXB). (From Swanson, 1991).

of the members of the phycocyanin family. Phycoerythrins fall into two classes. The first of these includes proteins previously designated as C-, B- or R-phycoerythrins, with two bilins on the α-subunit and three on the β-subunit. Class II phycoerythrins, isolated so far only from marine cyanobacteria *Synechococcus* sp., carry three bilins on each α- and β-subunit.

The structures of three C-phycocyanins (Schirmer et al., 1986, 1987; Duerring et al., 1991), a phycoerythrocyanin (Duerring et al., 1990), and a B-phycoerythrin (Ficner et al., 1992; Ficner and Huber, 1993) have been determined by X-ray crystallography. The three-dimensional structures of all of these phycobiliproteins are very similar.

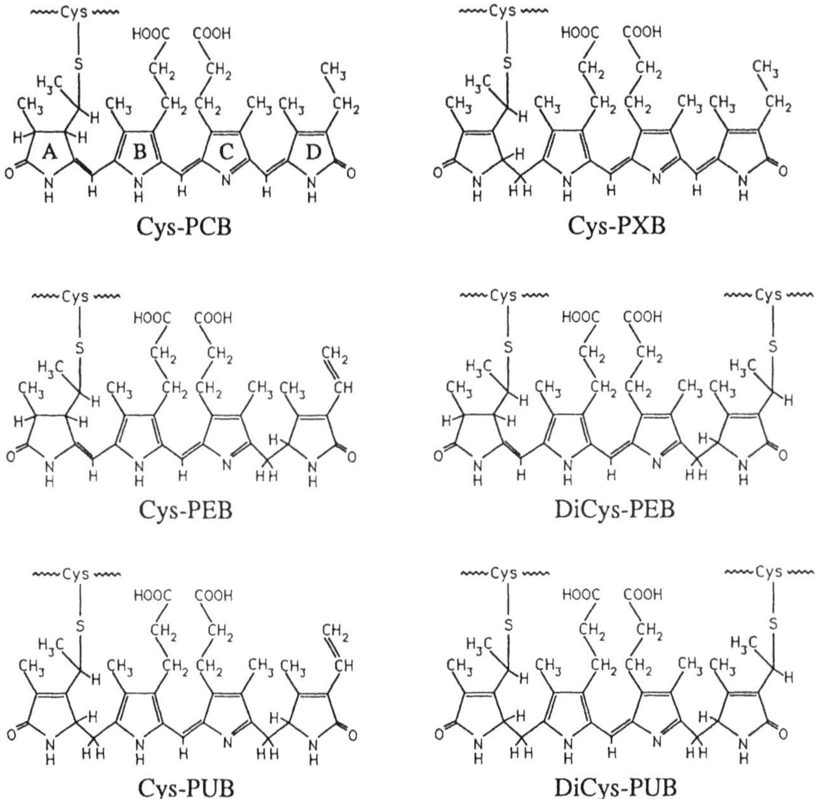

Figure 2. Structures of the cysteine-linked bilins found as prosthetic groups on cyanobacterial or red algal phycobiliproteins. Cys-PCB, Cys-PXB, Cys-PEB, and Cys-PUB designate phycocyanobilin, phycobiliviolin, phycoerythrobilin, and phycourobilin, respectively, each linked through a single thioether bond at C3′ to a cysteinyl residue. DiCys-PEB and DiCys-PUB designate phycoerythrobilin and phycourobilin, respectively, each linked through two thioether bonds at C3′ and C18′.

III. BILINS: THE LIGHT-HARVESTING CHROMOPHORES

Four different isomeric bilin prosthetic groups have been found attached to cyanobacterial and red algal phycobiliproteins: phycocyanobilin (PCB), phycoerythrobilin (PEB), phycobiliviolin (cryptoviolin; PXB), and phycourobilin (PUB) (Figure 2). The variation in structure between peptide-linked bilins is not limited to the differences within the tetrapyrrole moiety of the bilin itself. Additional chemical variation arises from the differences in the number of bonds to the protein and stereochemistry at the carbon atom(s) on the bilin that is (are) part of the thioether linkage (Figure 3).

Each bilin is linked to the polypeptide by a thioether linkage at C3′ of the bilin to a cysteinyl residue. The absolute configuration at the C3′ may be either *R* or *S*. At any given attachment site, only one stereoisomer is seen. However, both occur in the phycobiliproteins. For example, for the PCB groups linked to C-phycocyanin at α-84 and β-82, the configuration is *R*, whereas for the PCB at β-155 the configuration is *S* (Duerring et al., 1991). In contrast, in phycoerythrocyanin, the configuration at C3′ for all three bilins appears to be *R* (Duerring et al., 1990, 1991). The number of thioether linkages between the bilin and the protein provides another source of structural variation. In phycoerythrins, all bilins are linked through a single thioether linkage at C3′, except for the bilin attached nearest to the amino-terminus of the β-subunit. For this bilin, whether PEB or PUB, the linkage is through both C3′ and C18′ to cysteinyl residues β-50 and β-61 (Figures 1 and 2).

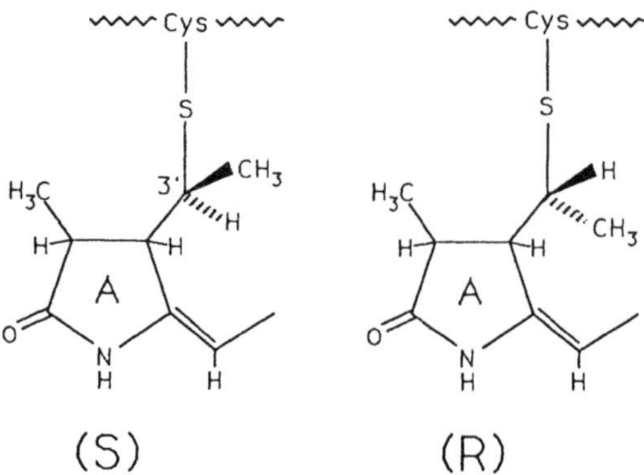

Figure 3. PCB bound to C-phycocyanin at α-84 and β-82 has the (*R*)-configuration at C3′, whereas PCB bound at β-155 has the (*S*)-configuration at C3′. (Data from Duerring et al., 1991).

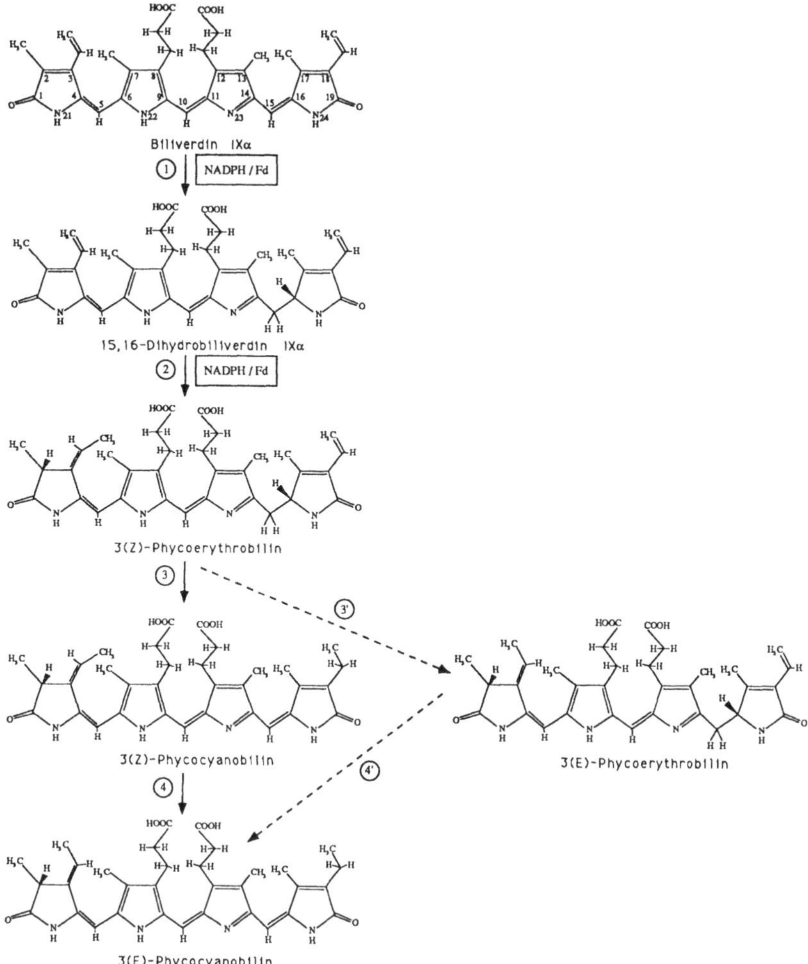

Figure 4. Biosynthesis of 3(*E*)-phycocyanobilin from biliverdin IXα. (Data from Beale and Cornejo, 1992a,b,c).

IV. BIOSYNTHESIS OF THE PHYCOBILINS

The pathway of phycobilin biosynthesis has been established in the thermoacidophilic unicellular rhodophyte *Cyanidium caldarium. C. caldarium* phycobiliproteins contain exclusively phycocyanobilin. Early studies with greening *C. caldarium* cells established that heme was a precursor of phycocyanobilin (for reviews see Brown et al., 1990; Beale and Weinstein, 1991). The first step in the conversion of heme to phycocyanobilin is the formation of biliverdin IXα in a heme oxygenase-catalyzed reaction (Troxler et al., 1979; Cornejo and Beale, 1988). The latter is converted to 3(*E*)-phycocyanobilin by two sequential ferredoxin-mediated two-electron reduction steps followed by isomerization reactions, as illustrated in Figure 4 (Beale and Cornejo, 1991a,b,c). A surprising feature of the biosynthetic pathway is the formation of PEB as a precursor to PCB.

V. BILIN ADDITION TO APOPHYCOBILIPROTEINS

The factors that govern the attachment of specific bilins at particular cysteinyl residues in phycobiliproteins have yet to be defined. In an organism with a complicated phycobilisome, such as a marine unicellular cyanobacterium with two types of phycoerythrins, R-phycocyanin II, bilin-bearing rod linkers, and several bilin-bearing core polypeptides, one of three bilins (PCB, PEB, or PUB) is attached at each of at least 20 different cysteinyl residues. There are no obvious patterns in the linear amino acid sequences surrounding the cysteinyl residues at these attachment sites that correlate with the presence of particular bilins.

The mechanism of attachment of PCB to C-phycocyanin has received considerable attention, and the results obtained with this protein offer a glimpse of the factors that may be involved in the attachment of bilins in general. C-Phycocyanin carries three PCB groups at α-84, β-82, and β-155. Arciero et al. (1988a,b,c) examined the possibility that the addition of PCB to C-phycocyanin was autocatalytic. *In vitro* studies on PCB addition to *Synechococcus* sp. PCC7002 apophycocyanin, expressed from the cloned genes in *E. coli*, showed that PCB would add specifically and rapidly at α-Cys84 and β-Cys82, but not at β-Cys155. At β-82, the products consisted of a mixture of the normal PCB adduct and a more highly oxidized mesobiliverdin adduct (with an extra double bond between C2 and C3, absent from PCB). Only the mesobiliverdin adduct was recovered from the α-84 site. These results argued strongly against the autocatalytic mechanism of bilin addition. In contrast, analogous experiments on bilin addition to apophytochrome support the autocatalytic mechanism for this biliprotein (Wahleithner et al., 1991; Cornejo et al., 1992).

The *Synechococcus* sp. PCC7002 phycocyanin operon has six open reading frames, *cpcBACDEF*. *CpcBACD* code for β^{PC}, α^{PC}, L_R^{33}, and $L_R^{8.5}$, respectively (Bryant, 1989; for nomenclature see Glazer, 1985). *CpcE* and *cpcF* do not encode structural components of the phycobilisome, but these two genes are cotranscribed with *cpcBACD* on low-abundance transcripts. The sequences of the *cpcE* and *cpcF* genes predict proteins of 268 and 205 residues, respectively (Zhou et al., 1992). The role of the *cpcE* and *cpcF* genes was examined by construction of interposon insertion mutations in each of these genes, as well as an interposon deletion mutation affecting both genes (Zhou et al., 1992). All three mutant strains showed a similar phenotype. These strains accumulated only about 10% of the wild-type level of C-phycocyanin. The β-subunit of this phycocyanin was indistinguishable from that of the wild type in size and in the structure of the PCB adducts at β-82 and β-155; however, addition of PCB to the α-subunit to form the wild-type holopolypeptide did not take place. Most (> 90%) of the α-subunit in the mutants was recovered as the apopolypeptide, and the balance as a mixture of unnatural bilin adducts (Swanson et al., 1992). The mutants formed wild-type levels of normal allophycocyanin, which carries PCB at α-84 and β-84 (Figure 1). These results argued for a direct specific role of CpcE and CpcF proteins in PCB addition to the apo-α-subunit.

Synchococcus sp. PCC7002 *cpcE* and *cpcF* were cloned and expressed in *E. coli* (Fairchild et al., 1992; Zhou et al., 1992). The recombinant CpcE and CpcF proteins were shown to catalyze the addition of PCB to *Synechococcus* sp. PCC7002 apo-α-phycocyanin to form a holo-α-phycocyanin indistinguishable from the natural product (Fairchild et al., 1992). Both CpcE and CpcF were required for the addition reaction and the roles of the individual proteins are yet to be established. A particularly interesting observation was the finding that CpcE plus CpcF were able to catalyze the transfer of bilin from native phycocyanin (the αβ C-phycocyanin heterodimer) to apo-α-phycocyanin.

Genes homologous to *Synechococcus* sp. PCC7002 *cpcE* and *cpcF* are present downstream from the genes encoding β^{PC} and α^{PC} in *Anabaena* sp. PCC7120 (Belnap and Haselkorn, 1987), *Calothrix* sp. PCC7601 (Mazel et al., 1988), *Pseudoanabaena* sp. PCC7409 (Bryant, 1991), and *Synechococcus* sp. WH8020 (Wilbanks, 1992; Wilbanks and Glazer, 1993). Tandeau de Marsac et al. (1988, 1990, 1991) have shown that pigmentation mutants of *Calothrix* sp. PCC7601, resulting from spontaneous insertion of endogenous IS elements (IS*701* and IS*703*) into the *cpcF* gene, produce the same phenotype as that described above for the *Synechococcus* sp. PCC7002 *cpcE* or *cpcF* mutants generated by interposon mutagenesis. It is likely therefore that the *cpcE* and *cpcF* gene homologs have equivalent functions in all of these organisms.

In studies of the *Synechococcus* sp. PCC7002 *cpcE* or *cpcF* mutants, pseudorevertants of both strains were observed to arise at high frequency (Swanson et al., 1992; Zhou et al., 1992). Analysis of the phycocyanin from a *cpcE* pseudorevertant, which produced a near wild-type level of phycocyanin with the α-subunit carrying PCB, revealed a single amino acid substitution, α^{PC}-Tyr129→Cys. Proof that this point mutation accounted fully for the altered phenotype was obtained as follows: A mutated *cpcA* gene (encoding α^{PC}) containing this substitution was constructed by site-directed mutagenesis, and transformed together with *cpcB* into a *Synechococcus* sp. PCC7002 *cpcBAC* deletion strain containing an insertionally inactivated *cpcE*. This strain was found to produce high levels of phycocyanin, and the majority of the α^{PC} carried PCB at α-Cys84 (Swanson et al., 1992). α-Tyr129 is conserved in all phycocyanins sequenced to date, forms part of the α-84 bilin binding site, and lies within 5 Å of α-Cys84. One interpretation of the phenotype seen in the pseudorevertant is that an enzyme system responsible for PCB addition at one of the other seven

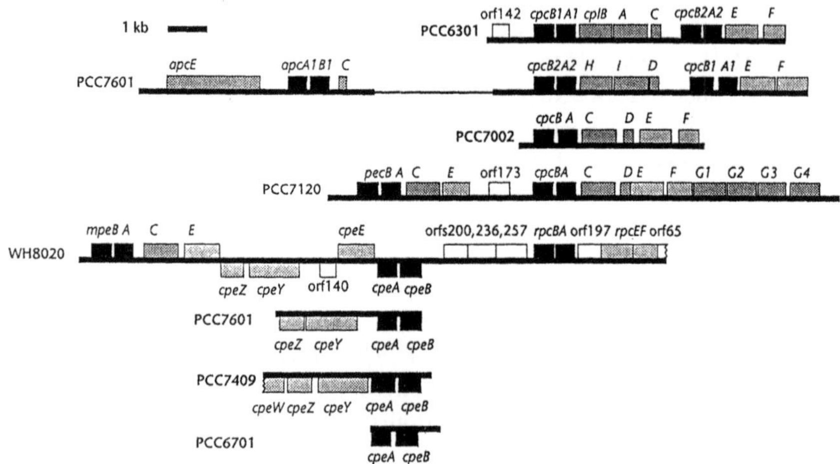

Figure 5. Phycobiliprotein gene organization in some cyanobacteria. Each heavy line represents sequence(s) for a particular strain, indicated in bold type by its Woods Hole or Pasteur Culture Collection number; the lighter line refers to a distance inferred from restriction mapping. Boxes above the line represent genes and open reading frames translated from left to right; boxes below the line are those translated from the complementary strand. Black boxes represent the genes for α- and β-subunits; dark gray boxes are those for linker polypeptides; light gray boxes are those of open reading frames with homologs in other phycobiliprotein gene clusters. All other open reading frames are represented by empty boxes. Clusters from different species have been aligned at the 5′ end of either the *cpcB* or *cpeB* gene. The sources of data are as follows: *Synechococcus* sp. PCC6301, Lind, 1988; *Calothrix* sp. (*Fremyella diplosiphon*) PCC7601 *apc* and *cpc* clusters, Lomas et al., 1987; Grossman et al., 1988; Mazel et al., 1988; Tandeau de Marsac et al., 1988; *cpe* cluster, Mazel et al., 1988; Tandeau de Marsac et al., 1988; *Synechococcus* sp. (*Agmenellum quadruplicatum*), Bryant, 1991; *Anabaena* sp. PCC7120, Belnap and Haselkorn, 1987; Eberlein and Kufer, 1990; Bryant et al., 1991; Swanson et al., 1992; *Synechococcus* sp. WH8020, Wilbanks, 1992; *Pseudanabaena* sp. PCC7409, Dubbs and Bryant, 1991.

phycobiliprotein PCB attachment sites in *Synechococcus* sp. PCC7002 is able to catalyze such addition to the apo-α^{PC}-Tyr129→Cys mutant subunit.

Phycobiliprotein gene organization in several cyanobacteria is illustrated in Figure 5. Open reading frames homologous to the *cpcE* and *cpcF* class of genes are associated with phycobiliproteins other than phycocyanin. For example, in *Anabaena* sp. PCC7120, *pecE* is found downstream of *pecBA* (encoding the α- and β-subunits of phycoerythrocyanin) (Swanson et al., 1992). In *Calothrix* sp. PCC7601 and in *Pseudanabaena* sp. PCC7409, *cpcE* and *cpcF* homologues *cpeY* and *cpeZ* lie upstream of *cpeBA* (which encode the α- and β-subunits of C-phycoerythrin) (Dubbs and Bryant, 1991). Thus, there are several genes whose products are candidates for roles in bilin attachment and/or phycobilisome assembly.

VI. GENERAL FEATURES OF PHYCOBILISOME STRUCTURE

Phycobilisomes invariably have two distinct morphological domains: a core which rests on the thylakoid membrane, and an array of rods that radiate away from the core (Mörschel et al., 1977; Bryant et al., 1979; Glazer et al., 1979; Wehrmeyer, 1983; see diagrammatic representation in Figure 6). Phycobilisomes from different organisms differ with respect to the details of core organization, and in the number and phycobiliprotein composition of the rods.

Allophycocyanin is the major phycobiliprotein of the phycobilisome core. In addition, each of the basal cylinders of the core of hemidiscoidal phycobilisomes contains allophycocyanin B (α^{APB}), a polypeptide homologous to the α-subunit of allophycocyanin, that functions as one of the two terminal energy acceptors in the phycobilisome. A copy of a PCB-bearing polypeptide of unknown function, homologous to the β-subunit of allophycocyanin ($\beta^{16-18.5}$), is present in each of the basal core cylinders. Two copies of a PCB-bearing large linker polypeptide (L_{CM}; see Figures 6 and 7 and the discussion below) are also present in the core. This polypeptide functions as the other terminal energy acceptor (Glazer, 1985).

The rods in all phycobilisomes contain one or another type of phycocyanin (Swanson et al., 1991). In some instances, either one or two different phycoerythrins are present as well (Ong et al., 1991). Less frequently, phycoerythrocyanin is present in the rod substructures. In no instance has an organism been found to contain both phycoerythrin and phycoerythrocyanin (Bryant, 1982). Phycocyanin forms the section of the rods proximal to the core, while phycoerythrins and phycoerythrocyanin form the distal portions of the rods (Gantt et al., 1976; Bryant et al., 1979; Glazer, 1982).

A family of polypeptides of 27 to 36 kDa, called "linker polypeptides", is responsible for the assembly of the phycobiliprotein components of the rods into discrete disc-shaped complexes, [e.g., $(\alpha\beta)_6L_R^{33}$ or $(\alpha\beta)_6L_{RC}^{29}$ in Figure 6] and for connecting these complexes in an ordered manner to each other to form the rod structure and to attach it to the phycobilisome core (Glazer, 1982; Gingrich et al., 1983; Glazer et al., 1983; Bryant, 1988).

Hemidiscoidal phycobilisomes in certain cyanobacteria have a higher degree of structural complexity. The phycobilisomes of *Mastigocladus laminosus* and *Anabaena* sp. PCC7120 are reported to have eight rods, with allophycocyanin forming the core-proximal segment in two of these, phycocyanin the middle segment, and phycoerythrocyanin occupying the distal position (Glauser et al., 1992a). Moreover, while most of the organisms with hemidiscoidal phycobilisomes have a single gene encoding the rod–core linker polypeptide (L_{RC}), *Anabaena* sp. and *M. laminosus* have genes (*cpcG1-cpcG4*) encoding four different L_{RC} polypeptides (Belnap and Haselkorn, 1987; Bryant et al., 1991). The hemidiscoidal eight-rod phycobilisomes of these organisms have been shown to contain at least three of these polypeptides (Bryant et al., 1991; Glauser et al., 1992b). Preliminary evidence indicates that each of the four L_{RC} polypeptides may attach two of the rods

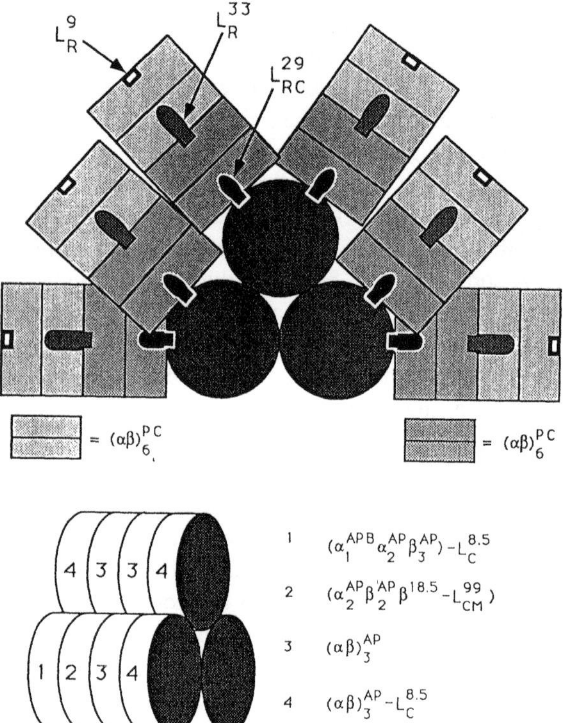

Figure 6. Schematic representation of the location of the components of the hemidiscoidal *Synechococcus* sp. PCC7002 phycobilisome. The rod substructures of this simple phycobilisome contain C-phycocyanin as the only phycobiliprotein. The composition of the phycobiliprotein-linker polypeptide complexes which make up the rod substructure is shown. The rod substructures are "capped" by a small linker polypeptide, L_R^9. Allophycocyanin is the major phycobiliprotein component of the core complexes. The arrangement shown for the core complexes is that suggested by Anderson and Eiserling (1986). The peripheral allophycocyanin complexes of each of the three core cylinders are "capped" by a small linker polypeptide $L_C^{8.5}$. The abbreviations PC, AP, and APB are used for the phycobiliproteins phycocyanin, allophycocyanin, and allophycocyanin B, respectively, and α^{AP} and β^{AP}, for the α- and β-subunits of these proteins. Subscripts indicate the number of copies of each component. Linker polypeptides are abbreviated L, with the superscript denoting the size in kilodaltons, and a subscript that specifies the location of the polypeptide: R, rod substructure; RC, rod–core junction; C, core substructure; CM, core–membrane junction (data from Bryant, 1991; for nomenclature, see Glazer, 1985).

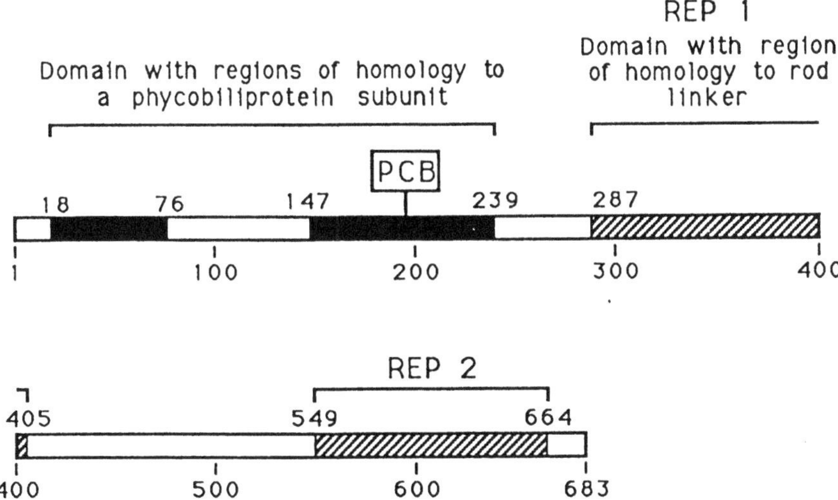

Figure 7. Schematic representation of the core–membrane linker (L_{CM}) of the *Synechococcus* sp. PC6301 phycobilisome. The phycobilisomes of *Synechococcus* sp. PCC6301 have the simplest core organization among hemidiscoidal phycobilisomes and the smallest L_{CM} (683 residues; 72,400 kDa). Their core is made up of two cylinders. There are two copies of the L_{CM} per phycobilisome (Glazer et al., 1983). In their model of the organization of the core complexes of the *Synechococcus* sp. PCC6301 phycobilisome, Capuano et al. (1991) propose that the *rep 1* and *rep 2* domains (which are homologous to rod linker polypeptides) each organize one of the core hexameric assemblies (see Figure 8). The amino-terminal domain of the L_{CM} is homologous to a phycobiliprotein subunit, and carries a phycocyanobilin. The phycocyanobilins on the L_{CM} polypeptides and those on the two α^{APB} subunits function as the terminal energy acceptors in the phycobilisome.

specifically to one of four different binding sites in the core. In support of this view, it has been shown that the *M. laminosus* phycobilisome rod–core complex $(\alpha\beta)_6^{PC}L_{RC}^{29.5}\cdot(\alpha\beta)_3^{AP}L_C^{8.9}$ (where $L_{RC}^{29.5}$ is the product of gene *cpcG2*) could be formed *in vitro* from the subcomplexes $(\alpha\beta)_3^{PC}L_{RC}^{29.5}$ and $(\alpha\beta)_3^{AP}L_C^{8.9}$, whereas no such complex could be obtained when the phycocyanin subcomplex was substituted by one containing the L_{RC} encoded by *cpcG3*. Moreover, the same $(\alpha\beta)_6^{PC}L_{RC}^{29.5}\cdot(\alpha\beta)_3^{AP}L_C^{8.9}$ complex was isolated from partially dissociated *M. laminosus* phycobilisomes (Glauser et al., 1992).

A large linker polypeptide (L_{CM}, ranging in size from 75 to 120 kDa, depending on the type of phycobilisome) organizes the allophycocyanin complexes within the phycobilisome core. L_{CM} has two or more repeats of a domain homologous in sequence to the rod linker polypeptides (ca. 30 kDa), and a phycobiliprotein domain which carries a terminal acceptor bilin (Figure 7). From the complete amino acid

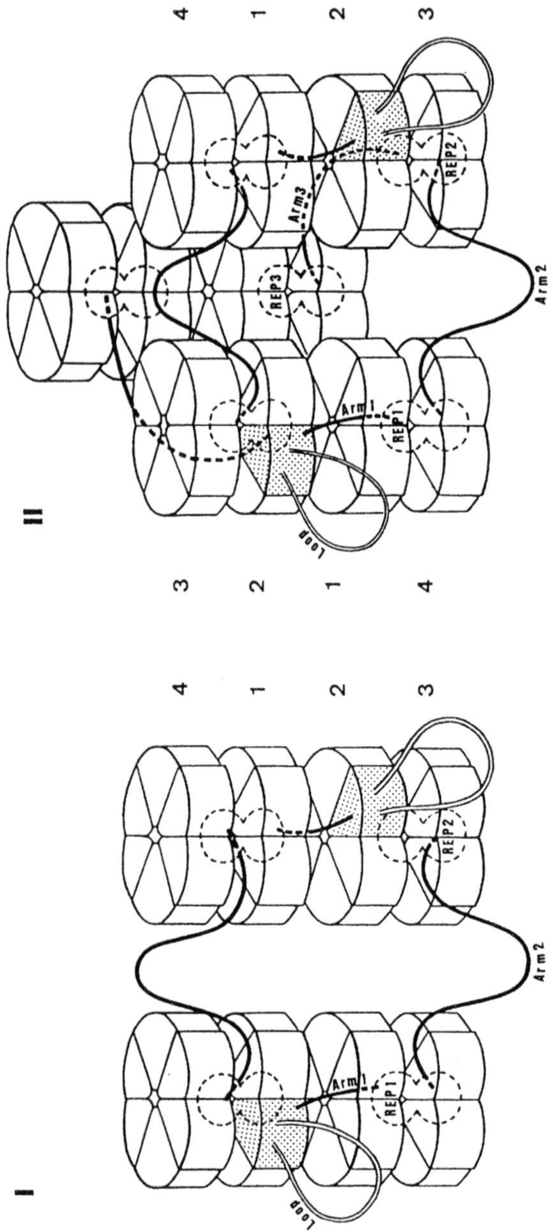

I
II
Loop
Arm1
Arm2
Arm3
REP1
REP2
REP3
3
2
1
4

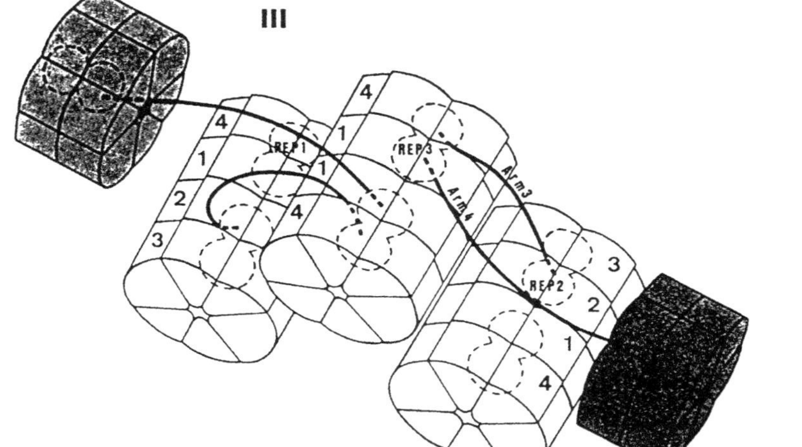

Trimer composition :

complex 1 $(\alpha^{AP}, \beta^{AP})_3$

complex 2 $(\alpha^{AP}, \beta^{AP})_2\ \beta^{18.3}\ L_{CM}$

complex 3 $(\alpha^{APB}, \alpha^{AP}{}_2, \beta^{AP}{}_3)\ L_C$

complex 4 $(\alpha^{AP}, \beta^{AP})_3\ L_C$

Figure 8. Schematic representation of the organization of the core complexes in phycobilisomes with bicylindrical and tricylindrical cores illustrating the proposed disposition of the domains of the two L_{CM}s. The L_{CM}s of phycobilisomes with a bicylindrical core have two *rep* domains (e.g., see Figure 7), while those of phycobilisomes with a tricylindical core have three (e.g., *Synechococcus* sp. PCC7002) or four (*Calothrix* sp. PCC7601) *rep* domains. The composition of the four types of trimeric disc assemblies is indicated in the figure. The phycobiliprotein domain of each of the two L_{CM}s (see Figure 7) is *cross-hatched* in diagram **I** and **II**. *Hollow lines* represent the extra sequences (designated loop) inserted into the phycobiliprotein homology domain of L_{CM} (see Figure 5); *solid lines* indicate the sequences that lie between the different domains of the L_{CM} (designated *Arms* in **I–III**); and the *two overlapping circles* represent the *rep* domains of the L_{CM}. For clarity, the domains of only one of the two L_{CM}s have been labeled. (**I**) Bottom view of the two-cylinder core found in *Synechococcus* sp. PCC 6301 phycobilisomes in which the L_{CM} has two *reps.* (**II**) Bottom view of the tricylindrical cores found in the phycobilisomes of *Synechococcus* sp. PCC7002 in which the L_{CM} has three *reps.* (**III**) Top view, of a tricylindrical core of the *Calothrix* sp. PCC7601 phycobilisome showing, in addition, two phycocyanin hexamers (*shaded*) that would represent the core-proximal elements of two of the six rods that emanate from the core. The four *reps* of the two L_{CM}s are drawn as well the *Arm3* and *Arm4* regions. Capuano et al. (1991) suggest that *rep4* serves as an L_{RC} to link a core-proximal phycocyanin hexamer to the top cylinder of the core. (Reprinted with permission from Capuano et al., 1991).

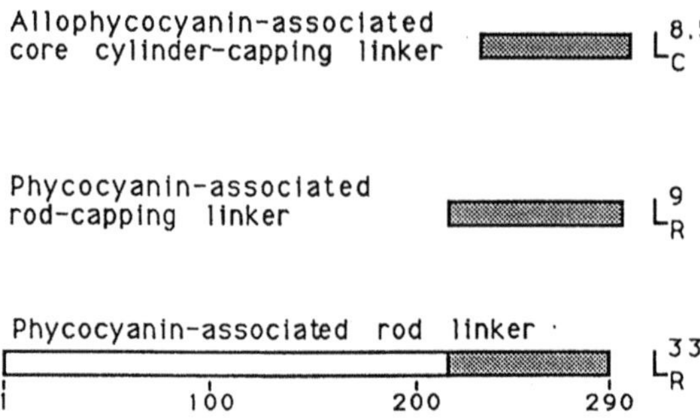

Figure 9. The small rod-capping and core-capping polypeptides of the phycobilisome are partially homologous to the carboxyl-terminal domain of the phycocyanin-associated rod linker polypeptides. The alignments shown are those for polypeptides from *Synechococcus* sp. PCC 7002 (Data from Bryant, 1991).

sequences of the large core linkers in the phycobilisomes of *Synechococcus* sp. PCC7002 (Bryant, 1991), *Cyanophora paradoxa* (Bryant, 1988), *Calothrix* sp. PCC7601 (Houmard et al., 1990), and *Synechococcus* sp. PCC6301 (Capuano et al., 1991), and by analogy to the known sequences and roles of the rod linkers, a model has been proposed for the organization of the components of the core domain. The details of this model are presented in Figure 8.

Small linker polypeptides of 8 to 9 kDa are associated with the terminal discs of the rod substructures and with the peripheral allophycocyanin trimeric complexes of the core (Figure 6). These polypeptides show homology to the carboxyl terminal portions of the rod linker polypeptides (Figure 9), and are believed to function in the termination of the assembly of the rod and core structures (de Lorimier et al., 1990).

VII. PHYCOBILIPROTEINS: SPECTROSCOPIC PROPERTIES

In native phycobiliproteins, the bilins show the following ranges of absorption maxima: PCB, 600–650 nm; PXB, 568 nm; PEB, 535–568 nm; and PUB, 495 nm. The dominant features of the absorption spectrum of a native phycobiliprotein are determined by the chemical structure of the bilins it carries. However, the native protein environment modulates the spectroscopic properties of the bilins in ways that are important to the determination of the pathways of energy transfer in phycobilisomes. For example, C-phycocyanin and allophycocyanin both carry only PCB groups. The absorption maximum of C-phycocyanin is at 620 nm, whereas that of allophycocyanin is at 650 nm. This difference in the absorption spectra ensures directional energy transfer from phycocyanin to allophycocyanin.

The modulation of the spectroscopic properties of the bilins within native phycobiliproteins is achieved in several ways. These linear tetrapyrroles are held in rigid extended conformations. In such conformations the bilins possess strong long-wavelength absorption bands. It is possible that in some phycobiliproteins, exciton interactions between bilins also modify their spectra. The pronounced differences between the spectra of allophycocyanin and C-phycocyanin (see above; Yeh et al., 1986) might be accounted for by strong interactions between the bilins in the latter protein.

The details of the protein environment about each bilin also affect its spectroscopic properties. The influence of the protein environment is dramatically demonstrated by the modulation of the spectroscopic properties of the bilin at β-82 in various phycobiliproteins by posttranslational modification. Each phycobiliprotein (with rare exceptions) contains a conserved unique modified amino acid residue, γ-N-methylasparagine at β-72 (Klotz et al., 1986; Klotz and Glazer, 1987; Rümbeli et al., 1987). From the X-ray crystal structures of the phycocyanins of *Synechococcus* sp. PCC7002 and *M. laminosus*, it is seen that the methyl group approaches closely ring B of the bilin at β-82 (Duerring et al., 1988). As discussed below, the β-82 bilin acts as the terminal energy acceptor within phycocyanins and phycoerythrins. Isolation of mutants lacking the protein asparagine methylase allowed comparison of the spectroscopic properties of "methylated" and "unmethylated" phycocyanins and phycobilisomes (Swanson and Glazer, 1990). The results showed that the presence of the methyl group of the γ-*N*-methylasparagine at β-72 shifts the absorption spectrum of the β-82 PCB in C-phycocyanin to the red, thus contributing to its function as a terminal acceptor. This fine tuning of the spectroscopic properties of the β-82 PCB contributes significantly to the efficiency of directional energy transfer in intact phycobilisomes (Swanson and Glazer, 1990).

VIII. ENERGY FLOW IN PHYCOBILISOMES

Excitation energy absorbed by the phycobilisomes is transferred with an efficiency exceeding 95% to the reaction centers (Porter et al., 1978). The gross organization of phycobiliproteins in phycobilisomes (see below) ensures thermodynamically favored directional energy transfer towards the core irrespective of the location at which the absorption of the photon takes place.

Energy transfer within individual phycobiliprotein hexamers is likewise directional. Spectroscopic measurements on C-phycocyanin crystals show that the bilin at β-82 is the terminal acceptor chromophore in this protein. Direct measurements, as well as calculations based on the crystal structures of C-phycocyanins, indicate that energy absorbed by any bilin within a hexamer is transferred very rapidly to the terminal acceptor bilin (Schirmer and Vincent, 1987; Siebzenrübl et al., 1987; Sauer and Scheer, 1988). Analyses of the locations of donor and acceptor bilins in a variety of phycocyanins and phycoerythrins indicate that the bilin at β-82 is conserved as the terminal acceptor in all of these proteins (Ong and Glazer, 1991; Figure 10). The bilin at β-82 occupies a location near the center of a phycobilipro-

PHYCOCYANINS

	α-84	β-82	β-155
C-Phycocyanin	PCB	PCB	PCB
Phycoerythrocyanin	PXB	PCB	PCB
R-Phycocyanin	PCB	PCB	PEB
R-Phycocyanin II	PEB	PCB	PEB
WH8501 Phycocyanin	PUB	PCB	PCB

PHYCOERYTHRINS

	α-75	α-83	α-140	β-50,61	β-82	β-159
C-PE		PEB	PEB	PEB	PEB	PEB
B-PE		PEB	PEB	PEB	PEB	PEB
R-PE		PEB	PEB	PUB	PEB	PEB
WH8020 PE(I)		PEB	PEB	PEB	PEB	PEB
WH8020 PE(II)	PUB	PEB	PEB	PUB	PEB	PEB
WH8103 PE(I)		PEB	PUB	PUB	PEB	PEB
WH8103 PE(II)	PUB	PUB	PUB	PUB	PEB	PEB
WH8501 PE(I)		PUB	PUB	PUB	PEB	PUB
WH8501 PE(II)		PUB	PUB	PUB	PEB	PUB

Figure 10. Location of terminal acceptor bilin in phycocyanins and phycoerythrins. For each protein, bilins which must serve as donor chromophores are *outlined*. Terminal energy acceptor bilins are shown in *larger boldface font*. The sites of attachment of the bilins in WH8020 and WH8103 PE(I) and PE(II) are from Ong and Glazer (1991), those in WH8501 PE(I) are from Swanson et al., 1991. The sources of data for the other sequences are as follows: C-PE, Sidler et al., 1986; B-PE, Lundell et al., 1984; Sidler et al., 1989; and R-PE, Klotz and Glazer, 1985. Residue numbering is based on the DNA sequence encoding WH8020 PE(II) (Wilbanks et al., 1991).

tein hexamer. Modeling based on the crystal structures of phycobiliproteins of bilin locations in other phycobiliproteins indicates that the donor bilins are located towards the periphery of the hexamers (Figure 11).

In summary, the picture of energy transfer in the phycobilisome is as follows. Energy absorbed by the donor chromophores in a biliprotein hexamer is rapidly transferred to the β-82 acceptor bilins. The absorption spectra of the acceptor bilins are influenced by interaction with specific linker polypeptides which occupy a central cavity within the hexamer such that the β-82 chromophores of consecutive phycocyanin disks (proceeding towards the core) lie at decreasing energy levels.

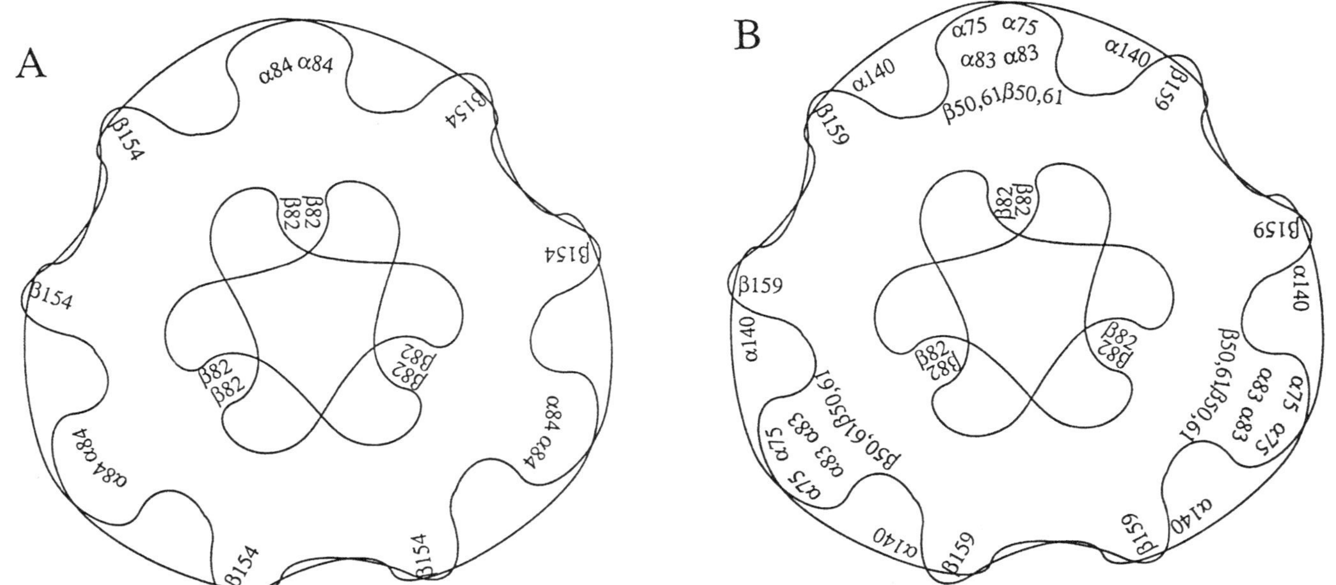

Figure 11. Schematic representation of chromophore positions in hexameric phycobiliproteins. Bilins are represented by the number of their cysteine attachment site(s) (e.g., the bilin at α-Cys140 is designated α140). The diagrams show a superposition of two trimers. (**A**) Hexamer of phycocyanin based on the structure of Schirmer et al., 1986. (**B**) A model of the WH8020 phycoerythrin II hexamer built onto the disc in A (Wilbanks, 1992).

The energy transferred to the allophycocyanin in the core is then rapidly localized on the terminal acceptor polypeptides α^{APB} and L_{CM}. Exclusive of intradisk transfers in the phycobiliprotein hexamers of the rods, the energy transfer pathway within phycobilisomes is:

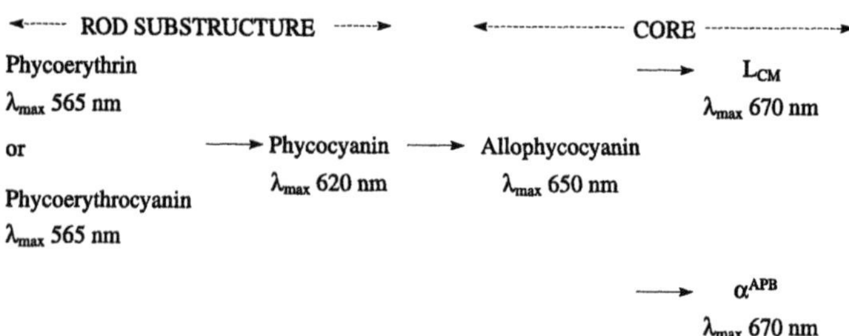

where only the long wavelength absorption maxima of each component are given.

IX. CONSTITUTIVE ADAPTATIONS IN PHYCOBILISOME COMPOSITION TO AMBIENT LIGHT QUALITY

Photosynthesis at low light intensity is limited by the rate at which excitation energy is delivered to the reaction centers. This rate is in turn dependent on the absorption cross-section of the light-harvesting antenna complexes for the available radiation. Comparison of the absorption properties of phycobilisomes of freshwater and marine organisms offers an excellent example of constitutive adaptation of the phycobiliprotein composition of the antenna complex to the ambient light quality. Adaptation by marine cyanobacteria and red algae to blue–green light is achieved by maximum utilization of PUB and PEB as light-harvesting groups within the phycobilisome and by utilizing a minimum number of PCB chromophores.

The phycobilisomes of marine cyanobacteria and red algae are rich in phycoerythrins. These phycoerythrins carry both PEBs (λ_{max} 530–565 nm) and PUBs (λ_{max} 495 nm). The PUB:PEB ratio in the phycoerythrins of open ocean unicellular cyanobacteria is particularly high. It is evident from inspection of Figure 12 that the phycobilisome of a marine cyanobacterium such as *Synechocystis* sp. WH8103 is well suited to the efficient absorption of those wavelengths of light best transmitted through water. The major phycoerythrin (phycoerythrin II) of *Synechocystis* sp. WH8103 phycobilisomes has a PUB:PEB ratio of 2:1 and carries six PUB + PEB chromophores per $\alpha\beta$. C-Phycoerythrin, the phycoerythrin present in many freshwater and soil cyanobacteria, carries five PEB per $\alpha\beta$. At 495 nm, the absorption coefficient of phycoerythrin II is almost 10-fold higher than that of C-phycoerythrin. C-Phycocyanin, the major phycobiliprotein component of most freshwater and soil cyanobacterial phycobilisomes, has very little absorption at 495 nm.

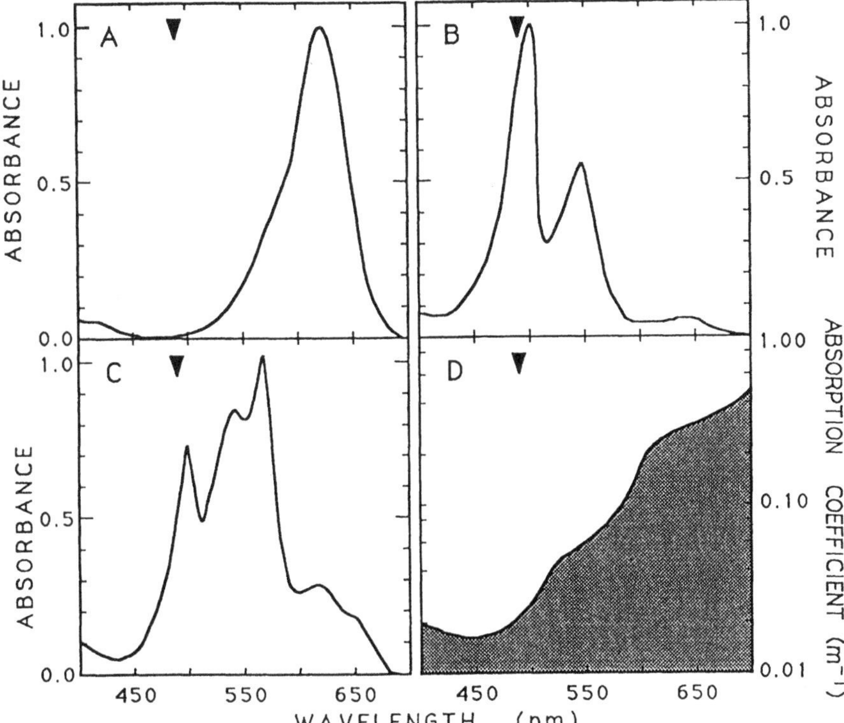

Figure 12. Comparison of the absorption spectra of phycobilisomes to that of the absorption spectrum of water. *Arrowheads* indicate 490 nm in each frame. (**A**) Phycobilisomes from the freshwater unicellular cyanobacterium *Synechococcus* sp. PCC6301. (**B**) Phycobilisomes from the marine unicellular cyanobacterium *Synechococcus* sp. WH8103. (**C**) Phycobilisomes from the marine red alga *Porphyridium cruentum.* (**D**) Absorption coefficient of water per meter as a function of wavelength. Note that **A–C** are plotted on a linear scale, whereas **D** is on a logarithmic scale.

The phycocyanins of freshwater and soil cyanobacteria always carry three PCBs, whereas those of the open ocean cyanobacteria and red algae retain the β-82 PCB, but frequently replace one or both of the other PCB groups with PEB or PUB (see Figure 1).

X. MODULAR CHANGES IN PHYCOBILISOME STRUCTURE: INTENSITY ADAPTATION

As a general rule, the phycobiliprotein and chlorophyll *a* content per cell in cyanobacteria and red algae decreases with increase in the irradiance (Allen, 1968;

Waaland et al., 1974; Jahn et al., 1984; Kana and Glibert, 1987). In some organisms, such as the cyanobacterium *Microcystis aeruginosa* (Raps et al., 1985), or the red alga *Griffthsia pacifica* (Waaland et al., 1974), the decrease in phycobiliprotein content is due entirely to a decrease in the number of phycobilisomes. In other organisms, such as *Synechococcus* sp. PCC6301, at high light intensity both the number of phycobilisomes decreases and the phycocyanin content per phycobilisome decreases (Yamanaka and Glazer, 1981). The decrease in the phycocyanin content per phycobilisome in cells grown at high light intensity reflects the presence of phycobilisomes with rod substructures with one to two hexamers of C-phycocyanin per rod. Phycobilisomes from cells grown at low light intensity have rod substructures with two to four hexamers of C-phycocyanin per rod.

Kana and Glibert (1987) have presented similar results on the effect of variation in irradiance on the light-harvesting components of the photosynthetic apparatus in the marine unicellular cyanobacterium, *Synechococcus* sp. WH7803 (Table 1). Phycoerythrin forms the distal portions of the rod substructures of *Synechococcus* sp. WH7803 phycobilisomes. Increase in irradiance results in a decrease in the total amount of both phycoerythrin and phycocyanin and a sharp decrease in the ratio of phycoerythrin to phycocyanin (from 14.4 at 30 $\mu E\ m^{-2}\ s^{-1}$ to 3.4 at 700 $\mu E\ m^{-2}\ s^{-1}$). These results indicate a loss of phycobilisomes and decrease in the size of the remaining phycobilisomes with increase in the intensity of irradiation.

Table 1. Effect of Variation in Irradiance on the Light-Harvesting Components of the Photosynthetic Apparatus in *Synechococcus* sp. WH7803[a]

Ig[b] ($\mu E\ m^{-2}\ s^{-1}$)	*Growth Rate* (d^{-1})	*PE/PC*[c]	*PE*[d]	*PC*[e]	*Chl a*[f]	*C*[g]
			(fg cell^{-1})			
30	0.44	14.4	143	9.9	3.7	270
50	0.77	13.3	120	9.0	5.0	252
100	1.21	13.1	81	6.2	3.4	237
160	1.40	11.7	56	4.8	3.1	231
400	1.71	5.3	20	3.8	2.4	250
700	1.72	3.4	7.2	2.1	1.4	195
1330	1.77	2.0	6.1	3.0	1.4	230
2000	1.97	2.6	8.5	3.2	1.3	266

Notes: [a]Data from Kana and Gilbert, 1987.
[b]Irradiances.
[c]PE to PC weight ratio.
[d]Phycoerythrin.
[e]Phycocyanin.
[f]Chlorophyll *a*.
[g]Cell carbon concentrations.

XI. MODULAR CHANGES IN PHYCOBILISOME STRUCTURE: CHROMATIC ADAPTATION

Chromatic adaptation—the response of the photosynthetic apparatus to changes in light quality—has been intensively studied in cyanobacteria for over 100 years. The most intensive molecular genetic and biochemical studies of this phenomenon have been performed with *Calothrix* sp. PCC7601 (*Fremyella diplosiphon*) (for reviews, see Grossman et al., 1988; Tandeau de Marsac et al., 1988; Grossman, 1990; Tandeau de Marsac, 1991).

In *Calothrix* sp. PCC7601, chromatic adaptation is now well understood at both the level of gene expression and phycobilisome structure. The genes for α- and β-subunits of phycoerythrin and for two phycoerythrin-associated linker polypeptides, L_R^{35} and $L_R^{35.5}$, are expressed in green but not in red light. Genes for the α- and β-subunits of a constitutive phycocyanin (PC_c) and for its associated rod–core linker, L_{RC}^{31}, are expressed in both red and green light. A second set of genes encoding the α- and β-subunits of an inducible phycocyanin (PC_i) and for its associated rod linkers, $L_R^{37.5}$ and L_R^{39}, and the rod-capping linker $L_R^{9.7}$, are expressed in red light. In consequence, the composition of the rods of *Calothrix* sp. PCC7601 phycobilisomes from cells grown in red and green light is as follows (Tandeau de Marsac et al., 1988; Grossman, 1990):

Red Light

$$\text{CORE}–(\alpha^{PCc}\beta^{PCc})L_{RC}^{31}–(\alpha^{PCi}\beta^{PCi})L_R^{37.5}–(\alpha^{PCi}\beta^{PCi})L_R^{39}–L_R^{9.7}$$

where the order and relative number of $(\alpha^{PCi}\beta^{PCi})L_R^{37.5}$ and $(\alpha^{PCi}\beta^{PCi})L_R^{39}$ discs are assigned arbitrarily.

Green Light

$$\text{CORE}–(\alpha^{PCc}\beta^{PCc})L_{RC}^{31}–(\alpha^{PE}\beta^{PE})L_R^{35.5}–(\alpha^{PE}\beta^{PE})L_R^{35}–(\alpha^{PE}\beta^{PE})L_R^{35}$$

where the order and relative number of $(\alpha^{PE}\beta^{PE})L_R^{35.5}$ and $(\alpha^{PE}\beta^{PE})L_R^{35}$ discs are assigned arbitrarily.

It is seen that the chromatic adaptation response results in a substitution of two hexameric phycocyanin complexes, which make up the distal portion of the rod substructures in red light-grown cells, by three phycoerythrin hexameric complexes in green light-grown cells. The rest of the phycobilisome structure is unaltered in red and green light.

XII. PHYCOBILISOMES: A STORE OF FIXED NITROGEN

Cyanobacteria deal with temporary deprivation in the supply of fixed nitrogen by degrading phycobiliproteins (Allen and Smith, 1969). The cyanobacterial photosystem I complex has a light-harvesting chlorophyll *a* antenna of 130 to 140 pigment molecules, while about 40 chlorophyll *a* are associated with each photosystem II reaction center. In addition to the chlorophyll *a* antenna, phycobilisomes provide 200 or more bilin chromophores per photosystem II reaction center. Under conditions of nitrogen deprivation, retention of the light-harvesting chlorophyll *a* antenna complexes allows the cyanobacteria to utilize the phycobilisome as a store of fixed nitrogen while only partially compromising the ability of the cells to generate energy and reducing power through photosynthesis.

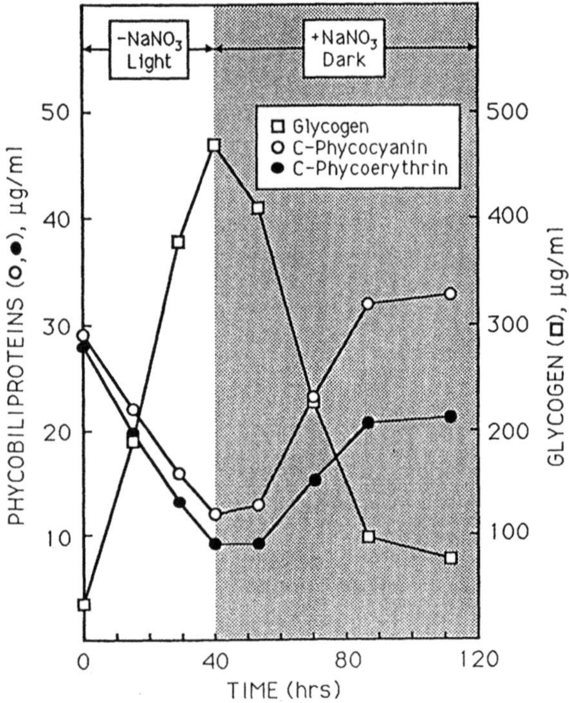

Figure 13. Effect of nitrate deprivation in the light and nitrate replenishment in the dark on the levels of phycoerythrin, phycocyanin, and glycogen measured in extracts of sonicated cells of *Synechocystis* sp. PCC6701. (Based on Figure 1 of Tandeau de Marsac et al., 1980).

As soon as the cells have exhausted their reserves of fixed nitrogen, the genes encoding the phycobiliproteins and the associated linker polypeptides are no longer transcribed (reviewed in Bryant, 1991). In the early stages of nitrogen starvation, the phycobiliprotein components that make up the distal portions of the phycobilisome rods are degraded more rapidly than entire phycobilisomes. Thus, phycobilisome antenna function is partially conserved in the early stages of phycobiliprotein breakdown. On prolonged nitrogen deprivation, the phycobilisomes are almost quantitatively degraded. Resynthesis of phycobiliprotein does not take place until a source of fixed nitrogen is again available.

The effectiveness of this adaptation in dealing with temporary shortage of fixed nitrogen is well illustrated in a study by Tandeau de Marsac et al. (1980) on *Synechocystis* sp. PCC6701 (Figure 13). Cells were partially depleted of their phycobiliproteins by deprivation of fixed nitrogen in the light. Continued photosynthesis during the starvation period led to a rapid increase in the glycogen content to a level representing up to 70% of the total cellular dry weight. Addition of nitrate to the medium allowed the cells to resynthesize their phycobiliproteins in the dark at the expense of rapid breakdown of the glycogen.

XIII. PHYCOBILISOMES: RESPONSE TO SULFUR DEPRIVATION

The uniform response of cyanobacterial strains to nitrogen deprivation is a decrease in phycobiliprotein content. In contrast, the responses to sulfur deprivation are idiosyncratic. For *Synechococcus* sp. PCC6301, sulfur deprivation leads to a rapid degradation of phycobiliproteins (Schmidt et al., 1982; Wanner et al., 1986). For *Synechococcus* sp. PCC7002, sulfur deprivation arrests growth, but does not lead to breakdown of phycobiliproteins (Bryant, 1988).

An extraordinary response is seen in *Calothrix* sp. PCC 7601. This organism has three phycocyanin operons: *cpc1*, *cpc2*, and *cpc3* (Tandeau de Marsac et al., 1988). When *Calothrix* sp. PCC7601 is grown in a medium with normal levels of sulfate (300 μM) the operon *cpc1* is expressed regardless of light quality with production of the α- and β-subunits of PC_c and its associated linker polypeptides. Expression of *cpc2* is triggered by exposure to red light with the production of PC_i and its associated linker polypeptides (see discussion of chromatic adaptation, above). The third operon, *cpc3*, is transcribed at a very low level independent of light quality (Mazel et al., 1988). However, *cpc3* is highly expressed at low concentrations of sulfate (15 μM) in the culture medium (Mazel and Marlière, 1989) in white light, whereas the *cpc1, cpc2*, and *cpe* (encoding phycoerythrin) operons are switched off under conditions of sulfur limitation.

Operon *cpc3* encodes the α- and β-subunits of an unusual phycocyanin (PC_s) and its three associated linker polypeptides. These five genes differ from their counterparts in the *cpc1* and *cpc2* operons in that they are free of codons specifying Met and Cys, except for the five Met initiation codons and the three Cys codons specifying the Cys residues at PCB attachment sites. Sequence analysis of PC_s reveals that the amino-terminal methionine is cleaved posttranslationally from both subunits. Interestingly, alignment of PC_s with PC_c and PC_i reveals that removal of the amino-terminal Met residues of the α- and β-subunits of PC_s puts their sequences in register with those of the corresponding subunits of PCc and PC_i. The allophycocyanin subunits of *Calothrix* sp. PCC7601 are expressed constitutively irrespective of sulfate limitation or light quality. As in the case of the PC_s subunits, the allophycocyanin subunits have protruding initiation Met residues when their sequences are aligned with those of other phycobiliproteins. These Met residues are cleaved from the subunits posttranslationally under all growth conditions. The α- and β-subunits of phycoerythrin, which are high in sulfur-containing residues, are not made under conditions of sulfur limitation. By minimizing the requirement for sulfur-containing amino acids by these means, *Calothrix* sp. PCC7601 is able to synthesize its phycobiliprotein antenna when sulfate is limited.

Mazel and Marlière (1989) point out that *Calothrix* sp. PCC7601 as a freshwater cyanobacterium faces an environment low in sulfate, and comment that other cyanobacteria (such as *Anabaena* sp. PCC7120 and *Synechococcus* sp. PCC6301) isolated from freshwater also have phycocyanins and allophycocyanins low in sulfur amino acids relative to the corresponding phycobiliproteins from organisms isolated from habitats with high-sulfate contents, such as seawater or acidic hot springs.

ACKNOWLEDGMENTS

The research described from the author's laboratory was supported by the National Institute of General Medical Sciences grant GM28994, the National Science Foundation grant DMB8816727, and a grant from the Lucille P. Markey Charitable Trust. The author is indebted to Gary J. Wedemayer, Sigurd M. Wilbanks, and Ronald V. Swanson for help in the preparation of the illustrations for this chapter, and thanks Donald A. Bryant, Nicole Tandeau de Marsac, Jean Houmard, and J. Clark Lagarias for communicating results prior to publication.

NOTE ADDED IN PROOF

Several excellent reviews, relevant to subjects covered here, appeared after submission of this manuscript. These include a detailed description of the molecular mechanisms that underly the adaptation of cyanobacteria to environmental stimuli (Tandeau de Marsac and Houmard, 1993), a review (that emphasizes molecular genetics) on the influence of environmental conditions on phycobilisome structure and turnover (Grossman et al., 1993), and an account of the biosynthesis of the phytochromobilin-bearing plant photoreceptor phytochrome (Terry et al., 1993).

REFERENCES

Allen, M.M. (1968). Photosynthetic membrane system in *Anacystis nidulans*. J. Bacteriol. 96, 836–841.

Allen, M.M. & Smith, A.J. (1969). Nitrogen chlorosis in blue-green algae. Arch. Microbiol. 69, 114–120.

Anderson, L.K. & Eiserling, F.A. (1986). Asymmetrical core structure in phycobilisomes of the cyanobacterium *Synechocystis* 6701. J. Mol. Biol. 191, 441–451.

Arciero, D.M., Bryant, D.A., & Glazer, A.N. (1988a). *In vitro* attachment of bilins to apophycocyanin. I. Specific covalent adduct formation at cysteinyl residues involved in phycocyanobilin binding in C-phycocyanin. J. Biol. Chem. 263, 18343–18349.

Arciero, D.M., Bryant, D.A., & Glazer, A.N. (1988b). *In vitro* attachment of bilins to apophycocyanin. II. Determination of the structures of tryptic bilin peptides derived from the phycocyanobilin adduct. J. Biol. Chem. 263, 18350–18357.

Arciero, D.M., Bryant, D.A., & Glazer, A.N. (1988c). *In vitro* attachment of bilins to apophycocyanin. III. Properties of the phycoerythrobilin adduct. J. Biol. Chem. 263, 18358–18363.

Beale, S.I. & Cornejo, J. (1991a). Biosynthesis of the phycobilins. Ferredoxin-mediated reduction of biliverdin catalyzed by extracts of *Cyanidium caldarium*. J. Biol. Chem. 266, 22328–22332.

Beale, S.I. & Cornejo, J. (1991b). Biosynthesis of the phycobilins. 3(Z)-Phycoerythrobilin and 3(*Z*)-phycocyanobilin are intermediates in the formation of 3(*E*)-phycocyanobilin from biliverdin IXα. J. Biol. Chem. 266, 22333–22340.

Beale, S.I. & Cornejo, J. (1991c). Biosynthesis of the phycobilins. 15,16-Dihydrobiliverdin IXα is a partially reduced intermediate in the formation of phycobilins from biliverdin IXα. J. Biol. Chem. 266, 22341–22345.

Beale, S.I. & Weinstein, J.D. (1991). In: Biosynthesis of Tetrapyrroles (Jordan, P.M., Ed.), pp. 155–235. Elsevier, Amsterdam.

Belnap, W.R. & Haselkorn, R. (1987). Cloning and light regulation of expression of the phycocyanin operon of the cyanobacterium *Anabaena*. EMBO J. 6, 871–874.

Brown, S.B., Houghton, J.D., & Vernon, D.I. (1990). Biosynthesis of phycobilins. Formation of the chromophore of phytochrome, phycocyanin, and phycoerythrin. J. Photochem. Photobiol. B. Biology 5, 3–23.

Bryant, D.A. (1982). Phycoerythrocyanin and phycoerythrin: Properties and occurrence in cyanobacteria. J. Gen. Microbiol. 128, 835–844.

Bryant, D.A. (1987). The cyanobacterial photosynthetic apparatus: comparison to those of higher plants and photosynthetic bacteria. Can. Bull. Fish. Aquat. Sci. 214, 423–500.

Bryant, D.A. (1988). In: Light-Energy Transduction in Photosynthesis: Higher Plant and Bacterial Models (Stevens, S.E., Jr. & Bryant, D.A., Eds.), pp. 62–90. The American Society of Plant Physiologists, Rockville, MD.

Bryant, D.A. (1991). In: Cell Culture and Somatic Cell Genetics of Plants. The Molecular Biology of Plastids and Mitochondria (Bogorad, L. & Vasil, I.K., Eds.), Vol. 7B, pp. 255–298. Academic Press, New York.

Bryant, D.A. (1992). Puzzles of chloroplast ancestry. Current Biology 2, 240–242.

Bryant, D.A., Guglielmi, N., Tandeau de Marsac, N., Castets, A.-M., & Cohen-Bazire, G. (1979). The structure of cyanobacterial phycobilisomes: A model. Arch. Microbiol. 110, 61–75.

Bryant, D.A., Stirewalt, V.L., Glauser, M., Frank, G., Sidler, W., & Zuber, H. (1991). A small multigene family encodes the rod-core linker polypeptides of *Anabaena* sp PCC 7120 phycobilisomes. Gene 107, 91–99.

Capuano, V., Braux, A.-S., Tandeau de Marsac, N., & Houmard, J. (1991). The "anchor polypeptide" of cyanobacterial phycobilisomes. Molecular characterization of the *Synechococcus* sp. PCC6301 *apcE* gene. J. Biol. Chem. 266, 7239–7247.

Conley, P.B., Lemaux, P.G., Lomax, T.L., & Grossman, A.R. (1986). Genes encoding major light-harvesting polypeptides are clustered on the genome of the cyanobacterium *Fremyella diplosiphon*. Proc. Natl. Acad. Sci. USA 83, 3924–3928.

Cornejo, J. & Beale, S.I. (1988). Algal heme oxygenase from *Cyanidium caldarium*. Partial purification and fractionation into three required protein components. J. Biol. Chem. 263, 11915–11921.

Cornejo, J., Beale, S.I., Terry, M.J., & Lagarias, J.C. (1992). Phytochrome assembly. The structure and biological activity of 2(*R*),3(*E*)-phytochromobilin derived from phycobiliproteins. J. Biol. Chem. 267, 14790–14798.

de Lorimier, R., Bryant, D.A., & Stevens, S.E., Jr. (1990). Genetic analysis of a 9-kDa phycocyanin-associated linker polypeptide. Biochim. Biophys. Acta 1019, 29–41.

Dubbs, J.M. & Bryant, D.A. (1991). Molecular cloning and transcriptional analysis of the *cpeBA* operon of the cyanobacterium *Pseudanabaena* species PCC7409. Mol. Microbiol. 5, 3073–3085.

Duerring, M., Huber, R., & Bode, W. (1988). The structure of γ-N-methylasparagine in C-phycocyanin from *Mastigocladus laminosus* and *Agmenellum quadruplicatum*. FEBS Lett. 236, 167–170.

Duerring, M., Schmidt, G.B., & Huber, R. (1991). Isolation, crystallization, crystal structure analysis, and refinement of constitutive C-phycocyanin from the chromatically adapting cyanobacterium *Fremyella diplosiphon* at 1.66 Å resolution. J. Mol. Biol. 217, 577–592.

Duerring, M., Huber, R., Bode, W., Ruembeli, R., & Zuber, H. (1990). Refined three-dimensional structure of phycoerythrocyanin from the cyanobacterium *Mastigocladus laminosus* at 2.7 Å. J. Mol. Biol. 211, 633–644.

Eberlein, M. & Kufer, W. (1990). Genes encoding both subunits of phycoerythrocyanin, a light-harvesting biliprotein from the cyanobacterium *Mastigocladus laminosus*. Gene 94, 133–136.

Fairchild, C.D., Zhao, J., Zhou, J., Colson, S.E., Bryant, D.A., & Glazer, A.N. (1992). Phycocyanin α-subunit phycocyanobilin lyase. Proc. Natl. Acad. Sci. USA 89, 7017–7021.

Ficner, R. & Huber, R. (1993). Refined crystal structure of phycoerythrin from *Porphyridium cruentum* at 0.23-nm resolution and localization of the gamma subunit. Eur. J. Biochem. 218,103–106.

Ficner, R., Lobeck, K., Schmidt, G., & Huber, R. (1992). Isolation, crystallization, crystal structure analysis and refinement of B-phycoerythrin from the red alga *Porphyridium sordidum* at 2.2 angstrom resolution. J. Mol. Biol. 228, 935–950.

Gantt, E. (1980). Structure and function of phycobilisomes: Light-harvesting complexes in red and blue-green algae. Intern. Rev. Cytol. 66, 45–80.

Gantt, E. (1990). In: Biology of the Red Algae (Cole, C.M. & Sheath, R.G., Eds.), pp. 203–219. Cambridge University Press, Cambridge.

Gantt, E., Lipschultz, C.A., & Zilinskas, B. (1976). Further evidence for a phycobilisome model from selective dissociation, fluorescence emission, immunoprecipitation, and electron microscopy. Biochim. Biophys. Acta 430, 375–388.

Gingrich, J.C., Lundell, D.J., & Glazer, A.N. (1983). Core substructure in cyanobacterial phycobilisomes. J. Cell. Biochem. 22, 1–14.

Glauser, M., Bryant, D.A., Frank, G., Wehrli, E., Sidler, W., & Zuber, H. (1992a). Phycobilisome structure in the cyanobacteria *Mastigocladus laminosus* and *Anabaena* sp. PCC7120: a new model. Eur. J. Biochem. 205, 907–915.

Glauser, M., Sidler, W., & Zuber, H. (1992b). Isolation, characterization, and reconstitution of phycobiliprotein rod-core linker polypeptide complexes from the phycobilisome of *Mastigocladus laminosus*. Photochem. Photobiol. 57, 344–351.

Glazer, A.N. (1981). In: The Biochemistry of Plants. Photosynthesis (Hatch, M.D., & Boardman, N.K., Eds.), Vol. 8, pp. 51–96. Academic Press, New York.

Glazer, A.N. (1982). Phycobilisomes: structure and dynamics. Ann. Rev. Microbiol. 36, 173–198.

Glazer, A.N. (1985). Light-harvesting by phycobilisomes. Ann. Rev. Biophys. Biophys. Chem. 14, 47–77.

Glazer, A.N. (1989). Light guides. Directional energy transfer in a photosynthetic antenna. J. Biol. Chem. 264, 1–4.

Glazer, A.N., Lundell, D.J., Yamanaka, G., & Williams, R.C. (1983). The structure of a "simple" phycobilisome. Ann. Microbiol. (Inst. Pasteur) 134B, 159–180.

Glazer, A.N., Williams, R.C., Yamanaka, G., & Schachman, H.K. (1979). Characterization of cyanobacterial phycobilisomes in zwitterionic detergents. Proc. Natl. Acad. Sci. USA 76, 6162–6166.

Grossman, A.R. (1990). Chromatic adaptation and the events involved in phycobilisome biosynthesis. Plant, Cell, and Environment 13, 651–666.

Grossman, A.R., Lemaux, P.G., Conley, P.B., Bruns, B.U., & Anderson, L.K. (1988). Characterization of phycobiliprotein and linker polypeptide genes in *Fremyella diplosiphon* and their regulated expression during complementary chromatic adaptation. Photosynthesis Res. 17, 23–56.

Grossman, A.R., Schaefer, M.R., Chiang, G.G., & Collier, J.L. (1993). The phycobilisome, a light-harvesting complex responsive to environmental conditions. Microbiol. Rev. 57, 725–749.

Houmard, J., Capuano, V., Colombano, M.V., Coursin, T., & Tandeau de Marsac, N. (1990). Molecular characterization of the terminal energy acceptor of cyanobacterial phycobilisomes. Proc. Natl. Acad. Sci. USA 87, 2152–2156.

Jahn, W., Steinbiss, J., & Zetsche, K. (1984). Light intensity adaptation of phycobiliprotein content of the red alga *Porphyridium*. Planta 161, 536–539.

Kana, T.M. & Glibert, P.M. (1987). Effect of irradiances up to 2000 μE m^{-2} s^{-1} on marine *Synechococcus* WH7803-II. Photosynthetic responses and mechanisms. Deep-Sea Research 34, 497–516.

Klotz, A.V. & Glazer, A.N. (1985). Characterization of the bilin attachment sites in R-phycoerythrin. J. Biol. Chem. 260, 4856–4863.

Klotz, A.V. & Glazer, A.N. (1987). γ-N-Methylasparagine in phycobiliproteins. Occurrence, location, and biosynthesis. J. Biol. Chem. 262, 17350–17355.

Klotz, A.V., Leary, J.A., & Glazer, A.N. (1986). Post-translational modification of asparaginyl residues. Identification of β-71 γ-*N*-methylasparagine in allophycocyanin. J. Biol. Chem. 261, 15891–15894.

Lind, L.K. (1988). Cloning and characterization of genes for light-harvesting antenna polypeptides in *Synechococcus* 6301. Ph.D. thesis, University of Umeå.

Lomax, T.L., Conley, P.B., Schilling, J., & Grossman, A.R. (1987). Isolation and characterization of light-regulated phycobilisome linker polypeptide genes and their transcription as polycistronic mRNA. J. Bacteriol. 169, 2675–2684.

Lundell, D.J., Glazer, A.N., DeLange, R.J., & Glazer, A.N. (1984). Bilin attachment sites in the α- and β-subunits of B-phycoerythrin. Amino acid sequence studies. J. Biol. Chem. 259, 3580–3592.

Mazel, D. & Marlière, P. (1989). Adaptive eradication of methionine and cysteine from cyanobacterial light-harvesting proteins. Nature (London) 341, 245–248.

Mazel, D., Houmard, J., & Tandeau de Marsac, N. (1988). A multigene family in *Calothrix* sp. PCC 7601 encodes phycocyanin, the major component of the cyanobacterial light-harvesting antenna. Mol. Gen. Genet. 211, 296–304.

Mörschel, E., Koller, K.-P., Wehrmeyer, W., & Schneider, H. (1977). Biliprotein assembly in the disc-shaped phycobilisomes of *Rhodella violacea*. I. Electron microscopy of phycobilisomes *in situ* and analysis of their architecture after isolation and negative staining. Cytobiologie 16, 118–129.

Mörschel, E. & Rhiel, E. (1987). Electron Microscopy of Proteins. 6. Membraneous Structures (Harris, J.R. & Horne, R.W., Eds.), pp. 209–254. Academic Press, London.

Ong, L.J. & Glazer, A.N. (1991). Phycoerythrins of marine unicellular cyanobacteria. I. Bilin types and locations and energy transfer pathways in *Synechococcus* spp. phycoerythrins. J. Biol. Chem. 266, 9515–9527.

Porter, G., Tredwell, C.J., Searle, G.F.W., & Barber, J. (1978). Picosecond time resolved energy transfer in *Porphyridium cruentum*. Part I. In the intact alga. Biochim. Biophys. Acta 501, 232–245.

Raps, S., Kycia, J.H., Ledbetter, M.C., & Siegelman, H.W. (1985). Light intensity adaptation and phycobilisome composition of *Microcystis aeruginosa*. Plant Physiol. 79, 983–987.

Rümbeli, R., Suter, F., Wirth, M., Sidler, W., & Zuber, H. (1987). Isolation and localization of N-methylasparagine in phycobiliproteins from the cyanobacterium *Mastigocladus laminosus*. Biol. Chem. Hoppe-Seyler 368, 1401–1406.

Sauer, K. & Scheer, H. (1988). Excitation transfer in C-phycocyanin. Förster transfer rate and exciton calculation based on new crystal structure data for C-phycocyanins from *Agmenellum quadruplicatum* and *Mastigocladus laminosus*. Biochem. Biophys. Acta 936, 157–170.

Schirmer, T., Bode, W., & Huber, R. (1987). Refined three-dimensional structures of two cyanobacterial phycocyanins at 2.1 and 2.5 Å resolution. J. Mol. Biol. 196, 677–695.

Schirmer, T. Huber, R., Schneider, M., Bode, W., Miller, M., & Hackert, M. (1986). Crystal structure analysis and refinement at 2.5 Å of hexameric C-phycocyanin from the cyanobacterium *Agmenellum quadruplicatum*. J. Mol. Biol. 188, 651–676.

Schirmer, T. & Vincent, M.G. (1987). Polarized absorption and fluorescence spectra of single crystals of C-phycocyanin. Biochim. Biophys. Acta 893, 379–385.

Schmidt, A., Erdle, I., & Köst, H.P. (1982). Changes of C-phycocyanin in *Synechococcus* 6301 in relation to growth on various sulfur compounds. Z. Naturforsch. 37, 870–876.

Sidler, W., Kumpf, B., Rüdiger, W., & Zuber, H. (1986). The complete amino acid sequence of C-phycoerythrin from the cyanobacterium *Fremyella diplosiphon*. Biol. Chem. Hoppe-Seyler 367, 627–642.

Sidler, W., Kumpf, B., Suter, F., Klotz, A.V., Glazer, A.N., & Zuber, H. (1989). The complete amino acid sequence of the α- and β-subunits of B-phycoerythrin from the rhodophytan alga *Porphyridium cruentum*. Biol. Chem. Hoppe-Seyler 370, 115–124.

Siebzenrübl, S., Fischer, R., & Scheer, H. (1987). Chromophore assignment in C-phycocyanin from *Mastigocladus laminosus*. Z. Naturforsch. Sect. C. Biosci. 42, 258–262.

Stanier, R.Y. & Cohen-Bazire, G. (1977). Phototrophic prokaryotes: the cyanobacteria. Ann. Rev. Microbiol. 31, 225–274.

Stanier, R.Y., Sistrom, W.R., Hansen, T.A., Whitton, B.A., Castenholz, R.W., Pfennig, N., Gorlenko, V.N., Kondratieva, E.N., Eimhjellen, K.E., Whittenbury, R., Gherna, R.L., & Trüper, H.G. (1978). Proposal to place the nomenclature of the cyanobacteria (blue-green algae) under the rules of the International Code of Nomenclature of Bacteria. Int. J. Syst. Bact. 28, 335–336.

Swanson, R.V. (1991). Studies on posttranslational modifications of phycobiliproteins. Ph.D. thesis, University of California, Berkeley.

Swanson, R.V. & Glazer, A.N. (1990). Phycobiliprotein methylation. Effect of γ-N-methylasparagine residue on energy transfer in the phycobilisome. J. Mol. Biol. 214, 787–796.

Swanson, R.V., de Lorimier, R., & Glazer, A.N. (1992). Genes encoding the phycobilisome rod substructure are clustered on the *Anabaena* chromosome: Characterization of the phycoerythrocyanin operon. J. Bacteriol. 174, 2640–2647.

Swanson, R.V., Ong, L.J., Wilbanks, S.M., & Glazer, A.N. (1991). Phycoerythrins of marine unicellular cyanobacteria. II. Characterization of phycobiliproteins with unusually high phycourobilin content. J. Biol. Chem. 266, 9528–9534.

Swanson, R.V., Zhou, J., Leary, J.A., Williams, T., de Lorimier, R., Bryant, D.A., & Glazer, A.N. (1992). Characterization of phycocyanin produced by *cpcE* and *cpcF* mutants and identification of an intergenic suppressor of the defect in bilin attachment. J. Biol. Chem. 267, 16146–16154.

Tandeau de Marsac, N. (1991). In: Cell Culture and Somatic Cell Genetics of Plants. The Molecular Biology of Plastids and Mitochondria (Bogorad, L. & Vasil, I.K., Eds.), Vol. 7B, pp. 417–446. Academic Press, New York.

Tandeau de Marsac, N. & Houmard, J. (1993). Adaptation of cyanobacteria to environmental stimuli: new steps towards molecular mechanisms. FEMS Microbiol. Rev. 104, 119–190.

Tandeau de Marsac, N., Castets, A.-M., & Cohen-Bazire, G. (1980). Wavelength modulation of phycoerythrin synthesis in *Synechocystis* sp. 6701. J. Bacteriol. 142, 310–314.

Tandeau de Marsac, N., Mazel, D., Damerval, T., Guglielmi, G., Capuano, V., & Houmard, J. (1988). Photoregulation of gene expression in the filamentous cyanobacterium *Calothrix* sp. PCC 7601: light-harvesting complexes and cell differentiation. Photosynthesis Res. 18, 99–132.

Tandeau de Marsac, N., Mazel, D., Capuano, V., Damerval, T., & Houmard, J. (1990). In: Molecular Biology of Membrane-Bound Complexes in Phototrophic Bacteria (Drews, G. & Dawes, E.A., Eds.), pp. 143–153. Plenum Press, New York.

Terry, M.J., Wahleithner, J.A., & Lagarias, J.C. (1993). Biosynthesis of the plant photoreceptor phytochrome. Arch Biochem. Biophys. 306, 1–15.

Troxler, R.F., Brown, A.S., & Brown, S.B. (1979). Bile pigment synthesis in plants. Mechanism of ^{18}O incorporation into phycocyanobilin in the unicellular rhodophyte *Cyanidium caldarium*. J. Biol. Chem. 254, 3411–3418.

Waaland, J.R., Waaland, S.D., & Bates, G. (1974). Chloroplast structure and membrane composition in the red alga *Griffithsia pacifica*. Regulation by light intensity. J. Phycol. 10, 193–199.

Wahleithner, J.A., Li, L., & Lagarias, J.C. (1991). Expression and assembly of spectrally active recombinant holophytochrome. Proc. Natl. Acad. Sci. USA 88, 10387–10391.

Wanner, G., Henkelmann, G., Schmidt, A., & Köst, H.P. (1986). Nitrogen and sulfur starvation of the cyanobacterium *Synechococcus* 6301. An ultrastructural, morphometrical, and biochemical comparison. Z. Naturforsch. 41c, 741–750.

Wehrmeyer, W. (1983). In: Photosynthetic Prokaryotes: Cell Differentiation and Function (Papageorgiou, C.C. & Packer, L., Eds.), pp. 1–22. Elsevier Biomedical, New York.

Wilbanks, S.M. (1992). Adaptive variation in phycoerythrins. Ph.D. thesis, University of California, Berkeley.

Wilbanks, S.M. & Glazer, A.N. (1993). Rod structure of a phycoerythrin-II-containing phycobilsome. I. Organization and sequence of the gene cluster encoding the major phycobiliprotein rod components in the genome of marine *Synechococcus* sp. WH8020. J. Biol. Chem. 268, 1226–1235.

Wilbanks, S.M., de Lorimier, R., & Glazer, A.N. (1991). Phycoerythrins of marine unicellular cyanobacteria. II. Sequence of a class II phycoerythrin. J. Biol. Chem. 266, 9535–9539.

Yamanaka, G. & Glazer, A.N. (1981). Dynamic aspects of phycobilisome structure: modulation of phycocyanin content of *Synechococcus* phycobilisomes. Arch Microbiol. 130, 23–30.

Yeh, S.W., Glazer, A.N., & Clark, J.H. (1986). Control of bilin transition dipole moment direction by macromolecular assembly: Energy transfer in allophycocyanin. J. Phys. Chem. 90, 4578–4580.

Zhou, J., Gasparich, G.E., Stirewalt, V.L., de Lorimier, R., & Bryant, D.A. (1992). The *cpcE* and *cpcF* genes of *Synechococcus* sp. PCC 7002: Construction and phenotypic characterization of interposon mutants. J. Biol. Chem. 267, 16138–16145.

Zuber, H. (1987). In: The Light Reactions (Barber, J., Ed.), pp. 197–259. Elsevier Biomedical, Amsterdam.

PHOTOPROTECTION AND PHOTOINHIBITORY DAMAGE

W. S. Chow

Advances in Molecular and Cell Biology
Volume 10, pages 151–196.

ISBN: 1-55938-710-6

I. INTRODUCTION

Photosynthesis occurs with maximum efficiency in light-limiting conditions, when the absorption of about 9 to 10 photons leads to the evolution of one O_2 molecule in a diverse range of nonstressed plants performing C_3 photosynthesis (Björkman and Demmig, 1987). On the other hand, plants exposed to full sunlight absorb more photons than are needed for photosynthesis at a maximum rate, thereby losing photosynthetic efficiency. A leaf exposed to 2000 μmol photons m^{-2} s^{-1}, for example, absorbs about 1700 μmol photons m^{-2} s^{-1}. If the light-saturated rate of photosynthesis corresponds to 40 μmol O_2 m^{-2} s^{-1}, and the evolution of each O_2 molecule requires 10 photons, then only 400 μmol photons m^{-2} s^{-1} are needed to sustain the maximum photosynthetic rate. That is, approximately one-quarter of the absorbed photons are usefully employed for photochemistry; the remaining photons are dissipated as heat, and a small proportion (approximately 1%) as chlorophyll *a* fluorescence.

The *controlled* dissipation of surplus photons as heat represents an important process that helps to protect the photosynthetic apparatus, for a debilitating effect on photosynthetic capability may occur when more photons are absorbed than can be utilized or dissipated in an orderly manner. This debilitating effect of excess visible light or photoinhibitory damage is observed whenever "the rates of transfer of excitation energy from light-harvesting pigment assemblies (the antenna) to photochemical reaction centers, are in excess of the rates of transfer of excitation energy from the reaction centers to the electron-transfer chain (the transducers)" (Osmond, 1981).

Photosynthetic organisms, being largely immobile in their immediate light environment, have developed various strategies to prevent photoinhibitory damage. At the whole-plant level, leaf orientation is important in determining the amount of radiation intercepted (Ludlow and Björkman, 1984; Öquist and Huner, 1991). An extreme example is *Eucalyptus pauciflora* (snow gum), which has isobilateral leaves growing in a vertical position (Kirschbaum and Farquhar, 1984). At the cellular level, the spatial distribution of chloroplasts in the cytoplasm may be an important strategy (Chow et al., 1988). At the macromolecular level, protection against photoinhibitory damage can sometimes be brought about via protein phosphorylation (whereby a phosphorylated portion of the light-harvesting antenna is detached from the photosystem II complex) in pea thylakoids (Horton and Lee, 1985; Habash and Baker, 1987); however, this is not a universal protective mechanism, since it does not occur in wheat thylakoids (Habash and Baker, 1987) possibly because of concurrent inhibition of light-saturated electron-transport capacity by protein phosphorylation in wheat thylakoids (Habash and Baker, 1990). At the substrate level, scavenging systems, which remove reactive oxygen generated during photosynthesis, appear to be important for protection against photoinhibitory damage (Asada and Takahashi, 1987; Richter et al, 1990b). When the supply of photosynthetically derived reductants exceeds demand, there is the potential within chloroplasts to reduce O_2 to the superoxide radical and H_2O_2 (the Mehler reaction). Provided the products of photosynthetic O_2 reduction are rapidly removed by enzymes to prevent toxicity, the Mehler reaction would maintain the linear electron-transport chain in a more oxidized state, thus helping to prevent photoinhibitory damage while sustaining a high rate of ATP production (Robinson, 1988). Under conditions of restricted CO_2 supply to photosynthesis, photorespiration (whereby CO_2 cycling helps to dissipate excess excitation energy) may also be important for counteracting photoinhibitory damage (Powles and Osmond, 1979).

Because damage to the photosynthetic apparatus may be caused by ultraviolet (see Bornman, 1989) as well as by visible light, Powles (1984) restricted the use of the term "photoinhibition" to the reduction of photosynthetic capacity, independent of photobleaching, induced by visible light in the wavelength band 400 to 700 nm. More recently, the term photoinhibition is sometimes used to include not only damage to the photosynthetic apparatus but also certain photoprotective mechanisms which reduce photosynthetic efficiency as a result of high-light treatments (e.g., Krause, 1988; Demmig-Adams and Adams, 1992a). Therefore, in this review, the term "photoinhibitory damage" is used to distinguish it from photoprotective aspects of photoinhibition.

This review will not cover all protective strategies, including those mentioned above, but will focus on some biochemical photoprotection strategies (listed in Section II) which are adopted by photosynthetic organisms in response to high light. The nature of photoinhibitory damage which results when photoprotective strategies fail will also be discussed, but in less detail, since the topic has been reviewed by others. Finally, there will be a brief discussion of recovery from photoinhibitory damage. This review represents a selection of aspects of a fundamental dilemma among photosynthetic organisms: How to maximize the efficiency of light capture and utilization in low light, and how to avoid the effects of too much light. (Anderson and Osmond, 1987; Barber and Andersson, 1991). There are a number of modern reviews on aspects of photoprotection (Krause, 1988; Demmig-Adams, 1990; Barber and Andersson, 1991; Demmig-Adams and Adams, 1992a,b) and on photoinhibitory damage (Osmond, 1981; Powles, 1984; Kyle and Ohad, 1986; Kyle et al., 1987; Krause, 1988; Barber and Andersson, 1991; Prasil et al., 1991), which may also be consulted.

II. PHOTOPROTECTION

A. Protection via ΔpH-Dependent Quenching of Excitation Energy

A pH difference, typically 3 pH units, is developed across the thylakoid membrane during illumination, such that the intrathylakoid space is acidic when the pH of the stroma of the chloroplast is near-neutral or slightly alkaline (Rottenberg et al., 1972). Such an "energized" state drives the formation of ATP from ADP and inorganic phosphate (Jagendorf, 1977).

The energized state, however, may have other roles in regulating the energy budget of chloroplasts. As mentioned above, chlorophyll *a* fluorescence emission (normally abbreviated to "chlorophyll fluorescence") represents a pathway of excitation-energy dissipation in the chloroplast. Thus, increased dissipation of excitation energy either by direct heat loss or photochemical conversion would constitute enhanced pathways in direct competition with fluorescence emission, leading to fluorescence quenching. A quenching of the yield of chlorophyll fluorescence associated with the energized state of chloroplasts was first reported by Murata and Sugahara (1969). This "energy-dependent" quenching is defined by the chlorophyll fluorescence non-photochemical quenching coefficient, q_{Ne}, (formerly called q_E), and is linearly related to the intrathylakoid H^+ concentration (Briantais et al., 1979). Mechanistically, q_{Ne} is thought to result from conformational changes within the light-harvesting pigment beds of photosystem II (PS II). These conformational changes, in turn, were assumed to arise by protonation of the intrathylakoid membrane surfaces (Krause et al., 1982; see also Krause and Weis, 1991) when counterions (principally Mg^{2+}) are exchanged for protons translocated into the intrathylakoid space during normal photosynthesis (Hind et al., 1974; Chow et al., 1976). More recently, these conformational changes are specifically proposed to involve aggregation of the light-harvesting complexes of PS II (LHC II, see below).

Chlorophyll fluorescence emission is itself a minor pathway of energy dissipation, and the energy-dependent quenching of chlorophyll fluorescence is thought to be a result of (not the cause of) increased thermal deactivation of the excited state of chlorophyll. Furthermore, it is thought to be a good indicator of the excitation status of chlorophyll. If thermal deactivation increases, one expects that it occurs at the expense of the quantum yield of photochemistry. Indeed, using intact spinach chloroplasts, Krause and Laasch (1987) demonstrated that the quantum yield of CO_2-dependent O_2 evolution decreased linearly with the increase in q_{Ne}. This diminution of quantum yield of O_2 evolution can be partly attributed to a more reduced state of Q_A (primary quinone acceptor in PS II) which typically accompanies an increase in q_{Ne} (Krause and Laasch, 1987). However, after allowing for the population of closed PS II reaction centers with reduced Q_A, Weis and Berry (1987) found that the quantum yield of *open* PS II reaction centers (estimated as Φ_s/q_p, where Φ_s is the steady-state quantum yield, and q_p is the photochemical quenching coefficient and a measure of oxidized Q_A) still decreased linearly with increase in

q_{Ne}. They proposed a model in which there is a ΔpH-dependent conversion of PS II from a state with high photochemical and fluorescence yields to a quenched state of low photochemical and fluorescence yields but high rates of thermal deactivation.

Whether the negative, linear relationship between Φ_s/q_p and q_{Ne} still holds at low irradiance is still open to question. Horton and Hague (1988) and Rees and Horton (1990) reported that the relationship was nonlinear in low light. Öquist and Chow (1992) observed a negative, near-linear relationship between the two parameters for a number of plant species grown under various conditions. In any case, a quenched state associated with the transthylakoid ΔpH gives rise to a loss of quantum yield of photochemistry in *open* PS II reaction centers.

A corollary is that such a quenched state, with efficient heat dissipation, should confer protection against photoinhibitory damage of chloroplasts. Indeed, Fork et al. (1986) proposed that the formation of the high-energy state, giving rise to an increased rate constant of radiationless transition in PS II reaction centers, could serve as a photoprotective mechanism in plants. Similarly, Krause and Behrend (1986) used uncouplers to show that photoinhibitory damage of isolated chloroplasts is exacerbated by decreased thermal deactivation, as expressed by ΔpH-dependent chlorophyll *a* fluorescence quenching. Ögren (1991) confirmed that leaves suffer greater photoinhibitory damage in the presence of the uncoupler, nigericin. Oxborough and Horton (1988) also found that energy-dependent quenching provided some protection against photoinhibitory damage, although their interpretation did not favor a direct effect of energy-dependent quenching on Φ_s/q_p. The mechanism of ΔpH-dependent quenching of excitation energy will be considered later in the light of more recent findings. Meanwhile, it is necessary to consider the photoprotective roles played by carotenoids in general, and the violaxanthin cycle in particular.

B. Protection via the Singlet State of Zeaxanthin

The protective role of carotenoids in preventing photooxidative damage is well known; without it, there would be no photosynthesis in the presence of oxygen (Cogdell, 1988). Carotenoids prevent photooxidative damage in two ways.

1. By scavenging the powerful oxidant, singlet O_2, which is formed via the reaction of triplet excited Chl with (ordinary) triplet O_2:

$$^{1}Chl + h\nu \rightarrow {}^{1}Chl^{*} \text{ (singlet excited Chl)}$$

$$^{1}Chl^{*} \rightarrow {}^{3}Chl^{*} \text{ (triplet excited Chl)}$$

$$^{3}Chl^{*} + {}^{3}O_2 \rightarrow {}^{1}Chl + {}^{1}O_2^{*} \text{ (singlet oxygen)}$$

$$^{1}O_2^{*} + {}^{1}Car \rightarrow {}^{3}O_2 + {}^{3}Car^{*} \text{ (triplet carotenoid)}$$

$$^3Car^* \rightarrow {}^1Car + heat$$

2. By quenching $^3Chl^*$ and so prevent singlet O_2 production:

$$^3Chl^* + {}^1Car \rightarrow {}^3Car^* + {}^1Chl$$

$$^3Car^* \rightarrow {}^1Car + heat$$

The protective role of carotenoids in detoxifying triplet Chl or singlet O_2 is depicted on the right-hand side of Figure 1.

The protection against photooxidative damage offered by carotenoids in general relates to coping with the effects arising from the formation of triplet chlorophyll. Under high-light conditions, however, chlorophyll molecules are continually raised to their singlet excited states, of which only a relatively small proportion undergo

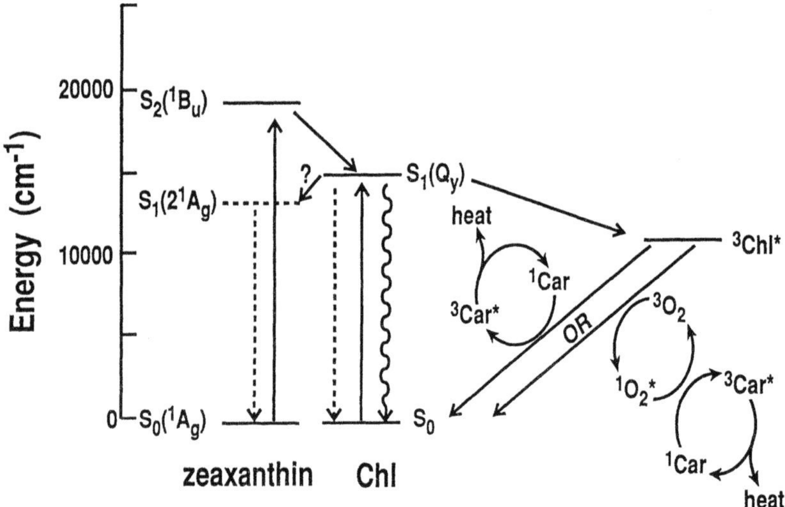

Figure 1. Low-energy excited states of zeaxanthin and Chl *a*, and the dissipation of the triplet excited state of Chl *a* directly or indirectly via carotenoids (Car). Energy transfer within or between molecules are indicated by *solid arrows*, the *vertical arrows* denoting excitation by light absorption. *Broken arrows* denote radiationless de-excitation, and the *wavy arrow* indicates fluorescence emission. The S_2 level of zeaxanthin corresponds to an absorption maximum at 515 nm in the solvent, CS_2, and the S_1 level (which cannot be reached by direct light absorption) is assumed to be 6000 cm^{-1} lower than S_2 (see Cosgrove et al., 1990). As the S_1 level of zeaxanthin appears to lie below the S_1 level of Chl *a*, energy transfer from the latter to the former is a strong possibility, although yet to be confirmed (indicated by '?'). The triplet state of Chl *a*, formed from S_1 by intersystem crossing, is dissipated either via carotenoid (^{1}Car) directly, or via triplet O_2, followed by an interaction of singlet O_2 with singlet carotenoid.

intersystem crossing to form the comparatively slow-decaying triplet states. Therefore, to avoid an excessive buildup of singlet chlorophyll, and the associated problem of electron transport not being able to keep pace with the generation of the most powerful oxidant in photosynthesis, $P680^+$ (Thompson and Brudvig, 1988), it is necessary for the excitation energy of singlet chlorophyll to be dissipated as rapidly as possible.

One way in which a quenching/deactivation of the singlet excited state of chlorophyll can occur is via energy transfer to zeaxanthin, followed by radiationless energy dissipation (Cosgrove et al., 1990; Demmig-Adams, 1990; Demmig-Adams and Adams, 1992b). For many years, it has been generally accepted that carotenoids, when acting as accessory light-harvesting pigments, transfer their excitation energy to chlorophyll. Therefore, the idea that energy may be transferred *from* the singlet state of chlorophyll *to* a singlet excited state of zeaxanthin appears novel and merits close examination. Before such an examination, however, some introductory remarks about the violaxanthin cycle and its role in nonradiative dissipation of excitation energy seem appropriate.

Following the first report of Sapozhnikov et al. (1957) that the level of violaxanthin in leaves could be reversibly altered by light and dark treatments, numerous studies have attempted to elucidate the function of these changes (e.g., Yamamoto, 1979), which are now known as the violaxanthin cycle or the xanthophyll cycle. Typically, when leaves are exposed to high irradiance, violaxanthin decreases, antheraxanthin increases transiently, and zeaxanthin increases, as de-epoxidation proceeds. The process is cyclic, because the level of violaxanthin recovers in the dark or in subdued light, when epoxidation of zeaxanthin and antheraxanthin occurs (Figure 2). While it was recognized that the violaxanthin cycle occurs in the thylakoid membrane and responds to light-induced proton translocation, its function remained unclear for a long time. Indeed, the observations that the forward and back reactions of the cycle are essentially dark reactions, despite their indirect regulation by light, led some to suggest that there was no direct role of the cycle in strictly light-dependent processes, such as photoprotection.

In recent years, however, the opposite conclusion has been reached by Demmig-Adams and co-workers, and Björkman and colleagues. Demmig et al. (1987) reported a massive formation of zeaxanthin and a decrease in violaxanthin, which was correlated with a decrease in variable chlorophyll fluorescence when leaves were exposed to high light in the absence of CO_2. The fluorescence changes were indicative of increased nonradiative energy dissipation, and the authors concluded that zeaxanthin may act as a quencher of fluorescence. Further studies by Demmig et al. (1988) established that the increase in zeaxanthin, on exposing *Nerium oleander* plants to high light and water stress, was correlated with an increase in the rate of nonradiative energy dissipation in the antenna chlorophyll which, in turn, caused a decrease in PS II photochemical efficiency.

If the formation of zeaxanthin enhances nonradiative energy dissipation, then acclimation to a high-irradiance environment should result in a greater pool of

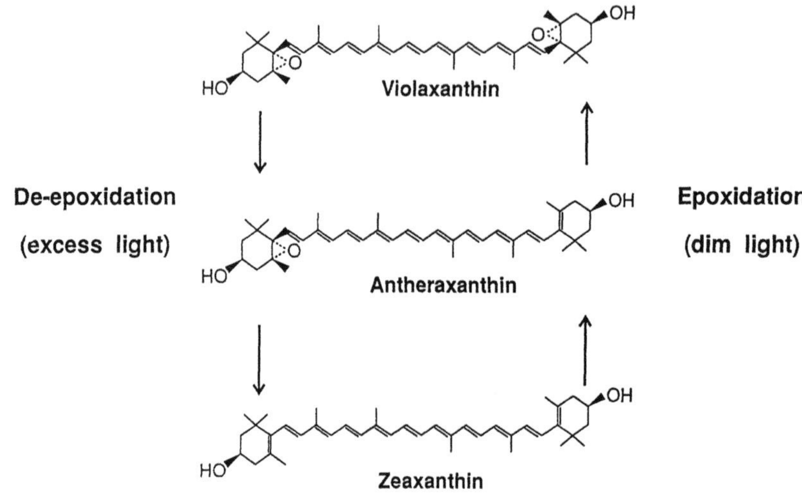

Figure 2. The violaxanthin cycle. The de-epoxidation of violaxanthin to zeaxanthin via antheraxanthin occurs in excess light, and is promoted by an acidic intrathylakoid pH, as well as a more reduced state of NADP. The epoxidation of zeaxanthin to violaxanthin via antheraxanthin occurs in limiting light or darkness.

violaxanthin-cycle components, as indeed was found in leaves of a number of different species (Thayer and Björkman, 1990; Demmig-Adams and Adams, 1992c). Moreover, a sudden transfer of plants from weak to strong light increased the violaxanthin-cycle components several fold (Demmig-Adams et al., 1989). The acclimation to high growth-irradiance is accompanied by an increased capacity for zeaxanthin formation. Conversely, under conditions where zeaxanthin formation is inhibited by dithiothreitol, a large part of the non-photochemical quenching of chlorophyll fluorescence, induced by excessive light, is eliminated (Bilger et al., 1989; Bilger and Björkman, 1990; Demmig-Adams et al., 1990). Furthermore, it has been shown photoacoustically that *in vivo* heat dissipation increases concomitantly with the conversion of violaxanthin to zeaxanthin in pea leaves exposed to strong light (Havaux et al., 1991a). While Havaux et al. (1991a) interpreted their results by ascribing a lipid-protection role to zeaxanthin in strong light, the possibility of zeaxanthin acting as a quencher of excitation energy of chlorophyll has not been ruled out. The above results together provide a strong case for a role of zeaxanthin in nonradiative dissipation of excitation, particularly in high-light environments (Demmig-Adams, 1990; Demmig-Adams and Adams, 1992a,b). What is the mechanism of the nonradiative energy dissipation process?

Carotenoids are intensely colored, due to strong absorption of light as molecules undergo a transition from the ground state S_o (^1A_g) to an excited state S_2 (^1B_u) (Figure 1). An achievement in recent years has been the discovery that below S_2

lies an optically forbidden singlet state, S_1 (2^1A_g), which cannot be reached directly from the ground state by light absorption (reviewed by Truscott, 1990; Koyama, 1991). This "hidden" state was first suggested by high-resolution optical spectroscopic examination of polyenes (with at least four conjugated double bonds) related to the visual chromophore (Christensen and Kohler, 1976) and by a Raman excitation experiment with β-carotene (Thrash et al., 1977). The S_1 state was initially suggested to lie at approximately 3500 cm^{-1} below S_2, at a level above the lowest singlet excited state of chlorophyll. This property has been widely cited in discussions of the antenna function of carotenoids in photosynthesis (Thrash et al., 1979; Siefermann-Harms, 1985) where energy is transferred from carotenoids to chlorophyll. Recent picosecond transient absorption studies on β-carotene have also been interpreted to support a relatively small $S_2 - S_1$ energy difference (Hashimoto and Koyama, 1989a) which implies that S_1 lies above the lowest singlet excited state of chlorophyll.

However, this relatively small $S_2 - S_1$ energy difference for carotenoids is not consistent with trends noted in shorter polyenes (Snyder et al., 1985). For example, a comparison of unsubstituted polyene hydrocarbons, with 4 to 7 conjugated double bonds, showed that the energy levels of both S_2 and S_1 decreased with an increase in the number of double bonds but, more importantly, that the $S_2 - S_1$ energy difference increased with conjugation length, reaching 6300 cm^{-1} in this series. For longer polyenes with more than 8 or 9 double bonds (e.g., β-carotene and zeaxanthin, each with 11 conjugated double bonds), the S_1 state is not easily detected in standard fluorescence or absorption experiments. This technical difficulty has hampered the exact location of the S_1 state in carotenoids of biological interest. Nevertheless, Cosgrove et al. (1990) estimated the $S_2 - S_1$ energy difference to be approximately 5500–6500 cm^{-1} for polyenes of conjugation length comparable to β-carotene. For zeaxanthin, the S_2 state lies just under 20,000 cm^{-1}, which corresponds to an absorption maximum at 515 nm in the solvent, CS_2; thus, the S_1 level could be at (or lower than) 14,000 cm^{-1}. This is less than 15,000 cm^{-1}, the energy level for the lowest singlet excited state of chlorophyll, corresponding to an absorption maximum of 660 nm (see Figure 1). Assuming that this estimation of S_1 for zeaxanthin is correct, energy transfer from chlorophyll to zeaxanthin is a definite possibility.

If energy is transferred from chlorophyll to zeaxanthin, how does the latter dissipate the energy as heat? A relatively large $S_2 - S_1$ energy difference for zeaxanthin implies a small $S_1 - S_o$ energy gap which, according to the energy-gap law (Cosgrove et al., 1990), increases the rate constant for nonradiative dissipation of the S_1 state. If zeaxanthin is an efficient sink for excitation energy, one would expect its S_1 state to be rapidly deactivated by radiationless transition, implying that S_1 is short-lived. Indeed, transient absorption (Hashimoto and Koyama, 1989a) and resonance Raman (Hashimoto and Koyama, 1989b) experiments indicate that the lifetime of the S_1 state of carotenoids is approximately 10 ps, which is of the same order of magnitude as the intrinsic charge transfer time of approximately 2.7

ps in the PS II reaction center (Schatz et al., 1988) and 2.6 ps at 283 K in the photosynthetic bacterial reaction center (Chan et al., 1991).

If zeaxanthin accepts and dissipates energy transferred from singlet chlorophyll, then violaxanthin, formed from zeaxanthin by epoxidation in limiting light or darkness (Demmig-Adams, 1990), should not possess this property; otherwise the quantum efficiency of light-limited photosynthesis will be less than maximal due to excessive heat dissipation. Upon epoxidation of zeaxanthin to form violaxanthin, the most obvious change in chemical structure is the decrease of conjugation length from 11 to 9 double bonds. Since the $S_2 - S_1$ energy difference decreases with decreasing conjugation length while the energy of S_2 is simultaneously raised (Snyder et al., 1985), it is possible that S_1 in violaxanthin is raised to a level above that of S_1 in chlorophyll, so that violaxanthin acts as an antenna pigment in transferring energy to chlorophyll. Confirmation of these speculations must await experiments which more directly and accurately locate the S_1 state in these xanthophylls. Presently, it is a tantalizing possibility that chloroplasts have a mechanism of reversibly changing the conjugation length between 9 and 11 double bonds, and this regulates nonradiative dissipation of excitation energy. Furthermore, the stereochemistry and hydrophobicity of the two ends of the violaxanthin formed by epoxidation of zeaxanthin may be so altered as to hinder the transfer of energy from chlorophyll to violaxanthin, even if violaxanthin had a lower S_1 energy level. This speculation, too, remains to be substantiated.

C. Protection via Zeaxanthin-Promoted Aggregation of LHC II

Recently, Horton et al. (1991) suggested a new hypothesis of an efficient pathway for nonradiative dissipation of excitation energy. This pathway is thought to arise from the aggregation of the light-harvesting complexes of PS II (LHC II) upon acidification of the intrathylakoid lumen. In this hypothesis, zeaxanthin potentiates the aggregation of LHC II, while violaxanthin is postulated to prevent aggregation by an unspecified mechanism. The decrease of fluidity in the peripheral region of the hydrophobic core of isolated pea thylakoid membranes, associated with the light-induced conversion of violaxanthin to zeaxanthin, is qualitatively consistent with the promotion of LHC II aggregation by zeaxanthin (Gruszecki and Strzalka, 1991). Because the de-epoxidase which converts violaxanthin to zeaxanthin has optimal activity at pH 5 (Hager, 1969), lumenal acidity would have the double action of promoting LHC II aggregation both directly and indirectly by regulating zeaxanthin formation.

Horton et al. (1991) hypothesized that the electronic state of chlorophyll in the aggregated state is modified dramatically, and heat dissipation is favored over fluorescence emission or energy transfer. In this scheme, chlorophyll molecules themselves are mainly responsible for the nonradiative deactivation of their own singlet states. Zeaxanthin, while it promotes the aggregated state, may not be directly involved in the quenching of the excitation energy; it may merely serve as a "quenching amplifier" (Rees et al., 1989; Noctor et al., 1991).

However, the possible roles of zeaxanthin both as a quenching amplifier and as a direct quencher of excitation energy (via its S_1 state) need not be mutually exclusive. Indeed, the spectrofluorimetric data of Ruban et al. (1991) suggest that both possibilities exist: the presence of zeaxanthin not only amplified energy-dependent quenching (with a peak at 700 nm), but also quenched PS II fluorescence in the absence of an energized state. If zeaxanthin plays both roles, it is all the more important in conferring photoprotection at high irradiances.

D. Protection via Electron Cycling Around PS II

So far, we have considered dissipation of excitation energy as heat by ΔpH- and/or zeaxanthin-controlled thermal deactivation probably occurring in the antenna pigment beds of PS II. If this "front line of defense" fails, more excitation arrives at the PS II reaction center than can be utilized in orderly electron transfer to generate reducing equivalents for CO_2 fixation. Under such circumstances, excessive reduction on the acceptor side and excessive oxidation on the donor side of the PS II reaction center would be expected, unless other protective strategies come into play.

One such strategy is an electron cycle operating around PS II (Falkowski et al., 1986) or in the simplest case, a back-reaction which reverses charge separation. Electron cycling around PS II has been suggested in several studies to protect the photosynthetic apparatus in high irradiance. It is favored by Horton and co-workers (Oxborough and Horton, 1988; Noctor and Horton, 1990; Rees and Horton, 1990) as a mechanism whereby the quantum yield of open reaction centers can be changed independently of energy-dependent non-photochemical quenching (q_{Ne}).

An obvious possibility for protection afforded by electron cycling around PS II is the re-reduction of $P680^+$, thereby shortening the lifetime of the radical cation. As discussed by Thompson and Brudvig (1988), $P680^+$, required for oxidizing water, is the most powerful oxidant in photosynthesis. It is also the only component capable of oxidizing the antenna chlorophyll of PS II. Thompson and Brudvig (1988) demonstrated the oxidation of chlorophyll at the expense of the Mn site during illumination at low temperatures. They further showed that photooxidized chlorophyll was reduced by cyt b_{559}. Since cyt b_{559} itself can be photoreduced via plastoquinol (Whitmarsh and Cramer, 1978), Thompson and Brudvig (1988) proposed that the electron cycle functions as shown in Figure 3. By reducing Chl_z^+ (the accessory chlorophyll linking the antenna and PS II reaction center) and thus removing a strong oxidant, cyt b_{559} is thought to play an important role in preventing photoinhibitory damage.

Using the isolated D1/D2/cytochrome b_{559} PS II reaction center complex in flash absorption studies, Telfer et al. (1991) also concluded that cyt b_{559}, reduced by DBMIB in anaerobic conditions, was able to shorten the lifetime of $P680^+$ to such

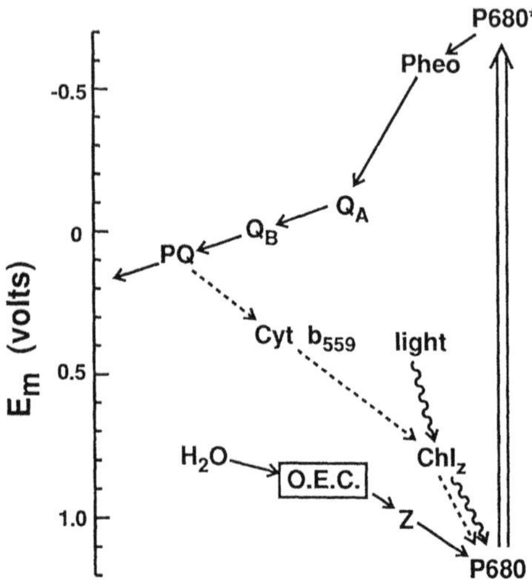

Figure 3. Pathways of energy and electron transfer in PS II, including a possible cyt b_{559}-dependent mechanism for protection against photoinhibition, as proposed by Thompson and Brudvig (1988). *Wavy arrows* indicate energy transfer. The *vertical arrow* symbolizes the excitation of P680 by a photon (intramolecular electron transfer). The usual electron transport pathway is shown by the *solid arrows*, depicting electron transfer from H_2O to the oxygen-evolving complex (O.E.C.), redox-active tyrosine (Z), P680, pheophytin (Pheo), Q_A, Q_B, the PQ pool, and subsequent electron carriers. Chl_z mediates energy transfer to P680. It is susceptible to oxidation by the most powerful oxidant in photosynthesis, $P680^+$, but could be maintained in a reduced state by cyt b_{559} which is in turn reduced by the PQ pool (*broken arrows*). Partially drawn after Blankenship and Prince (1985).

an extent that P680 itself was not photobleached by steady state illumination. On the other hand, when oxygen competed with DBMIB as an electron acceptor, cyt b_{559} reduction was hampered, and the lifetime of $P680^+$ was increased. Under these conditions, $P680^+$ oxidized β-carotene and Chl_z in the reaction center (Telfer et al., 1991). These authors, therefore, suggested that the major role of β-carotene in the PS II reaction center is to protect against photodynamic damage to P680, just as cyclic electron transfer via b_{559} plays a role in the protection of P680. The protection of P680 seems essential since, if $P680^+$ is allowed to accumulate, it may form a triplet state which then results in the formation of highly oxidizing singlet oxygen (Durrant et al., 1990).

In leaf discs, Canaani and Havaux (1990) observed the photooxidation of cyt b_{559} using red light of irradiance above 550 μmol photons m^{-2} s^{-1}. By considering the effects of an ADRY reagent (which accelerates deactivation of the oxygen-evolving system) and water or mild heat stress on the extent and kinetics of cyt b_{559}

photooxidation, they concluded that the electron cycle included the reduction of Z^+, a redox-active tyrosine on D1, by cyt b_{559}.

Whatever the detailed mechanism of cyclic electron flow around PS II, it is consistent with an earlier report that cyclic electron transfer around PS II occurs under high irradiance (Falkowski et al., 1986). These authors concluded that about 15% of the electrons cycle around PS II at saturating irradiance. They further proposed that this cycling is mediated by cyt b_{559}, and may prevent photoinhibitory damage. If indeed cyt b_{559} reverses the photooxidation of Chl_z or reduces the powerful oxidant Z^+ thereby preventing subsequent events of photoinhibitory damage, then the function of cyt b_{559} in the smallest PS II reaction center complex capable of photoreducing pheophytin (Barber et al., 1987; Nanba and Satoh, 1987; Akabori et al., 1988) is readily understood.

E. Protection via "Long-Term Downregulated" PS II Reaction Centers

After a photoinhibitory treatment, some PS II reaction centers lose functional activity (see later), and excitation energy arriving at those reaction centers does not result in useful charge separation or O_2 evolution. Functional PS II reaction centers are conveniently assayed by repetitive flashes in leaf discs (Jursinic and Pearcy, 1988; Chow et al., 1989a,b; Hart and Stemler, 1990). Even when the flash intensity is saturating, photoinhibited PS II complexes do not evolve O_2; this suggests the loss of reaction-center function (Chow et al., 1989b; Öquist et al., 1992a), rather than increased thermal dissipation of excitation from the antenna, in agreement with the interpretations of Ögren and Öquist (1984) as well as Cleland and Melis (1987).

Accompanying the loss of reaction-center function is the quenching of variable chlorophyll fluorescence. The non-photochemical quenching of chlorophyll fluorescence associated with photoinhibitory damage, measured as the quenching parameter q_{Ni}, reverses very slowly after the photoinhibitory treatment ceases (Osmond 1989, where $q_{Ni} = q_I$) and depends on protein synthesis (see below). Other quenching parameters reverse more rapidly. In order of increasing speed (Horton and Hague, 1988), these are: q_{Nt} (quenching due to protein phosphorylation), and q_{Ne} (quenching due to the energized state).

Although the non-photochemical quenching parameters can be resolved kinetically, they are all functionally equivalent during steady-state photosynthesis; they all contribute to quenching of chlorophyll fluorescence and, by implication, to the dissipation of excitation energy. Indeed, Horton and Hague (1988) found, over a wide range of high irradiance where q_p is small, that the sum $q_N = q_{Ne} + q_{Ni} + q_{Nt}$ is constant in isolated barley protoplasts. Öquist et al. (1992a) also reported that, in leaf discs of pea and *Tradescantia* grown in low irradiance, or pea grown in moderate light, q_{Ne} for control leaf discs was equal to $q_{Ne} + q_{Ni}$ measured in previously photoinhibited leaf discs when both types of leaf discs were briefly exposed to light at 1700 μmol photons $m^{-2}\ s^{-1}$ to assay these parameters. Thus, photoinhibition gives rise to a long-term (slowly reversible) quenching of chlorophyll fluorescence which is functionally indistinguishable from quenching due to the energized state; in both cases the net effect is enhanced dissipation of excitation energy. Indeed, the increased heat emission in photoinhibited cotyledons of *Raphanus sativus* has been directly demonstrated by the photoacoustic method (Buschmann, 1987).

By analogy with the idea that quenching associated with the energized state affords photoprotection at high irradiance, Krause (1988) and Öquist et al. (1992a) proposed that a photoinhibited PS II reaction center, by virtue of q_{Ni}, also dissipates excess excitation energy, and may protect neighboring and connected PS II reaction complexes from permanent damage.

Thus, photoinhibition is not necessarily entirely disadvantageous for the leaf: it may represent a mechanism for long-term regulation of PS II (Horton et al., 1987). Indeed, stable "long-term downregulation" of PS II via photoinhibition is found in both terrestrial (Ögren, 1988) and aquatic (Neale, 1987; Henley et al., 1991) environments where environmental factors (e.g., temperature, drought, nutrient deficiency) predispose plants towards photoinhibition (Osmond, 1981). In particular, the effects of photoinhibition are prevalent at low temperatures (Öquist et al., 1987), as exemplified by field studies of coniferous trees (Lundmark et al., 1988; Ottander and Öquist, 1991), Antarctic mosses (Post et al., 1990), snow gum (Ball et al., 1991), spinach (Somersalo and Krause, 1990), and maize (Long et al., 1990; Ortiz-Lopez et al., 1990). Photoinhibitory effects are also exacerbated when high salinity conditions prevail (Neale and Melis, 1989; Mishra et al., 1991; Sharma and Hall, 1991), or when high temperature and water stress are superimposed on high-light stress (Adams et al., 1987). They are also observed under conditions where high light appears to be the only stress factor, as in midday depression of photosynthesis in aquatic (Neale, 1987) and terrestrial (Ögren, 1988) systems, in the shade-adapted plant *Tradescantia albiflora* grown in full sunlight (Adamson et al., 1991; Chow et al., 1991a), or in understorey Pacific silver fir following clearcutting (Tucker et al., 1987). In all cases, photoinhibition is prevalent over a long period, and any protection that can be derived from the downregulation of some PS II reaction centers would obviously enable other PS II reaction centers to survive to carry out photosynthesis.

This form of photoprotection, however, requires that the downregulated PS II reaction complex be connected to neighboring and functional PS II complexes from which excess excitation energy can be transferred and dissipated as heat. More experiments are needed to confirm this suggested form of photoprotection, but two lines of evidence are consistent with this hypothesis. First, leaves of the chlorophyll *b*-less barley mutant are more prone to photoinhibitory damage than wild-type barley (Leverenz, Öquist and Wingsle, personal communication), presumably because of the relatively poor grana formation (Goodchild et al., 1966) and hence poor connectivity between PS II complexes. The greater susceptibility of the barley mutant to photoinhibitory damage is surprising because it exists despite the small antenna size of PS II (Ghirardi et al., 1986) and enhanced energy spillover from PS

II to PS I; these properties should mitigate photoinhibitory damage. There is a second line of circumstantial evidence supporting a protective role for downregulated PS II reaction centers: When protein synthesis is blocked, low-light-grown *Tradecantia albiflora* leaves are less prone to photoinhibitory damage than pea grown in low or moderate light (Öquist, Anderson, Hossack-Smith, and Chow, submitted). This difference may be due to better overall connectivity between photoinhibited PS II reaction complexes and neighboring, functional PS II complexes in *Tradescantia*, arising from the formation of large granal stacks in low-growth light (Adamson et al., 1991). If connectivity between PS II complexes (reviewed by Williams, 1977) is a prerequisite for this form of photoprotection against photoinhibitory damage, then one can add another facet to the significance of thylakoid stacking (Anderson, 1982; Anderson and Osmond, 1987).

The nature of the downregulated state of PS II, and the mechanism whereby a downregulated PS II reaction center dissipates excitation energy as heat remain to be determined; possibly, charge recombination or electron cycling around the reaction center complex may be involved.

F. Protection Conferred by Long-Term Photosynthetic Acclimation to Light

Long-term acclimation of the photosynthetic apparatus to light has received considerable attention, particularly in relation to the irradiance in the growth environment (Boardman, 1977; Wild, 1979; Björkman, 1981; Anderson, 1986; Anderson and Osmond, 1987; Anderson et al., 1988). Although most studies have concentrated on steady-state differences between plants grown in various irradiances, there is no doubt that the response of a leaf to a step change in growth irradiance is highly dynamic. A substantial response usually occurs within a day or two following a transfer to a new light regime, and is complete in approximately 7 days. For example, Grahl and Wild (1975) showed that the cytochrome *f* content in *Sinapis alba* thylakoids increased by 150% when the plants were transferred to a higher light environment. Davies et al. (1986) also observed a doubling of latent ATPase activity (indicative of the abundance of ATP synthase) in tomato chloroplasts 2 days after a step increase in growth irradiance, while Besford (1986) measured a doubling of leaf soluble protein and maximum Rubisco activity in the same experimental system within 7 days.

The above studies refer to the overall response of a leaf to different growth irradiances. Within any leaf, there is a light gradient correlated with depth of tissue and concomitantly, a gradation in the abundance of photosynthetic components; thus, chloroplasts located in palisade cells near the exposed leaf surface have a greater abundance of photosynthetic components compared with spongy tissue chloroplasts (Terashima and Inoue, 1985). Remarkably, such a gradation along the depth into the tissue, as represented by electron-transport capacity and the Chl *a*/ Chl *b* ratio, reversed within a few days following the inversion of a leaf in the light

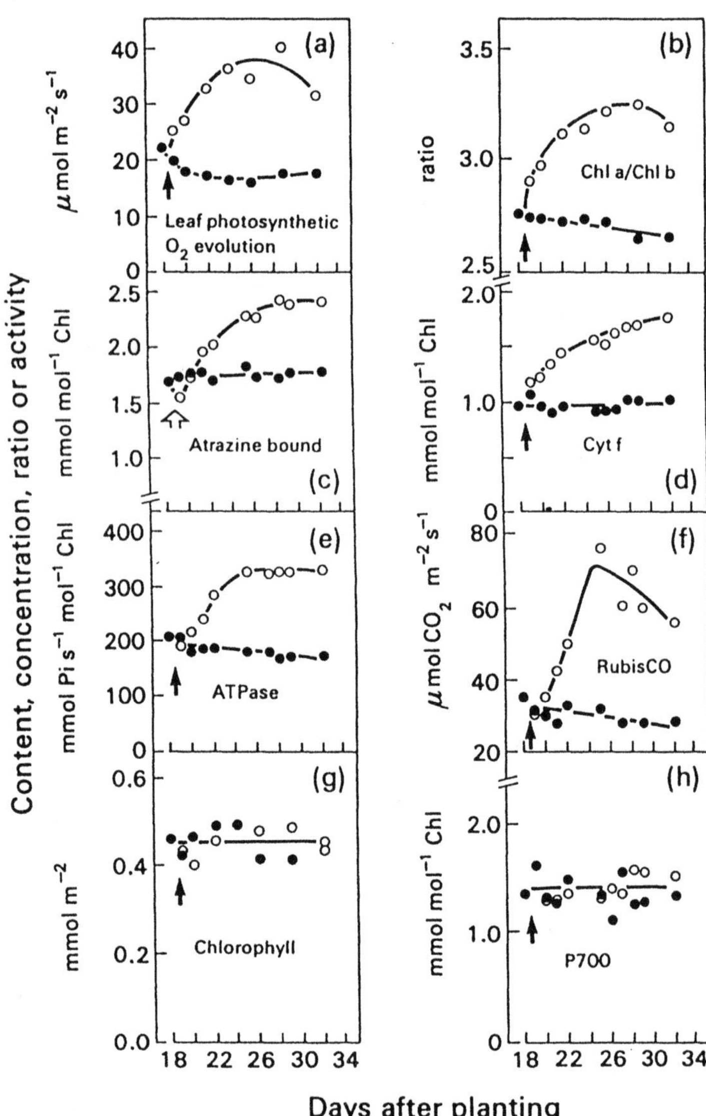

Figure 4. The dynamic acclimation of the photosynthetic apparatus of the third leaf from the base of pea grown in low irradiance (60 μmol photons m^{-2} s^{-1}) for 19 days, and then transferred to moderate irradiance (390 μmol photons m^{-2} s^{-1}). (•) Control low-light pea. (o) Pea acclimated to moderate light. *Solid arrows* indicate the time of transfer. The *open arrow* in (**c**) points to a temporary decrease in atrazine-binding capacity which was accompanied by a temporary decrease of quantum yield of O_2 evolution (data not shown). The figure is redrawn from Chow and Anderson (1987a,b).

(Terashima and Takenaka, 1986), again demonstrating the dynamic nature of photosynthetic acclimation to growth irradiance.

Chow and Anderson (1987a,b) attempted a detailed time-course study of changes in photosynthetic components in fully expanded pea leaves following a step increase from low to moderate growth irradiance (the latter equivalent to 20% of full sunlight). The results are summarized in Figure 4 (a)–(h). Following the transfer to moderate light, the light- and CO_2-saturated rate of photosynthesis per unit leaf area was doubled within 7 days (a). This was accompanied by an increase in the Chl *a*/Chl *b* ratio (b), without any detectable change in the total chlorophyll content per unit leaf area (g). The amounts of PS II assayed as atrazine-binding sites (c), cytochrome b_6/f complex assayed as cytochrome *f* (d), ATP synthase assayed as latent ATPase activity (e), and Rubisco protein assayed as the maximum enzyme activity (f) all increased, albeit to different extents. In contrast, the amount of PS I, assayed as P700, remained unaffected (h), presumably because PS I does not limit electron transport even in the high-light environment.

The above changes in the photosynthetic apparatus all contribute to an overall higher capacity for photosynthesis. Thus, fully acclimated leaves utilize the higher irradiance more effectively and can also better avoid photoinhibitory damage. This is partly because "a high photosynthetic capacity at high-light intensities can be expected to divert a greater fraction of the excitation energy to be used in photosynthesis, and less excess energy would thus be left at the reaction center to cause its inactivation" (Björkman, 1972). In contrast, in the transition period before the peas were fully acclimated, a transient decrease in the quantum yield of O_2 evolution in limiting light occurred over 2 days immediately after the light transfer. Subsequently, the quantum yield recovered to the control value (Chow and Anderson, unpublished). Similar results have been obtained with *Ulva* (Henley et al., 1991). The transient depression of quantum yield was almost certainly caused by photoinhibitory damage which occurred prior to the full adjustment of the photosynthetic apparatus in response to the elevated irradiance.

That photosynthetic acclimation to increased irradiance helps to avoid photoinhibitory damage is further exemplified by *Alocasia*. This plant is capable of sustained growth on the floor of a rainforest where the average irradiance is less than 0.5% of full sunlight (Björkman and Ludlow, 1972). Because of its remarkable shade tolerance, *Alocasia* is sometimes assumed to be a typical obligate shade species. However, it can grow in open, sunny areas. Indeed, in a controlled growth environment, *Alocasia* chloroplasts acclimated well to irradiances up to at least 30% of full sunlight, by modulating the amounts of Rubisco, cytochrome b_6/f complex, and PS II, and by changing the Chl *a*/Chl *b* ratio (Chow et al., 1988). Consequently, *Alocasia* leaves did not exhibit any symptoms of photoinhibitory damage at growth irradiances up to at least 30% of full sunlight.

Given that photosynthetic acclimation to a high-light environment protects against photoinhibitory damage, a corollary is that a species which does not readily modulate its photosynthetic components would be more prone to damage by high light. An example of such a species is *Tradescantia albiflora*, another shade tolerant plant, which is incapable of modulating its light-harvesting components on changing the growth irradiance. Thus, the PS II/PS I reaction center ratio is constant at different growth irradiances (Chow et al., 1991a). In addition, upon transfer to a higher light environment, there is a noticeable lag before any detectable increase in Rubisco (1 day) and cytochrome *f* (2 days). Consistent with such an inefficient acclimation of the photosynthetic apparatus, *Tradescantia* is easily photoinhibited (Chow et al., 1991a).

G. Does a Small Light-Harvesting Antenna Protect PS II Against Photoinhibitory Damage?

Upon acclimation to increased growth irradiance, the size of each PS II light-harvesting antenna is generally observed to decrease. Thus N_α, the number of chlorophyll molecules transferring excitation to each PS II_α reaction center, was 620 for *Chlamydomonas reinhardtii* grown under 47 μmol photons m^{-2} s^{-1}, but 460 for cells grown under 400 μmol photons m^{-2} s^{-1} (Neale and Melis, 1986). Similarly, the antenna size of PS II decreased from 400 to 130 Chl when *Chlorella* was grown under increasing irradiances (Ley and Mauzerall, 1982), and the difference in antenna size could be further emphasized by growth under extreme irradiance conditions (Ley, 1986). Changes in PS II antenna size in response to the light environment have been reviewed by Melis (1991).

In addition, a qualitative assessment of the antenna size of PS II as a function of growth irradiance can be obtained by treating the light-harvesting Chl *a/b*-protein complexes $LHCP^{1+2+3}$ as serving PS II, and CPa as the core Chl *a*-protein complex of PS II (Leong and Anderson, 1984). The Chl ratio, $LHCP^{1+2+3}$/CPa, decreased approximately twofold over the irradiance range used to grow pea plants (Leong and Anderson, 1984). Similarly, a twofold difference in the above ratio (obtained in a different electrophoretic gel system) was also observed in pumpkin plants grown in high or low light (Tyystjärvi et al., 1991).

It has been generally assumed that a smaller PS II antenna size offers protection against photoinhibitory damage. Given that high-light-grown pumpkin plants have a decreased PS II antenna size, Tyystjärvi et al. (1991) set out to compare the susceptibility to photoinhibitory damage of thylakoids isolated from plants grown in high or low light. Perhaps surprisingly, they found that photoinhibitory damage of isolated thylakoids in the absence of added electron acceptors is independent of the size of the light-harvesting antenna of PS II. A similar observation was made by us when pea was grown in moderate or low light: In the presence of chloramphenicol, an inhibitor of chloroplast–protein synthesis, leaf discs from plants grown in moderate light were more easily photoinhibited than low-light leaf discs, even though moderate-light leaves had a greater capacity for zeaxanthin-dependent mitigation of photoinhibitory damage (Öquist, Anderson, McCaffery, and Chow, submitted). Thus, the apparent lack of protection against photoinhibitory damage offered by a small PS II antenna requires an explanation. Conversely, one may ask whether a large antenna associated with PS II may confer characteristics which are not necessarily disadvantageous but instead limit excitation trapping and/or charge separation in high light.

An answer to this question may lie in the actual rate of light absorption by PS II as distinct from its physical antenna size. As pointed out by Melis et al. (1987), PS II is handicapped in terms of light absorption because of the substantial amounts of Chl *b* contained in its light-harvesting antenna. It is known that PS II_α with its large light-harvesting antenna is segregated in the granal stacks (Anderson and Melis, 1983). In low growth-irradiance, there is an increase in Chl *b* relative to Chl *a*, and granal stacks are larger; therefore, grana stacks, particularly in low-light chloroplasts, constitute regions of high pigment density, giving rise to a "sieve effect", whereby light absorption by PS II is actually lower than PS I (Melis et al., 1987). A greater sieve effect in low-light chloroplasts, due to larger grana stacks than in high-light chloroplasts, could lead to less excitation energy per unit time reaching each PS II reaction center for charge separation.

An additional or alternative answer to the above question may relate to the tight coupling of antenna pigments with the PS II reaction center. In their kinetic and energetic model for primary processes in PS II, Schatz et al. (1988) assume that the reaction center of PS II constitutes a shallow trap for an exciton which is delocalized over the complete antenna system, so that an exciton may visit the reaction center several times before being trapped. Because of this tight coupling, the apparent rate constant for charge separation in a PS II reaction center with a coupled antenna is smaller than the intrinsic rate constant for an isolated reaction center by a factor of N, the number of equivalent chlorophyll molecules coupled to the reaction center. Therefore, a large antenna (as in low-light chloroplasts) may lead to a small apparent rate constant for charge separation. Consequently, at any given excitation density in the antenna, there would be a lower rate of charge separation to form $P680^+ Pheo^-$; thus, large PS II antennae may not necessarily be disadvantageous in high light.

Indeed, it would seem that characteristics associated with good thylakoid stacking (viz., connectivity of PS II complexes, the sieve effect, and large PS II antenna giving rise to a small apparent rate constant for charge separation) should confer tolerance against photoinhibitory damage. Could it be that here lies an ecological significance of grana formation? If so, a shade-acclimated leaf, with extensive thylakoid stacking and associated enhancement of the characteristics mentioned above, may be better able to withstand potential photoinhibitory damage during long sunflecks than a leaf with poor grana formation, all else being equal.

III. PHOTOINHIBITORY DAMAGE

A. Background

Injury to the photosynthetic apparatus of green plants on prolonged exposure to intense light has been known for a long time (e.g., Ewart, 1896). Since there have been a number of reviews on photoinhibitory damage (Osmond, 1981; Powles, 1984; Kyle and Ohad, 1986; Kyle et al., 1987; Krause, 1988; Barber and Andersson, 1991; Prasil et al., 1992), this topic will be discussed in limited detail here.

In the early stages of a photoinhibitory treatment, photoprotective phenomena may occur before damage is evident. For instance, a photoinhibitory treatment initially induced a decrease in quantum yield of light-limited O_2 evolution with little or no loss of light-saturated photosynthetic capacity in leaves (Walker and Osmond, 1986) and protoplasts (Horton et al., 1987). Similarly, the cold-induced decline of light-limited quantum yield of O_2 in snow gum leaves occurred without a simultaneous decline in light- and CO_2-saturated photosynthetic capacity (Ball et al., 1991). As discussed by Osmond and Chow (1988) and Ball et al. (1991), a decrease of light-limited quantum yield without a loss of light-saturated photosynthetic capacity is probably an indication of a photoprotective mechanism which deflects excess excitation, thereby lowering the quantum yield.

Prolonged illumination with strong light, however, will eventually lead to photoinhibitory damage. Furthermore, damage to PS II is the most frequent result of photoinhibition (Critchley, 1988a). Light-induced malfunction of PS II would not lead to effective charge separation, hence the number of functional PS II complexes detected by the O_2 yield per single-turnover repetitive flash (Chow et al., 1989b, 1991b) would be decreased. The maximum efficiency of PS II photochemistry, assayed as F_v/F_m, would also be lowered if the absorbed light does not lead to useful photochemistry. Similarly, the overall quantum efficiency of O_2 evolution or CO_2 fixation would also be decreased. Figure 5 shows the effects of photoinhibitory treatment on these parameters, and their correlations in low-light grown *Tradescantia* or pea grown in low or moderate light. These correlations arise primarily because of light-induced malfunction in PS II. Since declines in F_v/F_m and quantum yield of O_2 evolution are each correlated with the loss of functional PS II complexes, they are also correlated with each other. Indeed a linear correlation between F_v/F_m and the quantum yield of O_2 evolution has been reported from a number of laboratories (Demmig and Björkman, 1987; Leverenz and Öquist, 1987; Adams and Osmond, 1988; Henley et al., 1991). Such correlations justify the use of the chlorophyll fluorescence parameter F_v/F_m as a rapid and convenient assay of photoinhibitory damage, provided photoprotective mechanisms have been allowed to relax. However, these correlations of F_v/F_m with the quantum yield of O_2 evolution are merely empirical: the former signal originating only from the layers of chloroplasts in a leaf nearest to the incident excitation light, and the latter measured for the leaf piece as a whole. Therefore, one should always establish that

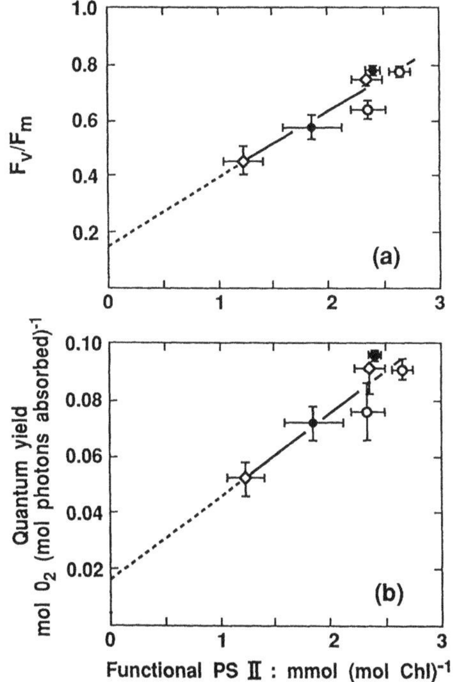

Figure 5. Correlation of the maximum efficiency of PS II photochemistry (F_v/F_m) and the quantum yield of O_2 evolution with the amounts of functional PS II reaction-center complexes in control and photoinhibited leaves of *Tradescentia* (◊) and pea (•) grown under 50 μmol photons m^{-2} s^{-1}, as well as pea (o) grown under 300 μmol photons m^{-2} s^{-1}. The leaf discs were photoinhibited by exposure to 1700 μmol photons m^{-2} s^{-1} for 4 hr at 22 °C, while floating on water. The graphs are replotted from Öquist et al. (1992a).

the relationship holds in each experimental system before accepting the assay of photoinhibitory damage based entirely on chlorophyll fluorescence measurements.

B. The Onset of Photoinhibitory Damage

During illumination at high irradiance, photosynthetic cells and tissues utilize a variety of photoprotective mechanisms to cope with the effects of excessive excitation energy, as discussed above. When these mechanisms cannot adequately protect the photosynthetic apparatus from the excessive visible radiation, photoinhibitory damage ensues. The onset of photoinhibitory damage seems to occur when the primary quinone acceptor is continually reduced beyond a certain extent during illumination. A measure, albeit it a nonlinear one (Havaux et al., 1991b), of the oxidation state of Q_A is given by the chlorophyll fluorescence quenching parameter q_P (van Kooten and Snel, 1990). Ögren (1991) observed, after a standard high-light treatment of willow trees, that decrease of F_v/F_m begins when q_P is maintained at less than 0.5 to 0.65, depending on growth conditions and leaf age. Similarly, Öquist et al. (1992a) found that inhibition of F_v/F_m began at $q_P \leq 0.56$ for nine species grown under diverse light conditions. Thus, the onset of photoinhibitory damage seems to occur when about half of the PS II reaction centers are closed.

Given the close relationship between the oxidation state of Q_A and the onset of photoinhibitory damage, any photosynthetic organism which can maintain Q_A at an oxidation level corresponding to $q_P \geq 0.6$ at high irradiance will obviously have greater tolerance to high light. A major component of frost-hardening appears to be an acclimation which keeps Q_A more oxidized under a given irradiance and low temperature. For example, the increased capacity of frost-hardened winter rye to keep Q_A oxidized under given light and temperature conditions can account for its higher resistance to photoinhibition compared with the nonhardened control (Öquist et al, 1992b). Because of the requirement of a highly oxidized state of Q_A (high q_P) in averting photoinhibition on the one hand and the importance of the photoprotective role of energy-dependent quenching of excitation energy (high q_N) on the other, it is reassuring that the best correlation was obtained when the extent of photoinhibitory damage of willow leaves of different ages treated in different ways was expressed in terms of both q_P and q_N in a multiple regression (Ögren, 1991).

C. Nature of the Damage to PS II

Chlorophyll Fluorescence Indicators

When q_P is maintained below some threshold value (ca. 0.6), with a fraction of the PS II reaction centers closed, photoinhibitory damage may result from the continued charge separation in those reaction centers with reduced Q_A. The nature and mechanisms of the ensuing damage to PS II has been extensively reviewed recently by Prasil et al. (1991); therefore, only a relatively brief survey will be presented here.

An early detectable change in the course of photoinhibitory damage is an increase in F_o, the chlorophyll fluorescence yield under conditions where PS II reaction centers are expected to be open (Krause, 1988; Krause and Weis, 1991). The increase in F_o may be as much as 80% (Ögren and Öquist, 1984; Franklin et al., 1992). It is further enhanced if the high-light treatment is given at low temperature (Greer and Laing, 1989; Kirilovsky et al., 1990b). It can also be enhanced under anaerobic conditions (Satoh, 1971; Krause et al., 1985; Kirilovsky and Etienne, 1991). The mechanism causing an increase in F_o during photoinhibition remains to be elucidated, but it is a prolonged effect, consistent with its correlation with q_{Ni}. It may be related to a decreased efficiency of excitation energy trapping (Ögren and

Öquist, 1984), a destabilization of Q_B^- which, in turn, enhances the back flow of electrons to Q_A (Ohad et al., 1990), a slower reoxidation of Q_A^- (Kirilovsky et al., 1990a), trapping of Q_A in a negatively-charged stable state (Setlik et al., 1990), or the accumulation of inactive permanently-closed PS II reaction centers which are damaged on the acceptor side (Nultsch et al., 1990). Presumably, increased F_o reflects some perturbation within the architecture of PS II, resulting in enhanced fluorescence emission; such enhanced emission is quite distinct from photoprotective quenching of excess excitation leading to decreased F_o (Krause, 1988; Franklin et al., 1992).

Accompanying the increase in F_o during a photoinhibitory treatment is a decrease in F_m as well as in F_v/F_m. Again, as in the case of F_o increase, various reasons for the decrease in F_m have been proposed, including the formation of protonated Q_A^- (Setlik et al., 1990) or modified Q_A (Cleland and Critchley, 1985), double reduction of Q_A (Styring et al., 1990), an increase in the lifetime of the quenching species $P680^+$ (Ohad et al., 1990), formation of $Pheo^-$ as a quenching species (Nedbal et al., 1986), and the formation of chlorophyll fluorescence quenchers within the light-harvesting pigment bed of PS II (Demmig et al., 1987; Rees et al., 1989). In the case of weak-light photoinhibition of NH_2OH-extracted PS II membranes, the loss of variable fluorescence, and hence of F_m, has been tentatively attributed to a carotenoid cation (Blubaugh et al., 1992). In general, the loss of F_m seems to be a slower process relative to the rise in F_o; however, this may be due to a photoprotective quenching of F_o which curtails the net increase in F_o (Franklin et al., 1992).

PS II Redox Sites Damaged by Photoinhibition

In most photoinhibitory treatments, irradiances comparable to or greater than full sunlight have been applied to low-light acclimated plants or leaves, or their isolated chloroplasts/thylakoids. Under these conditions, various sites of PS II inhibition have been reported either on the acceptor side or in the reaction center itself. Kyle et al. (1984) proposed that PS II electron transport in the early phase of photoinhibition is blocked solely at the Q_B-protein site, to which the secondary quinone acceptor binds. On the other hand, Tytler et al. (1984) reported that in the cyanobacterium *Microcystis aeruginosa* photoinhibition resulted in damage to the reaction center itself. The interpretations in these two studies, however, are complicated by uncertainty concerning the site at which silicomolybdate acts as an electron acceptor (see Critchley, 1988a) and the highly variable results that can be obtained with this electron acceptor (Nedbal et al., 1986). Further support for alterations at the Q_B site was obtained via measurements of thermoluminescence, the emission of light due to charge recombination between the S states and either Q_A^- or Q_B^-. In the initial stage of *in vivo* photoinhibition of *Chlamydomonas*, alteration of the thermoluminescence signal due to charge recombination between Q_B^- and the S states preceded the loss of that between Q_A^- and the S states (Ohad et al., 1988;

Ohad et al., 1990; Shochat et al., 1990); further, the temperature at which charge recombination occurs in the cells was downshifted, indicating destabilization of Q_B^-. However, no shift in the temperature of the thermoluminescence emission was observed in isolated spinach chloroplasts under anaerobic conditions, even though F_0 was more than doubled (Kirilovsky and Etienne, 1991), and this has been confirmed in other studies with isolated spinach (Vass et al., 1988) or pea (Farineau, 1990) chloroplasts.

A different primary site of photoinhibitory damage was suggested by Cleland and Critchley who proposed that photoinhibition led to inactivation of the PS II reaction center (Cleland and Critchley, 1985; Critchley, 1988a), possibly through a modification of the geometry of the reaction center (Cleland, 1988). Melis and co-workers (Cleland and Melis, 1987; Demeter et al., 1987) also reported that photoinhibition damaged PS II reaction center function, specifically the primary charge separation between P680 and pheophytin, in isolated spinach chloroplasts. However, Allakhverdiev et al. (1987), using PS II particles isolated from pea chloroplasts by digitonin–Triton X-100 fractionation, concluded that pheophytin reduction remains unimpaired. Indeed, impairment of primary charge separation is by far the slowest process (Nedbal et al., 1990). Possibly, photoinhibition first inhibits electron transfer from Q_A to Q_B, and is followed by secondary damage, which results in a low-fluorescent photoinactive state of the PS II reaction center (Kirilovsky et al., 1990a). In a similar interpretation, Styring et al. (1990) concluded that photoinhibition blocks electron transfer between pheophytin and Q_A, probably by impairment of the function of Q_A, while the primary charge separation reaction is still operational. According to these authors, such an impairment of the function of Q_A could occur via a double reduction of Q_A, which then leaves its sites.

In the studies on photoinhibitory damage to PS II redox sites described above, relatively high irradiances are applied to photosynthetic tissues, cells, and isolated chloroplasts or thylakoids, most of which initially possess active water-oxidizing ability. On the other hand, if PS II is impaired on the donor side beforehand, the sensitivity to photoinhibitory damage is increased by 2 to 3 orders of magnitude and can be observed in weak light. Impairment of PS II on the donor side, by Tris treatment or Cl^- depletion of isolated chloroplasts (Theg et al., 1986; Eckert et al., 1991; Jegerschold and Styring, 1991), or by NH_2OH treatment of leaf segments, chloroplasts (Callahan et al., 1986), or PS II membranes (Blubaugh and Cheniae, 1990; Blubaugh et al., 1992), can lead to weak-light photoinhibitory damage. In this type of photoinhibitory damage, the initial and the maximal variable chlorophyll fluorescence levels, measured in the presence of DCMU and the reductant, NH_2OH, are not significantly affected (Callahan and Cheniae, 1985; Blubaugh et al., 1992), although D1 degradation occurs rapidly under anaerobic conditions and strong light (Jegerschöld and Styring, 1991). The site of damage after weak-light inhibition is not at the Q_B locus or on the water-splitting complex, although still on the donor side of PS II (Callahan et al., 1986). More recent work by Blubaugh et al. (1992) indicated no impairment of charge separation from P680 via pheophytin

(Pheo) to the first stable electron acceptor, Q_A. Rather, when the donor side is impaired, the order of susceptibility of PS II components to photodamage is Chl/carotenoid > Z > D >> P680/Pheo/Q_A, where Z and D are redox-active tyrosines of the D1 and D2 polypeptides, respectively. Incidentally, Blubaugh et al. (1992) favor the view that the causative agent for photodamages is $P680^+$, in contrast to Thompson and Brudvig's suggestion that the formation of Chl^+ (on oxidation of antenna Chl by $P680^+$) is the immediate causative agent of photoinhibitory damage.

Two types of photoinhibitory damage have been described above: photodamage on the acceptor side resulting in failure to perform a stable charge separation, and that on the donor side blocking electron transfer to $P680^+$. They need not be mutually exclusive. Indeed, Eckert et al. (1991) observed, on irradiation of PS II membrane fragments with visible light, that photoinhibitory damage occurs at two different sites with different quantum yields: the dominating site of damage depends on the functional status of PS II and the prevailing irradiance. In any case, given that several redox components are on, or bound between D1 and D2, any modification on a particular component would be expected to have ramifications on electron transfer at other steps in the complex. It therefore seems unlikely that there will be general agreement on one particular site as the main target for photoinhibitory damage.

D. Turnover of the D1 Protein

Among thylakoid membrane proteins, the D1 protein within the PS II complex has the highest rate of synthesis and degradation (reviewed by Kyle, 1984). This turnover rate is dependent on light, being low in the dark and 50 to 80 times faster than other photosynthetic membrane proteins at high irradiances (Mattoo et al., 1984; Ohad et al., 1984). It has been suggested that the damage to the D1 protein, leading to turnover of the D1 polypetide, may occur as a natural consequence of its normal function in quinone reduction, or in the presence of oxygen radicals generated after quinone reduction (Arntzen et al., 1984).

Because of its enhanced rate of turnover in high light, the D1 protein has been studied extensively in relation to photoinhibition. In *Chlamydomonas* cells, the primary target of attack in high light seemed to be on the D1 protein: under the experimental conditions used, plastoquinone radicals accumulate at the Q_B site and destroy the Q_B-binding peptide (Kyle et al., 1984; Ohad et al., 1984). On the other hand, Arntz and Trebst (1986) reported that illumination of spinach thylakoid membranes under strictly anaerobic conditions and in the absence of an electron acceptor, inactivates PS II without any degradation of D1. The question still remains, however, whether the inactivation of PS II and degradation of D1 were tightly correlated under *aerobic* conditions. With spinach *thylakoids*, strong illumination induced the loss of PS II activity without loss of [^{35}S]methionine-prelabeled D1 (Cleland and Critchley, 1985; Cleland, 1988; unlike the case of *leaf discs*, Critchley et al., 1992) or of DCMU-binding sites or immunoassayable D1 (Cleland et al., 1990). On the other hand, photoinhibition of spinach leaf discs at 25 °C induced a parallel loss of functional PS II reaction centers, quantum yield of O_2 evolution, and atrazine-binding sites (Osmond and Chow, 1988). Similarly, strong illumination of isolated spinach thylakoids at 20 °C gave a parallel loss of linear electron-transport capacity, F_v/F_m, and atrazine-binding sites (Richter et al., 1991a). On the other hand, photoinhibition at a lower temperature (10 °C) brought about differential changes in these parameters; under such conditions, the loss of functional PS II reaction centers preceded that of herbicide-binding sites (Chow et al., 1989b).

At low temperature, the retardation of D1 degradation could be caused by at least two factors.

1. The first factor is that the lateral migration of a damaged PS II complex to the site of repair, which consists of the degradation of D1 followed by insertion of newly synthesized D1, is slow at low temperatures. Kettunen et al. (1991) demonstrated that illumination of intact pumpkin leaves with high light led to severe photoinactivation of PS II, but there was no net degradation of D1. Instead, a modified form of D1, termed D1*, with slightly lower electrophoretic mobility, was produced at the expense of the original form of D1. D1* could be detected only in the appressed thylakoid membrane regions, the main sites of photoinhibitory damage (Cleland et al., 1986; Mäenpää et al., 1987). Presumably, D1* is degraded in some way in stroma lamellae, where the degradation and insertion of the newly synthesized D1 into PS II are tightly coupled (Kettunen et al., 1991), whereas D1* in appressed membranes may actually be protected against degradation (Aro et al., 1992). Although it has been suggested that D1 degradation may occur in appressed thylakoid membranes (Hundal et al., 1990), the evidence seems only circumstantial. These authors based their interpretation on the loss of D1 in inside-out vesicles derived from appressed membranes of photoinhibited thylakoids, but the possibility was not to be ruled out that D1 could have been lost during the preparation of the inside-out vesicles when the granal structure was greatly perturbed. Given the uncertainty in the interpretation of the above study, the conclusions of Kettunen et al. (1991) seem to be valid. If so, the lateral migration of D1 from appressed to nonappressed membranes is a necessary step prior to D1 degradation. Indeed, Adir et al. (1990) concluded from experiments with *Chlamydomonas* that PS II translocates from the site of damage in the appressed to the nonappressed domain, where the D1 precursor protein is translated and replaces the degraded D1 protein. Virgin et al. (1990) also observed lateral migration of PS II subunits from appressed to nonappressed regions of isolated thylakoids, although the exact composition of the migrating subunits differs in these two studies. In the nonappressed regions of thylakoid membranes, a new copy of D1 is inserted and, once processed, the D1 protein retranslocates to the appressed region (Mattoo and Edelman, 1987). Specifically, PS II_β, activated from a Q_B-nonreducing to a Q_B-reducing state, could translocate from nonappressed regions to appressed regions (Neale and Melis,

1991). Thus, the lateral migration of PS II subunits could be part of the repair cycle (see Guenther and Melis, 1990) following photoinhibitory damage (Hundal et al., 1990), and is temperature-dependent.

2. The second factor and possibly more important effect of temperature on the degradation of D1, relates to the temperature dependence of enzymic processes. It is now generally agreed that, while the triggering of D1 for proteolytic attack requires light in thylakoid membranes (Ohad et al., 1985; Aro et al., 1990), in the isolated PS II core complex (Virgin et al., 1990), and in the isolated D1/D2/Cyt *b*-559 reaction center complex of PS II (Shipton and Barber, 1991), degradation of D1 proceeds readily in the dark (Aro et al., 1990). In particular, Misra et al. (1991) demonstrated with synthetic substrates and proteinase inhibitors that the isolated D1/D2/Cyt *b*-559 reaction center complex of PS II possesses a serine-type endopeptidase activity, which is an intrinsic property of the PS II reaction-center proteins. Virgin et al. (1991) have confirmed the existence of a serine-type protease as an integral part of the isolated spinach PS II core. Such a protease activity is obviously subject to the usual control by temperature; the degradation of D1 protein is therefore severely retarded below 7 °C, while PS II electron transport is much more sensitive to high light and low temperature (Aro et al., 1990).

An intriguing question regarding the intrinsic protease activity of the PS II reaction center is why no degradation of $D1^*$ seems to occur in the appressed membranes, and why $D1^*$ seems to be promptly degraded on lateral migration to nonappressed regions (Kettunen et al., 1991). For the protease activity to occur, could it be that certain conformational changes in $D1^*$ are necessary and that such changes are restrained by the ordered structure in grana? This possibility can be tested by a comparison of the degradation of $D1^*$ in both stacked and unstacked thylakoids.

A question related to the proteolytic breakdown of $D1^*$ in the stromal lamellae is the driving force responsible for the lateral migration of the photoinhibited PS II complex from appressed to nonappressed regions. At present, the origin of such a driving force is unknown. One possibility is that in a photoinhibited PS II core complex a tendency is generated for certain components (including D1 itself) to protrude out of the membrane surface. Energetically, for the system as a whole, it would be more favorable for the PS II core complex to migrate to nonappressed membranes rather than for the protruding components(s) to remain in the narrow partition gap between two appressed membranes. Such a mechanism could operate in a "conveyor belt" manner (Critchley, 1988b) to transfer PS II complexes from granal stacks to stromal lamellae. On lateral migration to stroma-facing membranes, proteolysis could occur in D1 followed by its replacement by insertion, perhaps in a direction perpendicular to the plane of the membrane of newly synthesized D1. Such an idea, however, remains speculative. Undoubtedly, the further study of the degradation and insertion of D1 will provide interesting findings.

Apart from external conditions, such as temperature and light, which influence D1 breakdown, there may also exist a difference in the intrinsic resistance of PS II to photoinhibitory damage between two forms of D1 which may be present in the reaction center complex of *Synechococcus* sp. The D1 protein in this cyanobacterium is coded for by the *psbA* multigene family containing the genes *psbAI, psbAII* and *psbAIII*. The mutant R2S2C3, possessing *psbAI* as the only active gene, produces Form I of the D1 protein in low- and moderate-light conditions. Another mutant R2K1, having *psbAII* and *psbAIII* as the active D1 genes, produces Form II of the D1 protein under high light (Schaefer and Golden, 1989a,b). In the presence of the protein-synthesis inhibitor, streptomycin, which blocks the PS II repair cycle, the mutant R2K1 was more resistant to photoinhibitory damage (Krupa et al., 1991). This indicates that Form II may have a higher intrinsic resistance of PS II to photoinhibitory damage. Whether the two forms of D1 protein have different degradability, thus accounting for the differential susceptibility to photoinhibitory damage, remains to be established.

IV. RECOVERY FROM PHOTOINHIBITORY DAMAGE

A. Dependence on Chloroplast–Protein Synthesis

The decline in the quantum yield of photosynthesis during and after photoinhibitory treatment may have various contributing factors, some of which relate to energy-dissipative processes (photoprotection), and others to malfunction of PS II (photoinhibitory damage) (Demmig-Adams and Adams, 1992a; Franklin et al., 1992). If one defines photoinhibitory damage in a narrow sense to describe damage to PS II due to excess visible light, and ignores the closing of PS II reaction centers that can be rapidly reactivated without replacement of the D1 protein (Kirilovsky and Etienne, 1991), then a feature which clearly distinguishes photoinhibitory damage from photoprotection is the requirement for chloroplast–protein synthesis during recovery from photoinhibition. Specifically, a requirement for synthesis of the D1 protein was shown to be necessary for recovery from photoinhibitory damage in *Chlamydomonas* (Kyle et al., 1984; Ohad et al., 1984), *Anacystis* (Samuelsson et al., 1985), and was indicated in pea (Ohad et al., 1985) and beans by studies based on chlorophyll fluorescence measurements (Greer et al., 1986). Furthermore, the difference in the ability to recover from photoinhibitory damage was attributed to different rates of the PS II repair cycle in high-light and low-light grown *Anacystis* (Samuelsson et al., 1987; Lönneborg et al., 1988), presumably due to different capacities for protein synthesis. In a study with *Lemna gibba*, Huse and Nilsen (1989) used chloramphenicol to block the synthesis of the D1 protein, and demonstrated that D1 synthesis is also important for recovery of photosynthesis from photoinhibitory damage in this higher plant.

When a second environmental stress interferes with protein synthesis, recovery from photoinhibitory damage is expected to be slow. In an investigation of the interaction between salt stress and photoinhibition on photosynthesis in barley and sorghum, Sharma and Hall (1991) suggested that salinity stress exerts adverse effects on cellular metabolic activities by predisposing the photochemical apparatus to photoinhibitory damage and hampering recovery by restraining the turnover of the D1 protein by an unspecified mechanism. Their suggestion is consistent with that of Ball and co-workers who proposed that salinity-induced potassium deficiency causes the loss of functional PS II in mangrove (Ball et al., 1987) and spinach (Chow et al., 1990) leaves. It is known that K^+ is required for protein synthesis in isolated intact chloroplasts (Fish and Jagendorf, 1982b); if the supply of K^+ is inadequate, protein synthesis would be impaired, resulting in a net loss of those proteins, particularly D1, with a high rate of natural turnover.

Although D1 synthesis is important for recovery from photoinhibitory damage, it is not clear to what extent the newly synthesized pool of D1 is related to the functional state of PS II. Huse and Nilsen (1989) observed a correlation between the recovery in photosynthesis and the relative amount of synthesized 32-kDa protein, when the irradiance during recovery was in the range 20 to 210 μmol photons m^{-2} s^{-1}. However, when the irradiance was 0 or 1000 μmol photons m^{-2} s^{-1}, the extent of recovery of photosynthesis was much greater than that of D1 resynthesis. Similarly, assaying D1 by the number of atrazine- or DCMU-binding sites, Chow et al. (1989b) found, within 10 h after cessation of a photoinhibitory treatment, that there was a substantial recovery in the number of functional PS II complexes, as well as the F_v/F_m ratio of chlorophyll fluorescence measured at room temperature or 77 K; however, there was no detectable increase in the number of herbicide-binding sites. Clearly, the nature of the newly synthesized D1 and its incorporation into PS II to form functional units is complex (Wettern, 1986) and worthy of further study during recovery from photoinhibition.

B. Dependence on Temperature and Irradiance

Until a few years ago, the ability of higher plants to recover from photoinhibitory damage was not well understood. To better understand the recovery process in higher plants, Greer et al. (1986) studied the effects of temperature and irradiance on recovery in *Phaseolus vulgaris*. Their results showed that no recovery occurred below 15 °C and maximum recovery at 30 °C. The threshold temperature for recovery, however, varies with species. For example, in contrast to *Phaseolus* (Greer et al., 1986), slow recovery occurred at 10 °C in two maize hybrids (Greer and Hardacre, 1989) and at 5 °C in barley (Greer et al., 1991). Nevertheless, the higher rate of recovery at a higher, but not excessive, temperature is consistent with the involvement of enzymic, temperature-dependent protein synthesis. Thus, in kiwifruit, the rate constant for recovery of the F_v/F_m ratio was increased by a factor of 2.9, 2.75, and 1.7 when the temperature was raised from 20 to 25 °C for plants grown under 300, 700, and 1300 μmol photons m^{-2} s^{-1}, respectively (Greer and Laing, 1988). Conversely, on lowering the temperature during a photoinhibitory treatment, the extent of damage was increased, probably because protein synthesis was hampered (Horton et al., 1987).

However, the dependence of recovery from photoinhibitory damage on temperature may also involve temperature effects at steps in the process other than protein synthesis. At low temperature, for example, degradation of the D1 protein of photoinhibited reaction centers may be retarded (Chow et al., 1989b; Gong and Nilsen, 1989). In addition, low temperature may affect both the removal of the products of D1 degradation and the insertion of newly synthesized D1 into reaction centers being repaired.

Another way to gain an insight into the recovery process is to manipulate the light regime under which recovery from photoinhibitory damage takes place. While slow recovery does occur in darkness in *Lemna gibba* (Skogen et al., 1986; Gjertsen and Nilsen, 1987), *Phaseolus vulgaris* (Greer et al., 1986), and kiwifruit (Greer and Laing, 1988), and despite a reported case of pea where light seemed to be unimportant for optimal recovery (Bhogal and Barber, 1987), it is generally found that light promotes recovery (Chaturvedi et al., 1985).

A notable feature of light-promoted recovery is, however, that optimal recovery is achieved with low irradiances. Skogen et al. (1986) demonstrated that a low irradiance of 10 μmol photons m^{-2} s^{-1} was sufficient for optimal recovery in high-light grown *Lemna gibba*: irradiances in excess of about 500 μmol m^{-2} s^{-1} retarded recovery. Another feature of light-promoted recovery is its dependence on the spectral distribution of the light. At the same photon irradiance of 115 μmol m^{-2} s^{-1}, monochromatic light (700 nm) which excites PS I was as effective as white xenon light, and both were more effective than monochromatic light at 477 or 662 nm. Skogen et al. (1986) discussed possible ways in which PS I may be involved in the recovery process; these include the supply of ATP for protein synthesis via cyclic electron flow around PS I, and the light-reactivation of certain photosynthetic enzymes that had been adversely affected by photoinhibition.

Perhaps a simpler explanation for the low-irradiance requirement of optimal recovery from photoinhibitory damage is the need to maintain an optimal stromal pH for protein synthesis. In isolated intact chloroplasts, the establishment of a maximum stroma pH of about 7.8 saturates at quite low irradiances (ca. 50 μmol photons m^{-2} s^{-1}; Heldt et al., 1973). This stromal pH is close to the optimal pH range (7.6 to 7.9) for protein synthesis observed when polysomes bound to washed pea thylakoids are supplied with amino acids (Bhaya and Jagendorf, 1984). The optimum pH range appears quite critical, with considerable inhibition of protein synthesis above and below these values (Bhaya and Jagendorf, 1984). The mechanism by which stromal pH controls translation by thylakoid-bound polysomes from pea chloroplasts appears to be related to the binding of stromal ribosomes into thylakoid polysomes; up to 74% more RNA was thylakoid-bound at pH 8.3 than at pH 7 (Hurewitz and Jagendorf, 1987), presumably because differences in thylakoid surface electrostatic properties at the two pHs give rise to different binding affinities.

Light increases the number and activity of ribosomes bound to pea chloroplast thylakoids *in vivo.* Consistent with the present idea of light regulation of stromal pH being closely related to recovery from photoinhibitory damage, the light-induced increase of ribosome binding was found by Fish and Jagendorf (1982a) to occur at much lower irradiances than required to half-saturate photosynthesis: a level of about 100 μmol photons m^{-2} s^{-1} (but not 25 μmol m^{-2} s^{-1}) was adequate for producing the maximum increase of bound ribosomes within a standard illumination period of 30 min. In terms of the present hypothesis on the role of low light in recovery from photoinhibitory damage, the ineffectiveness of an irradiance of 25 μmol photons m^{-2} s^{-1} in promoting ribosome binding may seem to be an anomaly. However, it is possibly not valid to strictly compare different systems, especially since the ribosomal work is performed with very young seedlings at 8 to 9 days after planting, when all the leaflets are still folded or when the first pair of leaves are just unfolding. In folded leaves pointing vertically, the "average" irradiance perceived by the "average" chloroplast may be too low when the (incident) irradiance is 25 μmol photons m^{-2} s^{-1}.

C. Recovery from Photoinhibitory Damage and from UV-B Stress

Although photoinhibition refers to the effects of high visible light on photosynthesis, natural sunlight may be inhibitory because of both its UV-B (Bornman, 1989) and visible components. A similarity between the damage caused by excessive visible light and UV-B radiation (280–320 nm) is that PS II is a common target affected by both stresses; in each case, damage to PS II is largely responsible for the loss of quantum yield of photosynthesis.

However, a closer scrutiny shows that the time-courses of recovery from the two stresses may be quite different. After photoinhibition at visible wavelengths for a few hours, leaves generally recover within a few hours, provided temperature and light are optimal for protein synthesis (see above). To our knowledge, there is little or no published information on recovery from UV-B irradiation. Our preliminary data (Chow and Anderson, unpublished) show that the maximal deleterious effects of supplementary UV-B in a growth cabinet on the light- and CO_2-saturated photosynthetic capacity, light-limited quantum yield of O_2 evolution, the number of functional PS II reaction centers, and the F_v/F_m ratio of pea leaves took about 40 h to develop after the cessation of the supplementary of UV-B treatment. It was only after this long lag period that net recovery was evident, as exemplified by the plot of quantum yield of O_2 evolution as a function of time in Figure 6a.

A number of reasons may explain why the deleterious UV-B effects continue to develop well after the cessation of the treatment, and why the recovery is so slow (days) compared with that from short-term photoinhibitory damage (hours). First, UV-B photons, being more energetic than those of visible light, are more likely to

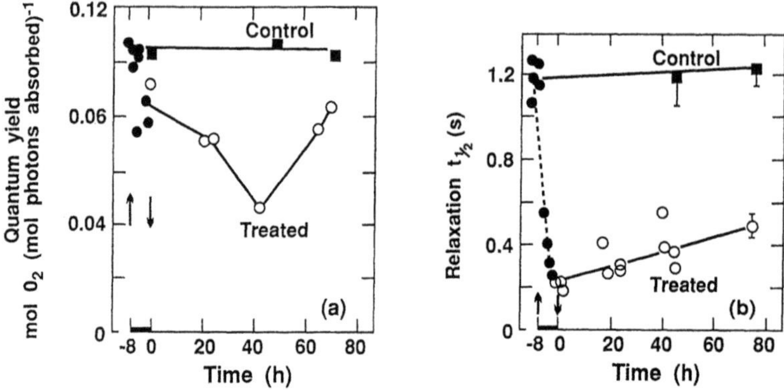

Figure 6. The time-course of changes in the light-limited quantum yield of O_2 evolution on the basis of absorbed photons (**a**) and in the half-time for relaxation of the flash-induced electrochromic shift at 515 nm (**b**) in pea leaves, during (•) and after (o) an 8-hr supplementary UV-B treatment of pea plants in a growth cabinet. The visible irradiance was 150 μmol m^{-2} s^{-1} (12 h/day). The spectral irradiance of UV-B was 50 and 220 mW m^{-2} nm^{-1} at 297 and 313 nm, respectively. *Upward arrows* indicate UV-B radiation on, and *downward arrows* UV-B radiation off. Note the continued loss of quantum yield of O_2 evolution for about 40 h following cessation of supplementary UV-B treatment before net recovery was evident. (Data of Chow, Strid, and Anderson, unpublished).

damage cellular membranes and denature nucleic acids, enzymes and other proteins. Once damaged, these cellular components may not be readily repaired.

Second, the structural integrity of thylakoid membranes was disrupted by supplementary UV-B, as indicated by the electrochromatic shift signal (Figure 6b). Following flash-induced charge separation in functional PS II and PS I reaction centers, an electric field is generated across the thylakoid membrane, altering the absorbance of antenna pigments by an electrochromic shift at 515 nm (Junge and Witt, 1968). The subsequent relaxation of the electric field occurs via ion movements across the thylakoid membrane, with a half-time ($t_{1/2}$) that is dependent upon the membrane permeability to ions. Figure 6b shows that the relaxation $t_{1/2}$ was drastically decreased from 1.2 s in the control to about 0.2 s after 8 h of supplementary UV-B irradiation. Even at 80 h after the cessation of UV-B treatment of pea plants, $t_{1/2}$ recovered only to a limited extent, suggesting a sustained leakiness of the thylakoid membrane to ions: this implies that ATP synthesis could have been partly uncoupled from electron transport. A limitation of ATP supply would have serious consequences for a system in need of repair, since the general synthesis, translocation (Lubben et al, 1988) and assembly of proteins (Ellis, 1987) are ATP-dependent.

Third, we have shown that the levels of *cab* (Jordan et al., 1991), *rbc*S, and *rbc*L (Jordan et al., 1992) transcripts were promptly depressed by UV-B, and that the recovery of these transcripts was extremely slow. The synthesis of the corresponding proteins could thus be impaired, resulting in a deterioration of photosynthetic performance.

Finally, a diversion of cellular activities towards the synthesis of enzymes and flavonoids for defence against UV-B (Chappell and Hahlbrock, 1984) must constitute significant competition for scarce resources when photosynthesis is impaired. Perhaps, for all of these reasons, it is not surprising that recovery from UV-B damage is much slower than from photoinhibitory damage.

V. CONCLUDING REMARKS

As one surveys the literature on photoprotection and photoinhibitory damage, one is overwhelmed by the amount of information which has accumulated on the subject, particularly in the past decade. One is also led to ask why such a concentrated effort has been focused on what is a mere part of photosynthesis.

The enthusiasm and interest directed towards this area of research probably arises from the vast spatial and temporal scales over which the phenomena of photoprotection and photoinhibitory damage occur, ranging from vegetation ecosystem to molecular complexes, and from long-term vegetation dynamics to rapid photophysical and photochemical events at photochemical reaction centers (Osmond, 1989). Not only are photoprotective strategies and photoinhibitory damage of ecological significance, but they also provide fertile territory in which the photosynthetic process and apparatus can be studied from the vantage point of several disciplines.

Studying about photoprotection and photoinhibitory damage has taught us a great deal about molecular processes of photosynthesis: (1) the mechanisms whereby excess excitation energy is dissipated or utilized, (2) the damage to the redox sites when defence strategies fail for one reason or another, and (3) mechanisms of subsequent repair required to sustain function. Nevertheless, there is still a great deal to be learned about photoprotection and photoinhibitory damage. The exact function of the violaxanthin cycle in controlled heat dissipation needs to be clarified. The long-term downregulated state of PS II, with its apparent ability for nonradiative dissipation of excitation energy, is a relatively recent idea which needs further investigation. The proteolytic degradation of D1 and other proteins, and the regulation of this enzymic process certainly merit in-depth study since they may ultimately have a strong bearing on the stability of the reaction center in the presence of strong light. Similarly, the mechanism of replacement of degraded D1 by newly synthesized D1 is only beginning to be understood. Even more puzzling is the reason for the fast turnover of D1. As suggested by Critchley et al. (1992), key questions include: Is the rapid turnover of D1 a fatal flaw in PS II leading to damage that needs constant repair? Alternatively, does D1 turnover have other regulatory roles? Unraveling the mechanisms of photoprotection and photoinhibitory damage will, almost certainly, reveal much not only about the protective and inhibitory events, but also about the molecular processes of photosynthesis itself.

ACKNOWLEDGMENTS

I am very grateful to Drs. Jan Anderson, Marilyn Ball, Bob Furbank, Steve Grace, Catherine Lovelock, Barry Osmond, Cecilia Sundby-Emanuelsson, Tom Wydrzynski, and Mrs. Steph McCaffery for their constructive comments on the manuscript, and to Prof. Gunnar Öquist for enlightening discussions during and after his visit to Canberra. I also wish to thank the Australian Research Council for the award of a National Research Fellowship which has partly led to the writing of this review.

ABBREVIATIONS

ADP:	adenosine diphosphate
ATP:	adenosine triphosphate
Car:	carotenoid
Chl:	chlorophyll
cyt:	cytochrome
D:	redox-active tyrosine on D2
D1:	herbicide/Q_B-binding protein in PS II, *psbA* gene product
D2:	Q_A-binding protein in PS II, *psbD* gene product
DBMIB:	2,5-dibromo-3-methyl-6-isopropyl-*p*-benzoquinone
DCMU:	3′-(3,4-dichlorophenyl)-1,1-dimethylurea
F_o:	chlorophyll fluorescence when PS II reaction centers are open
F_m:	maximum chlorophyll fluorescence
F_v:	variable chlorophyll fluorescence
LHC II:	light-harvesting complexes of PS II
OEC:	oxygen-evolving complex
Pheo:	pheophytin
P680:	primary electron donor in PS II
P700:	primary electron donor in PS I
PS:	photosystem
Q_A:	primary quinone electron acceptor in PS II
Q_B:	secondary quinone electron acceptor in PS II
q_{Ne}, q_{Ni}, and q_{Nt}:	coefficient for non-photochemical quenching of chlorophyll fluorescence due to the energized state, photoinhibitory damage, and protein phosphorylation, respectively
q_P:	coefficient for photochemical quenching
Rubisco:	ribulose 1,5-bisphosphate carboxylase/oxygenase

S_0: S_1, S_2, singlet states of molecules
UV-B: ultraviolet-B radiation
Z: redox-active tyrosine on D1.

REFERENCES

Adams, W.W., III & Osmond, C.B. (1988). Internal CO_2 supply during photosynthesis of sun and shade grown CAM plants in relation to photoinhibition. Plant Physiol. 86, 117–123.

Adamson, H.Y., Chow, W.S., Anderson, J.M., Vesk, M., & Sutherland, M. (1991). Photosynthetic acclimation of *Tradescantia albiflora* to growth irradiance: Morphological, ultrastructural and growth responses. Physiol. Plant. 82, 353–359.

Adir, N., Shochat, S., & Ohad, I. (1990). Light-dependent D1 protein synthesis and translocation is regulated by reaction center II. Reaction center II serves as an acceptor for the D1 precursor. J. Biol. Chem. 265, 12563–12568.

Akabori, K., Tsukamoto, H., Tsukihara, J., Nagatsuka, T., Motokawa, O., & Toyoshima, Y. (1988). Disintegration and reconstitution of photosystem II reaction center core complex. I. Preparation and characterization of three different types of subcomplex. Biochim. Biophys. Acta 932, 345–357.

Allakhverdiev, S.I., Setlikova, E., Klimov, V.V., & Setlik, I. (1987). In photoinhibited photosystem II particles pheophytin photoreduction remains unimpaired. FEBS Lett. 226, 186–190.

Anderson, J.M. (1982). The significance of grana stacking in chlorophyll *b*-containing chloroplasts. Photobiochem. Photobiophys. 3, 225–241.

Anderson, J.M. (1986). Photoregulation of the composition, function and structure of thylakoid membranes. Ann. Rev. Plant Physiol. 37, 93–136.

Anderson, J.M. & Melis, A. (1983). Localization of different photosystems in separate regions of chloroplast membranes. Proc. Natl. Acad. Sci. USA 80, 745–749.

Anderson, J.M. & Osmond, C.B. (1987). Shade-sun responses: Compromises between acclimation and photoinhibition. In: Photoinhibition (Kyle, D.J., Osmond, C.B., & Arntzen, C.J., Eds.). Topics in Photosynthesis, Vol. 9, pp. 1–38. Elsevier, Amsterdam.

Anderson, J.M., Chow, W.S., & Goodchild, D.J. (1988). Thylakoid membrane organization in sun/shade acclimation. Aust. J. Plant Physiol. 15, 11–26.

Arntz, B. & Trebst, A. (1986). On the role of the Q_B protein of PS II in photoinhibition. FEBS Lett. 194, 43–49.

Arntzen, C.J., Kyle, D.J., Wettern, M., & Ohad, I. (1984). Photoinhibition: A consequence of the accelerated breakdown of the apoprotein of the secondary acceptor of photosystem II. In: Biosynthesis of the Photosynthetic Apparatus. Molecular Biology, Development and Regulation (Hallick, R., Staehelin, L.A., & Thornber, J.P., Eds.). UCLA Symposium Series No. 14, pp. 313–324. A. R. Liss, New York.

Aro, E.-M., Hundal, T., Carlberg, I., & Andersson, B. (1990). *In vitro* studies on light-induced inhibition of photosystem II and D1-protein degradation at low temperatures. Biochim. Biophys. Acta 1019, 269–275.

Aro, E.-M., Kettunen, R., & Tyystjärvi, E. (1992). ATP and light regulate D1 protein modification and degradation: Role of $D1^*$ in photoinhibition. FEBS Lett. 297, 29–33.

Asada, K. & Takahashi, M. (1987). Production and scavenging of active oxygen in photosynthesis. In: Photoinhibition (Kyle, D.J., Osmond, C.B., & Arntzen, C.J., Eds.). Topics in Photosynthesis, Vol. 9, pp. 227–287. Elsevier, Amsterdam.

Ball, M.C., Chow, W.S., & Anderson, J.M. (1987). Salinity-induced potassium deficiency causes loss of functional photosystem II in leaves of the grey mangrove, *Avicennia marina*, through depletion of the atrazine-binding polypeptide. Aust. J. Plant Physiol. 14, 351–361.

Ball, M.C., Hodges, V.S., & Laughlin, G.P. (1991). Cold-induced photoinhibition limits regeneration of snow gum at tree line. Functional Ecol. 5, 663–668.

Barber, J., Chapman, D.J., & Telfer, A. (1987). Characterization of a PS II reaction center isolated from the chloroplasts of *Pisum sativum*. FEBS Lett. 220, 67–73.

Barber, J. & Andersson, B. (1992). Too much of a good thing: Light can be bad for photosynthesis. Trends Biochem. Sci. 17, 61–66.

Besford, R.T. (1986). Changes in some Calvin cycle enzymes of the tomato during acclimation to irradiance. J. Exp. Bot. 37, 200–210.

Bhaya, D. & Jagendorf, A.T. (1984). Optimal conditions for translation by thylakoid-bound polysomes from pea chloroplasts. Plant. Physiol. 75, 832–838.

Bhogal, M. & Barber, J. (1987). Photoinhibition and recovery in intact leaves of *Pisum sativum* grown in high and low light intensity. In: Progress in Photosynthesis Research (Biggins, J., Ed.), Vol. 4, pp. 91–94. Martinus Nijhoff, Dordrecht.

Bilger, W., Björkman, O., & Thayer, S.S. (1989). Light-induced spectral absorbance changes in relation to photosynthesis and the epoxidation state of xanthophyll cycle components in cotton leaves. Plant Physiol. 91, 542–551.

Bilger, W. & Björkman, O. (1990). Role of the xanthophyll cycle in photoprotection elucidated by measurements of light-induced absorbance changes, fluorescence and photosynthesis in leaves of *Hedera canariensis*. Photosynth. Res. 25, 173–185.

Björkman, O. (1972). Photosynthetic adaptation to contrasting light climates. Carnegie Inst. Wash. Yearbook 71, 82–85.

Björkman, O. & Ludlow, M.M. (1972). Characterization of the light climate on the floor of a Queensland rainforest. Carnegie Inst. Wash. Year Book 71, 85–94.

Björkman, O. (1981). Responses to different quantum flux densities. In: Physiological Plant Ecology I. Responses to the Physical Environment (Lange, O.L., Nobel, P.S., Osmond, C.B., & Ziegler, H., Eds). Encycl. Plant Physiol. New Series, Vol. 12A, pp. 57–107. Springer-Verlag, Berlin.

Björkman, O. & Demmig, B. (1987). Photon yield of O_2 evolution and chlorophyll fluorescence characteristics at 77 K among vascular plants of diverse origins. Planta 170, 489–504.

Blankenship, R.E. & Prince, R.C. (1985). Excited-state redox potentials and the Z scheme of photosynthesis. Trends Biochem. Sci. 10, 382–383.

Blubaugh, D.J. & Cheniae, G.M. (1990). Kinetics of photoinhibition in hydroxylamine-extracted photosystem II membranes: Relevance to photoactivation and sites of electron donation. Biochem. 29, 5109–5118.

Blubaugh, D.J., Atamian, M., Babcock, G.T., Golbeck, J.H., & Cheniae, G.M. (1992). Photoinhibition of hydroxylamine-extracted photosystem II membranes: Identification of the sites of photodamage. Biochem., in press.

Boardman, N.K. (1977). Comparative photosynthesis of sun and shade plants. Ann. Rev. Plant Physiol. 28, 355–377.

Bornman, J.F. (1989). Target sites of UV-B radiation in photosynthesis of higher plants. J. Photochem. Photobiol. B. Biol. 4, 145–158.

Briantais, J.-M., Vernotte, C., Picaud, M., & Krause, G.H. (1979). A quantitative study of the slow decline of chlorophyll *a* fluorescence in isolated chloroplasts. Biochim. Biophys. Acta 548, 128–138.

Buschmann, C. (1987). Induction kinetics of heat emission before and after photoinhibition in cotyledons of *Raphanus sativus*. Photosynth. Res. 14, 229–240.

Callahan, F.E. & Cheniae, G.M. (1985). Studies on the photoactivation of the water-oxidizing enzyme. I. Processes limiting photoactivation in hydroxylamine-extracted leaf segments. Plant Physiol. 79, 777–786.

Callahan, F.E., Becker, D.W., & Cheniae, G.M. (1986). Studies on the photoactivation of the water-oxidizing enzyme. II. Characterization of weak light photoinhibition of PS II and its light-induced recovery. Plant Physiol. 82, 261–269.

Canaani, O. & Havaux, M. (1990). Evidence for a biological role in photosynthesis for cytochrome *b*-559—a component of photosystem II reaction center. Proc. Natl. Acad. Sci. 87, 9295–9299.

Chan, C.-K., DiMagno, T.J., Chen, L.X.-Q., & Norris, J.R. (1991). Mechanism of the initial charge separation in bacterial photosynthetic reaction centers. Proc. Natl. Acad. Sci. USA. 88, 11202–11206.

Chappell, J. & Hahlbrock, K. (1984). Transcription of plant defense genes in response to UV light of fungal elicitor. Nature 311, 76–78.

Chaturvedi, R., Haugstad, M.K., Nilsen, S., & Skogen, D. (1985). Photoinhibition of photosynthesis: Effect of light and selective excitation of photosystems on recovery. Photosynthetica 19, 382–387.

Chow, W.S., Wagner, G., & Hope, A.B. (1976). Light-dependent redistribution of ions in isolated spinach chloroplasts. Aust. J. Plant Physiol. 3, 853–861.

Chow, W.S. & Anderson, J.M. (1987a). Photosynthetic responses of *Pisum sativum* to an increase in irradiance during growth. I. Photosynthetic activities. Aust. J. Plant Physiol. 14, 1–8.

Chow, W.S. & Anderson, J.M. (1987b). Photosynthetic responses of *Pisum sativum* to an increase in irradiance during growth. II. Thylakoid membrane components. Aust. J. Plant Physiol. 14, 9–19.

Chow, W.S., Qian, L., Goodchild, D.J., & Anderson, J.M. (1988). Photosynthetic acclimation of *Alocasia macrorrhiza* (L.) G. Don to growth irradiance: Structure, function and composition of chloroplasts. Aust. J. Plant Physiol. 15, 107–122.

Chow, W.S., Hope, A.B., & Anderson, J.M. (1989a). Oxygen per flash from leaf discs quantifies photosystem II. Biochim. Biophys. Acta 973, 105–108.

Chow, W.S., Osmond, C.B., & Huang, L.K. (1989b). Photosystem II function and herbicide binding sites during photoinhibition of spinach chloroplasts *in vivo* and *in vitro*. Photosynth. Res. 21, 17–26.

Chow, W.S., Ball, M.C., & Anderson, J.M. (1990). Growth and photosynthetic responses of spinach to salinity: Implications of K^+ nutrition to salt tolerance. Aust. J. Plant Physiol. 17, 563–578.

Chow, W.S., Adamson, H.Y., & Anderson, J.M. (1991a). Photosynthetic acclimation of *Tradescantia albiflora* to growth irradiance: Lack of adjustment of light-harvesting components and its consequences. Physiol. Plant. 81, 175–182.

Chow, W.S., Hope, A.B., & Anderson, J.M. (1991b). Further studies on quantifying photosystem II *in vivo* by flash-induced oxygen yield from leaf discs. Aust. J. Plant Physiol. 18, 397–410.

Christensen, R.L. & Kohler, B.E. (1976). High-resolution optical spectroscopy of polyenes related to the visual chromophore. J. Phys. Chem. 80, 2197–2200.

Cleland, R.E. (1988). Molecular events of photoinhibitory inactivation in the reaction center of photosystem II. Aust. J. Plant Physiol. 15, 135–150.

Cleland, R.E. & Critchley, C. (1985). Studies on the mechanism of photoinhibition in higher plants. II. Inactivation by high light of photosystem II reaction center function in isolated spinach thylakoids and O_2 evolving particles. Photobiochem. Photobiophys. 10, 83–92.

Cleland, R.E., Melis, A., & Neale, P.J. (1986). Mechanism of photoinhibition: Photochemical reaction center inactivation in system II of chloroplasts. Photosynth. Res. 9, 79–88.

Cleland, R.E. & Melis, A. (1987). Probing the events of photoinhibition by altering electron-transport activity and light-harvesting capacity in chloroplast thylakoids. Plant Cell Environ. 10, 747–752.

Cleland, R.E., Ramage, R.T., & Critchley, C. (1990). Photoinhibition causes loss of photochemical activity without degradation of D1 protein. Aust. J. Plant Physiol. 17, 641–651.

Cogdell, R. (1988). The function of pigments in chloroplasts. In: Plant Pigments. (Goodwin, T.W., Ed.), pp. 183–230.

Cosgrove, S.A., Guite, M.A., Burnell, T.B., & Christensen, R.L. (1990). Electronic relaxation in long polyenes. J. Phys. Chem. 94, 8118–8124.

Critchley, C. (1988a). The molecular mechanism of photoinhibition—facts and fiction. Aust. J. Plant Physiol. 15, 27–41.

Critchley, C. (1988b). The chloroplast thylakoid membrane system is a molecular conveyor belt. Photosynth. Res. 19, 265–276.

Critchley, C., Russel, A.W., & Bolhar-Nordenkampf, H.R. (1992). Rapid protein turnover in photosystem II: Fundamental flaw or regulatory mechanism? Photosynthetica 27, 183–190.

Davies, E.C., Chow, W.S., LeFay, J.M., & Jordan, B.R. (1986). Acclimation of tomato leaves to changes in light intensity: Effects on the function of the thylakoid membrane. J. Exp. Bot. 37, 211–220.

Demeter, S., Neale, P.J., & Melis, A. (1987). Photoinhibition: Impairment of the primary charge separation between P-680 and pheophytin in photosystem II of chloroplasts. FEBS Lett. 214, 370–374.

Demmig-Adams, B. (1990). Carotenoids and photoprotection in plants: A role for the xanthophyll zeaxanthin. Biochim. Biophys. Acta 1020, 1–24.

Demmig, B. & Björkman, O. (1987). Comparison of the effect of excess light on chlorophyll fluorescence (77 K) and photon yield of O_2 evolution in leaves of higher plants. Planta 171, 171–184.

Demmig-Adams, B. & Adams, W.W., III (1992a). Photoprotection and other responses of plants to high light stress. Annu. Rev. Plant Physiol. Plant Mol. Biol. 43, 599–626.

Demmig-Adams, B. & Adams, W.W., III (1992b). The xanthophyll cycle. In: Carotenoids in Photosynthesis (Young, A. & Britton, G., Eds.). Springer, Berlin/Heidelberg, in press.

Demmig-Adams, B. & Adams, W.W., III (1992c). Carotenoid composition in sun and shade leaves of plants with different life forms. Plant, Cell Environ. 15, 411–419.

Demmig, B., Winter, K., Krüger, A., & Czygan, F.-C. (1987). Photoinhibition and zeaxanthin formation in intact leaves. Plant Physiol. 84, 218–224.

Demmig, B., Winter, K., Krüger, A., & Czygan, F.-C. (1988). Zeaxanthin and the heat dissipation of excess light energy in *Nerium oleander* exposed to a combination of high light and water stress. Plant Physiol. 87, 17–24.

Demmig-Adams, B., Winter, K., Winkelmann, E., Krüger, A., & Czygan, F.-C. (1989). Photosynthetic characteristics and the ratios of chlorophyll, β-carotene, and the components of the xanthophyll cycle upon a sudden increase in growth light regime in several plant species. Botanica Acta 102, 319–325.

Demmig-Adams, B., Adams, W.W., III., Heber, U., Neimanis, S., Winter, K., Krüger, A., Czygan, F.-C., Bilger, W., & Björkman, O. (1990). Inhibition of zeaxanthin formation and of rapid changes in radiationless energy dissipation by dithiothreitol in spinach leaves and chloroplasts. Plant Physiol. 92, 293–301.

Durrant, J.R., Giorgi, L.B., Barber, J., Klug, D.R., & Porter, G. (1990). Characterization of triplet states in isolated photosystem II reaction centers: Oxygen quenching as a mechanism for photodamage. Biochim. Biophys. Acta 1017, 167–175.

Eckert, H.-J., Geiken, B., Bernarding, J., Napiwotzki, A., Eichler, H.-J., & Renger, G. (1991). Two sites of photoinhibition of the electron transfer in oxygen evolving and Tris-treated PS II membrane fragments from spinach. Photosynth. Res. 27, 97–108.

Ellis, J. (1987). Proteins as molecular chaperones. Nature 328, 378–379.

Ewart, A.J. (1896). On assimilatory inhibition in plants. J. Linn. Soc. 31, 364–461.

Falkowski, P.G., Fujita, Y., Ley, A., & Mauzerall, D. (1986). Evidence for cyclic electron flow around photosystem II in *Chlorella pyrenoidosa*. Plant Physiol. 81, 310–312.

Farineau, J. (1990). Photochemical alterations of photosystem II induced by two different photoinhibitory treatments in isolated chloroplasts of pea. A luminescence study. Biochim. Biophys. Acta. 1016, 357–363.

Fish, L.E. & Jagendorf, A.T. (1982a). Light-induced increase in the number and activity of ribosomes bound to pea chloroplasts *in vivo*. Plant Physiol. 69, 814–825.

Fish, L.E. & Jagendorf, A.T. (1982b). High rates of protein synthesis by isolated chloroplasts. Plant Physiol. 70, 1107–1114.

Fork, D.C., Bose, S., & Herbert, S.K. (1986). Radiationless transitions as a protection mechanism against photoinhibition in higher plants and algae. Photosynth. Res. 10, 327–333.

Franklin, L.A., Levavasseur, G., Osmond, C.B., Henley, W.J., & Ramus, J. (1992). Two components of onset and recovery during photoinhibition of *Ulva rotundata*. Planta 186, 399–408.

Ghirardi, M.L., McCauley, S.W., & Melis, A. (1986). Photochemical apparatus organization in the thylakoid membrane of *Hordeum vulgare* wild-type and chlorophyll *b*-less chlorina f2 mutant. Biochim. Biophys. Acta 851, 331–339.

Gjertsen, K. & Nilsen, S. (1987). Photoinhibition and recovery in intact leaves of *Pisum sativum* grown in high and low light intensity. In: Progress in Photosynthesis Research (Biggins, J., Ed.), Vol. 4, pp. 87–90. Martinus Nijhoff, Dordrecht.

Goodchild, D.J., Highkin, H.R., & Boardman, N.K. (1966). The fine structure of chloroplasts in a barley mutant lacking chlorophyll *b*. Exp. Cell Res. 43, 684–688.

Gong, H. & Nilsen, S. (1989). Effect of temperature on photoinhibition of photosynthesis, recovery and turnover of the 32 kD chloroplast protein in *Lemna gibba*. J. Plant Physiol. 135, 9–14.

Grahl, H. & Wild, A. (1975). Studies on the content of P700 and cytochromes in *Sinapis alba* during growth under two different light intensities. In: Environmental and Biological Control of Photosynthesis (Marcelle, R., Ed.), pp. 107–113. Dr W. Junk, The Hague.

Greer, D.H., Berry, J.A., & Björkman, O. (1986). Photoinhibition of photosynthesis in intact bean leaves: Role of light and temperature, and requirement for chloroplast-protein synthesis during recovery. Planta 168, 253–260.

Greer, D.H. & Laing, W.A. (1988). Photoinhibition of photosynthesis in intact kiwifruit (*Actinidia deliciosa*) leaves: Effect of light during growth on photoinhibition and recovery. Planta 175, 355–363.

Greer, D.H. & Laing, W.A. (1989). Photoinhibition of photosynthesis in intact kiwifruit (*Actinidia deliciosa*) leaves: Effect of growth temperature on photoinhibition and recovery. Planta 180, 32–39.

Greer, D.H. & Hardacre, A.K. (1989). Photoinhibition of photosynthesis and its recovery in two maize hybrids varying in low temperature tolerance. Aust. J. Plant Physiol. 16, 189–198.

Greer, D.H., Ottander, C., & Öquist, G. (1991). Photoinhibition and recovery of photosynthesis in intact barley leaves at 5 and 20 °C. Physiol. Plant. 81, 203–210.

Gruszecki, W.I. & Strzalka, K. (1991). Does the xanthophyll cycle take part in the regulation of fluidity of the thylakoid membrane? Biochim. Biophys. Acta 1060, 310–314.

Guenther, J.E. & Melis, A. (1990). The physiological significance of photosystem II heterogeneity in chloroplasts. Photosynth. Res. 23, 105–109.

Habash, D.Z. & Baker, N.R. (1987). Can protein phosphorylation alleviate photoinhibition of thylakoid photochemical activities? In: Progress in Photosynthesis Research (Biggins, J., Ed.), Vol. IV, pp. 75–78. Martinus Nijhoff, Dordrecht.

Habash, D.Z. & Baker, N.R. (1990). Demonstration of two sites of inhibition of electron transport by protein phosphorylation in wheat thylakoids. J. Exp. Bot. 41, 761–767.

Hager, A. (1969). Lichtbedingte pH-erniedrigung in einem chloroplasten-kompartiment als ursache der enzymatischen violaxanthin- $\rightarrow$ zeaxanthin-umwandlung; beziehungen zur photophosphorylierung. Planta 89, 224–243.

Hart, J.J. & Stemler, A. (1990). High light-induced reduction and low light-enhanced recovery of photon yield in triazine-resistant *Brassica napus* L. Plant Physiol. 94, 1301–1307.

Hashimoto, H. & Koyama, Y. (1989a). The C=C stretching Raman lines of β-carotene isomers in the S_1 state as detected by pump-probe resonance Raman spectroscopy. Chem. Phys. Lett. 154, 321–325.

Hashimoto, H. & Koyama, Y. (1989b). Raman spectra of all-trans-β-apo-8′-carotenal in the S_1 and T_1 states: A picosecond pump-and probe technique using ML-QS pulse trains. Chem. Phys. Lett. 162, 523–527.

Havaux, M., Gruszecki, W.I., Dupont, I., & Leblanc, R.M. (1991a). Increased heat emission and its relationship to the xanthophyll cycle in pea leaves exposed to strong light stress. J. Photochem. Photobiol. B: Biol. 8, 361–370.

Havaux, M., Strasser, R.J., & Greppin, H. (1991b). A theoretical and experimental analysis of the q_P and q_N coefficients of chlorophyll fluorescence quenching and their relation to photochemical and nonphotochemical events. Photosynth. Res. 27, 41–55.

Heldt, H.W., Werdan, K., Milovancev, M., & Geller, G. (1973). Alkalization of the chloroplast stroma caused by light-dependent proton flux into the thylakoid space. Biochim. Biophys. Acta 314, 224–241.

Henley, W.J., Levavasseur, G., Franklin, L., Osmond, C.B., & Ramus, J. (1991). Photoacclimation and photoinhibition in *Ulva rotundata* as influenced by nitrogen availability. Planta 184, 235–243.

Hind, G., Nakatani, H.Y., & Izawa, S. (1974). Light-dependent redistribution of ions in suspensions of chloroplast thylakoid membranes. Proc. Natl. Acad. Sci. USA 71, 1484–1488.

Horton, P. & Hague, A. (1988). Studies on the induction of chlorophyll fluorescence in isolated barley protoplasts. IV. Resolution of non-photochemical quenching. Biochim. Biophys. Acta 932, 107–115.

Horton, P. & Lee, P. (1985). Phosphorylation of chloroplast membrane proteins partially protects against photoinhibition. Planta 165, 37–42.

Horton, P., Lee, P., & Hague, A. (1987). Photoinhibition of isolated chloroplasts and protoplasts. In: Progress in Photosynthesis Research (Biggins, J., Ed.), Vol. 4, pp. 59–63. Martinus Nijhoff, Dordrecht.

Horton, P., Ruban, A.V., Rees, D., Pascal, A.A., Noctor, G., & Young, A.J. (1991). Control of the light-harvesting function of chloroplast membranes by aggregation of the LHCII chlorophyll-protein complex. FEBS Lett. 292, 1–4.

Hundal, T., Virgin, I., Styring, S., & Andersson, B. (1990). Changes in the organization of photosystem II following light-induced D1-protein degradation. Biochim. Biophys. Acta 1017, 235–241.

Hurewitz, J. & Jagendorf, A.T. (1987). Further characterization of ribosome binding to thylakoid membranes. Plant. Physiol. 84, 31–34.

Huse, H. & Nilsen, S. (1989). Recovery from photoinhibition: Effect of light and inhibition of protein synthesis of 32-kDa chloroplast protein. Photosynth. Res. 21, 171–179.

Jagendorf, A.T. (1977). Photophosphorylation. In: Photosynthesis. I. Photosynthetic electron transport and photophosphorylation (Trebst, A. & Avron, M., Eds.). Encyclo. Plant Physiol. New Series, Vol. 5, 307–337.

Jegerschöld, C. & Styring, S. (1991). Fast oxygen-independent degradation of the D1 reaction center protein in photosystem II. FEBS Lett. 280, 87–90.

Jordan, B.R., Chow, W.S., Strid, Å., & Anderson, J.M. (1991). Reduction in *cab* and *psbA* RNA transcripts in response to supplementary ultraviolet-B radiation. FEBS Lett. 284, 5–8.

Jordan, B.R., He, J., Chow, W.S., & Anderson, J.M. (1992). Changes in mRNA levels and polypeptide subunits of ribulose 1,5-bisphosphate carboxylase in response to supplementary ultraviolet-B radiation. Plant Cell Environ. 15, 91–98.

Junge, W. & Witt, H.T. (1968). On the ion transport system of photosynthesis: Investigation on a molecular level. Z. Naturforsch. 23b, 244–254.

Jursinic, P.A. & Pearcy, R.W. (1988). Determination of the rate limiting step for photosynthesis in a nearly isonuclear rapeseed (*Brassica napus* L.) biotype resistant to atrazine. Plant Physiol. 88, 1195–1200.

Kettunen, R., Tyystjärvi, E., & Aro, E.-M. (1991). D1 protein degradation during photoinhibition of intact leaves. FEBS Lett. 290, 153–156.

Kirilovsky, D. & Etienne, A.-L. (1991). Protection of reaction center II from photodamage by low temperature and anaerobiosis in spinach chloroplasts. FEBS Lett. 279, 201–204.

Kirilovsky, D., Ducruet, J.-M., & Etienne, A.-L. (1990a). Primary events occurring in photoinhibition in *Synechocystis* 6714 wild-type and an atrazine-resistant mutant. Biochim. Biophys. Acta 1020, 87–93.

Kirilovsky, D.L., Vernotte, C., & Etienne, A.L. (1990b). Protection from photoinhibition by low temperature in *Snechocystis* 6714 and in *Chlamydomonas reinhardtii*: Detection of an intermediary state. Biochem. 29, 8100–8106.

Kirschbaum, M.U.F. & Farquhar, G.D. (1984). Temperature dependence of whole-leaf photosynthesis in *Eucalyptus pauciflora* Sieb. ex Spreng. Aust. J. Plant Physiol. 11, 519–38.

Koyama, Y. (1991). Structures and functions of carotenoids in photosynthetic systems. J. Photochem. Photobiol. B: Biol. 9, 265–280.

Krause, G.H., Vernotte, C., & Briantais, J.-M. (1982). Photoinduced quenching of chlorophyll fluorescence in intact chloroplasts and algae. Biochim. Biophys. Acta 679, 116–124.

Krause, G.H. (1988). Photoinhibition of photosynthesis. An evaluation of damaging and protective mechanisms. Physiol. Plant. 74, 566–574.

Krause, G.H. & Weis, E. (1991). Chlorophyll fluorescence and photosynthesis: The basics. Ann. Rev. Plant Physiol. Plant Mol. Biol. 42, 313–349.

Krause, G.H. & Behrend, U. (1986). ΔpH-dependent chlorophyll fluorescence quenching indicating a mechanism of protection against photoinhibition of chloroplasts. FEBS Lett. 200, 298–302.

Krause, G.H. & Laasch, H. (1987). Energy-dependent chlorophyll fluorescence quenching in chloroplasts correlated with quantum yield of photosynthesis. Z. Naturforsch. 42c, 581–584.

Krause, G.H., Köster, S., & Wong, S.C. (1985). Photoinhibition of photosynthesis under anaerobic conditions studied with leaves and chloroplasts of *Spinacia oleracea* L. Planta 165, 430–438.

Krupa, Z., Öquist, G., & Gustafsson, P. (1991). Photoinhibition of photosynthesis and growth responses at different light levels in *psbA* gene mutants of the cyanobacterium *Synechococcus*. Physiol. Plant. 82, 1–8.

Kyle, D.J. (1984). The 32 000 Dalton Q_B protein of photosystem II. Photochem. Photobiol. 41, 107–116.

Kyle, D.J. & Ohad, I. (1986). The mechanism of photoinhibition in higher plants and green algae. In: Encyclopedia of Plant Physiology (Staehelin, L.A. & Arntzen, C.J., Eds.). New Series, Vol. 19, Photosynthesis III, pp. 468–475. Springer-Verlag, Berlin/Heidelberg.

Kyle, D.J., Ohad, I., & Arntzen, C.J. (1984). Membrane protein damage and repair: Selective loss of a quinone-protein function in chloroplast membranes. Proc. Natl. Acad. Sci. USA 81, 4070–4074.

Kyle, D.J., Osmond, C.B., & Arntzen, C.J. (1987). Photoinhibition. Topics in Photosynthesis (Barber, J., Ser. Ed.), Vol. 9. Elsevier, Amsterdam.

Leong, T.-Y. & Anderson, J.M. (1984). Adaptation of the thylakoid membranes of pea chloroplasts to light intensities. I. Study on the distribution of chlorophyll-protein complexes. Photosynth. Res. 5, 105–115.

Leverenz, J.W. & Öquist, G. (1987). Quantum yields of photosynthesis at temperatures between –2 °C and 35 °C in a cold-tolerant C_3 plant (*Pinus sylvestris*) during the course of one year. Plant, Cell Environ. 10, 287–295.

Ley, A.C. & Mauzerall, D. (1982). Absolute absorption cross sections for photosystem II and the minimum quantum requirement for photosynthesis in *Chlorella vulgaris*. Biochim. Biophys. Acta 680, 95–106.

Ley, A.C. (1986). Relationships among cell chlorophyll content, photosystem II light-harvesting and the quantum yield for oxygen production in *Chlorella*. Photosynth. Res. 10, 189–196.

Long, S.P., Farage, P.K., Groome, Q., Macharia, J.M.N., & Baker, N.R. (1990). Damage to photosynthesis during chilling and freezing, and its significance to the photosynthetic productivity of field crops. In: Current Research in Photosynthesis (Baltscheffsky, M., Ed.), Vol. IV, pp. 835–842. Kluwer Acad. Publ., Dordrecht.

Lönneborg, A., Kalla, S.R., Samuelsson, G., Gustafsson, P., & Öquist, G. (1988). Light-regulated expression of the *psbA* transcript in the cyanobacterium *Anacystis nidulans*. FEBS Lett. 240, 110–114.

Lubben, T.H., Theg, S.M., & Keegstra, K. (1988). Transport of proteins into chloroplasts. Photosynth. Res. 17, 173–194.

Ludlow, M.M. & Björkman, O. (1984). Paraheliotropic leaf movement in Siratro as a protective mechanism against drought-induced damage to primary photosynthetic reactions: Damage by excessive light and heat. Planta 161, 505–518.

Lundmark, T., Hällgren, J.-E., & Hedén, J. (1988). Recovery from winter depression of photosynthesis in pine and spruce. Trees 2, 110–114.

Mäenpää, P., Andersson, B., & Sundby, C. (1987). Difference in sensitivity to photoinhibition between photosystem II in the appressed and non-appressed thylakoid regions. FEBS Lett. 215, 31–36.

Mattoo, A.K., Hoffman-Falk, H., Marder, J.B., & Edelman, M. (1984). Regulation of protein metabolism: Coupling of photosynthetic electron transport to *in vivo* degradation of the rapidly metabolized 32-kDa protein of the chloroplast membranes. Proc. Natl. Acad. Sci. USA 81, 1380–1384.

Mattoo, A.K. & Edelman, M. (1987). Intramembrane translocation and posttranslational palmitoylation of the chloroplast 32-kDa herbicide-binding protein. Proc. Natl. Acad. Sci. USA 84, 1497–1501.

Melis, A. (1991). Dynamics of photosynthetic membrane composition and function. Biochim. Biophys. Acta 1058, 87–106.

Melis, A., Spangfort, M., & Andersson, B. (1987). Light-absorption and electron-transport balance between photosystem II and photosystem I in spinach chloroplasts. Photochem. Photobiol. 45, 129–136.

Mishra, S.K., Subrahmanyam, D., & Singhal, G.S. (1991). Interrelationship between salt and light stress on primary processes of photosynthesis. J. Plant Physiol. 138, 92–96.

Misra, A.N., Hall, S.G., & Barber, J. (1991). The isolated D1/D2/cyt *b*-559 reaction center complex of photosystem II possesses a serine-type endopeptidase activity. Biochim. Biophys. Acta 1059, 239–242.

Murata, N. & Sugahara, K. (1969). Control of excitation transfer in photosynthesis. III. Light-induced decrease of chlorophyll *a* fluorescence related to photophosphorylation system in spinach chloroplasts. Biochim. Biophys. Acta 189, 182–192.

Nanba, O. & Satoh, K. (1987). Isolation of a photosystem II reaction center consisting of D-1 and D-2 polypeptides and cytochrome *b*-559. Proc. Natl. Acad. Sci. USA 84, 109–112.

Neale, P.J. (1987). Algal photoinhibition and photosynthesis in the aquatic environment. In: Topics in Photosynthesis. Photoinhibition (Kyle, D.J., Osmond, C.B., & Arntzen, C.J., Eds.), Vol. 9, pp. 39–65. Elsevier, Amsterdam.

Neale, P.J. & Melis, A. (1986). Algal photosynthetic membrane complexes and the photosynthesis-irradiance curve: A comparison of light-adaptation responses in *Chlamydomonas reinhardtii* (chlorophyta). J Phycol. 22, 531–538.

Neale, P.J. & Melis, A. (1989). Salinity-stress enhances photoinhibition of photosynthesis in *Chlamydomonas reinhardtii*. J. Plant Physiol. 134, 619–622.

Neale, P.J. & Melis, A. (1991). Dynamics of photosystem II heterogeneity during photoinhibition: Depletion of PS II$_\beta$ from non-appressed thylakoids during strong-irradiance exposure of *Chlamydomonas reinhardtii*. Biochim. Biophys. Acta 1056, 195–203.

Nedbal, L., Masojídek, J., Komenda, J., Prasil, O., & Setlik, I. (1990). Three types of photosystem II photoinhibition. 2. Slow processes. Photosynth. Res. 24, 89–97.

Nedbal, L., Setlikova, E., Masojídek, J., & Setlik, I. (1986). The nature of photoinhibition in isolated thylakoids. Biochim. Biophys. Acta 848, 108–119.

Noctor, G. & Horton, P. (1990). Uncoupling titration of energy-dependent chlorophyll fluorescence quenching and photosystem II photochemical yield in intact pea chloroplasts. Biochim. Biophys. Acta 1016, 228–234.

Noctor, G., Rees, D., Young, A., & Horton, P. (1991). The relationship between zeaxanthin, energy-dependent quenching of chlorophyll fluorescence, and trans-thylakoid pH gradient in isolated chloroplasts. Biochim. Biophys. Acta 1057, 320–330.

Nultsch, W., Bittermsmann, E., Holzwarth, A.R., & Angel, G. (1990). Effects of strong light irradiation on antennae and reaction centers of the cyanobacterium *Anabaena variabilis*: A time-resolved fluorescence study. J. Photochem. Photobiol. (B) 5, 481–494.

Ögren, E. (1988). Photoinhibition of photosynthesis in willow leaves under field conditions. Planta 175, 229–236.

Ögren, E. (1991). Prediction of photoinhibition of photosynthesis from measurements of fluorescence quenching components. Planta 184, 538–544.

Ögren, E. & Öquist, G. (1984). Photoinhibition of photosynthesis in *Lemna gibba* as induced by the interaction between light and temperature. III. Chlorophyll fluorescence at 77 K. Physiol. Plant. 62, 193–200.

Ohad, I., Kyle, D.J., & Arntzen, C.J. (1984). Membrane protein damage and repair; removal and replacement of inactivated 32-kDa polypeptides in chloroplast membranes. J. Cell Biol. 99, 481–485.

Ohad, I., Kyle, D.J., & Hirschberg, J. (1985). Light-dependent degradation of the Q_B-protein in isolated pea thylakoids. EMBO J. 4, 1655–1659.

Ohad, I., Koike, H., Shochat, S., & Inoue, Y. (1988). Changes in the properties of RC II during the initial stages of photoinhibition as revealed by thermoluminescence measurements. Biochim. Biophys. Acta 933, 288–298.

Ohad, I., Adir, N., Koike, H., Kyle, D.J., & Inoue, Y. (1990). Mechanism of photoinhibition *in vivo*. A reversible light-induced conformational change of reaction center II is related to an irreversible modification of the D1 protein. J. Biol. Chem. 265, 1972–1979.

Öquist, G., Greer, D.H., & Ögren, E. (1987). Light stress at low temperature. In: Photoinhibition. Topics in Photosynthesis (Barber, J., Ser. Ed.), Vol. 9, pp. 67–87. Elsevier, Amsterdam.

Öquist, G. & Huner, N.P.A. (1991). Effects of cold acclimation on the susceptibility of photosynthesis to photoinhibition in Scots pine and in winter and spring cereals: A fluorescence analysis. Functional Ecol. 5, 91–100.

Öquist, G. & Chow, W.S. (1992). On the relationship between the quantum yield of photosystem II electron transport, as determined by chlorophyll fluorescence, and the quantum yield of CO_2-dependent O_2 evolution. Photosynth. Res. 33, 51–62.

Öquist, G., Chow, W.S., & Anderson, J.M. (1992a). Photoinhibition of photosynthesis represents a mechanism for the long-term regulation of photosystem II. Planta 186, 450–460.

Öquist, G., Hurry, V.M., Öquist, M.G., & Huner, N.P.A. (1992b). Differential resistance of frost-hardened and non- hardened winter rye to photoinhibition of photosynthesis is due to an increased capacity of frost-hardened rye to keep Q_A oxidized under similar light and temperature conditions. Photosynthetica 27, 231–235.

Ortiz-Lopez, A., Nie, G.Y., Ort, D.R., & Baker, N.R. (1990). The involvement of the photoinhibition of photosystem II and impaired membrane energization in the reduced quantum yield of carbon assimilation in chilled maize. Planta 181, 78–84.

Osmond, C.B. (1981). Photorespiration and photoinhibition. Some implications for the energetics of photosynthesis. Biochim. Biophys. Acta 639, 77–89.

Osmond, C.B. (1989). Photosynthesis from the molecule to the biosphere: A challenge for integration. In: Photosynthesis (Briggs, W. R., Ed.), pp. 5–17. A. R. Liss, New York.

Osmond, C.B. & Chow, W.S. (1988). Ecology of photosynthesis in the sun and shade: Summary and prognostications. Aust. J. Plant Physiol. 15, 1–9.

Ottander, C. & Öquist, G. (1991). Recovery of photosynthesis in winter stressed Scots pine. Plant, Cell Environ. 14, 345–349.

Oxborough, K. & Horton, P. (1988). A study of the regulation and function of energy-dependent quenching in pea chloroplasts. Biochim. Biophys. Acta 934, 135–143.

Post, A., Adamson, E., & Adamson, H. (1990). Photoinhibition and recovery of photosynthesis in Antarctic bryophytes under field conditions. In: Current Research in Photosynthesis (Baltscheffsky, M., Ed.), Vol. IV, pp. 635–638. Kluwer Acad. Publ., Dordrecht.

Powles, S.B. (1984). Photoinhibition of photosynthesis induced by visible light. Annu. Rev. Plant Physiol. 35, 15–44.

Powles, S.B. & Osmond, C.B. (1979). Photoinhibition of intact attached leaves of C3 plants illuminated in the absence of both carbon dioxide and of photorespiration. Plant Physiol. 64, 982–988.

Prasil, O., Adir, N., & Ohad, I. (1992). Dynamics of photosystem II: Mechanism of photoinhibition and recovery processes. In: Topics in Photosynthesis (Barber, J., Ed.), Vol. 11, pp. 295–348. Elsevier, Amsterdam.

Rees, D., Young, A., Noctor, G., Britton, G., & Horton, P. (1989). Enhancement of the ΔpH-dependent dissipation of excitation energy in spinach chloroplasts by light-activation: Correlation with the synthesis of zeaxanthin. FEBS Lett. 256, 85–90.

Rees, D. & Horton, P. (1990). The mechanisms of changes in photosystem II efficiency in spinach thylakoids. Biochim. Biophys. Acta 1016, 219–227.

Richter, M., Rühle, W., & Wild, A. (1990a). Studies on the mechanism of photosystem II photoinhibition I. A two-step degradation of D1-protein. Photosynth. Res. 24, 229–235.

Richter, M., Rühle, W., & Wild, A. (1990b). Studies on the mechanism of photosystem II photoinhibition II. The involvement of toxic oxygen species. Photosynth. Res. 24, 237–243.

Robinson, J.M. (1988). Does O_2 photoreduction occur within chloroplasts *in vivo*? Physiol. Plant. 72, 666–680.

Rottenberg, H., Grunwald, T., & Avron, M. (1972). Determination of ΔpH in chloroplasts. I. Distribution of [^{14}C] methylamine. Eur. J. Biochem. 25, 54–63.

Ruban, A.V., Rees, D., Noctor, G.D., Young, A., & Horton, P. (1991). Long-wavelength chlorophyll species are associated with amplification of high-energy-state excitation quenching in higher plants. Biochim. Biophys. Acta 1059, 355–360.

Samuelsson, G., Lönneborg, A., Rosenquist, E., Gustafsson, P., & Öquist, G. (1985). Photoinhibition and reactivation of photosynthesis in the cyanbacterium *Anacystis nidulans*. Plant Physiol. 79, 992–995.

Samuelsson, G., Lönneborg, A., Gustafsson, P., & Öquist, G. (1987). The susceptibility of photosynthesis to photoinhibition and the capacity for recovery in high and low light grown cyanobacteria, *Anasystis nidulans*. Plant Physiol. 83, 438–441.

Sapozhnikov, D.I., Krasovskaya, T.A., & Maevskaya, A.N. (1957). Change in the ratios of basic carotenoids of the plastids of green leaves under the action of light. Dokl. Akad. Nauk USSR, Botanical Sciences Sections, 113, 465–467.

Satoh, K. (1971). Mechanism of photoinactivation in photosynthetic systems. IV. Light-induced changes in the fluorescence transients. Plant Cell Physiol. 12, 13–27.

Schaefer, M.R. & Golden, S.S. (1989a). Light availability influences the ratio of two forms of D1 in cyanobacterial thylakoids. J Biol. Chem. 264, 7412–7417.

Schaefer, M.R. & Golden, S.S. (1989b). Differential expression of members of a cyanobacterial *psbA* gene family in response to light. J. Bacteriol. 171, 3973–3981.

Schatz, G.H., Brock, H., & Holzwarth, A.R. (1988). Kinetic and energetic model for the primary processes in photosystem II. Biophys. J. 54, 397–405.

Setlik, I., Allakhverdiev, S.I., Nedbal, L., Setlikova, E., & Klimov, V.V. (1990). Three types of photosystem II photoinhibition. 1. Damaging processes on the acceptor side. Photosynth. Res. 23, 39–48.

Sharma, P.K. & Hall, D.O. (1991). Interaction of salt stress and photoinhibition on photosynthesis in barley and sorghum. J. Plant Physiol. 138, 614–619.

Shipton, C.A. & Barber, J. (1991). Photoinduced degradation of the D1 polypeptide in isolated reaction centers of photosystem II: Evidence for an autoproteolytic process triggered by the oxidizing side of the photosystem. Proc. Natl. Acad. Sci. USA 88, 6691–6695.

Shochat, S., Adir, N., Gal, A., Inoue, Y., Mets, L., & Ohad, I. (1990). Photoinactivation of photosystem II and degradation of the D1 protein are reduced in a cytochrome b_6/f-less mutant of *Chlamydomonas reinhardtii*. Z. Naturforsch. 45c, 395–401.

Siefermans-Harms, D. (1985). Carotenoids in photosynthesis. I. Location in photosynthetic membranes and light-harvesting function. Biochim. Biophys. Acta 811, 325–355.

Skogen, D., Chaturvedi, R., Weidemann, F., & Nilsen, S. (1986). Photoinhibition of photosynthesis: Effect of light quality and quantity on recovery from photoinhibition in *Lemna gibba*. J. Plant Physiol. 126, 195–205.

Snyder, R., Arvidson, E., Foote, C., Harrigan, L., & Christensen, R.L. (1985). Electronic energy levels in long polyenes: $S_2 - S_0$ emission in all-trans-1,3,5,7,9,11,13,-tetradecaheptaene. J. Am Chem. Soc. 107, 4117–4122.

Somersalo, S. & Krause, G.H. (1990). Photoinhibition at chilling temperatures and effects of freezing stress on cold-acclimated spinach leaves in the field. A fluorescence study. Physiol. Plant. 79, 617–622.

Styring, S., Virgin, I., Ehrenberg, A., & Andersson, B. (1990). Strong light photoinhibition of electron transport in photosystem II. Impairment of the function of the first quinone acceptor, Q_A. Biochim. Biophys. Acta 1015, 269–278.

Telfer, A., De Las Rivas, J., & Barber, J. (1991). β-Carotene within the isolated photosystem II reaction center: Photooxidation and irreversible bleaching of this chromophore by oxidized P680. Biochim. Biophys. Acta 1060, 106–114.

Terashima, I. & Inonue, Y. (1985). Palisade tissue chloroplasts and spongy tissue chloroplasts in spinach: Biochemical and ultrastructural differences. Plant Cell Physiol. 26, 63–75.

Terashima, I. & Takenaka, A. (1986). Organization of photosynthetic system of dorsiventral leaves as adapted to the irradiation from the adaxial side. In: Biological Control of Photosynthesis (Marcelle, R., Clijsters, H., & Van Poucke, M., Eds.), pp. 219–230.

Thayer, S.S. & Björkman, O. (1990). Leaf xanthophyll content and composition in sun and shade determined by HPLC. Photosynth. Res. 23, 331–343.

Theg, S.M., Filar, L.J., & Dilley, R.A. (1986). Photoinactivation of chloroplasts already inhibited on the oxidizing side of photosystem II. Biochim. Biophys. Acta 849, 104–111.

Thompson, L.K. & Brudvig, G.W. (1988). Cytochrome *b*-559 may function to protect photosystem II from photoinhibition. Biochemistry 27, 6653–6658.

Thrash, R.J., Fang, H.L.-B., & Leroi, G.E. (1977). The Raman excitation profile spectrum of β-carotene in the preresonance region: Evidence for a low-lying singlet state. J. Chem. Phys. 67, 5930–5933.

Thrash, R.J., Fang, H.L.-B., & Leroi, G.E. (1979). On the role of forbidden low-lying excited states of light-harvesting carotenoids in energy transfer in photosynthesis. Photochem. Photobiol. 29, 1049–1050.

Truscott, T.G. (1990). The photophysics and photochemistry of the carotenoids. J. Photochem. Photobiol. B: Biol. 6, 359–371.

Tucker, G.F., Hinckley, T.M., Leverenz, J.W., & Jiang, S.-M. (1987). Adjustments of foliar morphology in the acclimation of understory Pacific silver fir following clearcutting. Forest Ecol. Management 21, 249–268.

Tytler, E.M., Whitelam, G.C., Hipkins, M.F., & Codd, G.A. (1984). Photoinactivation of photosystem II during photoinhibition in the cyanobacterium *Microcystis aeruginosa*. Planta, 160, 229–234.

Tyystjärvi, E., Koivuniemi, A., Kettenen, R., & Aro, E.-M. (1991). Small light-harvesting antenna does not protect from photoinhibition. Plant Physiol. 97, 477–483.

van Kooten, O. & Snel, J.F.H. (1990). The use of chlorophyll fluorescence nomenclature in plant stress physiology. Photosynth. Res. 25, 147–150.

Vass, I., Mohanty, N., & Demeter, S. (1988). Photoinhibition of electron transport activity of photosystem II in isolated thylakoids studied by thermoluminescence and delayed luminescence. Z. Naturforsch. 43c, 871–876.

Virgin, I., Ghanotakis, D.F., & Andersson, B. (1990). Light-induced D1-protein degradation in isolated photosystem II core complexes. FEBS Lett. 269, 45–48.

Virgin, I., Salter, A.H., Ghanotakis, D.F., & Andersson, B. (1991). Light-induced D1 protein degradation is catalyzed by a serine-type protease. FEBS Lett. 287, 125–128.

Walker, D.A. & Osmond, C.B. (1986). Measurement of photosynthesis *in vivo* with a leaf disc electrode: Correlations between light dependence of steady-state photosynthetic O_2 evolution and chlorophyll *a* fluorescence transients. Proc. R. Soc. Lond. B 227, 267–280.

Weis, E. & Berry, J.A. (1987). Quantum efficiency of photosystem II in relation to 'energy'-dependent quenching of chlorophyll fluorescence. Biochim. Biophys. Acta 894, 198–208.

Wettern, M. (1986). Localization of 32 000 dalton chloroplast protein pools in thylakoids: Significance in atrazine binding. Plant Sci. 43, 173–177.

Whitmarsh, J. & Cramer, W.A. (1978). A pathway for the reduction of cytochrome *b*-559 by photosystem II in chloroplasts. Biochim. Biophys. Acta 501, 83–93.

Wild, A. (1979). Physiologie der Photosynthese höherer pflanzen. Die Anpassung and Lichtbedingungen. Ber. Dtsch. Bot. Ges. 92, 341–364.

Williams, W.P. (1977). The two photosystems and their interactions. In: Primary Processes of Photosynthesis (Barber, J., Ed.). Topics in Photosynthesis, Vol. 2, pp. 99–147. Elsevier, Amsterdam.

Yamamoto, H.Y. (1979). Biochemistry of the violaxanthin cycle in higher plants. Pure & Appl. Chem. 51, 639–648.

MOLECULAR GENETIC MANIPULATION AND CHARACTERIZATION OF MUTANT PHOTOSYNTHETIC REACTION CENTERS FROM PURPLE NONSULFUR BACTERIA

Eiji Takahashi and Colin A. Wraight

Advances in Molecular and Cell Biology
Volume 10, pages 197–251.

ISBN: 1-55938-710-6

I. INTRODUCTION

Photosynthetic bacteria convert light energy into chemical energy by stabilizing the primary photochemical charge separation within a membrane-bound pigment–protein complex, called the photosynthetic reaction center (RC). The functional characteristics of the RC complex from purple nonsulfur bacteria such as *Rhodopseudomonas viridis* and *Rhodobacter sphaeroides* have been studied extensively (for recent reviews, see: Kirmaier and Holten, 1987; Parson, 1987; Feher et al., 1989; Boxer, 1990; Prince, 1990; Gunner, 1991). Activation of the RC by light results in the transfer of an electron from the excited singlet state of the primary donor (P), a dimer of bacteriochlorophyll (Bchl), to bacteriopheophytin (Bph), and then to the primary and secondary acceptor quinones, Q_A and Q_B, respectively. The role of a monomeric bacteriochlorophyll in the electron transfer from P to Bph is still controversial. In *Rps. viridis*, Q_A is menaquinone-9 and Q_B is ubiquinone-9, while both are ubiquinone-10 in *Rb. sphaeroides*. Two turnovers of the RC result in the double reduction of Q_B to form quinol (Q_BH_2), accompanied by the delivery of two protons through the protein to the Q_B head group.

Characterization of these bacterial RCs has been facilitated by their relative ease of isolation, as well as the stability of the isolated complexes. These factors also enabled the crystallization of the RCs to yield well-ordered crystals that have been analyzed by X-ray diffraction (Deisenhofer and Michel, 1989a; Allen and Feher, 1991; Michel, 1991). Deisenhofer, Michel, and co-workers determined the X-ray crystallographic structure of the RC complex from *Rps. viridis*, which became the first membrane-bound protein to have its three-dimensional structure established at atomic resolution (Deisenhofer et al., 1985; Michel et al., 1986b). Three groups have also solved the three-dimensional structure of RCs from *Rb. sphaeroides* (Allen et al., 1987a,b; Arnoux et al., 1990; Chang et al., 1991; El-Kabbani et al., 1991). From these studies, a great wealth of structural information on the RC has become available, including the arrangements of cofactors and protein-cofactor interactions, and this has enabled theoretical calculations of RC function based on the atomic structure (for reviews, see Freisner and Won, 1989; Boxer, 1990; Gunner, 1991).

In *Rb. sphaeroides*, the RC is composed of three protein subunits: L, M, and H. The L- and M-subunits form a quasisymmetrical heterodimer that binds all of the cofactors involved in photochemistry and subsequent charge separation. The *Rps. viridis* RC complex, in addition to the L-, M-, and H-subunits, also possesses a tightly bound cytochrome subunit containing four *c*-type hemes. The L- and M-subunits are partially homologous and many regions of the subunits are conserved among various bacterial species for which the RC genes have been sequenced, including *Rps. viridis* (Michel et al., 1986a), *Rb. sphaeroides* (Williams et al., 1983, 1984), *Rhodobacter capsulatus* (Youvan et al., 1984), *Rhodospirillum rubrum* (Belanger et al., 1988) and *Chloroflexus aurantiacus* (Ovchinnikov et al., 1988a,b; Shiozawa et al., 1989). The L- and M-subunits also show some homology with the D1 and D2 polypeptides of photosystem II (PS II) RCs of cyanobacteria, algae and higher plants, although the homology is more distant (Williams et al., 1986; Trebst, 1987; Komiya et al., 1988). Therefore, the bacterial RC serves as a useful guide for modeling structure/function relationships of PS II RCs whose three-dimensional structure has yet to be determined.

The structural details of the bacterial RCs came available just when the essential methodology for molecular engineering was being developed for purple bacteria. The methods for RC mutagenesis, involving cloning vectors for site-directed mutagenesis and expression systems in a RC^- (null) background, are now well established for *Rb. sphaeroides* (Farchaus and Oesterhelt, 1989; Paddock et al., 1989; Takahashi et al., 1990a) and *Rb. capsulatus* (Youvan et al., 1984; Bylina et al., 1989). The development of a RC^- strain for *Rps. viridis* has been hampered by difficulties in growing this species heterotrophically, but some advances in *Rps. viridis* RC mutant expression have been reported (Lang and Oesterhelt, 1989; Laußermair and Oesterhelt, 1992).

Site-directed mutagenesis studies of RCs have so far focused on the L- and M-subunits, since they bind all the cofactors involved in the light-induced charge separation processes. However, the H-subunit gene, *puhA*, of the RC has also been isolated and sequenced from *Rb. sphaeroides* (Donohue et al., 1986; Williams et al., 1986), *Rb. capsulatus* (Youvan et al., 1984), and *Rps. viridis* (Michel et al., 1985), and site-directed mutagenesis studies can be expected in the near future. In this chapter, we will address significant advances, made through mutagenesis of the L- and M-subunits, in understanding the roles of key residues in specific functions of the RC. The use of mutants and mutagenesis in the study of photosynthetic electron transport has been previously reviewed by Coleman and Youvan (1990) and by Diner et al. (1991a).

II. REACTION CENTER CHARACTERIZATION

A. Structure and Function

A casual examination of the RC atomic structure shows it to be quite symmetrically organized in terms of both protein and prosthetic groups, with a C_2 symmetry axis passing through the bacteriochlorophyll dimer (P) and perpendicular to what would be the plane of the membrane *in vivo*. From P, located near the periplasmic side of the membrane, two nearly identically arranged branches of Bchl, Bph, and Q span the L- and M-subunit complex, ending near the H-subunit on the cytoplasmic side of the membrane (Figure 1A). Following Allen et al. (1988a), these two branches of cofactors will be referred to as the A and B branches, depending on whether the terminal quinone is Q_A or Q_B. A nonheme ferrous iron is located on the C_2 axis, between the two quinones. Although the two branches are symmetrically arranged, functionally they are very asymmetrical. Transient absorption spectra of the Q_x band region of the optical absorption for *Rb. sphaeroides* at 5 K indicate that only the 545-nm absorbing Bph (identified as Bph_A) undergoes reduction and no transient Bph_B^- formation is observed (see Kirmaier and Holten, 1987). In *Rps. viridis* RCs trapped in the state P Bph_A^-, at 90 K, Bph_B^- forms with a quantum yield of 0.09 ± 0.06. By comparison with a P^* lifetime of 20 ps in RCs with Bph_A^- reduced (Holten et al., 1978), this corresponds to a rate of Bph_B^- formation of $\sim 5 \times 10^9$ s^{-1} ($\tau = 200$ ps). Since the normal rate of electron transfer to Bph_A at 90 K is about $(1\ ps)^{-1}$, the branching ratio of observed Bph_A/Bph_B reduction rates is at least 200 (Kellog et al., 1989). The controversial role of the monomeric $Bchl_A$ (see section II.C) does not substantially affect these conclusions. Thus, in spite of the symmetrical arrangement of chromophores, electron transfer occurs only through the A branch, to Q_B:

$$P^* \rightarrow (Bchl_A) \rightarrow Bph_A \rightarrow Q_A \rightarrow Q_B.$$

The functional asymmetry seemed to be reinforced by reports that removal of $Bchl_B$ by borohydride treatment did not significantly alter any electron transfer properties of RCs (Maróti et al., 1985). However, this result has been challenged recently with the suggestion that borohydride treatment causes a marked hypochromism of $Bchl_B$, of unknown origin, but not removal of the pigment (Struck et al., 1991).

B. Electron Transfer Theory

Almost all attempts to describe the early events of photosynthetic electron transfer have utilized nonadiabatic electron transfer theory, beginning with the following expression:

$$k_{et} = (4\pi^2/h)\, V^2\, FC \tag{1}$$

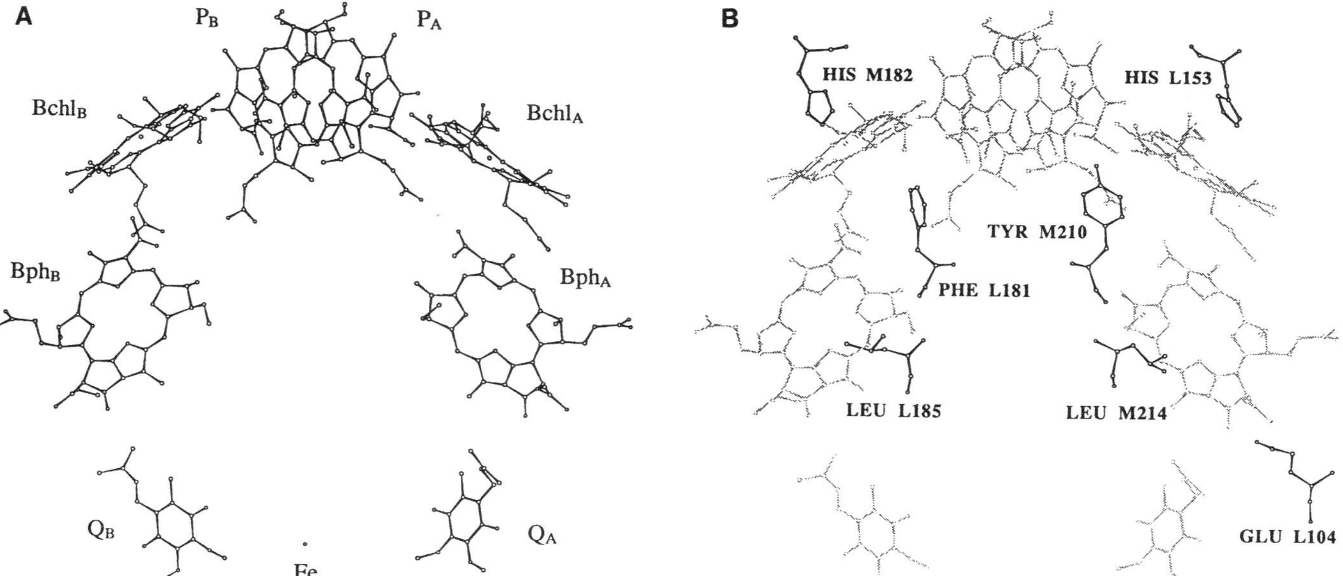

Figure 1. (**A**) Cofactors of the *Rhodobacter sphaeroides* reaction center. The phytyl side chains have been truncated after the first 5-carbon (isoprene) unit. Abbreviations: Bchl, bacteriochlorophyll; Bph, bacteriopheophytin; P, primary donor; Q, quinone. (X-ray crystallographic coordinates were kindly provided by J.P. Allen.) (**B**) Locations of some of the cofactor-associated amino acid residues (*solid lines*) that have been the targets of site-directed mutagenesis. The cofactors are shown with *light lines*. See text for details.

where V is the electronic matrix element, coupling the electron donor and acceptor species, and FC is the Franck–Condon factor, describing the accessibility of nuclear configurations compatible with electron transfer. V is proportional to the overlap of electronic wavefunctions of the donor and acceptor, and is the main source of the distance dependence for electron transfer. For nonadiabatic electron transfer, where the coupling is weak by definition, the distance is apt to be sufficiently large that the wavefunctions will follow an exponential decay. Thus, the overlap of donor and acceptor wavefunctions will similarly show an exponential dependence on distance:

$$V = V_0 e^{-\beta(r-r_0)} \tag{2}$$

where r_0 indicates, roughly, van der Waals contact. The value of β has been hotly debated on both theoretical and experimental grounds, but lies in the range of 0.6–1.5 $Å^{-1}$ in condensed media. A recent survey of rate processes in RCs and in artificial systems provided support for a value of 1.4 $Å^{-1}$ for transfers over 5 Å or more for donor–acceptor pairs not directly bonded to each other (Moser et al., 1992). This translates into a fall off in rate of 10-fold per 1.7 Å. The edge-to-edge distance between P and Bph_A is about 10 Å, corresponding to a rate for direct electron transfer of about $(0.5\ ns)^{-1}$, compared to the observed rate of $(1\ ps)^{-1}$, and it is clear that some mediation of the electron transfer must occur. In this context, it is not really debated whether $Bchl_A$ is involved, but only how—as a real intermediate or as a factor in the electronic overlap between P^* and Bph_A.

The Franck–Condon factor embodies the notion that no (nuclear) vibrational relaxation accompanies the "instantaneous" electron transfer. Thus, the energies of the reactant and product states must be equal at the moment of transfer. Attainment of this isoenergetic configuration is achieved by thermal fluctuations and vibrations. The necessary displacement from equilibrium of the reactant configuration, is determined by: (1) the net free energy of the reaction, ΔG_0 (the vertical displacement of reactant and product potential energy surfaces); (2) the shape of the reactant and product potential energy surfaces; and (3) their horizontal displacement, which indicates the degree of coupling between the electronic states and the nuclear vibrations represented by the potential energy surfaces. In the theoretical description introduced by Marcus, the latter two contributions are combined in the reorganization energy, λ. This is the work necessary to distort the reactant equilibrium nuclear configuration into that of the product, without transfer of the electron. The thermally weighted Franck–Condon factor is then given by:

$$\mathrm{FC} = (4\pi\lambda kT)^{-1/2} \exp[-\lambda(1 + \Delta G_0/\lambda)^2/4kT] \tag{3}$$

This relationship, in the context of Equation 1, predicts that the rate of reaction will be maximal when $\Delta G_0 = -\lambda$, and will fall off at smaller and larger values of ΔG_0. Demonstration of the fall off at highly exothermic values of ΔG_0 (the so-called "inverted" region) has been pursued in many systems, but has been convincingly demonstrated in only a few cases. When $\Delta G_0 = -\lambda$, the product potential energy curve cuts through the minimum of the reactant curve, and the reaction is "activationless". In fact, the theoretical temperature dependence is slightly negative, due to the preexponential factor of Equation 3. As the temperature is lowered, this corresponds to the settling of the reactant state thermal distribution into the lowest vibrational levels, which have the greatest overlap with the product state vibrational wavefunctions.

The presumption of a nonadiabatic process and the strict application of the Franck–Condon separation of electronic and nuclear factors may yet prove to be unsatisfactory but, at the present time, most theoretical descriptions of the primary charge transfer are restrained by them, and for our descriptive purposes this simple device will serve us well. However, a necessary further sophistication is the effect of vibrational quantization and nuclear tunneling on the temperature dependence of the electron transfer rate. The simplest theories strictly limit activationless behavior to when $\Delta G_0 = -\lambda$. However, if $\Delta G_0 \neq -\lambda$, an energy match may still be made between quantized vibrational levels of the reactant and product states. Although no classical overlap exists, nuclear tunneling between the two states can potentiate the electron transfer. In this way, apparent temperature independence of electron transfer can be extended to regions of ΔG_0 significantly different from $-\lambda$. In order to account for the smooth and gradual variation in the rates of electron transfer with changing ΔG_0, it is necessary to invoke coupling to more than one vibrational mode. Current descriptions of nonadiabatic electron transfer in RCs commonly use two or three, including a low-frequency, classical mode that acts as a smoothing function on the periodic matching of the quantized, higher frequency vibrations. It is likely, and is supported by molecular dynamic simulations, that a full description would require coupling to many vibrational modes highly dispersed throughout the protein (Creighton et al., 1988; Schulten and Tesch, 1991). One consequence of the coupling of high-frequency, quantized modes to the electron transfer is that the rate falls off very slowly, or not at all, in the inverted region (very exothermic values of ΔG_0).

C. Primary Events

The direct involvement of $BChl_A$ as a transient intermediate in primary charge separation leading to $P^+Bph_A^-$ was first proposed by Shuvalov et al. (1978) on the grounds of fleeting absorbance changes near 800 nm during the profile of a picosecond laser pulse. This was not supported by subsequent studies until recently when Zinth and co-workers (Holzapfel et al., 1989, 1990), using femtosecond excitation flashes, detected subpicosecond absorbance transients which they also attributed to $Bchl_A^-$ formation. At various wavelengths throughout the visible and near infrared, they could fit their data to two components with lifetimes of 0.9 and 3.3 ps at room temperature. They interpreted these kinetics in terms of a two-step mechanism, via $Bchl_A$, with the first step slow and the second step fast, thereby accounting for the low level of accumulation of $Bchl_A^-$. However, the kinetics of P^* decay, determined from the stimulated emission at 920 nm, were essentially monophasic, as previously reported by several other groups (Kirmaier et al, 1985; Woodbury et al., 1985; Breton et al., 1986; Martin et al., 1986; Wasielewski and Tiede, 1986; Fleming et al., 1988).

Until the reports from Zinth and co-workers, the persistent absence of experimental support for the direct involvement of $Bchl_A$ in the primary electron transfer led to the consideration of it as a virtual intermediate, acting to couple the donor and acceptor by "superexchange" (for review, see Friesner and Won, 1989). The effect is to change the value of V in Equation 1. Subpicosecond components in the early kinetics, comparable to that first reported by Holzapfel et al. (1989, 1990), have now been observed by others, but the controversy over the interpretation remains heated. Chan et al. (1991a), following the theoretical lead of Bixon et al. (1991), concluded that the two-step mechanism dominated at room temperature, but that the superexchange path contributed significantly (at least 50%) at low temperature. They also insisted on the need for reversibility in at least the first step (to $Bchl_A$).

Kirmaier and Holten (1990, 1991) argued that a conformational distribution in the RC population gave rise to kinetic complexity in several electron transfer processes, including $P^* \rightarrow P^+Bph_A^-$ and $P^+Bph_A^- \rightarrow P^+Q_A^-$. Conformational effect on the energetics of Bph_A were also proposed by Wraight and co-workers, to account for nonexponential kinetics of the $P^+Q_A^-$ and $P^+Q_B^-$ recombination kinetics, which proceed by repopulation of $P^+Bph_A^-$ in *Rps. viridis* (Shopes and Wraight, 1987a; Sebban and Wraight, 1989; Gao et al., 1991) and in *Rb. sphaeroides* with Q_A substituted by low-potential quinones (Gunner et al., 1986; Sebban, 1988).

Breton et al. (1988) observed a fast (0.4 ps) transient at low temperature and at various wavelengths including those expected of $Bchl_A$ involvement, but found the relative amplitudes to be dependent on the probe flash intensity and concluded that it was not directly related to electron transfer. More recently, also at low temperature, Martin and co-workers have observed both fast oscillatory phenomena, with periods of ≈0.5 and ≈2 ps (Vos et al., 1991), and very fast relaxation events, with time constants of 0.09 ps (Vos et al., 1992). The oscillations are indicative of vibrational coherence in the excited state, and are also seen in RCs from the D_{LL} mutant of *Rb. capsulatus*, which lacks Bph_A (Breton et al., 1990). The rapid relaxation, however, is not seen in the D_{LL} mutant RCs. Interestingly, a molecular dynamics simulation of the RC from *Rps. viridis* revealed a very fast (0.1 ps) dielectric relaxation of the protein matrix accompanying electron transfer, which persisted to low temperatures (Schulten and Tesch, 1991; Treutlein et al., 1992).

Although probably of importance to a detailed physical understanding of the primary events, none of these observations clearly confirm the existence of $Bchl_A^-$ as an intermediate. Furthermore, in all these studies no fast component was detectable in the lifetime of P^*. However, the technical improvements necessary to test the nonexponentiality of the kinetics finally revealed a previously unrecognized *slow* component in the P^* kinetics (Vos et al., 1992). The uncertainty in the baseline of absorption and stimulated emission measurements makes this determination difficult, but Du et al. (1992) have now shown this component with an amplitude of about 20% in spontaneous emission measured with a time resolution of 0.05 ps. At room temperature, P^* decays with lifetimes of 2.7 and 11 ps. This behavior fits well with the three-state scheme developed by Bixon et al. (1991) to analyze the data of Zinth and co-workers, but with very different rate constants:

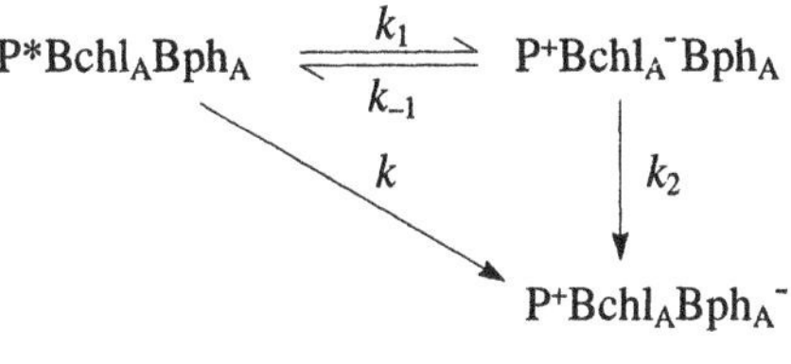

Scheme 1.

If the population of the intermediate state does not exceed 20%, the data of Du et al. (1992) place $P^+Bchl_A^-Bph_A$ within 25 meV of P*, above or below. Furthermore, the second step (k_2) is always the slowest. This is contrary to the view of Holzapfel et al. (1989). Over this free energy range, from exothermic to endothermic for the first step, the model yields the following values for the rate constants: $k = (8–6\ ps)^{-1}$, $k_1 = (6–10\ ps)^{-1}$, $k_{-1} = (12–6\ ps)^{-1}$; for k_2, the exothermic range yields values of $(12–16\ ps)^{-1}$, but if $P^+Bchl_A^-Bph_A$ lies above P^*, k_2 gets much slower.
Thus, the general picture is of the $P^+Bchl_A^-Bph_A$ state being populated to some extent, even if it lies somewhat above P^*, but with most of the electron transfer to $P^+Bchl_A\ Bph_A^-$ occurring by the superexchange pathway. Reversibility of the first step is a necessary condition for the observation of biphasic kinetics of P^* decay, but the other steps (k_2 and k) are considered irreversible because of the larger free energy drops (≈200 meV).

D. Asymmetry

The origin of the preferential electron transfer through the A-branch of the symmetrically arranged chromophores is one of the major unanswered questions concerning RC function. Several possibilities exist for determining the asymmetry of primary events, which can be controlled through nuclear Franck–Condon factors, and through the electronic coupling between electron donor and acceptor. The nuclear Franck–Condon factors are determined by the net energetics of the reaction and by vibrational coupling of the electronic transitions to the (protein) environment. A more favorable electrostatic gradient along the A-branch has been suggested by several estimates of varying crudity (Yeates et al., 1987; Michel-Beyerle et al., 1988; Treutlein et al., 1988), and the contributions of certain polar and ionizable residues are potentially susceptible to manipulation by mutagenesis. On the other hand, molecular dynamic simulations of primary events indicate the vibrational coupling to be extremely dispersed, with the main chain atoms contributing as much as side chain structures (Treutlein et al., 1992). The possibilities for more favorable electronic coupling between the electron donor and acceptor species on the active side largely reside in a few specific characteristics: (1) shorter distances between chromophores; (2) asymmetry in the excited state of P^*; and (3) the free energy of $Bchl_A$, allowing it to act either as a conventional intermediate in electron transfer, or as a medium of superexchange between P^* and Bph_A.

A detailed, but still highly approximate, calculation by Michel-Beyerle et al. (1988) using the *Rps. viridis* structure invoked some contribution from all these sources. However, the estimated branching ratio was only 12–25, compared to an experimental estimate of over 200 (Kellog et al., 1989). Since then, studies on site-directed mutants have provided substantial qualitative insight into this and other aspects of primary events, but a quantitative description is still elusive.

III. MUTANTS OF THE CHLORIN COFACTOR DOMAINS

A. Mutants of the P Region

His^{M202} and His^{L173}: The Heterodimer Mutants

The RC X-ray crystal structures of *Rps. viridis* and *Rb. sphaeroides* confirmed the earlier proposal of the dimeric nature of P (Norris and Katz, 1978), accounting for two of the four Bchls; the other two Bchls were found to be monomeric (Michel et al., 1986b; Yeates et al., 1988). All four Bchls have histidine residues available as an axial ligand to the central Mg^{2+} ions of the macrocycle. His^{L173} and His^{M202} (*R.c.* and *R.v.* His^{M200}) # are the axial ligands of P_A and P_B, respectively, the two Bchls of P (Figure 2). Replacement of either one of these residues with leucine, a hydrophobic residue, results in the incorporation of Bph in place of the normally associated Bchl, and the ratio of Bchl/Bph in the mutant RC is altered from 2 (in wild type RCs) to 1 (Bylina and Youvan, 1988; McDowell et al., 1990, 1991a). On the other hand, the substitution of histidine with glutamine does not alter the pigment composition. Evidently replacement of either of these histidines by a nonliganding residue results in the formation of a [Bchl–Bph] heterodimer. A double mutant of both histidines has not yet been reported.

Although not capable of photosynthetic growth, both His^{M202} → Leu and His^{L173} → Leu mutant RCs, referred to here as M202 (or M200) and L173 heterodimer mutants, are capable of undergoing primary charge separation. At room temperature, the measured lifetime of the excited state of the primary donor## in *Rb. sphaeroides* His^{M202} → Leu mutant RCs was approximately 18 ps, and the quantum yield of charge separation (formation of $D^+Bph_A^-$) was 40 % (McDowell et al. 1991a) compared to the wild type (Wt) values of 3.5 ps and ≈ 100 %, respectively.

The mutant data correspond to an electron transfer rate of $(45\ ps)^{-1}$. The reduction in quantum yield is larger than expected simply from the decrease in the primary charge transfer rate, and implies a significant rate of $(30\ ps)^{-1}$ for the nonproductive deexcitation of the excited heterodimer. For the $His^{L173} \rightarrow$ Leu mutant RCs, an even slower electron transfer rate of $(90\ ps)^{-1}$ was estimated, with a rate of $(80\ ps)^{-1}$ for decay to the ground state. An enhanced decay of P* in the heterodimer mutants can arise from stronger electronic coupling of the heterodimer excited states with the ground state (Warshel et al., 1988; McDowell et al., 1991a,b). With Q_A reduced, M200 heterodimer RCs from *Rb. capsulatus* were found to have a significantly decreased triplet yield, suggesting that the enhanced decay of P* was through the singlet channel (Bylina et al., 1990; Kolaczkowski et al., 1990).

It is worth noting that the non-photochemical lifetime of P^* in Wt RCs is not well known. The measured quantum yield (98%) and electron transfer rate of $(3.5\ ps)^{-1}$ imply a lifetime of 150 to 200 ps, but P^* in blocked RCs (with Bph_A reduced) has been reported to decay in about 20 ps (Holten et al., 1978). Although a direct involvement of the $Bchl_A$ monomer complicates the picture, on its own it does not help in the resolution of this discrepancy. A simple explanation is that the presence of Bph_A^- significantly perturbs the processes of P^* decay. Adding to the uncertainty, in reduced *Rb. sphaeroides* RCs, Schenck et al. (1981) observed a similar state to decay in 340 ps. It is of interest, therefore, that the rather exotic D_{LL} mutant from Youvan's lab, lacking Bph_A, exhibits a P^* lifetime of 190 ps (Breton et al., 1990). The mutation, in *Rb. capsulatus*, is a partial symmetrization, in which the D-helix of the M-subunit has been replaced by the D-helix of the L-subunit (Robles et al., 1990). In contrast to the monomer Bchls, the binding domains of the Bphs are provided by helices from both protein subunits.

The heterodimer mutants have provided an elegant test of ideas on the nature of the P^* excited state and, in particular, the possible role of a charge transfer contribution to P^* in the asymmetry of the primary events. An excellent qualitative view of the problem is given by McDowell et al. (1990). The lowest excited states of P are derived from mixing of four basis states, the two excited monomers, P_A^* and P_B^*, and two intradimer charge transfer states, $P_A^+P_B^-$ and $P_A^-P_B^+$. To a first order approximation, the two monomer excitations are expected to mix to give two excitonic components, and the two charge transfer states mix to yield two charge resonance states. The excitonic and charge resonance states then mix to give rise to four final configurations. The net charge transfer character of the resonance states and, hence, of the resultant states, depends on the energy difference between the two charge transfer basis states. Because of the inherent symmetry of the Wt structure, this is expected to be quite small. Even if the basis states are well matched to give strong mixing, the two charge transfer states will very nearly cancel. Stark effect measurements on the near infrared band in Wt RCs indicate that the difference between the dipole moments of the ground and excited states ($\Delta\mu_A$) is about 7 debye (DiMagno et al., 1990; Hammes et al., 1990). This is consistent with less than 20% charge transfer character for P^*.

Calculation of the electronic properties of P and P^* from the RC structures is dependent on the quality of the quantum chemical description of the "bare" dimer which is, at best, debatable. For *Rps. viridis*, a substantial asymmetry in the charge distribution of P^* has been estimated, with P_A positive and P_B negative (Plato et al., 1988). This was considered consistent with the observed asymmetry because the electronic overlap between P_B and $Bchl_A$ was calculated to be greater than that between P_A and $Bchl_A$. In *Rps. viridis*, but not in *Rb. sphaeroides*, P_B is actually (slightly) physically closer to $Bchl_A$ than is P_A. The EPR characteristics of the oxidized primary donor (P^+) and of the triplet (^3P) also support the idea that the primary donor of *Rps. viridis* RCs is quite asymmetric, compared to that of *Rb. sphaeroides* (Norris et al., 1989). However, even for *Rps. viridis*, the unreliability of this type of calculation is illustrated by the different result of Parson and Warshel (1987), who obtained the opposite polarity of the charge distribution from that of Plato et al. (1988). More extensive calculations by Parson et al. (1990a,b) on the *Rb. sphaeroides* structure showed the relative ordering of states distinguishing P_A^+ or P_B^+ to be very sensitive to the structural model used, especially the cutoff values for electrostatic interactions and the inclusion of dipole energies. In general, therefore, it seems that the charge transfer character of P^* is not yet well predicted by theory.

EPR and ENDOR spectroscopic studies of the cation radical (P^+ or D^+) in Wt and heterodimer RCs in solution (Huber et al., 1990) have now been extended to single crystals (Lendzian et al., 1990; Lous et al., 1990; Huber et al., 1992). This facilitated the assignment of the hyperfine couplings of the native P^+ from which the electron density map could be determined for the whole molecule. The spin density of the cation in the mutant RCs was essentially localized on the Bchl half of the heterodimer. From the assigned hyperfine couplings of the methyl groups, the spin density in the homodimer of *Rb. sphaeroides* (R26) was estimated to be distributed between P_A and P_B in a ratio of 2:1. Similar conclusions were reached by ENDOR and TRIPLE resonance spectroscopy of P^+ in single crystals of *Rb. sphaeroides* RCs and solution spectra of RCs from *Rb. sphaeroides* and *capsulatus, Rps. viridis* and *Rs. rubrum* (Lubitz et al., 1992). Molecular orbital calculations modelled the various observed asymmetries quite satisfactorily (Lenzian et al., 1993). Interestingly, different species exhibited significant differences in the asymmetry of the spin distribution, although all showed the same general trend of greater spin density on P_A. Dramatic effects were observed in mutants of *Rb. sphaeroides* in which the hydrogen bonding environment of P was changed viz. $Phe^{M197} \rightarrow His$, adding a hydrogen bond to the ring I acetyl group of P_B and $His^{L168} \rightarrow Phe$ and removing a hydrogen bond from P_A (Lubitz et al., 1992). The spin density distribution of P^+ thus appears to be a very sensitive probe of the environment and local structure of P^+.

The spin density distribution of P^+ is related to the charge transfer character of P^*, but the calculation is not sufficiently straightforward to clarify the picture for the native structure. However, the photoactivity of the M202 and L173 heterodimer mutants provides a strong indication that electronic asymmetry in the P^* excited state is not instrumental in determining the asymmetry of electron transfer. Upon excitation, the heterodimer RCs undergo a significant (>50%) intradimer charge transfer, as indicated by the appearance of a Bph "anion band" at 650 nm, in the subpicosecond timescale at both room temperature and 77 K (Kirmaier et al., 1989; McDowell et al., 1990, 1991b). The location of the negative charge on the Bph in the intradimer charge transfer state is consistent with the redox potential of the Bph/Bph^- couple, *in vitro*, which is about 300 mV more positive than $Bchl/Bchl^-$ (Fajer et al., 1975). Thus, the two heterodimer structures, corresponding to the M202 and L173 mutations, lead to charge transfer states of opposite polarity, but both exhibit similar activity along the normal A-branch of cofactors and neither of them leads to detectable charge separation along the B-branch.

The substitution of Bph for Bchl in the heterodimers is expected to cause a substantial increase in redox midpoint potential for D^+/D compared to P^+/P and, indeed, for the M202 and L173 mutant RCs, the E_m values are, respectively, 160 and 180 mV more positive (Davis et al., 1992). This places some strong constraints on the energetics of the electron transfer to Bph_A. Stark effect spectroscopy on the M202 heterodimer of *Rb. sphaeroides* shows the longest wavelength absorbance feature (935 nm at 77 K) to have $\Delta\mu_A \approx 15$ debye, which is smaller than expected from the apparent charge transfer character of the first detected transient state (DiMagno et al., 1990; Hammes et al., 1990; McDowell et al., 1990). It seems likely that this absorption band corresponds to D^*, a largely excitonic component with similar energy to P^* (890 nm at 77 K), and that rapid relaxation, in a few hundred femtoseconds, yields the intradimer charge transfer state, $D^{\pm}$, which is the first detectable transient state and has little oscillator strength of its own. The energy of $D^{\pm}$ is uncertain, but it cannot be much lower than the 935 nm feature of the absorption spectrum (1.35 eV) as it successfully leads to $D^+Bph_A^-$ at low temperature. Because of the raised E_m of D^+/D, the energy level of $D^+Bph_A^-$ is likely to be somewhat higher than that of $P^+Bph_A^-$, in Wt RCs, which has been placed at about 1.2 eV by measurements of delayed fluorescence (Woodbury and Parson, 1984).

Remarkably, therefore, the dramatic structural and redox perturbations of the heterodimer do not lead to large scale changes in the relative energetics of various states of the RCs. With this in mind, how is the 10- to 30-fold decrease in the primary electron transfer rate accounted for? Molecular dynamics simulations have suggested a reorganization energy (λ) for the primary charge separation of about 200 meV (Creighton et al., 1988; Parson et al., 1990a,b; Schulten and Tesch, 1991; Treutlein et al., 1992), and this is consistent with measured rates, free energies, and distances for the electron transfer (Moser et al., 1992). If the Wt RC is taken to have $\Delta G_o = -\lambda$ for the process, a shift in ΔG_o alone, within the confines described above, is unlikely to alter the rate by as much as is observed. Although λ for the heterodimer may be different, the rates remain roughly temperature-independent, so it cannot alter so much as to make the reaction highly activated.

We can also turn to the electronic coupling term, where distance is the dominant factor in determining biological electron transfer rates. Although the simple exponential dependence of Equation 2 is unlikely to be accurate at the short distances involved here, the β-factor of 1.4 $Å^{-1}$ would require a mighty shift (> 2Å) in closest approach to yield a slowing of 30-fold. It seems more reasonable to take the magnitude of the kinetic effects of the heterodimer mutants as reflecting the role of the monomer $Bchl_A$ in determining the electronic coupling between P^* and Bph_A.

In Wt RCs, the state $P^+Bchl_A^-$ is considered to be close to P^* in energy—possibly thermally accessible but at least coupled to the primary donor (P^*) and acceptor (Bph_A) so as to facilitate electron transfer through a superexchange mechanism. Roughly, the relative change in the rate via superexchange is equal to the inverse square of the relative change in ∂E, the gap between $P^+Bchl_A^-$ and the crossing point of the P^* and $P^+Bph_A^-$ curves. Thus, the 10- to 30-fold decrease in rate for the heterodimer RCs would imply a 3- to 6-fold increase in ∂E, if the process was dominated by the superexchange pathway. This would mean both a significant shift in the relative energy level of $D^+Bchl_A^-$ in the mutants and a rather small value of ∂E for the Wt. In fact, recent analyses of the complex primary kinetics suggest that $P^+Bchl_A^-$ in the Wt is nearly isoenergetic with P^* (Bixon et al., 1991; Du et al., 1992). Thus, we should expect $D^+Bchl_A^-$ to lie above D^* (or $D^{\pm}$), and the contribution of the two-step mechanism could significantly decrease in the mutants.

The behavior of the M202 and L173 heterodimer mutations shows that charge asymmetry in the primary donor excited state is not a necessary or sufficient factor for driving unidirectional electron transfer through the A branch. However, it must be borne in mind that evolution is able to operate on extremely small differences in reproductive fitness, and all sources of improvement in quantum yield are likely to be maximized. In fact, the introduction of an additional charge transfer characteristic in these mutants has a seriously deleterious effect on the quantum yield of electron transfer through enhanced coupling to the ground state leading to accelerated decay. Thus, the RC may be designed not so much to develop some feature of asymmetry in P^*, but actually to minimize it.

Control of the Redox Potential of P^+/P by Hydrogen Bonding

Partial symmetrization of the protein around P in *Rb. capsulatus* was found to cause a substantial change in the midpoint potential of P (Woodbury et al., 1990; Stocker et al., 1992; Taguchi et al., 1992). The specific causes of this effect were suggested to be alterations in the hydrogen bonding pattern to the macrocycle, including the ring I acetyl and ring V 9-keto groups. In particular, His^{L168} donates a hydrogen bond to the acetyl group of P_A, while the symmetry related residue, Phe^{M195}, lacks such an interaction with P_B. To test their involvement, single site mutations, $His^{L168} \rightarrow$ Phe and $Phe^{M195} \rightarrow$ His, were constructed and found to lower and raise the E_m of P, respectively (H.A. Murchison, N.W. Woodbury, J.P. Allen, and J.C. Williams, personal communication; Stocker et al., 1992).

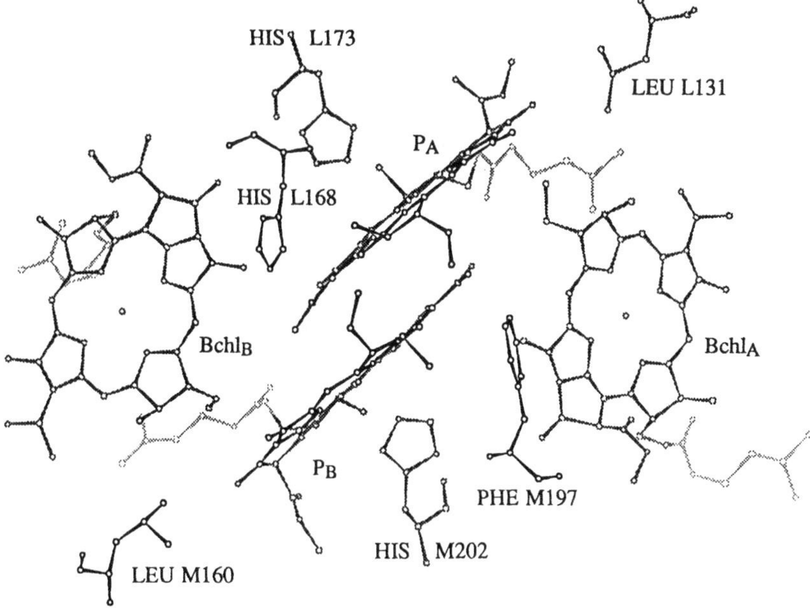

Figure 2. The primary donor bacteriochlorophylls, P_A and P_B, and the two monomer bacteriochlorophylls, $Bchl_A$ and $Bchl_B$, in the *Rb. sphaeroides* reaction center, shown with some symmetry-related amino acid residues targeted for mutagenesis. The phytyl side chains have been truncated after the first 5-carbon (isoprene) unit. *Various line densities* are used to indicate depth of field, with solid lines for the closest positions.

Williams et al. (1992) have further investigated this type of interaction, by introducing hydrogen bonding residues in symmetry related positions around P, in *Rb. sphaeroides* (Figure 2). Residues Leu^{L131} and Leu^{M160} were each altered to histidine, providing putative hydrogen bonds to the 9-keto groups of P_A and P_B, respectively. The substitutions caused increases in $E_m(P^+/P)$ of 80 and 55 mV, respectively, and decreases in the free energy gap between P^* and $P^+Bph_A^-$ of 60 and 25 meV (determined from the yield of delayed fluorescence).

In all these mutants at positions L131, L168, M160, and M197 (M195 in *R.c.*), the rate of the primary electron transfer from P^* to Bph_A varied according to a crude parabola of log(rate) versus free energy, with a maximum at about 130 meV (N.W. Woodbury et al., personal communication). As an indication of the reorganization energy of this process, this is in good agreement with other estimates and from molecular dynamic simulations (Parson et al., 1990a; Bixon et al., 1991; Moser et al., 1992; Treutlein et al., 1992). Although most of the points lay on the low side of the free energy maximum, the kinetics got faster at low temperature in all mutants. Activationless behavior is now generally expected from the inclusion of quantized, high-frequency vibrational modes coupled to electron transfer (Jortner, 1980;

Gunner et al., 1986; Gunner and Dutton, 1988), but it is perhaps surprising to observe this effect when the change in ΔG is such a substantial fraction of the value at the maximum, where $\Delta G = -\lambda$.

Tyr^{M210} and Phe^{L181}

The residue Tyr^{M210} (*R.c.* and *R.v.* Tyr^{M208}) attracted early attention as potentially important in the primary electron transfer. It is located close to P, $Bchl_A$, and Bph_A (see Figure 1B) and is conserved among many bacterial species, as well as in plants (Komiya et al., 1988). The residue structurally analogous to Tyr^{M210} on the B branch is Phe^{L181}, contributing to the symmetry-breaking between the two branches.

Electrostatic calculations, relevant to M208 in the *Rps. viridis* RC structure, showed $Bchl_A$ to be in a more positive environment than $Bchl_B$. Analysis of the net stability of the charge transfer states, $P^+Bchl_A^-$ and $P^+Bchl_B^-$, indicated that the free energy of $P^+Bchl_A^-$ was close to that of P^*, while $P^+Bchl_B^-$ was significantly above P^* (Parson et al., 1990a,b). "Mutation" of Tyr^{M208} to Phe, in the calculation, revealed a significant contribution of this residue in the interactions that lower the free energy of the $P^+Bchl_A^-$ state and, therefore, in facilitating electron transfer through the A branch, either through a real $Bchl_A^-$ intermediate or via superexchange.

Tyr^{M210} in *Rb. sphaeroides* has been mutagenized to Phe, Leu, and Ile (Gray et al., 1990a; Nagarajan et al., 1990; Wang et al., 1990). No large structural disturbances in the pigments surrounding the M210 mutation were detected in the linear dichroism (Gray et al., 1990a) and resonance–Raman spectra (Mattioli et al., 1991). However, a significant loss of Q_A during RC isolation and a red-shift in the Bph_B spectrum were observed for the mutant RCs, indicating transmission of structural alterations along the D-helix of the M-subunit, on which M210 is located. The lifetime of stimulated emission of P^*, a measure of primary electron transfer, was increased three- to sixfold in the mutants, compared to Wt at room temperature (Finkele et al., 1990; Gray et al., 1990b; Nagarajan et al., 1990). More significantly, Tyr^{M210} $\rightarrow$ Phe exhibited a flat temperature dependence, while Tyr^{M210} $\rightarrow$ Ile slowed down to $(50\ \text{ps})^{-1}$ at 150 K, and was constant at lower temperatures (Nagarajan et al., 1990). In a two-step process, this could imply the introduction of a small activation barrier ($E_{act} \approx 50$ meV) for $P^+Bchl_A^-$ formation, with nuclear tunneling accounting for temperature independence at low temperature. For a direct mechanism, involving $Bchl_A$ only through superexchange, the changes in temperature dependence for the three mutants would require a change in the level of $P^+Bph_A^-$ relative to P^*. This could easily arise from an effect of Tyr^{M210} on the E_m of P^+/P, which is sensitive to the hydrogen-bonding environment (Nagarajan et al., 1992; Stocker et al., 1992; Williams et al., 1992).

The slow kinetics of formation of $P^+Bph_A^-$ in the Tyr^{M210} $\rightarrow$ Ile, coupled with the high quantum yield (>95%), also allow estimates of a nonradiative decay time for P^* to the ground state of ≥ 300 ps at room temperature and ≥ 1 ns at low temperature (Nagarajan et al., 1990).

Mutants of Tyr^{M208} and of Phe^{L181} in *Rb. capsulatus* were investigated by Chan et al. (1991b). Double mutation, reversing the residues on M208 and L181 to Phe and Tyr, respectively, or single mutation of Tyr^{M208} to His, resulted in P^* lifetimes comparable to the Wt, while $Tyr^{M208} \rightarrow$ Thr gave a substantially slower decay time of 15 ± 3 ps. However, the mutant $Phe^{L181} \rightarrow$ Tyr, in which both M208 and L181 are Tyr, exhibited a P^* decay of 2.1 ± 0.3 ps at room temperature. This is faster than the Wt, but electron transfer is still exclusively through the A branch. Based on these results, Chan et al. (1991b) suggested that residues at M208 and L181 independently stabilize the primary donor through their aromaticity and hydrogen-bonding characteristics. Because of the symmetry of the two positions, the implication is that Tyr^{M208} stabilizes P^+ in $P^+Bchl_A^-$, rather than $Bchl_A^-$, and could, therefore, also affect the free energy level of $P^+Bph_A^-$. The following intriguing empirical relationship was also proposed:

$$1/\tau = 1/\tau_{M208} + 1/\tau_{L181}$$

On the basis of the available mutants, they obtained $\tau_{Phe} = 18.4$ ps and $\tau_{Tyr} = 4.2$ ps. Values for other mutations were predicted and the relationship awaits further testing.

The recent detection of biphasicity in the decay of P^*, in measurements of spontaneous emission, has not yet been integrated into the general picture. However, it is also seen in mutants of *Rb. capsulatus* (Du et al., 1992). The $Phe^{L181} \rightarrow$ Tyr mutant was very similar to the Wt, although a little faster, while $Tyr^{M208} \rightarrow$ Phe, and the double mutant $Tyr^{M208} \rightarrow$ Phe + $Phe^{L181} \rightarrow$ Tyr, both showed strongly biphasic decays, with a much slower (40 and 25 ps, respectively) and larger ($\approx$50 and 40%, respectively) slow component. Analysis in terms of Scheme 1 yielded small free energy gaps, with P^* 5–25 meV above $P^+Bchl_A^-$. Substitution of Tyr^{M208} with threonine caused a substantial slowing of both components of P^* spontaneous emission, with lifetimes of 120 and 70 ps. The corresponding free energy gap places $P^+Bchl_A^-$ 10 to 15 meV above P^* for this mutant. Interestingly, markedly biphasic kinetics had previously been resolved in RCs from *C. aurantiacus*, possibly because they are intrinsically slower (Feick et al., 1990; Becker et al., 1991).

In hole-burning studies of the near infrared absorption band, the longer lifetime of P^* in $Tyr^{M210} \rightarrow$ Phe mutant RCs of *Rb. sphaeroides* allowed separation of homogeneous and heterogeneous broadening mechanisms, but the mutation did not significantly affect the electronic character of P^* (Middendorf et al., 1991). The similarity in the electroabsorption spectrum (Stark effect) of the Wt and $Tyr^{M210} \rightarrow$ Phe mutant RCs also implied that the charge transfer state, $P^+Bchl_A^-$, does not mix strongly into P^*. (This can be extended to support the conclusion above that P^* has only a small contribution from charge transfer states of any sort.) They, therefore, suggested that the residue at M210/208 must exert its primary effect through the coupling with a product state. This would seem to be consistent with an effect via stabilization of P^+ in $P^+Bchl_A^-$, as suggested by Chan et al. (1991b).

The slower kinetics of the primary electron transfer observed for the $Tyr^{M210} \rightarrow$ Leu mutant, as well as other aliphatic substitutions, are consistent with the rate in native *C. aurantiacus* RCs (Kirmaier et al., 1986) where the equivalent residue is naturally leucine (Ovchinnikov et al, 1988b). To test this inference, it would be interesting to see how mutagenesis of this residue to tyrosine affects electron transfer rates in this thermophilic organism.

B. Mutants of the $Bchl_A$ and $Bchl_B$ Regions

His^{L153} and His^{M182}

The question of the participation of $Bchl_A$ in the electron transfer between P and Bph_A is largely an energetic one, and is mirrored by the possible role of $Bchl_B$ as guardian of the B-chain. In principle, we might expect these to be answerable through mutagenic manipulation of the properties of the monomer Bchls. The *Rps. viridis* RC structure shows His^{L153} and His^{M180} to be axial ligands to the central Mg^{2+} ions of $Bchl_A$ and $Bchl_B$, respectively (see Figure 1B; Michel et al., 1986b). In *Rb. sphaeroides*, Yeates et al. (1988) found asymmetry in the Mg^{2+} ligation, with His^{M182} (*R.v.* His^{M180}) as a ligand to Mg^{2+} of $Bchl_A$, but His^{L153} too far from $Bchl_B$, which appeared to be devoid of a protein-derived ligand. However, El-Kabbani et al. (1991) found His^{L153} to be involved in Mg^{2+} ligation in their *Rb. sphaeroides* RC structure, more in agreement with the *Rps. viridis* data and with the resonance–Raman study of Zhou et al. (1987).

In *Rb. capsulatus*, mutagenesis of $His^{L153} \rightarrow$ Ser or Thr, $His^{M180} \rightarrow$ Ser, as well as a double mutation of $His^{L153,M180} \rightarrow$ Ser, did not result in alteration of the RC pigment composition, and all isolated RCs were stable and photochemically active (Bylina et al., 1990), implying that Ser or Thr can function as axial ligands to the Mg^{2+} of Bchl without major perturbation of any essential functions of, for example, $Bchl_A$. Mutagenesis of $His^{L153} \rightarrow$ Arg or Leu, as well as $His^{M180} \rightarrow$ Arg or Leu resulted in RCs which were structurally unstable to varying degrees, but the His $\rightarrow$ Leu mutants were able to grow photosynthetically (Bylina et al., 1990). In *C. aurantiacus* the analogous residue to His^{M180} is Leu^{M173} (Ovchinnikov et al., 1988b). This difference is probably responsible for the presence of three Bphs in the *C. aurantiacus* RC (Blankenship et al., 1983) instead of the two found in Wt RCs of the *Rhodospirillaceae*. Thus, as also suggested by the heterodimer mutants of purple bacteria, mutation of His^{L153} or His^{M180} to Leu might be expected to result in the incorporation of Bph in place of Bchl, due to the loss of liganding potential for Mg^{2+}. Unfortunately, stable RCs could not be isolated from the $His^{L153} \rightarrow$ Leu mutant, for pigment and electron transfer analysis. However, EPR spectra of chromatophores from this mutant revealed the light-induced formation of stable P^+ at 80 K with a similar line width to Wt RCs, and a normal triplet signal in reduced preparations at 6 K (Bylina et al., 1990). Although these assays were performed under continuous illumination, the amplitudes of the triplet spectra, relative to the

RC concentration, was similar to that of Wt, implying that the quantum yield of radical formation ($P^+Bph_A^-$) was not greatly suppressed.

If, indeed, the His^{L153} → Leu mutation does result in the change $Bchl_A$ → Bph_A, the role of $Bchl_A$ in the electron transfer of Wt RCs would be well tested since the redox potential of Bph/Bph^- is significantly higher (≈300 mV) than Bchl/$Bchl^-$ (Fajer et al., 1975). The mutant state $P^+\mathit{Bph}_A^-$ should be at a substantially lower free energy than $P^+Bchl_A^-$, relative to P^*, and should certainly facilitate resolution of this state as a real intermediate. In fact, one might expect it to approach $P^+\mathit{Bph}_A Bph_A^-$ in energy, allowing charge recombination from $P^+\mathit{Bph}_A^-$ (which should be quite fast due to strong electronic coupling) to compete favorably with electron transfer to Bph_A and to Q_A, and leading to a severely impaired quantum yield. This is not supported by the EPR data or by the phototrophic competence of this mutant since, for example, the heterodimers mutants with primary quantum yields of 40 to 50% are not able to grow photosynthetically. Indeed, the functionality of all these mutant RCs at low temperature places strong constraints on the relative energetics of all intermediate states.

The His^{M180} → Leu mutation also exhibits photochemical activity at low temperature, producing essentially normal P^+ and triplet EPR signals in chromatophores (Bylina et al., 1990). Substitution of $Bchl_B$ by Bph, could have the long-sought effect of promoting "wrong way" electron transfer; that is, the higher redox potential of Bph compared to Bchl may make the state $P^+\mathit{Bph}_B^-$ accessible, either for direct involvement or for enhanced mixing in a superexchange mechanism. However, it should be noted that *C. aurantiacus* RCs do not appear to display B-side activity (Kirmaier et al., 1986; Kirmaier and Holten, 1987).

C. Mutants of the Bph_A and Bph_B Regions

Leu^M214 and Leu^L185

In the His^{M202} and His^{L173} mutants described above, removal of effective axial ligands to the Bchl Mg^{2+} ions resulted in the replacement of Bchl by Bph. The reverse of this mutagenesis strategy can cause the substitution of Bchl for Bph (see Figure 1B). The RC structures show Leu^{M214} and Leu^{L185} located over the macrocycles of Bph_A and Bph_B, respectively (Michel et al., 1986b; Yeates et al., 1988), and site-directed mutagenesis of Leu^{M214} → His, in *Rb. sphaeroides*, results in replacement of Bph_A by Bchl_A (designated β_L in the original work) (Kirmaier et al., 1991). This was evident in the ground state absorption spectrum and was confirmed by pigment composition assays. Although the redox properties of Bchl differ drastically from that of Bph, the primary electron transfer to Bchl_A occurs in 6.5 ps, forming $P^+Bchl_A\mathit{Bchl}_A^-$ with high quantum yield. Subsequent electron transfer to Q_A also occurred, but the quantum yield of $P^+Q_A^-$ formation was only ≈60 %, due mainly to a 10-fold increase in the rate of $P^+Bchl_A\mathit{Bchl}_A^-$ recombination (0.9 ns compared to 12 ns in Wt) but also to a threefold decrease in the forward electron transfer rate (600 ps compared to 220 ps).

The delayed fluorescence yield of the mutant indicated that the free energy of $P^+Bchl_A Bchl_A^-$ is ≈85 mV higher than $P^+Bchl_A Bph_A^-$, relative to P^* (Kirmaier et al., 1991). This is considerably less than the difference in E_m values for Bchl and Bph *in vitro*. Thus, the RC appears to accommodate the novel macrocycle with significant clamping of the redox properties. This can arise from solvation properties of the protein and from electronic interactions with other charge transfer states. It may be compared with the E_m shifts of 160 to 180 mV for the heterodimer primary donors, where coupling between the two halves of the dimer is expected to prevent the redox properties of either the Bchl or the Bph from being fully expressed.

The relatively small shift in free energy of $P^+Bchl_A Bchl_A^-$ can be reconciled with a large effect on the rate of recombination if the electronic coupling between $Bchl_A^-$ and P^+ is dependent on quantum mechanical mixing with higher states. Thus, any increase in the free energy of $P^+Bchl_A Bchl_A^-$ can substantially increase the superexchange interaction with $P^+Bchl_A^- Bchl_A$, which may have quite strong electronic coupling to the ground state. The magnitude of this effect is *not* expected to be the same for recombination and for the forward electron transfer, as the relevant states of the primary donor are not the same (P vs. P^*), differing by 1.45 eV in energy and probably with different potential surfaces (Kirmaier et al., 1991).

The comparative competence of this mutant, and its specific failing in terms of enhanced charge recombination of $P^+Bchl_A Bchl_A^-$, led Kirmaier et al. (1991) to suggest that Bph is necessary in this position to minimize electronic coupling to the ground state by lowering the energy of the native $P^+Bchl_A Bph_A^-$ state relative to $P^+Bchl_A^- Bph_A$, and thus to allow efficient electron transfer to Q_A. If this function is predominantly energetic, it may be possible, through further mutations, to "optimize" $Bchl_A$ to act almost like Bph_A (e.g., by adjusting its redox potential).

Mutagenesis of Leu^{L185} has not yet been reported in the literature, but a Leu^{L185} → His mutation might convert Bph_B to $Bchl_B$. In view of the inactivity of the B-branch, the expectations for this mutation are hard to assess.

Glu^{L104}

Following the publication of the *Rps. viridis* structure, one of the first residues to attract attention—as a possible contributor to the asymmetry of the RC electron transfer—was Glu^{L104}, which hydrogen bonds to the 9-keto group of ring V of Bph_A (see Figure 1B; Michel et al., 1986b; Yeates et al., 1988). A resonance–Raman study implicated the interaction between Bph_A and Glu^{L104} as being responsible for the partial double-bond character seen in the C_9–C_{10} bond of Bph_A (Bocian et al., 1987). The hydrogen-bonding interaction between the exchangeable proton of Glu^{L104} and Bph_A has also been investigated by ENDOR (Feher et al., 1988) and FTIR (Nabedryk et al., 1988). The structurally homologous residue to Glu^{L104} on the B-branch of the RC is Val^{M131} in *Rps. viridis* and *Rb. capsulatus*, which cannot form a hydrogen bond (Michel et al., 1986a,b). This asymmetry in the environments of the two Bphs prompted speculation that hydrogen bonding of Glu^{L104} to Bph_A may contribute to the unidirectionality of electron transfer through the A-branch. Glu^{L104} is also the only ionizable residue found near any of the six chromophores, other than the four histidines which serve as ligands to the four Mg^{2+} ions of the Bchls. The residue homologous to Glu^{L104} in the B-branch of *Rb. sphaeroides* is Thr^{M133}, which does hydrogen bond to Bph_B (Yeates et al., 1988), although presumably more weakly.

Mutagenic substitution of Glu^{L104} by Leu, Gln, and Lys was carried out in *Rb. capsulatus* by Bylina et al. (1988). The $Glu^{L104} \rightarrow$ Lys mutation resulted in the lack of RC assembly. The other two mutants were both photosynthetically competent and their 77 K absorption spectra showed that the interaction between Bph_A and Glu^{L104} contributes to the spectral red shift of the Q_x band of Bph_A relative to that of Bph_B in the Wt RC (Bylina et al., 1988), as originally suggested by Michel et al. (1986b). The Bph^- anion band, near 650 nm, was also blue-shifted in the mutants compared to the Wt. However, no change was observed in the high quantum yield or the directionality of electron transfer in the mutants. Also, linear dichroism spectroscopy indicated that Bph_A orientation in the protein, relative to the C_2 axis, is not affected significantly (Breton et al., 1989) and resonance–Raman spectroscopy showed that the other pigments also were not affected (Peloquin et al., 1990). Therefore, these mutations do not induce large structural changes in the RC.

Resonance–Raman studies showed that Wt and $Glu^{L104} \rightarrow$ Gln mutant RCs undergo a temperature-dependent conformational change, while $Glu^{L104} \rightarrow$ Leu does not (Peloquin et al., 1990, 1991). This was suggested to result from steric interaction or local conformational differences depending on the amino acid present at L104, and not due to electrostatic interactions. Local differences in the Bph_A environment may influence the $P^* \rightarrow Bph_A$ electron transfer kinetics which were 1.5 to 2-fold slower in the mutant RCs compared to the Wt . It remains to be seen how the temperature dependence of electron transfer rates are influenced by the conformational differences around Bph_A, depending on the substitutions made at L104.

IV. MUTANTS OF THE QUINONE DOMAINS

A. Mutants of the Q_A Region

In addition to fast electron transfer in the primary events, the RC provides opportunities for studying the coupling between electron and proton transfer, and the molecular basis of equilibrium protein–ligand interactions. The main focus, here, has been on the quinone electron acceptors of the RC, which interface the one-electron primary events with the two-electron redox chemistry of the quinone pool. The significant functional differences between Q_A and Q_B arise from the environment provided by the binding domains. The experimental amenity of the quinone binding sites to quinone replacement studies, especially in *Rb. sphaeroides* and *Rb. capsulatus* where the native Q_A and Q_B are chemically identical, and the demonstrated resilience of the RC to drastic amino acid substitutions by site-directed mutagenesis, makes the RC a potentially rich playground for exploring structure–function relationships in protein–cofactor associations.

Dutton and co-workers have built a substantial basis for quantitative structure–activity analysis of quinone function, by comparing the ability of many different quinones and quinone analogues to restore Q_A activity in quinone extracted RCs of *Rb. sphaeroides* (strain R26) and *Rps. viridis* (Gunner et al., 1985; Giangiacomo et al., 1990; Warncke and Dutton, 1990, 1992; Keske and Dutton, 1991; Keske et al., 1992). This approach provides information both on binding affinity and on *in situ* redox properties of the bound cofactor. The affinity of the binding site is defined by hydrogen bonding to the carbonyl oxygens (or other polar groups) of the quinone analogue, and by extensive van der Waals interactions. For a quinone with two carbonyls the binding free energy contributed by two hydrogen bonds is little different from that of a one-hydrogen bond to the single carbonyl of, for example, anthrone (Gunner et al., 1985; Warncke and Dutton, 1990). This suggests that the two carbonyl interactions of a quinone are mutually antagonistic. Whether this is of functional importance (e.g., in setting the redox potential) may be difficult to test since one of the hydrogen bonds is contributed from the main chain peptide nitrogen of Ala^{M260} (M258 in *R.v.* and *R.c.*).

Binding of quinones in the Q_A pocket causes a characteristic shift in redox potential from values determined in aprotic solvent (Prince et al., 1983,1986). Interestingly, when quinones of higher intrinsic redox potential are tested, the site appears to clamp the *in situ* potential at a maximum value, so the site-induced potential shifts decrease as the *in vitro* potentials get higher (Giangiacomo et al., 1990). The origin of this effect is not known, but it recalls the above discussion concerning the small shifts in redox potentials of the heterodimers and of $Bchl_A$. It is presumably significant that the binding affinity of the quinones is independent of the *in vitro* potential, so the *in situ* potential shift reflects the binding interactions of the semiquinone form (Giangiacomo et al., 1990).

Mutual antagonism also occurs between the head group and polyisoprene tail of ubiquinones as Q_A. Thus, although the presence of the tail contributes some additional net affinity, it is partially at the expense of head group binding (McComb et al., 1990). This implies that a small quinone does not occupy quite the same position in the Q_A site as the head group of a prenyl quinone.

Trp^{M252}

The X-ray structures of RCs immediately suggested Trp^{M252} (*R.v.* and *R.c.* Trp^{M250}) as a likely important amino acid residue in the Q_A region (Figure 3). It is located in the Q_A-binding pocket in van der Waals contact with Bph_A (Figure 2),

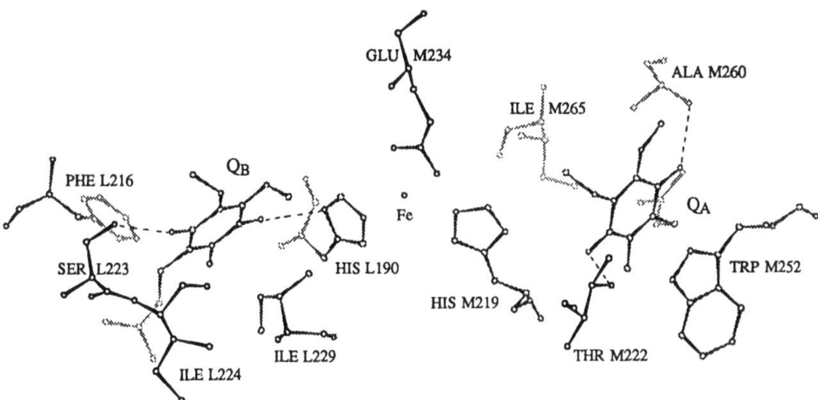

Figure 3. View of selected residues surrounding the two quinones, Q_A and Q_B, of the *Rb. sphaeroides* reaction center. The side chains of the quinones have been truncated to leave a single isoprene unit. Hydrogen bonds to the quinone carbonyls are shown by *dashed lines. Various line densities* are used to indicate depth, with *solid lines* for the closest positions.

and is conserved in all bacterial RC sequences determined to date, as well as in RCs of photosystem II (Trebst, 1987; Komiya et al., 1988). The symmetrically equivalent residue in the B-branch is Phe^{L216} located in the Q_B-binding pocket, but it is not in van der Waals contact with Bph_B. Because of its close contact with Q_A, alteration of Trp^{M252} has a substantial effect on quinone binding in the Q_A pocket, and on electron transfer.

$Trp^{M250/252}$ has been altered by site-directed mutagenesis in *Rb. capsulatus* and *Rb. sphaeroides* to several other residues (Coleman et al., 1990a; Stilz et al, 1990). In *Rb. capsulatus*, only substitutions which conserve the aromaticity of the residue; that is, $Trp^{M250} \rightarrow$ Phe or Tyr, permitted photosynthetic growth. Isolated RCs showed some photochemical activity, but the Q_A content was significantly depleted. In all other cases (Trp $\rightarrow$ Met, Thr, Leu, Glu, Val, Arg), Q_A loss was total, but photochemical activity (bleaching at 850 nm in continuous light) could be restored to the RCs by addition of menadione (menaquinone-0), a relatively tight-binding analogue of Q_A. Quinone-binding studies showed a wide range of apparent dissociation constants among the mutants. For example, replacement of Trp^{M250} by leucine resulted in a greater than 30-fold decrease in affinity for quinone compared to the Wt, while substitution with valine caused more than a 300-fold loss of affinity (Coleman et al., 1990a,b). This probably results from the loss of van der Waals, and especially π-orbital, overlap between Q_A, and the indole side chain of Trp.

In view of the markedly apolar nature of the Q_A-binding domain, it is remarkable that such substitutions as Glu and Arg lead to assembly of stable RCs that appear to retain some quinone affinity at the Q_A site. However, the nature of the assay (continuous illumination) must be borne in mind, as isolated RCs are capable of low quantum yield electron transfer to quinones and related acceptors that are not bound in the pocket, at all, but are presumably adhering to the RC in a nonspecific manner (K. Warncke and P.L. Dutton, personal communication).

Alteration of Trp^{M252} is also expected to affect electron transfer since Trp^{M252} is in a position to act as a bridging molecule between Bph_A and Q_A, enhancing the electronic coupling between them. The rate of electron transfer between Bph_A and Q_A at room temperature is ~ $5 \times 10^9\ s^{-1}$ and shows a slight negative temperature dependence (Kirmaier and Holten, 1987; Gunner and Dutton, 1988), presumably due in part to the activationless nature of the electron transfer process. The dependence of this step on temperature and on the free energy of the reaction has been extensively studied by Gunner and Dutton and co-workers using quinone analogues as Q_A to adjust the free energy level of the $P^+Q_A^-$ state (Woodbury et al., 1984; Gunner et al., 1986; Gunner and Dutton, 1988). Plato et al. (1989) have discussed the possible role of $Trp^{M250/252}$ in *Rps. viridis* and *Rb. sphaeroides* as a superexchange mediator in the electron transfer from Bph_A^- to Q_A. On the basis of quantum chemical calculations, using the X-ray structural information for both species, they concluded that superexchange is dominant over the direct electron transfer by as much as 10-fold.

Consistent with the expectations of Plato et al. (1989), $Trp^{M250} \rightarrow$ Leu mutant RCs from *Rb. capsulatus* exhibited a 16-fold decrease in $Bph_A^- \rightarrow Q_A$ electron transfer rate, although it remained activationless (Coleman et al., 1990b). In the framework of nonadiabatic electron transfer theory (rate $\propto V^2$), this implies a fourfold decrease in the electronic coupling between Bph_A^- and Q_A, provided that the energies of Bph_A^- and Q_A are themselves unchanged. In fact, the rate of $P^+Q_A^-$ recombination with naphthoquinone as Q_A was noticeably temperature-dependent, unlike the Wt. This suggests that the E_m of Q_A/Q_A^- is lowered by the mutation, so that some recombination occurs by thermal repopulation of $P^+Bph_A^-$.

Qualitatively similar behavior was seen in mutants of *Rb. sphaeroides*. Comparisons of $Bph_A^- \rightarrow Q_A$ electron transfer rates in $Trp^{M252} \rightarrow$ Tyr and Phe mutants revealed a three- to fourfold decrease in electron transfer rate in the mutants compared to Wt, corresponding to the decrease in the π electron density of the aromatic residues (Stiltz et al., 1990).

Thr^{M222}

In *Rps. viridis*, Thr^{M220} hydrogen bonds to the nitrogen atom of the indole ring of Trp^{M250} and thus could influence the orientation of the tryptophan parallel to Q_A. In *Rb. sphaeroides*, however, the equivalent residue, Thr^{M222}, is hydrogen bonded to the O2 carbonyl of the ubiquinone (see Figure 3). There is disagreement between the two independent X-ray refinements as to whether it makes an additional hydrogen bond to Trp^{M252} (Allen et al., 1988b; El-Kabbani et al., 1991).

Mutation of Thr^{M222} to Val in *Rb. sphaeroides* RCs did not affect the rate of the $Bph_A^- \rightarrow Q_A$ electron transfer rate (Stilz et al., 1990). The quinone affinity was lowered compared to the Wt, but not as much as in the Trp^{M252} mutants. This is consistent with the finding that Q_A analogues with a single carbonyl bind almost as strongly as those with two (Gunner et al., 1985; Warncke and Dutton, 1990). This might indicate that Q_A is "suspended" between two hydrogen bonds, either one of which can become stronger if the other is removed. Furthermore, in the absence of Thr^{M222}, Q_A may hydrogen bond to His^{M219}, as in *Rps. viridis* RCs.

Ile^{M265}

The contributions of van der Waals interactions in Q_A binding are well delineated by the work of Dutton's group, and are amenable to mutagenesis studies. Size limitations on effective Q_A analogues gave an outline of the binding pocket in good agreement with the X-ray structure. Subsequent analyses of solvation energies for various, relevant functional groups are now providing a quantitative description of the binding energetics (Keske et al., 1991). Substitutions of Ile^{M265} by Ser and Thr in *Rb. sphaeroides* caused readily interpretable changes in affinities for various quinones (see Figure 3). The affinity for a medium-sized species, such as 1,4-naphthoquinone, was decreased, consistent with weaker contact with the smaller mutant residues, but the capacity for binding larger quinones was enhanced (K. Warncke and E. Takahashi, unpublished). Interestingly, RCs from the carotenoid-containing strain Ga, parent to these mutants, showed tighter binding of naphthoquinone and less tolerance of larger quinones than R26. This is consistent with the general folklore that the presence of the carotenoid "stiffens" the RC structure.

The $Ile^{M265} \rightarrow$ Ser and Thr mutant RCs exhibited significant acceleration of the $P^+Q_A^-$ recombination reaction, which appeared to arise from unexpectedly large decreases ($\approx$ 100 mV) in the *in situ* redox potential of Q_A/Q_A^-, thereby stimulating recombination via $P^+Bph_A^-$ (K. Warncke and E. Takahashi, unpublished). At the present, early stage of this work, the analogue and mutational approaches are proving usefully complementary in structure–activity studies of quinone binding and function.

B. Mutants of the Q_B Region

Glu^{L212}, *Asp*^{L213}, *Ser*^{L223}, *etc.: Mutations Affecting Proton Transfer*

In the presence of secondary donors to re-reduce P^+, two turnovers of the RC result in the double reduction of Q_B accompanied by the uptake of two protons to form a quinol, Q_BH_2 (see below; Figure 5). Since the Q_B site is completely occluded from the solvent by the protein, the delivery of two protons from the aqueous phase to Q_B occurs by proton transfer through the protein. In principle, this could occur by diffusive penetration of the protein matrix or by a hopping mechanism involving

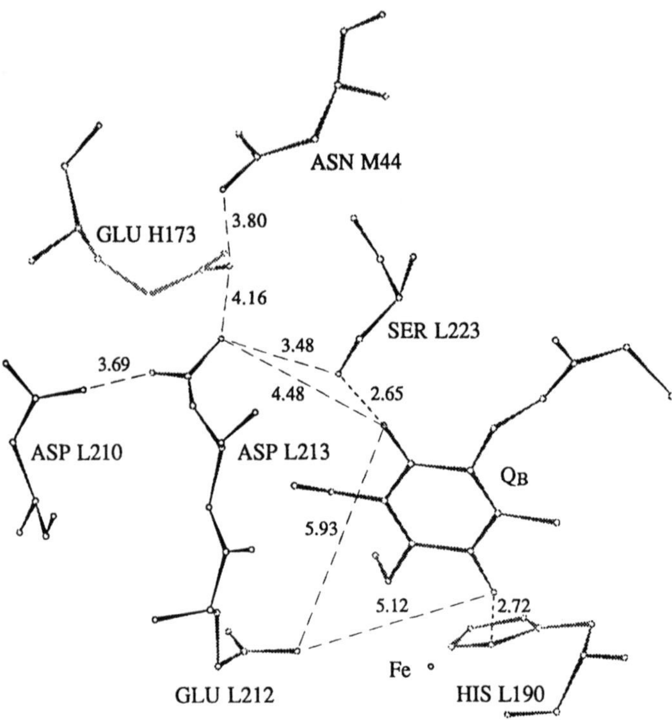

Figure 4. The Q_B binding site of the *Rb. sphaeroides* reaction center, showing residues implicated in protonation reactions. See text for details. The side chain of Q_B has been truncated after the first isoprene unit. Hydrogen bonds are shown by *short-dashed lines. Long-dashed lines* connect several selected residues with the distances given in angstroms. *Various line densities* are used to indicate depth, with *lighter lines* for the more distant entities.

polar ionizable residues. Examination of the X-ray structure of the RC reveals no simple "channels", but does show many ionizable residues from the L-, M-, and H-subunits in and around the Q_B site (Figure 4), in contrast to the Q_A site which contains only a few, weakly polar residues. Possible pathways of proton transfer to Q_B have been identified by Allen et al. (1988b) for *Rb. sphaeroides* and Deisenhofer and Michel (1989b) for *Rps. viridis*. Several water molecules are seen close to the Q_B site in the high resolution structure of *Rps. viridis* RCs (Deisenhofer and Michel, 1989a,b), and these may also participate in the proton transfer. The residues in close contact with the Q_B, including Glu^{L212}, Asp^{L213}, and Ser^{L223}, have been altered in recent site-directed mutagenesis studies on *Rb. sphaeroides* RCs. These studies have identified two terminal steps in the route by which protons ultimately arrive at Q_B (Figure 5).

Delivery of the first proton to Q_B, either in the state $Q_A^-Q_B^-$ or $Q_AQ_B^{2-}$, involves Asp^{L213}, probably in concert with Ser^{L223}. The hydroxyl group of Ser^{L223} is hydrogen-bonded to the O5 carbonyl oxygen of Q_B, and is also within hydrogen-bonding distance of the side chain of Asp^{L213}. Both the $Ser^{L223} \rightarrow$ Ala mutant (Paddock et al., 1990a) and $Asp^{L213} \rightarrow$ Asn mutant (Takahashi and Wraight, 1990) show normal rates of $P^+Q_A^-$ recombination, indicating the absence of any drastic structural changes due to the mutations. This is supported by the EPR spectrum of Q_B^- which is unaltered, perhaps surprisingly, even by the Ser^{L223} mutation. However, in both mutants the RC was blocked after two turnovers in the state $Q_A^-Q_B^-$ due to failure in the delivery of the first proton to Q_B. This results in inhibition of the second electron transfer.

Ser^{L223}, Asp^{L213} or water could all act as the direct proton donor to Q_B. However, mutation of Ser^{L223} to Thr or Asp results in good activity, while Asn does not, strongly implicating Ser^{L223} as the initial donor (Okamura and Feher, 1992). The role of Asp^{L213} may then be the rapid or concerted reprotonation of Ser^{L223} to permit net proton transfer to the quinone. Alternatively, since Asp^{L213} is always ionized ($pK \leq 6$ even in the presence of Q_B^-; Takahashi and Wraight, 1992a), it may serve to raise the p*K* of Ser-OH/OH_2^+ making it a functional proton donor. [The p*K* of asparagine, a nonfunctional substitute for Ser^{L223}, may be too low (approximately –6) for such facilitation to help.] In order to participate directly in proton donation on the second transfer, Asp^{L213} must be rapidly and transiently protonated by internal transfer in the $Q_A^-Q_B^-$ or $Q_AQ_B^{2-}$ states.

In fact, the pH dependence of Q_A/Q_B electron transfer events at low pH is complicated by the apparent interaction of several acidic groups in the vicinity of Q_B, including Asp^{L210}, Asp^{L213}, Arg^{L217}, and Glu^{H173} (see Shinkarev and Wraight, 1992). For example, Paddock et al. (1992) altered Asp^{L210} to asparagine and found the $Q_A^-Q_B \leftrightarrow Q_AQ_B^-$ equilibrium to be almost pH-independent below pH 7, but with no kinetic impairment of function. Thus, the charge state of Asp^{L210} contributes to the specific ionization properties of the cluster of acidic groups—the mutation presumably shifts an apparent p*K* out of range—but is not essential for proton delivery to Q_B.

For mutants with impaired proton delivery, Takahashi and Wraight (1991) found that the addition of small protonophores, such as azide, stimulated the transfer of the second electron in a concentration- and pH-dependent manner. This suggests that small molecules from the aqueous phase are accessible to a site close to the Q_B site.

The transfer pathway for the second proton may involve the residues Glu^{L212}, Asp^{L213}, and possibly His^{L190}. Paddock et al. (1989) altered Glu^{L212} to the nonionizable residue glutamine and found the mutation to essentially eliminate the alkaline pH dependences of the one-electron $Q_A^- Q_B \leftrightarrow Q_AQ_B^-$ equilibrium and the electron transfer rate. This essentially identifies Glu^{L212}, with an unusually high p*K* (≈9.6), as the residue that inhibits the first electron transfer when ionized. The mutation also exhibited a greatly reduced rate of transfer for the second proton to Q_B^{2-}, resulting in blockage of the RCs, after three turnovers, in the state $PQ_A^-Q_BH^-$. A double mutation, $Glu^{L212} \rightarrow$ Gln + $Asp^{L213} \rightarrow$ Asn, lacked any pH dependence of

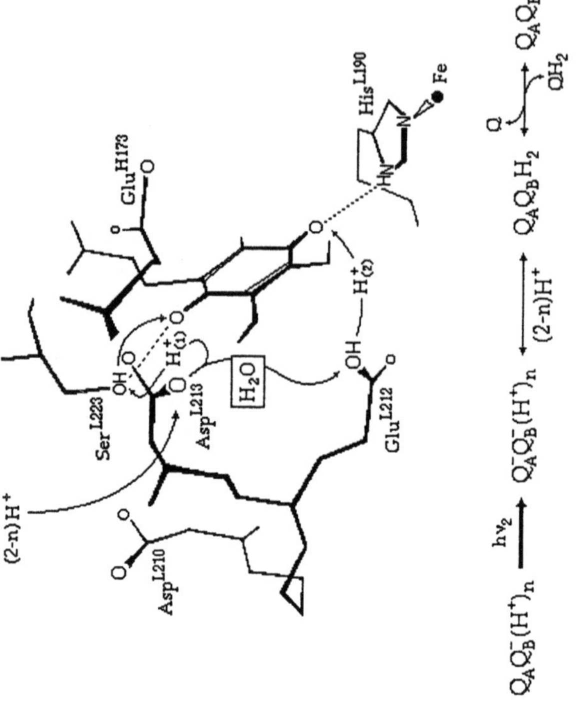

(2-n)H+
Ser L223
Asp L213
Asp L210
Glu L212
Glu H173
His L190
Fe
H2O
H+(1)
H+(2)
$Q_AQ_B^-(H^+)_n \xrightarrow{h\nu_2} Q_A^-Q_B^-(H^+)_n \xleftrightarrow[(2-n)H^+]{} Q_AQ_BH_2 \leftrightarrow Q_AQ_B$
Q
QH2

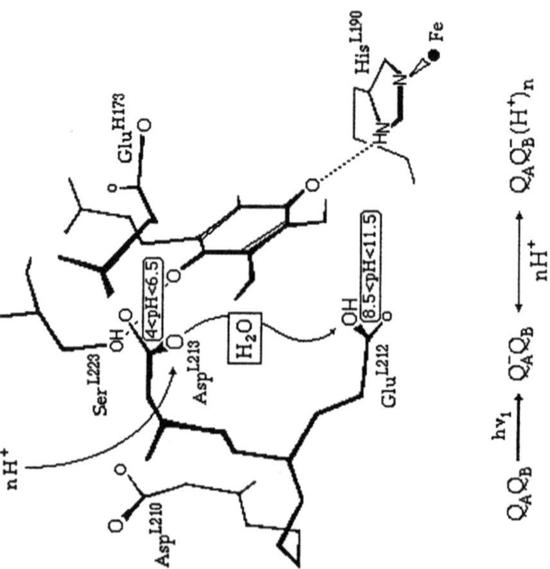

nH+
Ser L223
Asp L213
Asp L210
Glu L212
Glu H173
His L190
Fe
H2O
4<pH<6.5
8.5<pH<11.5
$Q_AQ_B \xrightarrow{h\nu_1} Q_A^-Q_B \xleftrightarrow[nH^+]{} Q_AQ_B^-(H^+)_n$

the $Q_A^-Q_B \leftrightarrow Q_AQ_B^-$ equilibrium and electron transfer rate, in good agreement with the behavior of the two single-site mutants (Takahashi and Wraight, 1992a).

Mutation of Glu^{L212} to aspartic acid caused a substantial decrease in the $Q_A^-Q_B \leftrightarrow Q_AQ_B^-$ equilibrium constant, apparently due to the pK of Asp^{L212} being much lower (< 6) than the native glutamic acid (Paddock et al., 1990b). The reason for such a dramatic shift in pK for a relatively conservative substitution is unknown, but it implies that the properties of the native Glu^{L212} are due to a fine balance of interactions such that substitution by a shorter aspartic acid residue results in a significant reorganization of the charge compensating interactions. Structural information on this mutant might be revealing.

Proton delivery to Q_B, in the formation of Q_BH_2, may be summarized as follows (Takahashi and Wraight, 1992a):

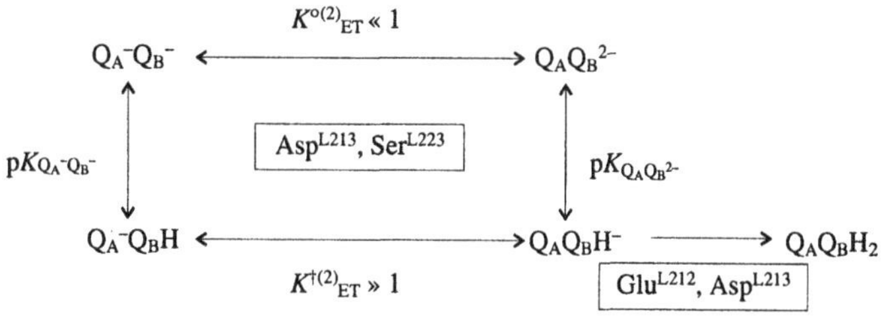

Scheme 2.

Figure 5. Protonation events accompanying the first and second electron transfers to Q_B. Hydrogen bonds to the quinone carbonyls are shown by *dashed lines*; proposed paths of proton conduction are shown by *solid lines*. Schemes for the flash-induced electron and proton transfers are shown beneath each figure. (*Left*) After the first flash ($h\nu_1$), sub-stoichiometric proton uptake occurs to satisfy changes in the ionization states of various groups, determined by pK shifts induced by the charge on Q_B^- (Maróti and Wraight, 1988; McPherson et al., 1988). At low pH (4 < pH < 6.5), the extent of H^+ bonding is largely determined by the pK shift and ionization state of Asp^{L213}, and possibly other groups in a cluster around it. At high pH (8.5 < pH < 11.5), the proton uptake is largely in response to the pK shift of Glu^{L212}. Asp^{L213} is suggested to be on the path of proton delivery to Glu^{L212}, probably with the involvement of water molecules in the pocket. (*Right*) After a second flash ($h\nu_2$), further proton uptake occurs with the formation of quinol. The quinone complex returns to its original state with exchange of the quinol for quinone in the Q_B pocket, and the average proton uptake is determined by the stoichiometry of quinol formation, 1 H^+ per e^-. Since nH^+ are bound on the first flash, the uptake on the second flash is $(2 - n)H^+$. The first proton to reach the Q_B head group is delivered via Asp^{L213} probably via Ser^{L223} and possibly with the involvement of bound water. The second proton is delivered to Q_BH^- from Glu^{L212}, probably via Asp^{L213} and bound water.

Two kinetic possibilities are implied whereby either electron or proton transfer is the first event in forming Q_B^{2-} or Q_BH^- as intermediate states, respectively (Wraight, 1979). Limiting values for the four constants have been derived from experiments on wild type (strain R26) (Kleinfeld at al., 1985) and mutant RCs (Takahashi and Wraight, 1992a):

$$pK_{Q_A{-}Q_B{-}} < 7.4,\ pK_{Q_AQ_B2{-}} \geq 10.7 \qquad K^{\dagger(2)}{}_{ET} \geq 30,\ K^{o(2)}{}_{ET} \leq 0.02.$$

The two routes are both characterized by an unfavorable initial equilibrium followed by a potentially favorable one, pulling the net electron transfer towards $Q_AQ_BH^-$. In the upper route, the high pK of the dianion, Q_B^{2-} [p$K \geq 13$, *in vitro* (Chambers, 1974; Morrison et al., 1982)], will allow this species to obtain a proton readily from Ser-OH_2^+ or neutral Asp. In the lower route, the electron transfer to the neutral semiquinone is expected to be facile.

The limiting parameters for proton delivery and the second electron transfer to Q_B in the $Glu^{L212} \rightarrow$ Gln mutant RCs were generally similar to those for Wt (R26) RCs, but a closer comparison indicated a decrease of 50 mV in the free energy drop from $Q_A^-Q_B^-$ / $Q_A^-Q_BH$ to $Q_AQ_B^{2-}$/$Q_AQ_BH^-$ (Takahashi and Wraight, 1992a). This could arise from a lesser ability of glutamine, compared to glutamic acid, to charge compensate Q_BH^- and/or Q_B^{2-} and is consistent with the observed weaker stabilization of Q_B^- in this mutant. Such effects could arise from specific shifts in the pKs; for example via hydrogen bonding, or from electrostatic influences on the charged species (Wraight, 1982).

Proton delivery is completed, and the electron transfer equilibrium is pulled over further with the uptake of a second proton from the medium via Glu^{L212} and probably Asp^{L213} as shown in Scheme 2.

Somewhat similar results have been obtained with mutants of *Rb. capsulatus*, but with interesting differences. Hanson et al. (1992; and D.M. Tiede, personal communication) have selected revertants of a double mutant of $Glu^{L212} \rightarrow$ Ala + $Asp^{L213} \rightarrow$ Ala, which is unable to grow phototrophically. The revertants fall into several categories:

1. Restoration of Asp^{L213}, alone, resulted in good photosynthetic activity, implying only a weak, kinetic dependence on Glu^{L212}, unlike *Rb. sphaeroides*.
2. Second site reversion at position M43 (Asn $\rightarrow$ Asp). This is particularly noteworthy since it generates the natural situation found in *Rps. viridis* and *R. rubrum* where L213 is Asn and M43 (or M44) is Asp.
3. Second site reversion at $Gly^{L225} \rightarrow$ Asp. This residue is located on the opposite side of the quinone from L213, but close to M43.
4. Second site reversion at $Arg^{M231} \rightarrow$ Leu. This unusual mutation may be expected to result in substantial structural changes. From the *Rb. sphaeroides* X-ray structure, the equivalent Arg^{M233} is involved in significant electrostatic

> interactions with several residues, most notably Glu^{H122}, but also Asp^{H170} and Glu^{H230}, among others. Apart from global structural changes, loss of the arginine may free up one or more acidic groups to function in the same general manner as Asp^{L213} or the other revertants at L225 and M43.

All of these revertants exhibit high levels of activity with good Q_B turnover, although the rates of the second electron transfer and proton delivery have not been established. However, the original double mutant was significantly less inhibited than the $Glu^{L212} \rightarrow$ Gln + $Asp^{L213} \rightarrow$ Asn double mutant in *Rb. sphaeroides*, indicating, perhaps, less structural specificity or greater flexibility in *Rb. capsulatus*.

Similar mutations have been reported in *Rb sphaeroides* (Okamura and Feher, 1992). An $Asp^{L213} \rightarrow$ Asn + $Asn^{M44} \rightarrow$ Asp double mutant has been constructed, and a second site reversion of the $Asp^{L213} \rightarrow$ Asn mutant was found to contain $Arg^{M233} \rightarrow$ Cys. Both showed high rates of proton transfer, and it was suggested that the main function of Asp^{L213} was in the electrostatic stabilization of protons in the interior of the protein rather than a specific proton donor role.

Together with the action of small weak acids in restoring proton delivery to Q_B-site mutants and the high resolution structure of the *Rps. viridis* RC, these results suggest to us a picture of the Q_B site as a significantly aqueous pocket surrounded by a "thin" proton impermeable shell. One or two residues in the shell (e.g., Asp^{L213} in *Rb. sphaeroides* and *capsulatus* and possibly Asp^{M43} in *Rps. viridis* and *Rs. rubrum*) act as chemically specific points of delivery for protons from the surrounding protein matrix, which provides many parallel paths for proton transfer and may be highly penetrated by water. Small structural perturbations, either as differences between species or as mutational alterations, can provide alternative paths of communication to Q_B, with water playing a significant role within the Q_B pocket per se. The purpose of the occluded nature of the Q_B pocket may be to limit the bulk nature of the water so that the binding and redox properties of Q_B can be fine-tuned; for example, to bind Q, Q^-, and QH_2 differentially (Wraight, 1982), and to facilitate exchange of the head group with the hydrophobic membrane phase.

A major difference between the $Asp^{L213} \rightarrow$ Asn and $Ser^{L223} \rightarrow$ Ala mutants in *Rb. sphaeroides* is the large increase in the $Q_A^-Q_B \leftrightarrow Q_AQ_B^-$ equilibrium in favor of Q_B reduction, and an altered pH dependence of the equilibrium in the former resulting from the change in charge upon substitution. The $Ser^{L223} \rightarrow$ Ala mutant does not show such behavior compared to the Wt at pH 7.5. Asp^{L213} is normally ionized and evidently presents an electrostatic restriction to the first electron transfer, as well as contributing significantly to the elevated p*K* of Glu^{L212} (Takahashi and Wraight, 1992a). This is consistent with the dominant contribution from electrostatics in the functional properties of Q_B (Shinkarev and Wraight, 1992). In an attempt to test this, Takahashi and Wraight (1992b) constructed a $Glu^{L212} \rightarrow$ Lys mutant with the expectation that the lysine would be positively charged and would very strongly stabilize Q_B^-. In fact, the mutant exhibited behavior quite similar to the $Glu^{L212} \rightarrow$ Gln mutant, with a pH-independent rate of transfer of the first electron, less stable Q_B^-, and almost normal rate of second electron transfer. This suggests that the lysine is neutralized either by deprotonation or by compensation. The former is unexpected on the basis of the elevated p*K* of the native Glu^{L212} in the same environment. The latter may not involve the most obvious residue, Asp^{L213}, because the net charge of the Q_B pocket would then be zero, as in the double mutant ($Glu^{L212} \rightarrow$ Gln + $Asp^{L213} \rightarrow$ Asn) which strongly stabilizes Q_B^- (Takahashi and Wraight, 1992a). On the other hand, formation of ionic interactions between alternative residues and charged Lys^{L212} would probably require significant structural perturbations.

Interestingly, an equivalent mutant (MAV2 = $Glu^{L212} \rightarrow$ Lys) has been isolated in *Rps. viridis*, as an atrazine resistance mutation (Ewald et al., 1990). Again contrary to expectation, but different from the *Rb. sphaeroides* mutant, the first electron transfer at pH 7.0 was more than 1000 times slower than in Wt (Leibl et al., 1992). Above about pH 9 the first electron transfer accelerated dramatically, possibly coincident with deprotonation of the lysine. The second electron transfer, on the other hand, is only 10 times slower than Wt and exhibits a similar pH dependence. The origin of the striking differences between these two mutants may lie in the distinction at position L213 (Asp in *Rb. sphaeriodes*, Asn in *Rps. viridis*).

The predominance of charged groups and electrostatic interactions in determining the properties of the Q_B site draws attention to an interesting feature of the RC structure in the quinone binding domains, as noted by M.R. Gunner (personal communication). Electrostatic calculations indicate that the number of ionized acidic groups is unusually large and can only be supported because of a substantial positive potential from main chain dipoles oriented roughly in parallel in a "picket fence" motif. This is especially evident in the interhelix loop *de* that forms the Q_B pocket.

Electrical Measurements of Acceptor Quinone Activity

The generation of a photovoltage due to charge movements can be detected in an oriented RC ensemble, as in a membrane bilayer (Schönfeld et al., 1979; Packham et al., 1980) or light gradient (Trissl, 1983). The major part of this voltage is due to electron transfer (e.g., from P to Q_A), and electron transfer from Q_A^- to Q_B is not electrogenic since the two quinones lie in the same plane (Feher and Okamura, 1984; Kaminskaya et al., 1986). However, transfer of protons to amino acid residues of the Q_B pocket of the RC can make a small contribution ($\leq 10\%$) to the net photovoltage. The magnitude of this contribution depends on the nature of the protolytic groups and reactions involved. Accompanying the first electron transfer, proton electrogenicity is only seen at 7 > pH > 8 (Drachev et al., 1990; Brzezinski et al., 1991a); that is, in those regions where groups close to Q_B (Glu^{L212} and Asp^{L213}/Asp^{L210}) have been implicated in proton binding. Near neutral pH, proton binding accompanying Q_B reduction is small and is thought to be widely dispersed (Maróti and Wraight, 1988; McPherson et al., 1988; Shinkarev and Wraight, 1992).

These assignments are well supported by the electrogenic responses of mutant RCs and chromatophores. Proton binding and electrogenicity are suppressed at alkaline pH in $Glu^{L212} \rightarrow Gln$ mutant RCs (Brzezinski et al., 1991a; Shinkarev et al., 1992). In the $Asp^{L213} \rightarrow Asn$ mutant, proton binding and electrogenicity are compressed into the neutral region (Brzezinski et al., 1991a) due to loss of Asp^{L213} and the resultant downward shift in the pK of Glu^{L212} (Takahashi and Wraight, 1992a).

The major electrogenic act of the acceptor quinone complex is the proton uptake associated with double reduction of Q_B to the quinol (Feher and Okamura, 1984; Kaminskaya et al., 1986; Drachev et al., 1990). In $Glu^{L212} \rightarrow Gln$ mutant RCs, the photovoltage is diminished by 50 to 70%, but retains a similar half time (100–200 μs at pH 7) to the wild type (Brzezinski et al., 1991b). This is consistent with the delivery of only one proton to Q_B/Q_B^{2-} through an unimpaired pathway. The anionic quinol, Q_BH^-, remains bound to the mutant RC until the second proton arrives on the time scale of about 1 s when its release can be detected in chromatophores by the electrogenic turnover of the cytochrome bc_1 complex (Shinkarev et al., 1992).

Brzezinski et al. (1992) have reported a small voltage change of opposite polarity to normal in the absence of Q_B. The behavior of this transient mirrored the electron transfer when Q_B was present (i.e., $\tau \approx 160$ μs at pH 7, pH-independent rate over the range $5 < pH < 9$), but decreased above pH ≈ 9. Furthermore, the rate of the transient was pH-independent in $Glu^{L212} \rightarrow Gln$ mutant RCs. The implication is that it represents a small movement of charge that may be required (rate limiting) for electron transfer to Q_B. One possibility is that it represents a movement of Q_A^-, perhaps transferring its hydrogen bond from Thr^{M222} to His^{M219}. The reason for the dependence on the charge state of Glu^{L212} is obscure, but it adds to the many examples of "crosstalk" between the two quinone binding sites. Structural heterogeneity in the hydrogen-bonding configuration of Q_A, between Thr^{M222} and His^{M219}, may also be a source of the biphasicity in the $Q_A^-Q_B \rightarrow Q_AQ_B^-$ electron transfer kinetics (Maróti and Wraight, 1990; Takahashi et al., 1992) and in the $P^+Q_A^-$ recombination kinetics at low temperature (Parot et al., 1987).

Herbicide-Resistant Mutants

Since the neutral quinone species, Q_B and Q_BH_2, are not tightly bound to the RC, inhibitor molecules can compete with the quinones of the membrane pool for the occupancy of the Q_B-binding pocket (Wraight, 1981; Stein et al., 1984). Competitive inhibitors of the Q_B site include *o*-phenanthroline and the commercially important s-triazine herbicides (e.g., atrazine and terbutryn) that block electron transfer from Q_A^- in both bacterial and photosystem II (PS II) RCs (Stein et al., 1984). The binding locations of terbutryn and *o*-phenanthroline within the Q_B pocket of the *Rps. viridis* RC have been determined by X-ray structural analysis

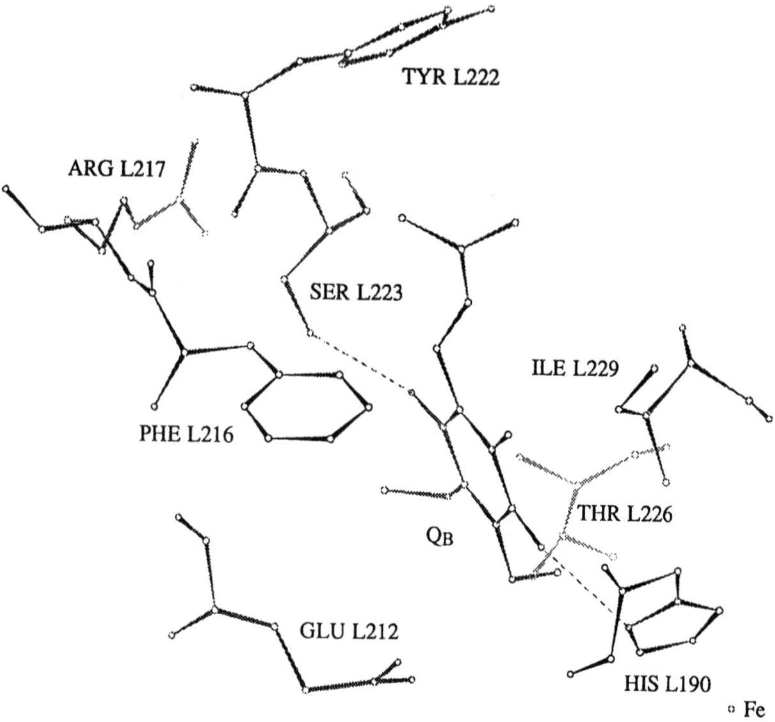

Figure 6. View of the Q_B-binding pocket in the *Rb. sphaeroides* reaction center with residues that confer herbicide resistance to the reaction center when mutated (except for His^{L190}). See text for details. Hydrogen bonds are shown by *dashed lines. Lighter lines* indicate greater depth.

(Figure 6; Michel et al., 1986b). Terbutryn is located at one end of the Q_B pocket and hydrogen bonded to the hydroxyl group of Ser^{L223} and the peptide nitrogen of Ile^{L224}, while *o*-phenanthroline is at the opposite end, closer to the iron atom, and hydrogen bonded to His^{L190}. Although the structure of the PS II RC is still not known, similarity in function (Crofts and Wraight, 1983) and amino acid sequence (Williams et al., 1986), as well as a similar mechanism of inhibitor action (Velthuys, 1981), provide strong indications of structural homology between the bacterial and PS II Q_B sites (Barber, 1987; Trebst, 1987). However, benzonitrile and urea inhibitors, most notably DCMU (or diuron), which inhibit electron transfer in PS II RCs, are not very effective inhibitors of bacterial RCs, indicating some differences between the Q_B pockets of the two types of RC.

Spontaneous and chemically induced herbicide-resistant mutants of several bacterial species have been isolated by selecting for photosynthetic growth in the presence of herbicides (Brown et al., 1984; Okamura, 1984; Stein et al., 1984). Sequence analyses have revealed most of the mutations to be located in the *de* loop between transmembrane helices D and E of the L-subunit which make up the Q_B-binding pocket. Some mutants contain double mutations in the Q_B loop, while others contain a second mutation in the M-subunit. Herbicide-resistant mutants, in general, show altered quinone and herbicide-binding properties, as well as altered electron transfer properties.

Interest in herbicide-resistant mutants stems from the use of bacterial RCs as a model system for PS II (Wraight, 1981; Stein et al., 1984) and the obvious agronomic significance of herbicide resistance, both as a scourge when found in weed species and potential blessing if it could be engineered into crop species without loss of photosynthetic capacity. The latter property is a common corollary of herbicide resistance. The general principles governing quinone binding and function, and some expectations of herbicide-resistance behavior, have been outlined by Wraight (1982) and Wraight and Shopes (1989).

An early attempt to explore the molecular nature of herbicide resistance was made on *Rb. capsulatus*. Bylina and Youvan (1987) performed extensive site-directed mutagenesis of Ile^{L229}, substituting Ile with 17 other residues (all except Phe and Pro). RC assembly was not affected in any of the mutants and seven of the mutants were capable of growing photosynthetically (Val, Ala, Leu, Met > Thr, Cys, Ser), including six mutants which were resistant to atrazine (Met > Leu, Thr > Ser, Ala, Cys). Characterization of these mutants was largely restricted to a herbicide-resistance growth assay that appeared to show some variation in resistance to various triazine herbicides (Bylina et al., 1989). Only the mutant $Ile^{L229} \rightarrow$ Met was clearly resistant to all five herbicides tested, which may be why this mutation is one of the most widely found spontaneous herbicide-resistant mutants (Gilbert et al., 1985; Schenck et al., 1986; Paddock et al., 1988; Takahashi, 1992). The X-ray structures of *Rps. viridis* and *Rb. sphaeroides* RCs show that the extra length of methionine compared to isoleucine would, indeed, intrude into the binding region of all occupants of the Q_B site, including quinone and inhibitors.

The growth assay used by Bylina et al. (1989) did not elucidate any functional characteristics, but this work on *Rb. capsulatus* did show $Phe^{L216} \rightarrow$ Leu, Thr, and Pro to induce triazine resistance, and revealed mutations at two residues not previously identified as causing herbicide resistance, $Thr^{L226} \rightarrow$ Ala and Met, and $Gly^{L228} \rightarrow$ Val, and Arg. Baciou et al. (1992) have found the $Thr^{L226} \rightarrow$ Ala mutant to exhibit greatly increased (25 ×) sensitivity to *o*-phenanthroline and a slightly enhanced affinity for ubiquinone (Q-10 and Q-6). Interestingly, *Rps. viridis* RCs, where alanine is the native residue at position L226, are much more sensitive to *o*-phenanthroline than Wt *Rb. sphaeroides* and *capsulatus*. Possibly threonine provides some steric hindrance to *o*-phenanthroline binding, either by its size, per se, or through its hydrogen-bonding propensities.

Several spontaneous triazine-resistant mutants of *Rps. viridis* have been isolated and characterized. Four of these, selected on terbutryn (Sinning et al., 1989a), have been identified as†: T1 = $Ser^{L223} \rightarrow$ Ala + $Arg^{L217} \rightarrow$ His, T3 = $Phe^{L216} \rightarrow$ Ser + $Val^{M263} \rightarrow$ Phe, T4 = $Tyr^{L222} \rightarrow$ Phe, and T6 = $Phe^{L216} \rightarrow$ Ser. All mutant RCs showed decreased affinity for terbutryn and *o*-phenanthroline compared to the Wt. The affinity for the native Q_B species, Q-9, decreased in the T3, T4, and T6 mutant RCs, but increased at least 10-fold in T1. This is a very unexpected result as the X-ray structure clearly implicates the serine hydroxyl in hydrogen bonding to the quinone carbonyl (Michel et al., 1986b), and in *Rb. sphaeroides* the single mutations, $Ser^{L223} \rightarrow$ Pro or $Ser^{L223} \rightarrow$ Ala, resulted in much lower quinone binding (Paddock et al., 1990a, 1991). However, in *Rps. viridis* the O5 carbonyl oxygen also appears to hydrogen bond to the peptide NH of L224, which may, therefore, be able to compensate for the loss of the serine. Furthermore, the fact that a single carbonyl is energetically as good as two in the Q_A site suggests that strengthening of the remaining hydrogen bond can compensate for the loss of one carbonyl. Such compensatory actions may be facilitated by the second mutation in this strain, $Arg^{L217} \rightarrow$ His. Difference Fourier analysis of X-ray diffraction data from T1 mutant RCs showed a shift in the Asn^{L213} side chain, allowing it to interact with His^{L217} (Sinning et al., 1990a). The histidine is small enough to rotate into position to hydrogen bond with Asn^{L213}, whereas the native arginine may be unable to. Thus, the second mutation may help maintain some structural integrity of the Q_B site, and the resulting minor structural alterations may contribute to the surprising increase in the affinity of Q-9 in this mutant.

None of these mutants showed marked effects ($\leq$ 3-fold) on the rate of $P^+Q_AQ_B^-$ backreaction, indicating similar values for the one-electron transfer equilibrium constant, K_{AB}. However, in the T1 mutant the pH dependence of the $P^+Q_B^-$ charge recombination revealed a novel p$K \approx 8.3$ influencing the energetics of the Q_A to Q_B electron transfer. This is presumably due to the mutated histidine at L217. The protonated histidine effectively stabilized Q_B^- by about 30 meV (Baciou et al., 1991), suggesting that Arg^{L217} (p$K > 11$) may contribute to the stabilization of Q_B^- in Wt RCs. As with the pK of Glu^{L212} in Wt *Rb. sphaeroides* RC, this pK is higher than the textbook value for histidine of $\approx$ 6.5.

Atrazine-selected mutants of *Rps. viridis* have also been reported (Ewald et al., 1990). One of these (MAV5) also contained the $Arg^{L217} \rightarrow$ His mutation in a double mutant combination with $Val^{L220} \rightarrow$ Leu. In the absence of an $Arg^{L217} \rightarrow$ His single mutation, comparison of the MAV5 and T1 mutants allows some extraction of the effects of the Ser^{L223} mutation, assuming that the $Val^{L220} \rightarrow$ Leu substitution does not have severe effects itself. Leibl et al. (1992) found the rate of the first electron transfer from Q_A^- to Q_B at pH 7.0 to be slightly (2–10 ×) slower in both mutants compared to Wt, while the second electron transfer was drastically impaired only in the T1 mutant (700-fold at pH 7.0, > 10^4-fold at pH 10.0). This is consistent with the behavior reported for the $Ser^{L223} \rightarrow$ Ala mutation in *Rb. sphaeroides* (Paddock et al., 1990a), and indicates that the effects of the $Arg^{L217} \rightarrow$ His and $Val^{L226} \rightarrow$ Leu substitutions on electron transfer are slight.

Mutants containing the $Phe^{L216} \rightarrow$ Ser mutation (T3 and T6) are likely to have altered herbicide-binding properties arising from the loss of hydrophobic interaction with Phe^{L216}, in addition to the change in the polarity of the Q_B pocket. In the T3 double mutant ($Phe^{L216} \rightarrow$ Ser + $Val^{M263} \rightarrow$ Phe), the redox energies of Q_A/Q_A^- and Q_B/Q_B^- were both altered (Baciou et al., 1991). The $P^+Q_A^-$ recombination rate was accelerated six- to sevenfold due to a smaller activation free energy via the $P^+Bph_A^-$ state. Because of the location of the second-site mutation in the Q_A pocket, this was ascribed to an effect on Q_A^- rather than Bph_A^-. Somewhat conflictingly, the altered pH dependence of this reaction in T3 mutant RCs was interpreted as implying a greater influence of Bph_A^- on an ionizable residue that was more affected by Q_A^- in the Wt.

In the T3 mutant, the Q_B site mutation, $Phe^{L216} \rightarrow$ Ser, caused a substantial stabilization of Q_B^- so that the net effect was a somewhat slower $P^+Q_B^-$ recombination rate compared to the Wt (Baciou et al., 1991). Assuming that the $P^+Bph_A^-$ level is unchanged, the effects of the mutations were to destabilize $P^+Q_A^-$ by 50 meV, stabilize $P^+Q_B^-$ by 40 meV, and increase the free energy drop between $Q_A^-Q_B$ and $Q_AQ_B^-$ by 90 meV.

The apparent harmony in the assessment of the T3 mutant is spoiled by the fact that the T6 single mutant ($Phe^{L216} \rightarrow$ Ser) exhibits quite different affinities for inhibitor binding (Sinning et al., 1989a). T6 is at least 10 times more resistant to terbutryn and *o*-phenanthroline than T3. Although this is based on I_{50} values, these were determined at a quinone concentration close to half saturation (Q_{50}), as recommended by Shopes and Wraight (1987b) and Wraight and Shopes (1989), making them directly comparable. An effect of the M263 mutation on the Q_B pocket is implied by this comparison.

The T4 mutant has attracted special interest. The mutation, $Tyr^{L222} \rightarrow$ Phe, eliminates a hydrogen bond between the hydroxyl group of the Tyr^{L222} side chain and the backbone carbonyl of Asp^{M43} (*R.s.* Asn^{M44}) causing substantial structural disturbances in the Q_B pocket (Sinning et al., 1990b). T4 mutant RCs were found to be very sensitive to the PS II herbicide DCMU (I_{50} = 8 μM), and somewhat sensitive to ioxynil (I_{50} = 100 μM), while Wt RCs showed high resistance (I_{50} > 10 mM) to these two herbicides (Sinning et al., 1989b). This is correlated with the global shift of the *de* loop, allowed by the loss of the Tyr^{L222}-Asp^{M43} hydrogen bond which loosens the Q_B pocket; the DCMU phenyl ring packs nearly parallel to Phe^{L216} (Sinning et al., 1990b). Also, the Phe^{L222} side chain rotates about 90° and would interfere with the *t*-butyl group of terbutryn. However, at the present level of resolution it is not really clear why these changes facilitate DCMU binding and decrease the affinity for *o*-phenanthroline, and also why the wild type cannot bind DCMU.

From the point of view of the bacterial RC as a model for PS II, the enhanced affinity of T4 mutant RCs for DCMU is considered suggestive, along with the initial observation that the Q_B^- EPR signal bears some resemblance to that of PS II (Sinning et al., 1989a). The mutant Q_A^- EPR signal was also substantially enhanced by the binding of terbutryn or *o*-phenanthroline in the Q_B pocket, somewhat as seen for the PS II Q_A^- signal (Rutherford et al., 1984), but quite different from Wt RCs. However, the mutant Q_A^- EPR signal was also largely unaffected by DCMU (Sinning et al., 1989b).

In spite of the significant changes in binding properties and structure of the Q_B site in T4 mutant RCs, the pH dependence of $P^+Q_B^-$ charge recombination rate constants for the T4 mutant and the Wt are very similar, suggesting that the charge distribution and electrostatic interactions around Q_B are not affected drastically (Baçiou et al., 1991). Similar findings were reported for a $Tyr^{L222} \rightarrow$ Gly mutation in *Rb. sphaeroides* with Q-10 as Q_B (Paddock et al., 1988, 1991). The Q_A^- to Q_B electron transfer rate was reduced by a factor of 4, but the $P^+Q_AQ_B^-$ charge recombination rates and equilibrium constants were unaffected, suggesting similar reductions in Q_B and Q_B^--binding affinities. This is a reasonable consequence of the type of structural disturbances caused by disruption of the hydrogen bond between Tyr^{L222} and Asn^{M44}, similar to those seen for the equivalent mutation in *Rps. viridis* (Sinning et al., 1990b).

In a $Tyr^{L222} \rightarrow$ Phe mutant of *Rb. sphaeroides*, Takahashi et al. (1990) found an altered pH dependence of binding of Q-0 (2,3-dimethoxy-5-methyl-1,4-benzoquinone, a water soluble analogue of the native ubiquinone, Q-10) as Q_B with p$K \approx$ 7.8, but little effect on the intrinsic one-electron transfer equilibrium (at saturating quinone concentrations). In wild type (R26) RCs, the pK for Q-10 binding is about 9.8 (C.A. Wraight, unpublished observations)—very similar to the pK for electron transfer attributed to Glu^{L212} (Paddock et al., 1989; Takahashi and Wraight, 1992a). Possibly the ionization state of Glu^{L212} governs both quinone binding and the electron transfer equilibrium at alkaline pH. Since the electron transfer pK is not greatly changed in the mutant RCs, this would suggest that Glu^{L212} in the mutant has a pK of 7.8 when the Q_B pocket is empty, but 9.8 when it is occupied. In the Wt, on the other hand, the pK would be similar (9.6–9.8) regardless of occupancy, perhaps because of a firmer pocket structure provided by the hydrogen bond between Tyr^{L222} and Asn^{M44}.

Paddock et al. (1988) analyzed RCs from three spontaneous terbutryn-resistant *Rb. sphaeroides* mutants identified by sequence analysis as[†]: IM(L229) = $Ile^{L229} \rightarrow$ Met, SP(L223) = $Ser^{L223} \rightarrow$ Pro, and YG(L222) = $Tyr^{L222} \rightarrow$ Gly. All three mutants displayed reduction in the binding of Q-0. The electron transfer rates for SP(L223) and YG(L222) mutant RCs were decreased at saturating levels of Q-0, while IM(L229) RCs showed a slight increase compared to the Wt. All three mutants were more resistant to terbutryn and *o*-phenanthroline compared to Wt RCs, but their relative degree of resistance for the two herbicides differed considerably. The Ser^{L223} and Tyr^{L222} mutants were more resistant to terbutryn than the Ile^{L229} mutant, while all three mutants were similarly resistant to *o*-phenanthroline. However, when the native ubiquinone, Q-10, was used for herbicide binding studies on the three mutants, the potential structure–function correlations became less clear, especially as the meaning of "resistance" was not clearly defined (Paddock et al., 1991). With Q-10 as Q_B, the order of resistance to terbutryn was: IM(L229) ≈ SP(L223) > YG(L222) > Wt. This is quite different from the order observed with Q-0 as Q_B SP(L223) ≈ YG(L222) > IM(L229) > Wt. In the case of *o*-phenanthroline, only the mutant IM(L229) was more resistant than the Wt, while the other two mutants were actually more sensitive. In contrast, with Q-0 as Q_B, the order was: YG(L222) > IM(L229) > SP(L223) > Wt. Since these mutants were selected for terbutryn resistance, it is not surprising that not all mutants are resistant to *o*-phenanthroline. However, Ile^{L229} makes strong van der Waals contact with *o*-phenanthroline (Michel et al., 1986b), and decreased affinity due to mutation of Ile^{L229} is expected.

The differences in herbicide binding in the presence of Q-0 or Q-10 presumably arise from the effect of the isoprene tail on Q-10 binding *relative* to Q-0 binding. Resistance, it must be remembered, arises when a mutation diminishes herbicide binding *relative* to quinone binding. Thus, the switching of positions of the YG(L222) and IM(L229) mutants when comparing terbutryn and *o*-phenanthroline resistance with Q-0 or Q-10 is fully consistent with the higher affinity for Q-0 in YG(L222) than in IM(L229) (Paddock et al., 1988, 1991). In the Q_A site of Wt (R26) RCs, the isoprene side chain appears to weaken the head-group binding, presumably realigning it, even though the net binding is enhanced (McComb et al., 1990). If a similar effect occurs in the Q_B site, the tail of Q-10 may position the quinone head group so as to make less contact with L229, thus desensitizing quinone binding to mutations at this residue and leading to greater terbutryn resistance with Q-10 relative to Q-0. Similarly, the enhanced sensitivity of the Ser^{L223} and Tyr^{L222} mutants to *o*-phenanthroline, when Q-10 is used, might arise if the Q-10 head group experiences more steric hindrance than Q-0 in the distal (Ser^{L223}) end of the mutant Q_B pocket, while *o*-phenanthroline binding is less altered. This is weakly supported by X-ray structural analysis of the T4 mutant ($Tyr^{L222} \rightarrow$ Phe) in *Rps. viridis* where a relatively global structural change is seen, resulting in a narrower Q_B pocket with enhanced steric interactions at the distal-end of the domain (Sinning et al., 1990b).

The mutation $Ser^{L223} \rightarrow$ Pro resulted in a twofold increase in the charge recombination rate for $P^+Q_B^-$, indicating a smaller free energy gap between $Q_A^-Q_B$ and $Q_AQ_B^-$ (Paddock et al., 1991). Since Ser^{L223} is hydrogen bonded to the quinone carbonyl group, the loss of this hydrogen bond may weaken Q_B binding in the mutant. If this effect were more pronounced for Q_B^- than for Q_B, the effect would be to lower the E_m of Q_B/Q_B^-, leading to the observed behavior of these mutant RCs. However, when Ser^{L223} was altered to Ala by site-directed mutagenesis, the equilibrium between $Q_A^-Q_B$ and $Q_AQ_B^-$ was not affected (Paddock et al., 1990a). Thus, the behavior of the SP(L223) mutant RC is not due to the specific loss of the serine hydroxyl. However, alternative hydrogen-bonding patterns to the main chain peptide nitrogens may also be lost due structural disturbances induced by the proline substitution.

Mutation of Ile^{L229} to Met caused the states $Q_A^-Q_B$ and $Q_AQ_B^-$ to be isoenergetic. This implies an even more distinct effect on Q_B^- binding compared to Q_B, and could arise, in part, from a shift of the quinone ring away from the divalent iron atom of the acceptor complex, due to steric interaction with the bulkier methionine side chain.

The Nonheme Iron (Fe^{II}) Ligands

The ferrous iron atom lies between Q_A and Q_B in asymmetric octahedral coordination provided by four histidines (L190, L230, M219, M266) and a glutamate (M234), which is bidentate (see Figures 3, 4, and 6). Two of the histidines, M219 and L190, provide hydrogen bonding to the carbonyl O2 of Q_A and Q_B, respectively. The iron atom is known not to be involved electronically in electron transfer from Q_A to Q_B (Debus et al., 1986; Buchanan and Dismukes, 1987), but it does modify the redox properties of the quinones and provides stability to the RC complex. Several divalent transition-metal ions can function in this position, and in some strains Mn appears to be preferentially bound (Rutherford et al., 1985). In the absence of any metal coordination, Q_A is susceptible to double reduction (Debus et al., 1986), implying enhanced accessibility to protons to promote reduction to the QH_2 state. Also, the rate of Q_A reduction by Bph_A^- is dramatically slowed (Kirmaier et al., 1986). The reason for this is unknown, but perhaps the simplest explanation is a structural perturbation that increases the distance involved.

Interest in the iron atom has been sparked by the apparent differences in its properties in bacterial RCs compared to PS II. In bacteria, the iron is not oxidizable at any potentials so far achieved, whereas in PS II it has been equated with an anomalous electron acceptor activity, with $E_{m,7} = 400$ mV, long known as Q_{400} (see Diner and Petrouleas, 1987). The physiological significance of this reversible oxidation–reduction is unknown, but it brings some characteristic variety to PS II behavior!

Also commonly charged to the iron in PS II is a central role in the so-called bicarbonate effect—an apparent dependence on bicarbonate for full activity of the quinone acceptor complex of PS II (see reviews by Blubaugh and Govindjee, 1988; Diner et al., 1991b). In fact, it has been reported that full activity is seen in the absence of bicarbonate provided that certain other inhibitory anions are not present, implying that the major part of the bicarbonate effect is to reverse this inhibition (Jursinic and Stemler, 1992). The phenomenology of the bicarbonate effect is diverse and complex, and at least two roles, and possibly two sites of action, have been ascribed to it: as a ligand to the iron atom, and as a participant in proton delivery to Q_B.

In the bacterial RC, where the iron is fully coordinated, no bicarbonate effects are seen (Shopes et al., 1989) and the iron cannot be oxidized. It was therefore suggested that the glutamate ligand (M232 in *R.c.* and *R.v.*, M234 in *R.s.*) might be missing in PS II, opening a position for bicarbonate to act (Michel and Deisenhofer, 1989). Action at, or very close to, the iron atom is evident from the significant effects that bicarbonate has on the EPR signals of Q_A^- and Q_B^- which are magnetically coupled to the iron (Vermaas and Rutherford, 1984; Hallahan et al., 1991; Nugent et al., 1992), and especially its ability to displace NO from the iron atom (Diner et al., 1991b). However, the analogy with Glu^{M234} has not been supported by mutation studies on the *Rb. sphaeroides* RC (Wang et al., 1992). Replacement of Glu^{M234} by glutamine, valine, or glycine had no significant effects on any functional aspects of the quinone complex and, even more surprising, had no effect on the Fe-Q_A^- EPR signal from pH 6 to 10. Minor perturbations of the rate of electron transfer from Bph_A^- to Q_A were observed. In view of the disturbance of the charge balance in this region caused by these substitutions, such passivity is quite unexpected. In a similar study, Williams et al. (1991) have substituted Glu^{M234} with aspartic acid. At neutral pH this mutant exhibited little difference form the wild type, but more marked changes were apparent at high pH (>10).

Preliminary results on the mutagenesis of some of the histidine ligands to the iron have also been reported (Williams et al., 1991). In contrast to the small effect upon mutation of the glutamate, substitution of any of three of the histidine ligands, L190, M219, or M266 by cysteine and/or glutamine caused rather dramatic effects including loss of the iron. The behavior of these mutants was very similar to that of iron-depleted Wt RCs described by Debus et al. (1986). In particular, Q_A^- to Q_B electron transfer was 2 to 3 times slower. Intriguingly, Q_A^- to Q_B electron transfer assayed as an electrochromic response of the RC pigments, especially Bph_A (Verméglio, 1977; Shopes and Wraight, 1985), appeared to be insensitive to herbicides, whereas other assays of quinone function, such as the rate of $P^+(Q_AQ_B)^-$ recombination, or cytochrome oxidation under continuous illumination, were sensitive. This is reminiscent of the small, flash-induced electrical transient described by Brzezinski et al. (1992), associated with Glu^{L212} but seen even in the absence of Q_B. Clearly some caution must be exercised in interpreting behavior based on indirect assays, like electrochromism.

V. CONCLUSIONS

In the last two years there has been an explosion of activity based on site-directed mutations of bacterial reaction centers from *Rb. sphaeroides* and *capsulatus*. In the very near future we can also expect the pace to pick up with *Rps. viridis* as working expression systems are now being developed (Laußermair and Oesterhelt, 1992). In some cases, the effects of the mutations have been dramatic and unexpected, including the heterodimer mutants, substitution of Bph_A by Bchl, and, at opposite extremes of functional disturbance, mutation of Asp^{L213} and of Glu^{M234}. Part of the unexpectedness of these and many other mutations is the capacity of the RC to accommodate radical changes, to assemble, and, to a large extent, to function. Many of these first-generation mutations have focussed on residues of obvious potential, requiring only rather superficial inspection of the structure and reaping rewards almost disproportionate to the effort! And, indeed, the rewards have been great. However, the future probably belongs to more thoughtful approaches, of which perhaps the first example is the carefully tailored mutants of Williams et al. (1992), for modifying the hydrogen bonding around P to perturb the redox properties and energetics of the intermediate states in quite subtle ways, thereby leaving the mechanism essentially intact.

NOTES

#Amino acid residues will be numbered according to the species in question, or according to the *Rb. sphaeroides* sequence when generalized. The numbering for homologous amino acid residues in other bacterial species will also be noted where appropriate (*R.s.* = *Rb. sphaeroides*, *R.v.* = *Rps. viridis*, *R.c.* = *Rb. capsulatus*, *C.a.* = *C. aurantiacus*, *R.r.* = *Rs. rubrum*).

##The primary donor of the heterodimer mutants will be designated D, c.f., P for the wild-type homodimer.

†For this section, abbreviations and acronyms for specific mutant strains will be given following the original authors. However, our preferred notation, for mutants of known sequence, is, e.g., L213DN, indicating L-subunit residue 213 altered from aspartic acid (D) to asparagine (N). This is very easily adapted to accommodate more complex notation for protein subunits, as in photosystem II: D1-264SA, i.e., D1-subunit residue 264 altered from serine (S) to alanine (A).

ABBREVIATIONS

$Bchl_A$, $Bchl_B$: monomer bacteriochlorophylls associated with the A and B branches, respectively, of reaction center cofactors.

Bchl: bacteriochlorophyll substituted for bacteriopheophytin in certain mutant reaction centers, due to the effects of site-directed mutagenesis.

Bph_A, Bph_B: bacteriopheophytins associated with the A and B branches, respectively, of reaction center cofactors.

Bph: bacteriopheophytin substituted for bacteriochlorophyll in certain mutant reaction center is due to the effects of site-directed mutagenesis.

P: "native" primary electron donor of the reaction center, a 'special pair' or dimer of bacteriochlorophyll; P_A, P_B—the monomer bacteriochlorophylls of the special pair, associated with the A or B branch; P^*—excited singlet state of the primary donor.

D: primary electron donor of certain mutant reaction centers composed of bacteriochlorophyll-bacteriopheophytin heterodimer as result of site-directed mutation; D^* —excited singlet state of the heterodimer primary donor, $D^{\pm}$ intradimer charge transfer state of the heterodimer primary donor.

Q_A, Q_B: primary and secondary quinones, respectively, of the reaction center; Q-0, Q-10, etc: ubiquinone-0 and ubiquinone-10, indicating the number of isoprene units in the side chain.

Q, Q^-, QH_2: quinone, (anionic) semiquinone and quinol forms of ubiquinone.

L, M, H: the core protein subunits of bacterial reaction centers, originally designated "light", "medium", and "heavy" on the basis of mobility in SDS polyacrylamide gel electrophoresis.

D1, D2: the core protein subunits of photosystem II reaction centers.

RC: reaction center.

Wt: wild type.

PS II: photosystem II of oxygenic photosynthesis.

DCMU: dichlorophenyl-dimethyl-urea.

I_{50}, Q_{50}: concentrations of inhibitor (I) or quinone (Q) that give half maximal inhibition of activity, respectively.

REFERENCES

Allen, J.P. & Feher, G. (1991). Crystallization of reaction centers from *Rhodobacter sphaeroides*. In: Crystallization of Membrane Proteins (Michel, H., Ed.), pp. 137–153. CRC Press, Boca Raton.

Allen., J.P., Feher, G., Yeates, T.O., Komiya, H., & Rees, D.C. (1987a). Structure of the reaction center from *Rhodobacter sphaeroides* R-26: the cofactors. Proc. Natl. Acad. Sci. USA 84, 5730–5734.

Allen., J.P., Feher, G., Yeates, T.O., Komiya, H., & Rees, D.C. (1987b). Structure of the reaction center from *Rhodobacter sphaeroides* R-26: the protein subunits. Proc. Natl. Acad. Sci. USA 84, 6162–6166.

Allen, J.P., Feher, G., Yeates, T.O., Komiya, H., & Rees, D.C. (1988a). Structure of the reaction center from *Rhodobacter sphaeroides* R-26 and 2.4.1. In: The Photosynthetic Bacterial Reaction Center (Breton, J. & Vermeglio, A., Eds.), pp. 5–11. Plenum, New York.

Allen, J.P., Feher, G., Yeates, T.O., Komiya, H., & Rees, D.C. (1988b). Structure of the reaction center from *Rhodobacter sphaeroides* R-26: protein-cofactor (quinones and Fe^{2+}) interactions. Proc. Natl. Acad. Sci. USA 85, 8487–8491.

Arnoux, B., Ducruix, A., Astier, C., Picaud, M., Roth, M., & Reiss-Husson, F. (1990). Towards the understanding of the function of *Rb sphaeroides* Y wild type reaction center: gene cloning, protein and detergent structures in the three-dimensional crystals. Biochimie 72, 525–530.

Baciou, L., Sinning, I., & Sebban, P. (1991). Study of Q_B^- stabilization in herbicide-resistant mutants from the purple bacterium *Rhodopseudomonas viridis*. Biochemistry 30, 9110–9116.

Baciou, L., Bylina, E.J., & Sebban, P. (1993). Study of wild type and genetically modified reaction centers from *Rhodobacter capsulatus*: structural comparison with *Rhodopseudomonas viridis* and *Rhodobacter sphaeroides*. Biophys. J. 65, 652–660.

Barber, J. (1987). Photosynthetic reaction centres: a common link. Trends Biochem. Sci. 12, 321–326.

Becker, M., Nagarajan, V., Middendorf, D., Parson, W.W., Martin, J.E., & Blankenship, R.E. (1991). Temperature dependence of the initial electron transfer kinetics in photosynthetic reaction centers of *Chloroflexus aurantiacus*. Biochim. Biophys. Acta 1057, 299–312.

Bélanger, G., Bérard, J., Corriveau, P., & Gingras, G. (1988). The structural genes coding for the L- and M-subunits of *Rhodospirillum rubrum* photoreaction center. J. Biol. Chem. 263, 7632–7638.

Bixon, M., Jortner, J., & Michel-Beyerle, M.-E. (1991). On the mechanism of the primary charge separation in bacterial photosynthesis. Biochim. Biophys. Acta 1056, 301–315.

Blankenship, R.E., Feick, R., Bruce, B.D., Kirmaier, C., Holten, D., & Fuller, R.C. (1983). Primary photochemistry in the facultative green photosynthetic bacterium *Chloroflexus aurantiacus*. J. Cell Biochem. 22, 251–261.

Blubaugh, D.J. & Govindjee (1988). The molecular mechanism of the bicarbonate effect at the plastoquinone reductase site of photosynthesis. Photosynthesis Res. 19, 85–128.

Bocian, D.F., Boldt, N.J., Chadwick, B.W., & Frank, H.A. (1987). Near-infrared-excitation resonance Raman spectra of bacterial photosynthetic reaction centers. Implications for path-specific electron transfer. FEBS Lett. 214, 92–96.

Boxer, S. G. (1990). Mechanisms of long-distance electron transfer in proteins: Lessons from photosynthetic reaction centers. Annu. Rev. Biophys. Biophys. Chem. 19, 267–299.

Breton, J., Martin, J.-L., Migus, A., Antonetti, A., & Orszag, A. (1986). Femtosecond spectroscopy of excitation energy transfer and initial charge separation in the reaction center of the photosynthetic bacterium *Rhodopseudomonas viridis*. Proc. Natl. Acad. Sci. USA 83, 5121–5125.

Breton, J., Martin, J.-L., Fleming, G.R., & Lambry, J.-C. (1988). Low-temperature femtosecond spectroscopy of the initial step of electron transfer in reaction centers from photosynthetic purple bacteria. Biochemistry 27, 8276–8284.

Breton, J., Bylina, E.J., & Youvan D.C. (1989). Pigment organization in genetically modified reaction centers of *Rhodobacter capsulatus*. Biochemistry 28, 6423–6430.

Breton, J., Martin J.-L., Lambry, J.-C., Robles, S.J., & Youvan, D.C. (1990). Ground state and femtosecond transient absorption spectroscopy of a mutant of *Rhodobacter capsulatus* which lacks the initial electron acceptor bacteriopheophytin. In: Reaction Centers of Photosynthetic Bacteria (Michel-Beyerle, M.-E., Ed.), pp. 293–302. Springer-Verlag, Berlin, New York.

Brown, A.E., Gilbert, C W., Guy, R., & Arntzen, C.J. (1984). Triazine herbicide resistance in the photosynthetic bacterium *Rhodopseudomonas sphaeroides*. Proc. Natl. Acad. Sci. USA 81, 6310–6314.

Brzezinski, P., Paddock, M.L., Rongey, S.H., Okamura, M.Y., & Feher, G. (1991a). Electrogenicity associated with proton uptake by Glu-L212 and Asp-L213 upon forming Q_B^- in bacterial RCs. Biophys. J. 59, 143a.

Brzezinski, P., Paddock, M.L., Rongey, S.H., Okamura, M.Y., & Feher, G. (1991b). Electrogenicity associated with proton uptake by Q_B^{2-} in bacterial RCs. Biophys. J. 59, 143a.

Brzezinski, P., Paddock, M.L., Messinger, A., Okamura, M.Y., & Feher, G. (1992). Electrogenic change after the formation of $D^+Q_A^-$ in bacterial reaction centers. Biophys. J. 61, A101.

Buchanan, S.K. & Dismukes, G.C. (1987). Substitution of Cu^{2+} in the reaction center diquinone electron acceptor complex of *Rhodobacter sphaeroides*: Determination of the metal-ligand coordination. Biochemistry 26, 5049–5055.

Bylina, E.J. & Youvan, D.C. (1987). Genetic engineering of herbicide resistance: Saturation mutagenesis of isoleucine 229 of the reaction center L subunit. Z. Naturforsch. 42c, 769–774.

Bylina, E.J. & Youvan, D.C. (1988). Directed mutations affecting spectroscopic and electron transfer properties of the primary donor in the photosynthetic reaction center. Proc. Natl. Acad. Sci. USA 85, 7226–7230.

Bylina, E.J., Kirmaier, C., McDowell, L., Holten, D., & Youvan D.C. (1988). Influence of an amino acid residue on the optical properties and electron transfer dynamics of a photosynthetic reaction center complex. Nature (London) 336, 182–184.

Bylina, E.J., Jovine, R.V.M., & Youvan, D.C. (1989). A genetic system for rapidly assessing herbicides that compete for the quinone binding site of photosynthetic reaction centers. Bio/technology 7, 69–74.

Bylina, E.J., Kolaczkowski, S.V., Norris, J.R., & Youvan, D.C. (1990). EPR characterization of genetically modified reaction centers of *Rhodobacter capsulatus*. Biochemistry 29, 6203–6210.

Chambers, J.Q. (1974). Electrochemistry of quinone. In: The Chemistry of the Quinoid Compounds (Patai, S., Ed.), pp. 737–792, Wiley Interscience, New York.

Chan, C.-K., DiMagno, T.J., Chen, L.X.-Q., & Norris, J.R. (1991a). Mechanism of the initial charge separation in bacterial photosynthetic reaction centers. Proc. Natl. Acad. Sci. USA 88, 11202–11206.

Chan, C.-K., Chen, L.X.-Q., DiMagno, T.J., Hanson, D.K., Nance, S.L., Schiffer, M., Norris, J.R., & Fleming., G.R. (1991b). Initial electron transfer in photosynthetic reaction centers of *Rhodobacter capsulatus* mutants. Chem. Phys. Lett. 176, 366–372.

Chang, C.-H., El-Kabbani, O., Tiede, D., Norris, J., & Schiffer, M. (1991). Structure of the membrane-bound protein photosynthetic reaction center from *Rhodobacter sphaeroides*. Biochemistry 30, 5352–5360.

Coleman, W.J. & Youvan, D.C. (1990). Spectroscopic analysis of genetically modified photosynthetic reaction centers. Annu. Rev. Biophys. Biophys. Chem. 19, 333–367.

Coleman, W.J., Bylina, E.J., & Youvan, D.C. (1990a). Reconstitution of photochemical activity in *Rhodobacter capsulatus* reaction centers containing mutations at tryptophan M-250 in the primary quinone binding site. In: Current Research in Photosynthesis (Baltscheffsky, M., Ed.), Vol. I, pp. 149–152. Kluwer Academic Publishers, Dordrecht, Netherlands.

Coleman, W.J., Bylina, E.J., Aumeier, W., Siegl, J., Eberl, U., Heckmann, R., Ogrodnik, A., Michel-Beyerle, M.-E., & Youvan D.C. (1990b). Influence of mutagenic replacement of tryptophan M250 on electron transfer rates involving primary quinone in reaction centers of *Rhodobacter capsulatus*. In: Reaction Centers of Photosynthetic Bacteria (Michel-Beyerle, M.-E., Ed.), pp. 273–282. Springer-Verlag, Berlin, New York.

Creighton, S., Hwang, J.-K., Warshel, A., Parson, W.W., & Norris, J. (1988). Simulating the dynamics of the charge separation process in bacterial photosynthesis. Biochemistry 27, 774–781.

Crofts, A.R. & Wraight, C.A. (1983). The electrochemical domain of photosynthesis. Biochim. Biophys. Acta 726, 149–185.

Davis, D., Dong, A., Caughey, W.S., & Schenck, C.C. (1992). Energetics of the oxidized primary donor in WT and heterodimer mutant reaction centers. Biophys. J. 61, A153.

Debus, R.J., Feher, G., & Okamura, M.Y. (1986). Iron-depleted reaction centers from *Rhodopseudomonas sphaeroides* R-26.1: Characterization and reconstitution with Fe^{2+}, Mn^{2+}, Co^{2+}, Ni^{2+}, Cu^{2+} and Zn^{2+}. Biochemistry 25, 2276–2287.

Deisenhofer, J., Epp, O., Huber, R., & Michel, H. (1985). Structure of the protein subunits in the photosynthetic reaction center of *Rhodopseudomonas viridis* at 3 Å resolution. Nature 318, 618–624.

Deisenhofer, J. & Michel, H. (1989a). The photosynthetic reaction center from the purple bacterium *Rhodopseudomonas viridis*. EMBO J. 8, 2149–2169.

Deisenhofer, J. & Michel, H. (1989b). High-resolution structures of photosynthetic reaction centers. Annu. Rev. Biophys. Chem. 20, 247–266.

DiMagno, T.J., Bylina, E.J., Angerhofer, A., Youvan, D.C., & Norris, J.R. (1990). Stark effect in wild-type and heterodimer-containing reaction centers from *Rhodobacter capsulatus*. Biochemistry 29, 899–907.

Diner, B.A. & Petrouleas, V. (1987). Q_{400}, the non-heme iron of the Photosystem II iron-quinone complex. A spectroscopic probe of quinone and inhibitor binding to the reaction center. Biochim. Biophys. Acta 895, 107–125.

Diner, B.A., Nixon, P.J., & Farchaus, J.W. (1991a). Site-directed mutagenesis of photosynthetic reaction centers. Curr. Opinion Struct. Biol. 1, 546–554.

Diner, B.A., Petrouleas, V., & Wendoloski, J.J. (1991b). The iron-quinone electron-acceptor complex of photosystem II. Physiol. Plant. 81, 423–436.

Donohue, T.J., Hoger, J.H., & Kaplan, S. (1986). Cloning and expression of the *Rhodobacter sphaeroides* reaction center H gene. J. Bacteriol. 168, 953–961.

Drachev, L.A., Mamedov, M.D., Mulkidjanian, A.Ya., Semenov, A. Yu., Shinkarev, V.P., & Verkhovsky, M.I. (1990). Electrogenesis associated with proton transfer in the reaction center protein of the purple bacterium *Rhodobacter sphaeroides*. FEBS Letts 259, 324–326.

Du, M., Rosenthal, S.J., Xie, X., DiMagno, T.J., Schmidt, M., Hanson, D.K., Schiffer, M., Norris, J.R., & Fleming, G.R. (1992). Femtosecond spontaneous-emission studies of reaction centers from photosynthetic bacteria. Proc. Natl. Acad. Sci. USA 89, 8517–8521.

El-Kabbani, O., Chang, C.-H., Tiede, D., Norris, J., & Schiffer, M. (1991). Comparison of reaction centers from *Rhodobacter sphaeroides* and *Rhodopseudomonas viridis*: Overall architecture and protein-pigment interactions. Biochemistry 30, 5361–5369.

Ewald, G., Wiessner, C., & Michel, H. (1990). Sequence analysis of four atrazine-resistant mutants from *Rhodopseudomonas viridis*. Z. Naturforsch. 45c, 459–462.

Fajer, J., Brune, D.C., Davis, M.S., Forman, A., & Spaulding, L.D. (1975). Primary charge separation in bacterial photosynthesis: Oxidized chlorophylls and reduced pheophytin. Proc. Natl. Acad. Sci. USA 72, 4956–4960.

Farchaus, J.W. & Oesterhelt, D. (1989). A *Rhodobacter sphaeroides pufL, M* and *X* deletion mutant and its complementation in *trans* with a 5.3 kb *puf* operon shuttle fragment. EMBO J. 8, 47–54.

Feher, G. & Okamura, M.Y. (1984). Structure and function of the reaction center from *Rhodopseudomonas sphaeroides*. In: Advances in Photosynthesis Research (Sybesma, C., Ed.), Vol. II, pp. 155–164, Martinus Nijhoff, Dordrecht.

Feher, G., Isaacson, R.A., & Okamura, M.Y. (1988). ENDOR of exchangeable protons of the reduced intermediate acceptor in reaction centers from *Rhodobacter sphaeroides* R-26. In: The Photosynthetic Bacterial Reaction Center: Structure and Dynamics (Breton, J. & Verméglio, A., Eds.), pp. 229–235. Plenum, New York.

Feher, G., Allen, J.P., Okamura, M.Y., & Rees, D.C. (1989). Structure and function of bacterial photosynthetic reaction centers. Nature 339, 111–116.

Feick, R., Martin, J.L, Breton, J., Volk, M., Scheidl, G., Langenbacher, T., Urbano, C., Ogrodnik, A., & Michel-Beyerle, M.-E. (1990). Biexponential charge separation and monoexponential decay of P^+H^- in reaction centers of *Chloroflexus aurantiacus*. In: Reaction Centers of Photosynthetic Bacteria (Michel-Beyerle, M.-E., Ed.), Springer Series in Biophysics, Vol. 6, pp. 181–188. Springer-Verlag, Berlin, New York.

Finkele, U., Lauterwasser, C., Zinth, W., Gray, K.A., & Oesterhelt, D. (1990). Role of tyrosine M210 in the initial charge separation of reaction centers of *Rhodobacter sphaeroides*. Biochemistry 29, 8517–8521.

Fleming, G.R., Martin, J.L., & Breton, J. (1988). Rates of primary electron transfer in photosynthetic reaction centers and their mechanistic implications. Nature (London) 333, 190–192.

Friesner, R.A. & Won, Y. (1989). Spectroscopy and electron transfer dynamics of the bacterial photosynthetic reaction center. Biochim. Biophys. Acta 977, 99–122.

Gao, J., Shopes, R.J., & Wraight, C.A. (1991). Heterogeneity of kinetics and electron transfer equilibria in the bacteriopheophytin and quinone electron acceptors of reaction centers from *Rhodopseudomonas viridis*. Biophys. Biochim. Acta 1056, 259–272.

Giangiacomo, K.M., Gunner, M.R., & Dutton, P.L. (1990). Q_A and Q_B site control of quinone electrochemistry in the photosynthetic reaction center from *Rhodobacter sphaeroides*. Biophys. J. 57, 566a.

Gilbert, C.W., Williams, J.G.K., Williams, K.A.L., & Arntzen, C.J. (1985). Herbicide action in photosynthetic bacteria. In: Molecular biology of the photosynthetic apparatus (Steinbeck, K.E., Bonitz, S., Arntzen, C.J., & Bogorad, L., Eds.), pp. 67–71. Cold Spring Harbor Laboratory, Cold Spring Harbor.

Gray, K.A., Farchaus, J.W., Wachtiveitl, J., Breton, J., & Oesterhelt, D. (1990a). Initial characterization of site-directed mutants of tyrosine M210 in the reaction center of *Rhodobacter sphaeroides*. EMBO J. 9, 2061–2070.

Gray, K.A., Farchaus, J.W., Wachtveitl, J., Breton, J., Finkele, U., Lauterwasse, C., Zinth, W., & Oesterhelt, D. (1990b). The role of tyrosine M210 in the initial charge separation in the reaction center of *Rhodobacter sphaeroides*. In: Reaction Centers of Photosynthetic Bacteria (Michel-Beyerle, M.-E., Ed.), Springer Series in Biophysics, Vol. 6, pp. 251–264. Springer-Verlag, Berlin, New York.

Gunner, M.R., Braun, B.S., Bruce, J.M., & Dutton, P.L. (1985). The characterization of the Q_A binding site of the reaction center of *Rhodopseudomonas sphaeroides*. In: Antennas and Reaction Centers of Photosynthetic Bacteria (Michel-Beyerle, M.-E., Ed.), pp. 298–305, Springer-Verlag, Berlin, New York.

Gunner, M.R., Robertson, D.E., & Dutton, P.L. (1986). Kinetic studies on the reaction center protein from *Rhodopseudomonas sphaeroides*: The temperature and free energy dependence of electron transfer between various quinones in the Q_A site and the oxidized bacteriochlorophyll dimer. J. Phys. Chem. 90, 3783–3795.

Gunner, M.R. & Dutton, P.L. (1988). Temperature and $-\Delta G^0$ dependence of the electron transfer from $BPh^{\bullet-}$ to Q_A in reaction center protein from *Rhodobacter sphaeroides* with different quinones as Q_A. J. Am. Chem. Soc. 111, 3400–3412.

Gunner, M.R. (1991). The reaction center protein from purple bacteria: Structure and function. Current Topics in Bioenergetics 16, 319–367.

Hallahan, B.J., Ruffle, S.V., Bowden, S.J., & Nugent, J.H.A. (1991). Identification and characterization of EPR signals involving Q_B semiquinone in plant photosystem II. Biochim. Biophys. Acta 1059, 181–188.

Hammes, S.L., Mazzola, L., Boxer, S.G., Gaul, D.F., & Schenck, C.C. (1990). Stark spectroscopy of the *Rhodobacter sphaeroides* reaction center heterodimer mutant. Proc. Natl. Acad. Sci. USA 87, 5682–5686.

Hanson, D.K., Baciou, L., Tiede, D.M., Nance, S.L. Schiffer, M., & Sebban, P. (1992). In bacterial reaction centers protons can diffuse to the secondary quinone by alternative pathways. Biochim. Biophys. Acta 1102, 260–265.

Holten, D.H., Windsor, M.W., Parson, W.W., & Thornber, J.P. (1978). Primary photochemical processes in isolated reaction centers of *Rhodopseudomonas viridis*. Biochim. Biophys. Acta 501, 112–126.

Holzapfel, W., Finkele, U., Kaiser, W., Oesterhelt, D., Scheer, H., Stilz, H.U., & Zinth, W. (1989). Observation of a bacteriochlorophyll anion radical during the primary charge separation in a reaction center. Chem. Phys. Lett. 160, 1–7.

Holzapfel, W., Finkele, U., Kaiser, W., Oesterhelt, D., Scheer, H., Stilz, H.U., & Zinth, W. (1990). Initial electron transfer in the reaction center from *Rhodobacter sphaeroides*. Proc. Natl. Acad. Sci. USA 87, 5168–5172.

Huber, M., Lous, E.J., Isaacson, R.A., Feher, G., Gaul, D., & Schenck, C.C. (1990). EPR and ENDOR studies of the oxidized donor in reaction centers of *Rhodobacter sphaeroides* Strain R-26 and two heterodimer mutants in which histidine M202 and L173 was replaced by leucine. In: Reaction Centers of Photosynthetic Bacteria (Michel-Beyerle, M.-E., Ed.), Springer Series in Biophysics, Vol. 6, pp. 219–228. Springer-Verlag, Berlin, New York.

Huber, M., Isaacson, R.A., Abresch, E.C., Feher, G., Gaul, D., & Schenck, C.C. (1992). ENDOR of the oxidized primary donor of the heterodimer mutants HL(M202) and HL(L173) in single crystals of reaction centers from *Rb. sphaeroides*. Biophys. J. 61, A104.

Jortner, J. (1980). Dynamics of the primary events in bacterial photosynthesis. J. Am. Chem. Soc. 102, 6676–6686.

Jursinic, P.A. & Stemler, A. (1992). High rates of photosystem II electron flow occur in maize thylakoids when the high-affinity binding site for bicarbonate is empty of all monovalent anions or has bicarbonate bound. Biochim. Biophys. Acta 1098, 359–367.

Kaminskaya, O.P., Drachev, L.A., Konstantinov, A.A., Semenov, A.Yu., & Skulachev, V.P. (1986). Electrogenic reduction of the secondary quinone acceptor in chromatophores of *Rhodospirillum rubrum*. FEBS Lett. 202, 224–228.

Kellog, E.C., Kolaczkowski, S., Wasielewski, M.R., & Tiede, D.M. (1989). Measurement of the extent of electron transfer to the bacteriopheophytin in the M-subunit in reaction centers of *Rhodopseudomonas viridis*. Photosynthesis Res. 22, 47–59.

Keske, J.M. & Dutton, P.L. (1991). Extraction and reconstitution of the Q_A site in *Rhodopseudomonas viridis*: Evidence for Q_A site similarities between *Rhodopseudomonas viridis* and *Rhodobacter sphaeroides*. Biophys. J. 59, 144a.

Keske, J.M., Bruce, J.M., & Dutton, P.L. (1991). A new method of solvation analysis: Applications to quinones. Z. Naturforsch. 45c, 430–435.

Kirmaier, C. & Holten, D. (1987). Primary photochemistry of reaction centers from the photosynthetic purple bacteria. Photosynthesis Res. 13, 225–260.

Kirmaier, C. & Holten, D. (1990). Evidence that a distribution of bacterial reaction centers underlies the temperature and detection-wavelength dependence of the rates of the primary electron-transfer reactions. Proc. Natl. Acad. Sci. USA. 87, 3552–3556.

Kirmaier, C. & Holten, D. (1991). An assessment of the mechanism of initial electron transfer in bacterial reaction centers. Biochemistry 30, 609–613.

Kirmaier, C., Holten, D., Feick, R., & Blankenship, R.E. (1983). Picosecond measurements of the primary photochemical events in reaction centers isolated from the facultative green photosynthetic bacterium *Chloroflexus aurantiacus*. Comparison with the purple bacterium *Rhodopseudomonas sphaeroides*. FEBS Lett. 158, 73–78.

Kirmaier, C., Parson, W.W., & Holten, D. (1985). Temperature and detection-wavelength dependence of the picosecond electron transfer kinetics measured in *Rhodopseudomonas sphaeroides* reaction centers. Resolution of new spectral and kinetic components in the primary charge separation process. Biochim. Biophys. Acta 810, 33–48.

Kirmaier, C., Blankenship, R.E., & Holten, D. (1986). Formation and decay of radical-pair state P^+I^- in *Chloroflexus aurantiacus* reaction centers. Biochim. Biophys. Acta 850, 275–285.

Kirmaier, C., Holten, D., Bylina, E.J., & Youvan, D.C. (1988). Electron transfer in a genetically modified bacterial reaction center containing a heterodimer. Proc. Natl. Acad. Sci. USA 85, 7562–7566.

Kirmaier, C., Bylina, E., Youvan, D.C., & Holten, D. (1989). Subpicosecond formation of the intradimer charge transfer state [$BChl^+_{LP}BPh^-_{MP}$] in reaction centers from the $His^{M200} \rightarrow$ Leu mutant of *Rhodobacter capsulatus*. Chem. Phys. Lett. 159, 251–257.

Kirmaier, C., Gaul, D., DeBey, R. Holten, D., & Schenck, C.C. (1991). Charge separation in a reaction center incorporating bacteriochlorophyll for photoactive bacteriopheophytin. Science 251, 922–927.

Kleinfeld, D., Okamura, M.Y., & Feher, G. (1985). Electron transfer in reaction centers of *Rhodopseudomonas sphaeroides*. II. Free energy and kinetic relations between the acceptor states $Q_A^-Q_B^-$ and $Q_AQ_B^{2-}$. Biochim. Biophys. Acta 809, 291–310.

Kolaczkowski, S.V., Bylina, E.J., Youvan, D.C., & Norris, J.R. (1990). Examination of the *Rhodobacter capsulatus* special pair in wild-type and heterodimer-containing reaction centers by time-resolved optically detected magnetic resonance. In: Molecular Biology of Membrane-Bound Complexes in Phototrophic Bacteria (Drews, G. & Dawes, E.A., Eds.), pp. 305–311. Plenum, New York.

Komiya, H., Yeates, T.O., Rees, D.C., Allen, J.P., & Feher, G. (1988). Structure of the reaction center from *Rhodobacter sphaeroides* R-26 and 2.4.1: Symmetry relations and sequence comparisons between different species. Proc. Natl. Acad. Sci. USA 85, 9012–9016.

Lang, F.S. & Oesterhelt, D. (1989). Gene transfer system for *Rhodopseudomonas viridis*. J. Bacteriol. 171, 4425–4435.

Laußermair, E. & Oesterhelt, D. (1992). A system for site-specific mutagenesis of the photosynthetic reaction center in *Rhodopseudomonas viridis*. EMBO J. 11, 777–783.

Leibl, W., Sinning, I., Ewald, G., Michel, H., & Breton, J. (1992). Initial characterization of the proton transfer pathway to Q_B in *Rps. viridis*: Electron transfer kinetics in herbicide resistant mutants. In: The Photosynthetic Bacterial Reaction Center II: Structure, Spectroscopy, and Dynamics (Breton, J. & Verméglio, A., Eds.), pp. 389–394. Plenum, New York.

Lendzian, F., Huber, M., Plato, M., Bumann, D., Lubitz, W., & Möbius, K. (1990). ENDOR and TRIPLE resonance investigation of the primary donor cation radical P_{865}^+ in single crystals of *Rhodobacter sphaeroides* R-26 reaction centers. In: Reaction Centers of Photosynthetic Bacteria (Michel-Beyerle, M.-E., Ed.), Springer Series in Biophysics, Vol. 6, pp. 57–68. Springer-Verlag, Berlin, New York.

Lendzian, F., Huber, M., Isaacson, R.A., Endeward, B., Plato, M., Bönigk, B., Möbius, K., Lubitz, W., & Feher, G. (1993). The electronic structure of the primary donor cation radical in *Rhodobacter sphaeroides* R26: ENDOR and TRIPLE resonance studies in single crystals of reaction centers. Biochim. Biophys. Acta 1183, 139–160.

Lous, E.J., Huber, M., Isaacson, R.A., & Feher, G. (1990). EPR and ENDOR studies of the oxidized primary donor in single crystals of reaction centers of *Rhodobacter sphaeroides* R-26. In: Reaction Centers of Photosynthetic Bacteria (Michel-Beyerle, M.-E., Ed.), Springer Series in Biophysics, Vol. 6, pp. 45–55. Springer-Verlag, Berlin, New York.

Lubitz, W., Rautter, J., Geßner, C., Lendzian, F., Allen, J.P., Murchison, H.A., Williams, J.C., & Woodbury, N.W. (1992). EPR and ENDOR studies of the primary donor cation radical in native and genetically modified bacterial reaction centers. In: The Photosynthetic Bacterial Reaction Center II: Structure, Spectroscopy, and Dynamics (Breton, J. & Verméglio, A., Eds.), pp. 99–107. Plenum, New York.

Maróti, P., Kirmaier, C., Wraight, C.A., Holten, D., & Pearlstein, R.M. (1985). Photochemistry and electron transfer in borohydride-treated photosynthetic reaction centers. Biochim. Biophys. Acta 810, 132–139.

Maróti, P. & Wraight, C.A. (1988). Flash-induced H^+ binding by bacterial reaction centers: Influences of the redox states of the acceptor quinones and primary donor. Biochim. Biophys. Acta 934, 329–347.

Maróti, P. & Wraight, C.A. (1989). Kinetic correlation between electron transfer and H^+-binding in reaction centers of photosynthetic bacteria *Rb. sphaeroides*. Biophys. J. 55, 182a.

Martin, J.-L., Breton, J., Hoff, A.J., Migus, A., & Antonetti, A. (1986). Femtosecond spectroscopy of electron transfer in the reaction center of the photosynthetic bacterium *Rhodopseudomonas sphaeroides* R-26: Direct electron transfer from the dimeric bacteriochlorophyll primary donor to the bacteriopheophytin acceptor with a time constant of 2.8 ± 0.2 psec. Proc. Natl. Acad. Sci. USA 83, 957–961.

Mattioli, T.A., Gray, K.A., Lutz, M., Oesterhelt, D., & Robert, B. (1991). Resonance Raman characterization of reaction centers bearing site-directed mutations at tyrosine M210. Biochemistry 30, 1715–1722.

McComb, J.C., Stein, R.R., & Wraight, C.A. (1990). Investigations on the headgroup substitution and isoprene side-chain length in the function of primary and secondary quinones of bacterial reaction centers. Biochim. Biophys. Acta 1015, 156–171.

McDowell, L.M., Kirmaier, C., & Holten, D. (1990). Charge-transfer and charge-resonance states of the primary electron donor in wild-type and mutant bacterial reaction centers. Biochim. Biophys. Acta 1020, 239–246.

McDowell, L.M., Gaul, D., Kirmaier, C., Holten, D., & Schenck, C.C. (1991a). Investigation into the source of electron transfer asymmetry in bacterial reaction centers. Biochemistry 30, 8315–8322.

McDowell, L.M., Kirmaier, C., & Holten, D. (1991b). Temperature-independent electron transfer in *Rhodobacter capsulatus* wild type and $His^{M200} \rightarrow$ Leu photosynthetic reaction centers. J. Phys. Chem. 95, 3379–3383.

McPherson, P.H., Okamura, M.Y., & Feher, G. (1988). Light-induced proton uptake by photosynthetic reaction centers from *Rhodobacter sphaeroides* R-26. I. Protonation of the one-electron states $D^+Q_A^-$, DQ_A^-, $D^+Q_AQ_B^-$, and $DQ_AQ_B^-$. Biochim. Biophys. Acta 934, 348–368.

Michel, H. (1991). Crystallization of Membrane Proteins. CRC Press, Boca Raton.

Michel, H. & Deisenhofer, J. (1989). Relevance of the photosynthetic reaction center from purple bacteria to the structure of photosystem II. Biochemistry 27, 1–7.

Michel, H., Weyer, K.A., Gruenberg, H., & Lottspeich, F. (1985). The "heavy" subunit of the photosynthetic reaction centre from *Rhodopseudomonas viridis*: Isolation of the gene, nucleotide and amino acid sequence. EMBO J. 4, 1667–1672.

Michel, H., Weyer, K.A., Gruenberg, H., Dunger, I., Oesterhelt, D., & Lottspeich, F. (1986a). The "light" and "medium" subunits of the photosynthetic reaction center from *Rhodopseudomonas viridis*: Isolation of the genes, nucleotide and amino acid sequence. EMBO J. 5, 1149–1158.

Michel, H., Epp, O., & Deisenhofer, J. (1986b). Pigment-protein interactions in the photosynthetic reaction center from *Rhodopseudomonas viridis*. EMBO J. 5, 2445–2451.

Michel-Beyerle, M.-E., Plato, M., Deisenhofer, J., Michel, H., Bixon, M., & Jortner, J. (1988). Unidirectionality of charge separation in reaction centers of photosynthetic bacteria. Biochim. Biophys. Acta 932, 52–70.

Middendorf, T.R., Mazzola, L.T., Gaul, D.F., Schenck, C.C., & Boxer, S.G. (1991). Photochemical hole-burning spectroscopy of a photosynthetic reaction center mutant with altered charge separation kinetics: Properties and decay of the initially excited state. J. Phys. Chem. 95, 10142–10151.

Morrison, L.E., Schelhorn, J.E., Cotton, T.M., Bering, C.L., & Loach, P.A. (1982). Electrochemical and spectral properties of ubiquinone and synthetic analogs: Relevance to bacterial photosynthesis. In: Function of Quinones in Energy Conserving Systems (Trumpower, B.L., Ed.), pp. 35–58. Academic Press, New York.

Moser, C.C., Keske, J.M., Warncke, K., Farid, R.S., & Dutton, P.L. (1992). Nature of biological electron transfer. Nature (London) 355, 796–802.

Nabedryk, E., Andrianambinintsoa, S., Mäntele, W., & Breton, J. (1988). FTIR spectroscopic investigations of the intermediary electron acceptor photoreduction in purple photosynthetic bacteria and green plants. In: The Photosynthetic Bacterial Reaction Center: Structure and Dynamics. (Breton, J. & Verméglio, A., Eds.), pp. 237–250. Plenum, New York.

Nagarajan, V., Parson, W.W., Gaul, D., & Schenck, C. (1990). Effect of specific mutations of tyrosine-(M)210 on the primary photosynthetic electron-transfer process in *Rhodobacter sphaeroides*. Proc. Natl. Acad. Sci. USA 87, 7888–7892.

Nagarajan, V., Parson, W.W., Davis, D., & Schenck, C.C. (1993). Kinetics and free energy gaps of electron-transfer reactions in *Rhodobacter sphaeroides* reaction centers. Biochemistry 32, 12324–12336.

Norris, J.R. & Katz, J.J. (1978). Oxidized bacteriochlorophyll as photoproduct. In: The Photosynthetic Bacteria. (Clayton, R.K. & Sistrom, W.R., Eds.), pp. 397–418. Plenum, New York.

Norris, J.R., Budil, D.E., Gast, P., Chang, C.-H., El-Kabbani, O., & Schiffer, M. (1989). Correlation of paramagnetic states and molecular structure in bacterial photosynthetic reaction centers: The symmetry of the primary electron donor in *Rhodopseudomonas viridis* and *Rhodobacter sphaeroides* R26. Proc. Natl. Acad. Sci. USA 86, 4335–4339.

Nugent, J.H.A., Doetschman, D.C., & Maclachlan, D.J. (1992). Characterization of the multiple EPR line shapes of iron-semiquinones in photosystem II. Biochemistry 31, 2935–2941.

Okamura, M.Y. (1984). On the herbicide site in bacterial reaction centers. In: Biosynthesis of the photosynthetic apparatus: Molecular biology, development and regulation. (Thornber, J.P., Staehelin, L.A., & Hallick, R.B., Eds.), pp. 381–390. Alan R. Liss, New York.

Okamura, M.Y. & Feher, G. (1992). Proton transfer in reaction centers from photosynthetic bacteria. Annu. Rev. Biochem. 61, 861–896.

Ovchinnikov, Yu.A., Abdulaev, N.G., Zolotarev, A.S., Shmukler, B.E., Zargarov, A.A., Kutuzov, M.A., Telezhinskaya, I.N., & Levina, N.B. (1988a). Photosynthetic reaction center of *Chloroflexus aurantiacus*. I. Primary structure of L-subunit. FEBS Lett. 231, 237–242.

Ovchinnikov, Yu.A., Abdulaev, N.G., Shmuckler, B.E., Zargarov, A.A., Kutuzov, M.A., Telezhinskaya, I.N., Levina, N.B., & Zolotarev, A.S. (1988b). Photosynthetic reaction center of *Chloroflexus aurantiacus*. Primary structure of M-subunit. FEBS Lett. 232, 364–368.

Packham, N.K., Packham, C., Mueller, P., Tiede, D.M., & Dutton, P.L. (1980). Reconstitution of photochemically active reaction centers in planar phospholipid membranes. Light induced electrical currents under voltage-clamped conditions. FEBS Lett. 110, 101–106.

Paddock, M.L., Rongey, S.H., Abresch, E.C., Feher, G., & Okamura, M.Y. (1988). Reaction centers from three herbicide-resistant mutants of *Rhodobacter sphaeroides* 2.4.1: Sequence analysis and preliminary characterization. Photosynthesis Res. 17, 75–96.

Paddock, M.L., Rongey, S.H., Feher, G., & Okamura, M.Y. (1989). Pathway of proton transfer in bacterial reaction centers: Replacement of glutamic acid 212 in the L-subunit by glutamine inhibits quinone (secondary acceptor) turnover. Proc. Natl. Acad. Sci. USA 86, 6602–6606.

Paddock, M.L., McPherson, P.H., Feher, G., & Okamura, M.Y. (1990a). Pathway of proton transfer in bacterial reaction centers: replacement of serine-L223 by alanine inhibits electron and proton transfers associated with reduction of quinone to dihydroquinone. Proc. Natl. Acad. Sci. USA 87, 6803–6807.

Paddock, M.L., Feher, G., & Okamura, M.Y. (1990b). pH dependence of charge recombination in RCs from *Rb. sphaeroides* in which Glu-L212 is replaced with Asp. Biophys. J. 57, 569a.

Paddock, M.L., Feher, G., & Okamura, M.Y. (1991). Reaction centers from three herbicide resistant mutants of *Rhodobacter sphaeroides* 2.4.1: Kinetics of electron transfer reactions. Photosynthesis Res. 27, 109–119.

Paddock, M.L., Juth, A., Feher, G., & Okamura, M.Y. (1992). Electrostatic effects of replacing Asp-L210 with Asn in bacterial RCs from *Rb. sphaeroides*. Biophys. J. 61, A153.

Parot, P., Thiery, J., & Verméglio, A. (1987). Charge recombination at low temperature in photosynthetic bacterial reaction centers: Evidence for two conformational states. Biochim. Biophys. Acta 893, 534–543.

Parson, W.W. (1987). The bacterial reaction center. In: Photosynthesis (Amesz, J., Ed.), pp. 43–61. Elsevier, Amsterdam.

Parson, W.W. & Warshel, A. (1987). Spectroscopic properties of photosynthetic reaction centers. 2. Application of the theory to *Rhodopseudomonas viridis*. J. Am. Chem. Soc. 109, 6152–6163.

Parson, W.W., Chu, Z.-T., & Warshel, A. (1990a). Electrostatic control of charge separation in bacterial photosynthesis. Biochim. Biophys. Acta 1017, 251–272.

Parson, W.W., Nagarajan, V., Gaul, D., Schenck, C.C., Chu, Z.-T., & Warshel, A. (1990b). Electrostatic effects on the speed and directionality of electron transfer in bacterial reaction centers: The special role of tyrosine M-208. In: Reaction Centers of Photosynthetic Bacteria (Michel-Beyerle, M.-E., Ed.), Springer Series in Biophysics, Vol. 6, pp. 239–249. Springer-Verlag, Berlin, New York.

Peloquin, J.M., Bylina, E.J., Youvan, D.C., & Bocian, D.F. (1990). Resonance Raman studies of genetically modified reaction centers from *Rhodobacter capsulatus*. Biochemistry 29, 8417–8424.

Peloquin, J.M., Bylina, E.J., Youvan, D.C., & Bocian, D.F. (1991). Effects of pigment-protein interactions on the conformation of the primary electron acceptor in *Rhodobacter capsulatus* reaction centers. Biochim. Biophys. Acta 1056, 85–88.

Plato, M., Möbius, K., Michel-Beyerle, M.-E., Bixon, M., & Jortner, J. (1988). Intermolecular electronic interactions in the primary charge separation in bacterial photosynthesis. J. Am. Chem. Soc. 110, 7279–7285.

Plato, M., Michel-Beyerle, M.-E., Bixon, M., & Jortner, J. (1989). On the role of tryptophan as a superexchange mediator for quinone reduction in photosynthetic reaction centers. FEBS Lett. 249, 70–74.

Prince, R.C., Dutton, P.L., & Bruce, J.M. (1983). Electrochemistry of ubiquinones, menaquinones and plastoquinones in aprotic solvents. FEBS Lett. 160, 273–276.

Prince, R.C., Lloyd-Williams, P., Bruce, J.M., & Dutton, P.L. (1986). Voltammetric measurements of quinones. In: Methods in Enzymology (Fleischer, S. & Fleischer, B., Eds.), Vol. 125, pp. 109–119. Academic Press, New York.

Prince, R.C. (1990). Bacterial photosynthesis: from photons to Δp. In: The Bacteria, Vol. XII. Bacterial Energetics (Krulwich, T.A., Ed.), pp. 111–149. Academic Press, New York.

Robles, S.J., Breton, J., & Youvan, D. (1990). Partial symmetrization of the photosynthetic reaction center. Science 248, 1402–1405.

Rutherford, A.W., Zimmermann, J.L., & Mathis, P. (1984). The effect of herbicides on components of the PS II reaction center measured by EPR. FEBS Lett. 165, 156–162.

Rutherford, A.W., Agalidis, I., & Reiss-Husson, F. (1985). Manganese-quinone interactions in the electron acceptor region of bacterial photosynthetic reaction centers. FEBS Lett. 182, 151–157.

Schenck, C.C., Parson, W.W., Holten, D., & Windsor, M.W. (1981). Transient states in reaction centers containing reduced bacteriopheophytin. Biochim. Biophys. Acta 635, 383–392.

Schenck, C.C., Sistrom, W.R., & Capaldi, R.A. (1986). Structure-function studies in reaction centers: Characterization of an herbicide resistance mutation in *Rhodopseudomonas sphaeroides*. Biophys. J. 49, 486a.

Schönfeld, M., Montal, M., & Feher, G. (1979). Functional reconstitution of photosynthetic reaction centers in planar lipid bilayers. Proc. Natl. Acad. Sci. USA 76, 6351–6355.

Schulten, K. & Tesch, M. (1991). Coupling of protein motion to electron transfer: Molecular dynamics and stochastic quantum mechanics study of photosynthetic reaction centers. J. Chem. Phys. 158, 421–446.

Sebban, P. (1988). pH effect on the biphasicity of the $P^+Q_A^-$ charge recombination kinetics in the reaction centers from *Rhodobacter sphaeroides*, reconstituted with anthraquinones. Biochim. Biophys. Acta 936, 124–132.

Sebban, P. & Wraight, C.A. (1989). Heterogeneity of the $P^+Q_A^-$ recombination kinetics in reaction centers from *Rhodopseudomonas viridis*: The effects of pH and temperature. Biochim. Biophys. Acta 974, 54–65.

Shinkarev, V.P., Takahashi, E., & Wraight, C.A. (1993). Flash-induced electric potential generation in wild type and L212EQ mutant chromatophores of *Rhodobacter sphaeroides*: $Q_B H_2$ is not released from L212EQ mutant reaction centers. Biochim. Biophys. Acta 1142, 214–216.

Shinkarev, V.P. & Wraight, C.A. (1993). Electron and proton transfer in the acceptor quinone complex of RCs of phototrophic bacteria. In: The Photosynthetic Reaction Center (Deisenhofer, J. & Norris, J., Eds.), pp. 375–387. Academic Press, New York.

Shiozawa, J.A., Lottspeich, F., Oesterhelt, D., & Feick, R. (1989). The primary structure of the *Chloroflexus aurantiacus* reaction-center polypeptides. Eur. J. Biochem. 180, 75–84.

Shopes, R.J. & Wraight, C.A. (1985). The acceptor quinone complex of *Rhodopseudomonas viridis* reaction centers. Biochim. Biophys. Acta 806, 348–356.

Shopes, R.J. & Wraight, C.A. (1987a). Charge recombination from the $P^+Q_A^-$ state in reaction centers from *Rhodopseudomonas viridis*. Biochim. Biophys. Acta 893, 409–425.

Shopes, R.J. & Wraight. C.A. (1987b). Herbicide-resistant reaction center mutants of *Rhodopseudomonas viridis*. In: Progress in Photosynthesis Research. (Biggins, J., Ed.), Vol. II, pp. 397–400. Martinus Nijhoff, Dordrecht.

Shopes, R.J., Blubaugh, D.J., Wraight, C.A., & Govindjee. (1989). Absence of a bicarbonate depletion effect in electron transfer between quinones in chromatophores and reaction centers of *Rhodobacter sphaeroides*. Biochim. Biophys. Acta 974, 114–118.

Shuvalov, V.A., Klevanik, A.V., Sharkov, A.V., Matveetz, Ju.A., & Krukov, P.G. (1978). Picosecond detection of BChl-800 as an intermediate electron carrier between selectively excited P_{870} and bacteriopheophytin in *Rhodospirillum rubrum* reaction centers. FEBS Lett. 91, 135–139.

Sinning, I., Michel, H., Mathis, P., & Rutherford, A.W. (1989a). Characterization of four herbicide-resistant mutants of *Rhodopseudomonas viridis* by genetic analysis, electron paramagnetic resonance, and optical spectroscopy. Biochemistry 28, 5544–5553.

Sinning, I., Michel, H., Mathis, P., & Rutherford, A.W. (1989b). Terbutryn resistance in a purple bacterium can induce sensitivity toward the plant herbicide DCMU. FEBS Lett. 256, 192–194.

Sinning, I., Koepke, J., Schiller, B., & Michel, H. (1990a). First glance on the three-dimensional structure of the photosynthetic reaction center from a herbicide-resistant *Rhodopseudomonas viridis* mutant. Z. Naturforsch. 45c, 455–458.

Sinning, I., Koepke, J., & Michel, H. (1990b). Recent advances in the structure analysis of *Rhodopseudomonas viridis* reaction center mutants. In: Reaction Centers of Photosynthetic Bacteria (Michel-Beyerle, M.-E., Ed.), Springer Series in Biophysics, Vol. 6, pp. 199–208. Springer-Verlag, Berlin, New York.

Stein, R.R., Castellvi, A.L., Bogacz, J.P., & Wraight, C.A. (1984) Herbicide-quinone competition in the acceptor complex of photosynthetic reaction centers from *Rhodopseudomonas sphaeroides*: A bacterial model for PS-II-herbicide activity in plants J. Cell. Biochem. 24, 243–359.

Stilz, H.U., Finkele, U., Holzapfel, W., Lauterwasser, C., Zinth, W., & Oesterhelt, D. (1990). Site-directed mutagenesis of threonine M222 and tryptophan M252 in the photosynthetic reaction center of *Rhodobacter sphaeroides*. In: Reaction Centers of Photosynthetic Bacteria (Michel-Beyerle, M.-E., Ed.), pp. 265–271. Springer-Verlag, Berlin, New York.

Stocker, J.W., Taguchi, A.K.W., Murchison, H.A., Woodbury, N.W., & Boxer, S.G. (1992). Spectroscopic and redox properties of Sym1 and (M)F195H: *Rhodobacter capsulatus* reaction center symmetry mutants which affect th e initial electron donor. Biochemistry 31, 10356–10362.

Struck, A., Müller, A., & Scheer, H. (1991). Modified bacterial reaction centers. 4. The borohydride treatment reinvestigated: Comparison with selective exchange experiments at binding sites $B_{A,B}$ and $H_{A,B}$. Biochim. Biophys. Acta 1060, 262–270.

Taguchi, A.K.W., Stocker, J.W., Alden, R.G., Causgrove, T.P., Peloquin, J.M., Boxer, S.G., & Woodbury, N.W. (1992). Biochemical characterization and electron transfer reactions of *sym1*, a *Rhodobacter capsulatus* reaction center symmetry mutant which affects the initial electron donor. Biochemistry 31, 10345–10355.

Takahashi, E. & Wraight, C.A. (1990). A crucial role for Asp^{L213} in the proton transfer pathway to the secondary quinone of reaction centers from *Rhodobacter sphaeroides*. Biochim. Biophys. Acta 1020, 107–111.

Takahashi, E. & Wraight, C.A. (1991). Small weak acids stimulate proton transfer events in site-directed mutants of the two ionizable residues, Glu^{L212} and Asp^{L213}, in the Q_B-binding site of *Rhodobacter sphaeroides* reaction centers. FEBS Lett. 283, 140–144.

Takahashi, E. (1992). Characterization of site-directed mutants in the Q_B binding site of the photosynthetic reaction center of *Rhodobacter sphaeroides*. Ph.D. Thesis, University of Illinois, Urbana-Champaign.

Takahashi, E. & Wraight, C.A. (1992a). Proton and electron transfer in the acceptor quinone complex of *Rhodobacter sphaeroides* reaction centers: Characterization of site-directed mutants of the two ionizable residues, Glu^{L212} and Asp^{L213}, in the Q_B binding site. Biochemistry 31, 855–866.

Takahashi, E. & Wraight, C.A. (1992b). Characterization of photosynthetic reaction center from *Rb. sphaeroides* with the site-directed mutation L212 Glu → Lys. Biophys J. 61, A104.

Takahashi, E., Maróti, P. & Wraight, C.A. (1990). Site-directed mutagenesis of *Rhodobacter sphaeroides* reaction center: The role of tyrosine L222. In: Current Research in Photosynthesis (Baltscheffsky, M., Ed.), Vol. I, pp. 169–172. Kluwer Academic Publishers, Dordrecht, Netherlands.

Takahashi, E., Maróti, P., & Wraight, C.A. (1992). Coupled proton and electron transfer pathways in the acceptor quinone complex of reaction centers from *Rhodobacter sphaeroides*. In: Electron and Proton Transfer in Chemistry and Biology (Müller, A., Ratajczak, H., Junge, W. and Diemann, E., Eds.), pp. 219–236. Elsevier Science Publ.., Amsterdam.

Trebst, A. (1987). The three-dimensional structure of the herbicide binding niche on the reaction center polypeptides of photosystem II. Z. Naturforsch. 42c, 742–750.

Treutlein, H., Schulten, K., Niedermeier, C., Deisenhofer, J., Michel, H., & DeVault, D. (1988). Electrostatic control of electron transfer in the photosynthetic reaction center of *Rhodopseudomonas viridis*. In: The Photosynthetic Bacterial Reaction Center: Structure and Dynamics (Breton, J. & Verméglio, A., eds.), pp. 369–378. Plenum, New York.

Treutlein, H., Schulten, K., Brünger, A.T., Karplus, M., Deisenhofer, J., & Michel, H. (1992). Chromophore-protein interactions and the function of the photosynthetic reaction center: A molecular dynamics study. Proc. Natl. Acad. Sci. USA 89, 75–79.

Trissl, H. (1983). Spatial correlation between primary redox components in reaction centers of *Rhodopseudomonas sphaeroides* measured by two electrical methods in nanosecond range. Proc. Natl. Acad. Sci. USA 80, 7173–7177.

Velthuys, B.R. (1981). Electron-dependent competition between plastoquinone and inhibitors for binding to photosystem II. FEBS Lett. 126, 277–281.

Vermaas, W.F.J. & Rutherford, A.W. (1984). EPR measurements on the effects of bicarbonate and triazme resistance on the acceptor side of photosystem II. FEBS Lett. 175, 243–248.

Verméglio, A. (1977). Secondary electron transfer in reaction centers of *Rhodopseudomonas sphaeroides*. Out-of-phase periodicity of two for the formation of ubisemiquinone and fully reduced ubiquinone. Biochim. Biophys. Acta 459, 516–524.

Vos, M.H., Lambry, J.-C., Robles, S.J., Youvan, D.C., Breton, J., & Martin, J.-L. (1991). Direct observation of vibrational coherence in bacterial reaction centers using femtosecond absorption spectroscopy. Proc. Natl. Acad. Sci. USA 88, 8885–8889.

Vos, M.H., Lambry, J.-C., Robles, S.J., Youvan, D.C., Breton, J., & Martin, J.-L. (1992). Femtosecond spectral evolution of the excited state of bacterial reaction centers at 10 K. Proc. Natl. Acad. Sci. USA 89, 613–617.

Wang, X., Gast, G., Kirmaier, C., Holten, D., & Wraight, C.A. (1990). Characterization of M210 (Tyr → Phe) mutant from the photosynthetic bacteria *Rb. sphaeroides*. Biophys. J. 57, 565a.

Wang, X., Cao, J., Maróti, P., Stilz, H.U., Finkele, U., Lauterwassse, C,. Zinth, W., Oesterhelt, D., Govindjee, & Wraight, C.A. (1992). Is bicarbonate in photosystem II the equivalent of the glutamate ligand to the iron atom in bacterial reaction centers? Biochim. Biophys. Acta 1100, 1–8.

Warncke, K. & Dutton, P.L. (1990). Resolution of *in situ* interaction energies between molecules and the reaction center Q_A site: Implications for electron transfer function. In: Current Research in Photosynthesis (Baltscheffsky, M., Ed.), Vol. I, pp. 157–160. Kluwer Academic Publishers, Dordrecht, Netherlands.

Warshel, A., Creighton, S., & Parson, W.W. (1988). Electron transfer pathways in the primary event of bacterial photosynthesis. J. Phys. Chem. 92, 2696–2701.

Wasielewski, M.R. & Tiede, D.M. (1986). Sub-picosecond measurements of primary electron transfer in *Rhodopseudomonas viridis* reaction centers using near-infrared excitation. FEBS Lett. 204, 368–372.

Williams, J.C., Steiner, L.A., Ogden, R.C., Simon, M.I., & Feher, G. (1983). Primary structure of the M subunit of the reaction center from *Rhodopseudomonas sphaeroides*. Proc. Natl. Acad. Sci. USA 80, 6505–6509.

Williams, J.C., Steiner, L.A., Feher, G., & Simon, M.I. (1984). Primary structure of the L subunit of the reaction center from *Rhodopseudomonas sphaeroides*. Proc. Natl. Acad. Sci. USA 81, 7303–7307.

Williams, J.C., Steiner, L.A., & Feher, G. (1986). Primary structure of the reaction center from *Rhodopseudomonas sphaeroides*. Proteins 1, 312–325.

Williams, J.C., Paddock, M.L., Feher, G., & Allen, J.P. (1991). Effects of iron ligand substitutions in reaction centers from *Rhodobacter sphaeroides*. Biophys. J. 59, 142a.

Williams, J.C., Alden, R.G., Murchison, H.A., Peloquin, J.M., Woodbury, N.W., & Allen, J.P. (1992). Effects of mutations near the bacteriochlorophylls in reaction centers from *Rhodobacter sphaeroides*. Biochemistry 31, 11029–11037.

Woodbury, N.W., Parson, W.W., Gunner, M.R., Prince, R.C., & Dutton, P.L. (1984). Radical-pair energetics and decay mechanisms in reaction centers containing anthraquinones, naphthoquinones or benzoquinones in place of ubiquinone. Biochim. Biophys. Acta 851, 6–22.

Woodbury, N.W.T. & Parson, W.W. (1984). Nanosecond fluorescence from isolated photosynthetic reaction centers of *Rhodopseudomonas sphaeroides*. Biochim. Biophys. Acta 767, 345–361.

Woodbury, N.W., Becker, M., Middendorf, D., & Parson, W.W. (1985). Picosecond kinetics of the initial photochemical electron-transfer reaction in bacterial reaction centers. Biochemistry 24, 7516–7521.

Woodbury, N.W., Taguchi, A.K., Stocker, J.W., & Boxer, S.G. (1990). Preliminary characterization of pAT-3, a symmetry enhanced reaction center mutant of *Rhodobacter capsulatus*. In: Reaction Centers of Photosynthetic Bacteria (Michel-Beyerle, M.-E., Ed.), Springer Series in Biophysics, Vol. 6, pp. 303–312. Springer-Verlag, Berlin, New York.

Wraight, C.A. (1979). Electron acceptors of bacterial photosynthetic reaction centers. II. H^+ binding coupled to secondary electron transfer in the quinone acceptor complex. Biochim. Biophys. Acta 548, 309–327.

Wraight, C.A. (1981). Oxidation-reduction physical chemistry of the acceptor quinone complex in bacterial photosynthetic reaction centers: Evidence for a new model of herbicide activity. Israel J. Chem. 21, 348–354.

Wraight, C.A. (1982). The involvement of stable semiquinones in the two-electron gates of plant and bacterial photosystems. In: Function of Quinones in Energy Conserving Systems (Trumpower, B.L., Ed.), pp. 182–192. Academic Press, New York.

Wraight, C.A. & Shopes, R.J. (1989). Quinone binding and herbicide activity in the acceptor quinone complex of bacterial reaction centers. In: Techniques and New Developments in Photosynthesis Research (Barber, J. & Malkin, R., Eds.), pp. 183–191. Plenum, New York.

Yeates, T.O., Komiya, H., Rees, D.C., Allen, J.P., & Feher, G. (1987). Structure of the reaction center from *Rhodobacter sphaeroides* R-26: Membrane-protein interactions. Proc. Natl. Acad. Sci. USA 84, 6438–6442.

Yeates, T.O, Komiya, H., Chirino, A., Rees, D.C., Allen, J.P., & Feher, G. (1988). Structure of the reaction center from *Rhodobacter sphaeroides* R-26 and 2.4.1: Protein-cofactor (bacteriochlorophyll, bacteriopheophytin, and carotenoid) interactions. Proc. Natl. Acad. Sci. USA 85, 7993–7997.

Youvan, D.C., Bylina, E.J., Alberti, M., Begusch, H., & Hearst, J.E. (1984). Nucleotide and deduced polypeptide sequences of the photosynthetic reaction-center, B875 antenna, and flanking polypeptides from *R. capsulata*. Cell 37, 949–957.

Zhou, Q., Robert, B., & Lutz, M. (1987). Intergeneric structural variability of the primary donor of photosynthetic bacteria: Resonance Raman spectroscopy of reaction centers from two *Rhodospirillum* and *Rhodobacter* species. Biochim. Biophys. Acta 890, 368–376.

PROTON-TRANSLOCATING NAD(P)-H TRANSHYDROGENASE AND NADH DEHYDROGENASE IN PHOTOSYNTHETIC MEMBRANES

J. Baz Jackson and Alastair G. McEwan

Advances in Molecular and Cell Biology
Volume 10, pages 253–286.

ISBN: 1-55938-710-6

I. INTRODUCTION

H^+-transhydrogenase (H^+-Thase) and NADH dehydrogenase are membrane proteins found in the inner membranes of mitochondria and many bacteria. Both use nicotinamide nucleotides as substrates and are coupled to the proton electrochemical gradient. In this article we shall briefly summarize recent findings relevant to the structure and mechanism of H^+-Thase and NADH-dehydrogenase and speculate on the function of cyanobacteria and the function of H^+-Thase in bacteria, including photosynthetic bacteria, and the putative NADH dehydrogenase in chloroplasts. Other recent discussions pertaining to NADH dehydrogenase can be found in (Weiss et al., 1991) and to H^+-Thase in (Rydstrom et al., 1987; Jackson, 1991).

II. H^+-TRANSHYDROGENASE

H^+-Thase is a proton pump (Figure 1). It can translocate protons against an electrochemical gradient (Δp) across a membrane using the free energy released during the transfer of reducing equivalents between NAD(H) and NADP(H). In bacteria, the sites for nucleotide binding face the cytoplasm. H^+-Thase can be envisaged as a *generator* of Δp; the transfer of [H^-] from NADPH to NAD^+ causes the translocation of H^+ from the cytoplasm to the outside of the cell generating a Δp with the same polarity as that produced by respiratory or photosynthetic electron transport. Alternatively, H^+-Thase can be a *consumer* of Δp; if the protonic potential is high enough outside the cell, the inward translocation of H^+ through the enzyme will cause the reduction of $NADP^+$ by NADH. The actual direction of the reaction depends on the relative magnitudes of:

$$\Delta G = n\mathrm{F}\,\Delta p$$

(where n is the H^+/H^-, see below) and the free energy of the [H^-] transfer reaction:

$$\Delta G = \Delta G^{o1} + RT\ln([\mathrm{NADPH}][\mathrm{NAD^+}]/[\mathrm{NADP^+}][\mathrm{NADH}])$$

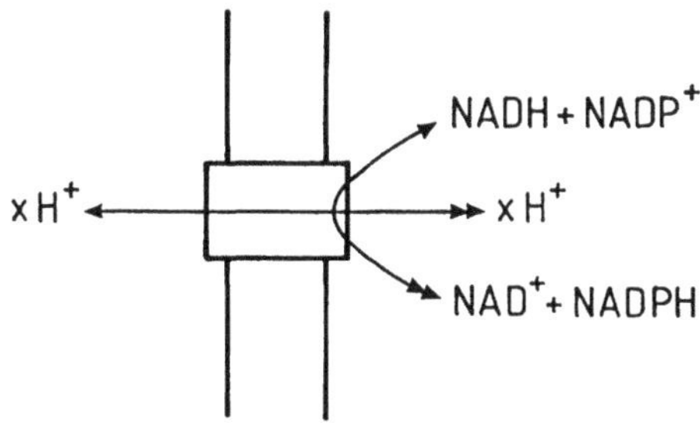

Figure 1. H^+-transhydrogenase as a proton pump.

A. Distribution of H^+-Transhydrogenase

H^+-Thase is found in the inner membranes of animal mitochondria (see Rydstrom and Hoek, 1988) and in many bacterial membranes (see Fisher and Earle, 1982). Some bacteria, however, possess a flavin-containing, "BB"-type transhydrogenase which does not pump protons (Rydstrom et al., 1987). There are no documented instances of wild-type (Wt) bacteria lacking both types of enzyme. It is perhaps unlikely that the BB enzyme can coexist with H^+-Thase in a single compartment of the same organism, although the condition could arise if the activities of the two enzymes were independently regulated. The situation remains to be established. The photosynthetic bacteria, *Rhodobacter capsulatus*, *Rhodobacter sphaeroides*, and *Rhodospirillum rubrum* have very active H^+-Thases, that is, relative to rates of ATP synthesis (Keister and Yike, 1966; Konings and Guillory, 1972; Cotton et al., 1987). Chloroplasts are reported to lack the enzyme (Vourdouw et al., 1983). Ferredoxin:$NADP^+$ oxidoreductase from chloroplasts has a low transhydrogenase activity, but this is not thought to be of physiological consequence (Voordouw et al., 1983). Plant mitochondria have transhydrogenase activity (Carlenor et al., 1988), but it is not yet clear whether the enzyme is proton-translocating or of the BB-type. Unfortunately, at present there is little information on the distribution of transhydrogenases in the cyanobacteria.

B. Structure of H^+-Transhydrogenase

The primary amino acid sequences of H^+-Thases from *E. coli* and bovine heart mitochondria have been determined from the nucleotide sequences of the gene and cDNA, respectively (Clarke et al., 1986; Yamaguchi et al., 1988). The *E. coli* enzyme comprises two polypeptides (alpha, M_r 53,906; beta, M_r 48,667) which, arranged contiguously, show strong homology with the single polypeptide of mitochondrial H^+-Thase, M_r 109,212 (Figure 2). On the basis of hydropathy plots there may be three large domains in the H^+-Thase (Figure 2). Domains I and III are

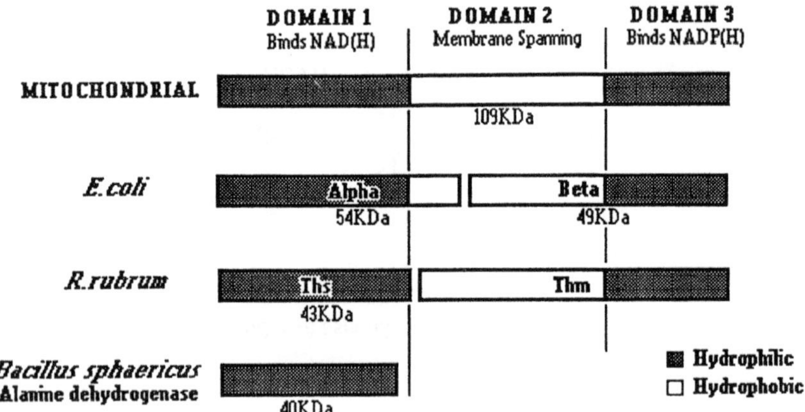

Figure 2. The domain and polypeptide composition of H^+-transhydrogenase. The *unshaded boxes* represent relatively hydrophilic regions and the *shaded boxes* represent hydrophobic regions, based on the sequences of the enzymes from mitochondria and *E. coli.*

relatively hydrophilic and probably mainly protrude from the membrane. Domain II is predominantly hydrophobic and could be composed of up to 14 transmembrane helical segments (Yamaguchi et al., 1988). H^+-Thase from *Rb. capsulatus* has a subunit composition similar to that in *E. coli*, and it also requires detergent for isolation and purification (Lever et al., 1991).

The enzyme from *Rhs. rubrum* is interestingly different. It can be separated into a water-soluble (Th_s) and a membrane-bound (Th_m) component. Individually, Th_s and Th_m have no transhydrogenase activity, but full activity can be reconstituted by mixing the two components (Fisher and Guillory, 1971). Purified Th_s (Cunningham et al., 1992b) has an M_r of 43,000 on SDS-PAGE and an N-terminal sequence similar to that of the α-subunit of the *E. coli* enzyme (and the corresponding region close to the N-terminal portion of the mitochondrial protein). This suggests that Th_s is soluble because the "break" with the remainder of the protein occurs, not as in *E. coli* in the hydrophobic region, but in the relatively hydrophilic Domain I (see Figure 2). The polypeptide composition of Th_m is not yet known.

C. Nucleotide Binding Sites on H^+-Transhydrogenase

Kinetic analysis indicates that there are separate sites on H^+-Thase for NAD^+/NADH and for $NADP^+$/NADPH (Hanson, 1979; Homyck and Bragg, 1979; Enander and Rydstrom, 1982; Lever et al., 1991). In the primary amino acid sequence of H^+-Thase from mitochondria and the α-subunit of the *E. coli* enzyme there is a segment showing strong homology with nucleotide-binding domains of other proteins. It was proposed that this is the NAD(H) binding site (Clarke et al., 1986; Yamaguchi et al., 1988). Support for this was obtained from experiments with dicyclohexylcarbodiimide (DCCD) and 5^1-[*p*-(fluorosulphonyl)benzoyl] adenosine (FSBA). These reagents inhibit mitochondrial transhydrogenase, apparently by interfering with NAD(H) binding (Phelps and Hatefi, 1984a,b, 1985). They react with glu and tyr residues, respectively, just downstream from the putative nucleotide-binding domain (Wakabayashi and Hatefi, 1987a,b).

In the presence of NADH to block binding into the NAD(H) site, FSBA reacted with another tyr residue (tyr-1006) close to the C-terminus of mitochondrial H^+-Thase. It is possible that this is part of the NADP(H) binding site (Wakabayashi and Hatefi, 1987b). However, the situation is not completely clear. It has been found that 8-azido-AMP, an analogue of NAD(H), also binds at tyr-1006 (Hu et al., 1988). Furthermore, the pattern of inhibition by DCCD seems rather complex. As well as the protective effect of NAD(H) mentioned above, NADP(H) can increase the degree of inhibition by DCCD of H^+-Thase from mitochondria, *E. coli* and *Rhs. rubrum* (Pennington and Fisher, 1981; Phelps and Hatefi, 1981; Clarke and Bragg, 1985; Palmer et al., 1992). In the latter enzyme this effect is confined to the membrane-bound component of the enzyme (Th_m) supporting the implication of an NADP(H) binding site within Domains II and III.

The conformational changes resulting from nucleotide binding and causing changes in the susceptibility to DCCD and FSBA are typical of the response to a number of other protein modification reagents (references cited in Yamaguchi et al., 1990; Jackson, 1991). Nucleotide-induced conformational changes are also evident in the reassembly of active H^+-Thase from the Th_s and Th_m of *Rhs. rubrum* (Cunningham et al., 1992a). The mechanistic significance of none of these conformational changes is understood but, in general, there are considerably more reported for NADP(H) than for NAD(H). Assuming that the above assignments of the nucleotide-binding sites are correct, it may be that during catalysis, Domain III is conformationally more mobile than Domain I.

D. The Mechanism of H^+-Transhydrogenase

Nucleotide substrates bind to H^+-Thase to give a ternary complex. At least under some conditions the substrate addition is random (Hanson, 1979; Homyck and Bragg, 1979; Enander and Rydstrom, 1982; Lever et al., 1991). In deenergized H^+-Thase (e.g., in the solubilized enzyme), nucleotide binding appears to reach equilibrium, suggesting that the interconversion of the ternary complexes (or a subsequent reaction) may be rate-limiting. H^- equivalents from NADH labeled in the 4A position of the nicotinamide ring are transferred to the 4B position on NADPH without exchanging with water (Lee et al., 1965). Therefore, it must be concluded either that the H^- transfer from one nucleotide to the other is direct, or that any involvement of the protein as an acceptor–donor of H^- must be confined to the ternary complexes. Substitution of the 4A H atom of the nicotinamide ring with ^2H did not have any effect on the rate of transhydrogenase (Fisher and Kaplan, 1973), suggesting that conformational changes in the ternary complex, not the hydride transfer step itself, are rate-limiting.

The ratio of protons translocated per hydride equivalent transferred between nucleotides (the H^+/H^- ratio) has been measured by several different procedures (reviewed in Palmer and Jackson, 1992a). The consensus value is 1.0, but a nonintegral value of 0.5 cannot be ruled out. In contrast with observations on ATP synthesis, there is little to detract from the simple view that the H^+-Thase is driven only by a delocalized proton electrochemical gradient—tests with both the mitochondrial and chromatophore system have yielded unique relationships between Δp and the rate of transhydrogenase independently of how the value of Δp is modified (Cotton et al., 1987; Persson et al., 1987; Petronilli et al., 1991). It seems unlikely that there are any significant localized interactions between electron transport components and H^+-Thase (Palmer et al., 1992b). This is also well supported by reconstitution experiments either with H^+-Thase alone (Rydstrom, 1979; Fisher and Earle, 1980) or together with bacteriorhodopsin or ATP synthase (Eytan et al., 1987a,b, 1990). In a recent article, observations on the anomalous uncoupler sensitivity of the transhydrogenase reaction in membrane vesicles from an overexpressing strain of *E. coli* were interpreted in terms of a localized proton pathway (Chang et al., 1992). However, "anomalous" results can also be explained in terms of delocalized coupling if the vesicle population was heterogeneous. $NADP^+$ reduction by NADH in vesicles possessing only H^+-Thase would be stimulated by uncoupler; in vesicles containing H^+-Thase and respiratory chains and/or ATPase the reaction would be inhibited by uncoupler.

The dependence, at a range of pH values, of the steady-state rate of the transhydrogenase reaction on membrane potential ($\Delta pH = 0$) when the nucleotide substrates are saturating provides a useful test of the coupling mechanism (Cotton et al., 1989). The simplest interpretation of the data is that, at low values, membrane potential increases the rate constant of a single reaction responsible for the charge translocation step through the enzyme but, at high values, another component becomes rate-limiting (Jackson, 1991).

This model also explains the existence of a rapid burst in the transhydrogenase reaction, measured with thio-$NADP^+$ and NADH, at the onset of illumination of chromatophores (Palmer and Jackson, 1990). On the basis that in deenergized chromatophores the nucleotide substrates reach equilibrium, and that turnover is limited by the interconversion of the ternary complexes, it is likely that the enzyme is loaded with thio-$NADP^+$ and NADH before the onset of illumination. During the burst a prompt increase in the rate constant of the charge translocation step in the interconversion of the ternary complexes leads to the rapid production of thio-NADPH on the enzyme. Subsequent slow release of the products is assumed to be the rate-limiting step during turnover in steady-state.

From experimental information currently available, there is little to choose between direct coupling models for H^+-Thase, in which proton translocation through the enzyme is directly coupled to protolytic reactions involved in H^- transfer (Mitchell, 1966; Skulachev, 1970; Mitchell, 1972), and indirect coupling (Skulachev, 1974; Rydstrom et al., 1981; Fisher and Earle, 1982; Yamaguchi and Hatefi, 1989). This point was discussed in detail (Jackson, 1991) but more structural and kinetic studies on H^+-Thase are urgently needed. However, it is worth emphasizing (1) that the acid–base chemistry of the nicotinamide nucleotides is very complex and provides some interesting possibilities for mechanisms of direct coupling, and (2) that the soluble dehydrogenases may provide useful models for the mechanism of H^+ exchange between the solvent and the catalytic center.

As discussed by Cunningham et al., 1992b, there is, in fact, strong homology between the first 400 residues of the α-subunit of H^+-Thase of *E. coli* (reported by Clarke et al., 1986) and the complete sequence of the single subunit of the soluble alanine dehydrogenases of *Bacillus stearothermophilus* and *B. sphericus* (reported by Kuroda et al., 1990): 30% sequence identity and 50% similarity between the H^+-Thase and the *Bacillus* enzymes. The homology is not confined to the nucleotide-binding sequence. The significance of this observation is not yet clear, but the soluble component of the *Rhs. rubrum* H^+-Thase, which probably corresponds to the same part of the *E. coli* α-subunit and which is capable of reconstituting transhydrogenase activity to depleted chromatophores, does not have any alanine dehydrogenase activity.

E. Function of H^+-Transhydrogenase

Although H^+-Thase was discovered more than 25 years ago, the function of the enzyme is still the subject of debate. Rydstrom and Hoekt (1988) have advanced the view that in animal mitochondria the enzyme has a role as a "buffer", counteracting either the oxidation of NAD(P)H or the depletion of Δp. The requirement for NADPH may be tissue-specific, but in general it is thought to be required for the degradation of xenobiotics or for the removal of damaging free radicals generated in the respiratory chain.

It is possible also in bacteria that H^+-Thase has several functions. Here, we shall discuss several plausible roles for the enzyme in bacteria. The view that emerges is that H^+-Thase is centrally involved in balancing the supply of NADH and NADPH for biosynthesis and detoxification in the cell. Its contribution is dependent on the supply of reduced nucleotides from catabolic dehydrogenases, and on the value of Δp. It is unlikely that H^+-Thase functions as a major generator of Δp, but it could play a role in "overflow metabolism".

Although the function is not known, it should be emphasized that H^+-Thase can proceed at high rates, at least *in vitro*. In chromatophores from photosynthetic bacteria rates of light-driven transhydrogenase are typically as much as 50% of those of light-driven ATP synthesis (Cotton et al., 1987). Thus, H^+-Thase can, in principle, be a substantial generator or consumer of Δp and could significantly influence the energy economy of the cell. It is also self-evident that the enzyme operates at the important interface between the nicotinamide nucleotide pools in the bacterial cytoplasm.

H^+-Thase in the Provision of Low-Potential Reductant

Early studies showed that the activity of membrane-bound transhydrogenase in *E. coli* was repressed (though not completely) when bacteria were transferred from an NH_4^+-containing to an amino acid-containing medium (Bragg et al., 1972; Liang and Houghton, 1981) and was de-repressed when cells were transferred from a rich medium to a glucose-minimal medium (Houghton et al., 1976). It was suggested that H^+-Thase serves to generate NADPH for amino acid biosynthesis and can therefore be partly dispensed with during growth in the presence of amino acids. There are some observations which cast doubt on this simple view. First, Gerolimatos and Hanson (1978) found that repression of transhydrogenase activity by individual amino acids was not particularly related to a requirement for NADPH in biosynthesis. Leucine, methionine, and alanine were the most effective at causing repression. Second, Csonka and Fraenkel (1977) found in an isotopic study that H^+-Thase is only a minor source, at most, of NADPH for amino acid biosynthesis in *E. coli*. Finally, a transhydrogenase insertion mutant of *E. coli* was able to grow aerobically at normal rates in minimal medium containing glucose, fructose, glycerol, or succinate (Hanson and Rose, 1980). These studies failed to reveal a phenotype for the mutant: only in strains with double mutations (e.g., in both H^+-Thase and glucose-6-phosphate dehydrogenase; the latter being considered as an alternative generator of NADPH) was there a decrease in the aerobic growth rate. It was concluded that H^+-Thase can contribute to the supply of NADPH under some conditions, but is not essential. Other NADP-linked enzymes, such as glucose-6-phosphate dehydrogenase, 6-phosphogluconate dehydrogenase, and isocitrate dehydrogenase could be equally or more important.

In discussions of the role of H^+-Thase in the provision of NADPH for biosynthesis in bacteria, it has not always been emphasized that, at large values of Δp, the nucleotide will be generated at a low redox potential relative to NAD(H). An important feature of H^+-Thase could be that it generates a very strong reducing agent (NADPH) for biosynthesis while maintaining the NAD^+ in an oxidized state to serve as an acceptor for the catabolic dehydrogenases. This is despite the fact that the standard potentials of NAD(H) and NADP(H) are similar. If H^+-Thase remains in equilibrium with Δp, then:

$$[\mathrm{NADPH}][\mathrm{NAD}^+]/[\mathrm{NADP}^+][\mathrm{NADH}] = \exp\{n\mathrm{F}\,\Delta p/RT\}$$

where n is the H^+/H^- ratio. Unfortunately, in bacteria, there is little available information about the redox potential of either of the nicotinamide nucleotide couples. There have been attempts to measure the total concentrations of NAD(H) and NADP(H) (e.g., Anderson and van Meyenburg, 1977; Vourdouw et al., 1983)

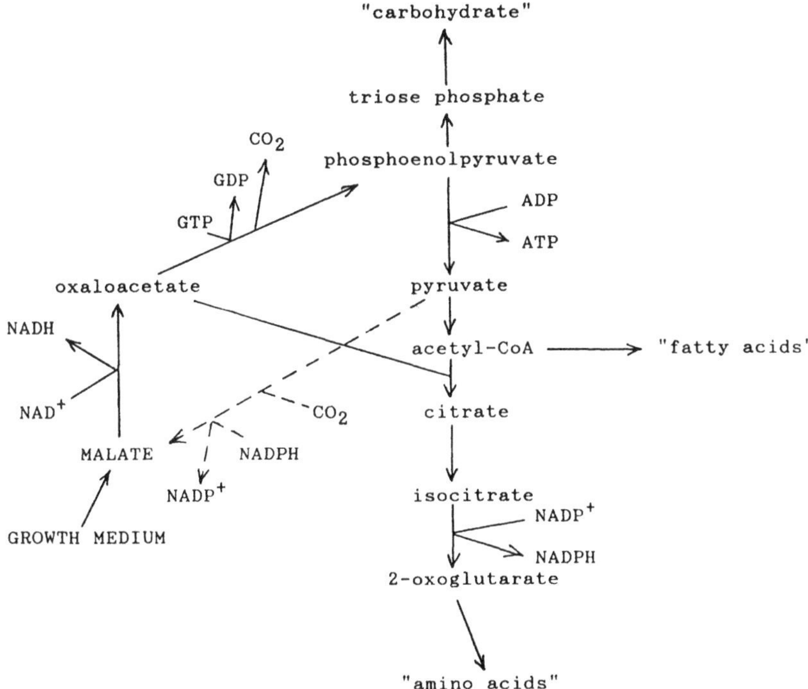

Figure 3. The central metabolic pathways of purple bacteria growing phototrophically on malate.

but it was recognized that the values were meaningless in a thermodynamic sense because the nucleotides would have been partly in an enzyme-bound form. Incidentally, this problem has been circumvented in animal cells by looking at metabolite couples thought to be in equilibrium with NAD(H) or NADP(H) (e.g., see Veech, 1987). If the redox potential of NAD^+/NADH in bacterial cytoplasm is in the region of –280 mV (similar to that found in the matrix of mitochondria and not inconsistent with the total NAD(H) found in the cytoplasm of *E. coli*), then for a Δp of 200 mV, the redox potential of the $NADP^+$/NADPH couple could be as low as –380 mV. Even though the free NAD(H) would be > 90% oxidized, the NADPH would be capable of causing significant reduction of other important low-potential redox couples in the cell.

In this context, H^+-Thase may be especially important in the anaerobic metabolism of *photosynthetic* bacteria growing on organic acids such as malate, succinate, lactate, or pyruvate as a source of carbon. Here, the first steps (Figure 3) involve the generation of NADH by NAD-linked dehydrogenases (e.g. Tayeh and Madigan, 1987). In the absence of oxygen, routes for substantial reoxidation of the NADH might be restricted (but see below). However, NADPH at low potential is required for amino acid and fatty acid biosynthesis, and for detoxification. Thus, the action of H^+-Thase would be to drive $NADP^+$ reduction by NADH using the energy of Δp. The maintenance of the NAD(H) pool at a high redox potential would permit continued turnover of carbon metabolism; the maintenance of the NADP(H) pool at low potential would provide the driving force for biosynthesis. Note that a soluble ("BB" type) transhydrogenase would not generate the NADP(H) at low potential relative to the NAD(H) nor effectively fulfill this purpose. The device could represent a central controlling feature of the energy metabolism of the cell since it would result in a tight coupling between the generation and utilization of biosynthetic precursors by way of the nicotinamide nucleotides; it would also be highly regulated by the value of Δp generated by photosynthetic electron flow. Note, however, that this is probably not the exclusive route for the production of NADPH in photosynthetic bacteria since other enzymes of intermediary metabolism (e.g. isocitrate dehydrogenase; Beatty and Gest, 1981; Leyland and Kelly, 1991) are at least partly NADP-linked. Depending on the precise requirement for NADH and NADPH and on the catabolic pathways employed for a particular substrate, there will be a different degree of involvement of H^+-Thase. In growing photosynthetic bacteria there may be a substantial requirement for NADH: (1) even during (anaerobic) growth on NH_4^+ there is a significant synthesis of energy storage polymers (Eidels and Preiss, 1970); and (2) even during growth on organic acids there can be a high rate of CO_2 fixation by way of a Calvin cycle using an NAD^+-linked triose phosphate dehydrogenase (Lascelles, 1960; Ormerod and Gest, 1962).

H^+-Thase as a Generator of Δp

It is possible, in principle, for the H^+-Thase in bacteria to *use* the redox energy of the NADP(H) and NAD(H) couples to *generate* Δp. Rydstrom and Hoek (1988) suggested that in animal cells H^+-Thase could function as a generator of Δp during anoxic stress. The same may be true in bacteria. Thus, during a period following withdrawal of the energy supply, NADPH (previously driven to a low potential by the action of H^+-Thase or the NADP-linked dehydrogenases or both, see above) and NAD^+ would serve as substrates for H^+-Thase and cause outward H^+ translocation. This could only maintain Δp during brief periods of energy depletion: the total capacity of the nicotinamide nucleotide pools to carry out the transhydrogenase reaction is not large [the concentrations of NAD(H) and NADP(H) in bacterial cytoplasm are approximately 10^{-3} M and 3×10^{-4} M, respectively; Lundquist and Olivera, 1971) and so for *sustained* turnover, continued reoxidation of NADH and rereduction of $NADP^+$ would be required.

It is conceivable that if the catabolic dehydrogenases of a chemoheterotrophic bacterium growing aerobically (or in the presence of an appropriate electron acceptor) were substantially NADP-linked, then the NADPH produced could serve to drive H^+-Thase with the formation of Δp. The NADH resulting from this reaction could then be used as an electron donor for NADH dehydrogenase (and a subsequent respiratory chain) with further development of Δp. It seems unlikely, however, that actively growing bacterial cells well-supplied with nutrients would use their NADPH both as a source of energy and as a source of reducing power for amino acid and fatty acid biosynthesis and for detoxification: The advantage of having two pools of nicotinamide nucleotide would be lost since the redox potentials of the NAD(H) and the NADP(H) would be locked into one another (by way of the relationship with Δp) with little independent flexibility for the control of NADPH-dependent biosynthetic metabolism.

H^+-Thase for the Oxidation of NADPH and Reduction of NAD^+

Of course, when operating in the direction of Δp generation, as just described, H^+-Thase is also causing oxidation of NADPH by NAD^+ and, in some conditions, this may be of physiological value. Consider the following. Typically (though perhaps not universally), when chemotrophic bacteria are growing aerobically and when phototrophic bacteria are growing anaerobically on NH_4^+ and an organic acid or a hexose, an NADP-specific dehydrogenase (e.g., ICDH) is included in the catabolic pathway (Ragland et al., 1966; Beatty and Gest, 1981). Presumably (as described above) these enzymes generate much of the NADPH for amino acid biosynthesis, and H^+-Thase operating as a Δp consumer may contribute to this. Now if bacteria are deprived of nitrogen, a common response is to divert metabolism into the production of storage compounds, such as glycogen or β-hydroxybutyrate (Dawes and Senior, 1973). The enzymes involved in these syntheses, triose phosphate dehydrogenase and acetoacetyl CoA reductase, prefer NADH to NADPH (Lascelles, 1960; Haywood et al., 1988; but see Ritchie et al., 1971), and thus transhydrogenase may be centrally involved in the metabolic switch. Following withdrawal of the nitrogen source, the NADPH concentration in the cells will rise, the NADH concentration will fall, and H^+-Thase will catalyze H^- transfer from NADPH to NAD^+. A soluble, nonenergy-linked transhydrogenase would serve the same purpose but, in contrast to H^+-Thase, would not contribute to the generation of Δp. Of course if H^+-Thase were operating under these conditions, then the value of Δp generated by electron transport would impose a limit on the mass action ratio, $[NADPH][NAD^+]/[NADP^+][NADH]$. If this is a determining factor it would mean that NADH-consuming pathways are favored at low Δp and NADPH-consuming pathways are favored at high Δp.

H^+-Thase as a "Valve" for the Dissipation of Δp

The phenomenon of "metabolic uncoupling" or "overflow metabolism" in bacteria is well established (Tempest and Neijssel, 1984). In combination with reactions that can proceed with both NAD- and NADP-linked enzymes, H^+-Thase could perform a role in the cyclical dissipation of Δp once a critical value has been exceeded.

In cells of *Rb. capsulatus* taken from exponential phase cultures, the ATP synthase is the major consumer of Δp (Clark et al., 1983). The same is probably true in other (non-photosynthetic) bacteria but this is difficult to demonstrate directly. If the cell is deprived of anabolic substrate(s), the rate of consumption of ATP will decrease, and therefore the phosphorylation potential will increase. This will tend to depress the rate of ATP synthesis and lead to an increase in the value of Δp. The extent of these changes is difficult to predict but a substantial increase in Δp might have undesirable effects on the cell membrane.

In fact, the major coupling reactions in the intact cell take place across a very narrow range of Δp. Thus, ATP synthesis (Clark et al., 1983) and, for example, K^+ transport (Golby et al., 1990) take place only above a rather high threshold: the major consumers of Δp are inactivated at low Δp. The physiological consequence of this is that bacterial growth has a marked threshold dependence on Δp (Taylor and Jackson, 1985). It is likely that bacteria have adapted to operate with Δp as large as possible so that they can, when necessary, generate a large phosphorylation potential to drive biosynthetic metabolism. The upper limit of Δp when mainly in the form of $\Delta\Psi$ would be the electrical breakdown value of the membrane, which for prolonged exposure is probably in the region of 200–400 mV (Zimmerman, 1982) and therefore quite close to the values of membrane potential that are recorded in intact bacterial cells (e.g., Clark and Jackson, 1981). Thus, any increase in the value of Δp for the reasons described above could cause breakdown of the membrane with serious consequences.

There are two strategies that the cell might employ to prevent this. An increase in Δp would tend to decrease the rate of electron flow as a result of "respiratory control", generally thought to be due to thermodynamic backpressure on the electron transport-driven proton pumps. For both respiratory (Cotton et al., 1981) and photosynthetic electron flow (Golby et al., 1990) in *Rb. capsulatus* this kind of control has been demonstrated by treating the cells with venturicidin to inhibit the ATP synthase. However, even with sufficient venturicidin to completely inhibit ATP synthesis, the depression of the electron transfer rate was only in the region of 50%—probably not enough in itself to substantially limit the increase in Δp. Furthermore, the current–voltage curves in the presence of venturicidin still revealed a pronounced diodic character. Interestingly, however, the steep increase in ionic current in the presence of venturicidin occurred at an appreciably higher Δp than in the absence of inhibitor (Golby et al., 1990). It is likely, therefore, that at elevated values the bacterial cell membrane operates a "valve" to allow dissipation of Δp. It is possible that, in combination with a pair of cytoplasmic dehydrogenases operating cyclically to oxidize NADPH and reduce NAD^+, H^+-Thase can serve just this purpose.

The requirements are that there should be both NAD-dependent and NADP-dependent dehydrogenases linking two metabolites in a cyclic manner. Equilibrium should favor the oxidized metabolite, but the equilibrium constant should not be very large. In photosynthetic bacteria a candidate system links malate and pyruvate (see Figure 3). In cells growing on malate or succinate, the major metabolic route of many species is probably through malate dehydrogenase (MDH) and phosphoenolpyruvate carboxykinase (PEPCK) (see e.g., Klemme, 1976). Carbohydrate biosynthesis then proceeds through the phosphoglycerates, and the carbon skeletons for amino acid and heme biosynthesis are generated by the action of pyruvate kinase (PK), pyruvate dehydrogenase, and the enzymes of the TCA "cycle". The equilibrium constant of the malic enzyme generally favors catalysis in the direction of malate $\rightarrow$ pyruvate *unless* the [NADPH]/[$NADP^+$] ratio is high ($K_{eq} = 34.4$ in the presence of 1 mM CO_2; Veech et al., 1969). It is proposed that if the rapid operation of H^+-Thase keeps the nicotinamide nucleotides close to equilibrium with Δp, then at high (but only at high) values of Δp the elevated ratios of [NADPH]/[$NADP^+$] tip the direction of the malic enzyme from pyr $\rightarrow$ malate. MDH continues to operate in the direction malate $\rightarrow$ oxaloacetate and is favored by the large ratio of NAD^+/NADH also generated by H^+-Thase. Under these conditions, the resulting cyclical operation of MDH, PEPCK, PK, and ME together with H^+-Thase would serve only to dissipate Δp.

In different organisms, alternative pairs of dehydrogenases might combine with H^+-Thase to produce this controlled dissipation of Δp. In some bacteria (and also in animal heart mitochondria) the NAD-dependent and NADP-dependent isocitrate dehydrogenases would be possible candidates; the equilibrium constant of the NADP-linked enzyme also generally favors reduction of $NADP^+$ (i.e., isocitrate $\rightarrow$ oxoglutarate) unless the [NADPH]/[$NADP^+$] ratio is driven (possibly by H^+-Thase) to a large value. It is also possible that single ICDH enzymes with dual coenzyme specificity (e.g., Leyland and Kelly, 1991) could by themselves carry out the cyclical consumption of the products of H^+-Thase. Although the equilibrium constant of the reaction is appropriate ($K_{eq} = 1.17 \times 10^3$ in the presence of 1 mM CO_2; Londesborough and Dalziel, 1968) there is little kinetic data available for bacterial ICDH operating in the direction oxoglutarate $\rightarrow$ isocitrate. In itself, this hypothesis does not invoke any allosteric control of the enzymes involved; it operates simply on the basis of thermodynamic potentials. However, it seems likely that other control mechanisms are overlaid on this one; for example, the isocitrate dehydrogenases are subject to very tight regulation by allosteric affectors.

Concluding Remarks on the Function of H⁺-Thase

Though there is little direct evidence, it seems likely that this active and potentially critical enzyme functions differently according to the needs of the bacterial cell. In actively growing aerobic cultures well-provided with substrates and an energy source, H^+-Thase probably plays a subsidiary role in the provision of NADPH for the biosynthesis of amino acids and fatty acids. In anaerobic photosynthetic cultures this role may be especially important if the demand for

NADH, produced by soluble dehydrogenases of intermediary metabolism, is low, and the demand for NADPH is high. However, if the requirement for NADPH decreases (e.g., if the nitrogen source is withdrawn), then the H^+-Thase could reverse direction and serve to generate NADH (and Δp) at the expense of NADPH, a situation which would favor the production of storage polymers. In other circumstances of biosynthetic limitation and energy surplus, in combination with pairs of NAD-linked and NADP-linked dehydrogenases operating cyclically, H^+-Thase could catalyze proton reuptake by bacterial cells and prevent a damaging buildup of the proton electrochemical gradient.

III. NADH DEHYDROGENASES

A. The Structure of Mitochondrial NADH Dehydrogenase

NADH dehydrogenase (complex 1) is the largest and most complicated energy-transducing enzyme of the mitochondrial respiratory chain. The enzyme functions as an NADH–ubiquinone oxidoreductase, but until recently there has been a rather limited understanding of its structure and even less was known about its catalytic mechanism (reviewed in Hatefi, 1985; Ragan, 1987). This situation has changed dramatically over the last few years, and several different approaches have coincided to produce new insights into the structure of complex 1 (reviewed in Weiss et al., 1991). A brief description of mitochondrial complex 1 is presented below and this provides a framework for a review of NADH dehydrogenases in photosynthetic membranes.

Hatefi et al. (1962) described the first isolation of complex 1 from bovine heart mitochondria and, apart from the enzyme from *Neurospora crassa* (Ise et al., 1985), it is the only mitochondrial NADH dehydrogenase which has been investigated in detail. Preparations of complex 1 contain between 25 and 30 distinct polypeptides (a recent report has suggested 26 polypeptides (Fearnley et al., 1989)). Complex 1 can be fragmented to generate two water-soluble complexes; the flavoprotein (FP) fragment and the iron protein (IP) fragment (Ragan, 1987). The FP fragment contains 3 polypeptides (molecular mass of 51, 24, and 10 kDa) and FMN in equimolar amounts (Galante and Hatefi, 1979; Ragan et al., 1982a). This subcomplex catalyzes electron transfer between NADH and a wide variety of water-soluble acceptors. The IP fragment contains 6 major polypeptides and about 9 to 10 atoms of iron (Heron et al., 1979; Ragan et al., 1982b). Together, the FP and IP fragments possess the majority of the iron–sulphur centers (centers N-1a, N-1b, N-3, and N-4) which have been assigned to complex 1 (Ragan, 1987). The remaining iron–sulphur centers, including center N-2, are located in the hydrophobic (HP) fragment which remains after resolution of complex 1 with perchlorate (Ohnishi et al., 1985). The HP fragment accounts for 70% of the total protein in complex 1 and is devoid of flavin (Ragan et al., 1982b). A remarkable series of experiments by Weiss et al.,

(1991) has led to further developments. They found that in *N. crassa* the inhibition of mitochondrial protein synthesis by chloramphenicol resulted in the synthesis of a small form of NADH dehydrogenase in place of the large form (Friedrich et al., 1989). This small form of NADH dehydrogenase is composed of 13 nuclear-encoded polypeptides, and contains FMN and all of the assigned iron–sulphur centers of complex 1 other than center N-2 (Wang et al., 1991). It appears to be composed of the FP and IP fragments and has a catalytic site for ubiquinone reduction which has relatively low affinity and is insensitive to the inhibitors piericidin A and rotenone (Friedrich et al., 1989). None of the subunits of the small form are found in the HP fragment of complex 1. Since the HP fragment has been thought to contain the high-affinity binding site for ubiquinone and the site of rotenone binding (Earley and Ragan, 1984), it has been proposed by Weiss et al. (1991) that the large form of NADH dehydrogenase is extended on the acceptor side of the small form. Support for this view has come from electron microscopy studies of single particles of the large and small forms and the HP fragment of complex 1 (Weiss et al., 1991). The large form of the enzyme has an L-shaped structure resembling a boot (Figure 4). The two arms of the large form can be distinguished and they appear to be

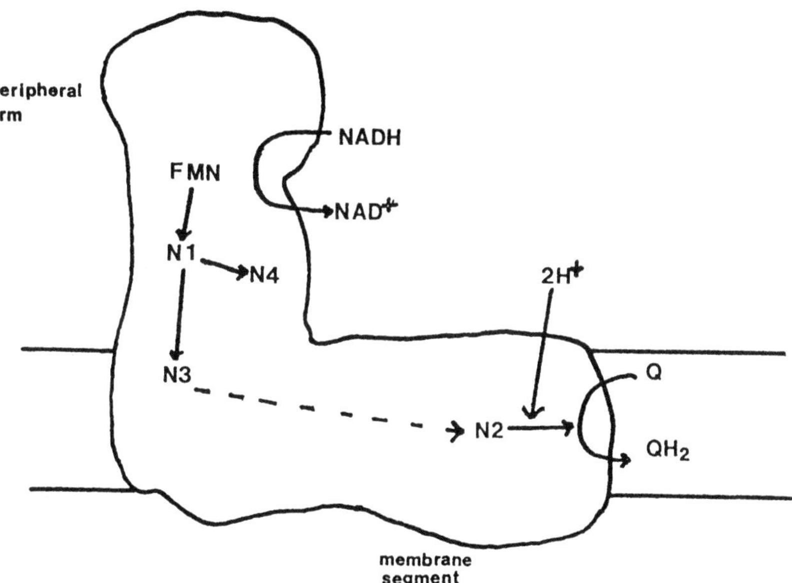

Figure 4. Model for electron transfer in mitochondrial complex 1. Electron flow from NADH to iron–sulphur center N-3 occurs within the peripheral arm of NADH dehydrogenase. Electron transfer from iron–sulphur center N-2 occurs within the membrane segment. Weiss et al. (1991) have proposed that an internal ubiquinone reduction–oxidation cycle is responsible for transfer of electrons between the two arms of NADH dehydrogenase. Modified from Weiss et al. (1991).

composed of the small form, which protrudes from the membrane (the peripheral arm) and a membrane-buried component, which corresponds to the HP fragment.

The model for complex 1 raises the question of how such a remarkable structure might have evolved. DNA and protein sequencing has provided an insight into this problem. Nine subunits of the peripheral arm of complex 1 (corresponding to the FP and IP fragments) have been sequenced and they all appear to be nuclear-encoded (reviewed in Weiss et al., 1991). Comparisons of deduced protein sequences have revealed that the 75, 51, and 24 kDa subunits of the peripheral arm are closely related to subunits of the water-soluble NAD$^+$-linked hydrogenase of *Alcaligenes eutrophus* (Pilkington et al., 1991a). Further DNA sequence analysis by Pilkington et al. (1991a) has indicated that the 51-kDa subunit probably contains a site of NADH binding; this is consistent with the results of photoaffinity-labeling studies (Chen and Guillory, 1981). The 49-kDa subunit, which is a component of the IP fragment, has been found to have a sequence homology to the product of the *orf* 5 gene of the *hyd* locii in *E. coli* (Bohm et al., 1990). The *hyd* gene cluster encodes components of formate hydrogen lyase. In addition, *orfs* 3 and 4 of the *hyd* gene cluster appear to have homology with the genes encoding subunits ND4 and ND1 of mitochondrial complex 1 (Bohm et al., 1990). The discovery of the mitochondrial ND subunits preceded the work described above and it is now known that seven of the subunits of the HP fragment of complex 1 are encoded by mitochondrial genes (Chomyn et al., 1985, 1986). The ND1 subunit is the most conserved of the ND genes and several lines of evidence indicate that it contains the binding site for ubiquinone (Friedrich et al., 1990) and rotenone (Earley et al., 1987). ND1 has also been shown to have sequence homology to the hydrophobic N-terminus of glucose dehydrogenase (Friedrich et al., 1990). Glucose dehydrogenase, located in the periplasm of bacteria, such as *Acinetobacter calcoaceticus* (Duine et al., 1982), functions as a glucose–ubiquinone oxidoreductase. We will return to DNA sequence comparisons in the next section which discusses the presence of *orfs* in the chloroplast genome that encode homologues of subunits of complex 1.

B. NADH Dehydrogenases in Photosynthetic Membranes

Purple Bacteria

The majority of species of purple photosynthetic bacteria are facultative anaerobes and most are capable of chemoheterotrophic growth in the dark (Ferguson et al., 1987). It follows that they should possess an active NADH dehydrogenase. The first evidence that NADH dehydrogenase was coupled to energy transduction came from Klemme (1969) who demonstrated that light-driven, energy-linked reverse electron flow from ubiquinol to NAD$^+$ was sensitive to uncouplers. Klemme (1969) also showed that the reaction was inhibited by rotenone, the classical inhibitor of mitochondrial complex 1. This observation has been confirmed by La Monica and

Marrs (1976), and more recently by Berks and Ferguson (1991) who reported that in membranes from *R. capsulatus* strain B10 5 to 7% of the NADH dehydrogenase activity was insensitive to rotenone and piericidin A. Berks and Ferguson (1991) also reported that the percentage of inhibitor-sensitive activity did not differ between membranes from cells grown photoheterotrophically or chemoheterotrophically. These data indicate that the dominant NADH dehydrogenase activity in membranes of *R. capsulatus* is an enzyme of the mitochondrial type, but the presence of a second NADH dehydrogenase cannot be excluded. Genetic evidence which is consistent with this view has come from Marrs et al. (1972) who isolated a mutant M-1 of *R. capsulatus* which was able to grow photoheterotrophically but not chemoheterotrophically with oxygen as an electron acceptor. In membranes from M-1 the NADH oxidase activity was only 13% of the activity in membranes from wild-type cells, and furthermore this residual activity was insensitive to rotenone (Marrs et al., 1972; Marrs and Gest, 1973). Studies of membranes using EPR spectroscopy (Zannoni and Ingledew, 1983) revealed that the mutant lacked two iron–sulfur centers, E_{m7} at –370 mV and –115 mV.

The close resemblance between the respiratory chains of mitochondria and those of *Rhodobacter* and the related non-photosynthetic bacterium *Paracoccus denitrificans* has been noted (John and Whatley, 1975). It is therefore not surprising that the NADH dehydrogenase of *R. capsulatus* is susceptible to inhibitors of mitochondrial complex 1. Further similarities have been revealed by EPR spectroscopy. Ohnishi and co-workers have identified two binuclear and three tetranuclear iron–sulfur centers corresponding to N-1a, N1-b, N-2, N-3, and N-4 of mitochondrial complex 1 in membranes of *P. denitrificans* (Meinhardt et al., 1987), and similar centers are reported to be present in *R. capsulatus* (Meinhardt et al., 1989). The high-potential iron–sulfur center which is absent in *R. capsulatus* M-1 (Zannoni and Ingledew, 1983) probably corresponds to center N-2.

Genetic evidence, inhibitor titrations, and EPR studies show that a mitochondrial-type NADH dehydrogenase is present in *R. capsulatus*, and a similar situation probably exists in related species. An advantage of studying the counterparts of mitochondrial respiratory enzymes from bacterial sources is that they often possess a comparatively simple polypeptide composition while retaining similar redox centers and mechanistic properties. The success of this approach is illustrated by studies of the cytochrome bc_1 complex of *R. capsulatus* (Davidson and Daldal, 1987) and *P. denitrificans* (Yang and Trumpower, 1983). The purification of a mitochondrial-type NADH dehydrogenase from bacteria would be of value. Most efforts have been directed towards the enzyme from *P. denitrificans*. Yagi (1986) purified an NADH dehydrogenase from membranes of P. dentrificans after solubilization using the chaotropic agent sodium bromide. The preparation exhibited NADH-ubiquinone oxidoreductase activity with ubiquinone-1 as an electron acceptor and was composed of 10 distinct polypeptides (6 major and 4 minor) as indicated by SDS PAGE (Yagi, 1986). The purified enzyme also contained FMN and multiple iron–sulphur centers, although no detailed analysis of the latter was carried out.

Immunological analysis has revealed that homologues of the 49-kDa polypeptides of the IP fragment and the 51- and 25-kDa polypeptides of the FP fragment are present in NADH dehydrogenase of *P. denitrificans* (George et al., 1986). Photoaffinity labeling of the *P. denitrificans* enzyme by Yagi and Dinh (1990) using NADH has confirmed that an approximate 50-kDa polypeptide contains the site of NADH binding. Xu et al. (1991) have cloned the gene encoding the 50-kDa polypeptide and it has homology with the NADH binding subunit of the FP fragment of complex 1. Yagi et al. (1991) have also reported that genes encoding homologues of the 24-kDa subunit (FP fragment) and the 75-, 49-, and 30-kDa polypeptides (IP fragment) are located in this gene cluster. Homologues of the ND1 subunit, the probable binding site for ubiquinone, and the ND5 subunit, which is thought to contain iron–sulfur center N-2 (Weiss et al., 1991), have also been identified (Yagi et al., 1991). The results of Yagi's group confirm that *P. denitrificans* possesses a mitochondrial-type NADH dehydrogenase, and it seems likely that a similar complex will be found in *Rhodobacter* species.

A notable feature of the preparations of bacterial NADH dehydrogenases is that they have lost sensitivity to rotenone (George and Ferguson, 1984; Yagi, 1986). This appears to correlate with the absence of polypeptides corresponding to the HP fragment of mitochondrial complex 1. In view of the model of Weiss et al. (1991) for mitochondrial complex 1 (see Figure 4), it seems probable that detergent solubilization of bacterial dehydrogenases may result in fragmentation of the enzyme into a subcomplex containing the FP and IP fragments (the peripheral arm of NADH dehydrogenase) and the HP fragment (the membrane segment).

In addition to mitochondrial-type NADH dehydrogenase other enzymes which catalyze the oxidation of NAD(P)H and reduction of non-physiological electron acceptors have been identified in membranes of *R. capsulatus*. Oshima and Drews (1981) have purified an NADH dehydrogenase which was composed of six identical 15-kDa subunits and contained FAD. The existence of simple flavoprotein dehydrogenases in membranes of bacteria is not without precedence; in *E. coli* an FAD-dependent NADH dehydrogenase which is composed of a single subunit (Jaworowski et al., 1981) is present in addition to a more complex energy-conserving NADH dehydrogenase (Matsushita et al., 1987). A third NADH dehydrogenase activity has been identified in detergent-solubilized membranes from *R. capsulatus*. This has been assigned to the dihydrolipoamide dehydrogenase subunit of the oxo–acid dehydrogenase complexes which appear to be associated with the membrane in this bacterium (B.C. Berks, A.G. McEwan, and S.J. Ferguson, unpublished observations). This latter observation highlights the caveat that not all "NADH dehydrogenases" which are active with non-physiological electron acceptors arise from respiratory enzymes.

Cyanobacteria

A single NADH dehydrogenase activity has been identified in membranes from *Anabaena variabilis* (Alpes et al., 1989). The NADH dehydrogenase has been purified and was found to contain one major 17-kDa polypeptide, although the presence of a minor 51-kDa polypeptide meant that the 17-kDa polypeptide could not be unambiguously identified as the NADH-oxidizing component. NADH dehydrogenase activity could be reconstituted from inactive enzyme using FAD, but not FMN. This suggests that the cyanobacterial NADH dehydrogenase is not related to the flavoprotein component of mitochondrial complex 1. If an NADH dehydrogenase resembling complex 1 exists in *A. variabilis*, then its activity must be very low or the enzyme must be unstable towards detergents.

Homologues of Subunits of Mitochondrial Complex 1 in Chloroplasts and Cyanobacteria

A surprising finding (Ohyama et al., 1986; Ohyama et al., 1988) has been that 7 *orfs* in the chloroplast genome of the liverworts *Marchantia polymorpha* and tobacco *Nicotiana tabacum* encode proteins with sequence homology to subunits ND1-ND6 and ND4-L of mitochondrial complex 1. The chloroplast genes in *M. polymorpha* are named *ndh*1-6 and *ndh*4-L (Ohyama et al., 1986; Matsubayashi et al., 1987), and *ndh*A-G in *N. tabacum* (Ohyama et al., 1988). The ND subunits of complex 1 are encoded by mitochondrial DNA, but recent observations by Walker and coworkers have established that homologues of some of the nuclear-encoded subunits of mitochondrial complex 1 are coded in chloroplast DNA. A homologue of the 49-kDa subunit of the IP fragment has strong homology with ORFs 392 and 393 in the chloroplast genomes of *M. polymorpha* and *N. tabacum*, respectively (Fearnley et al., 1989). The 30-kDa subunit of the IP fragment also appears to be homologous to ORF 169 in *M. polymorpha* and ORF 158 in *N. tabacum* (Pilkington et al., 1991b). There is no clear evidence that either of these *orfs* encode iron–sulfur proteins but recently, homology between the 23-kDa subunit of the HP fragment of complex 1 and the chloroplast-encoded FrxB has been established (Dupuis et al., 1991).

The *frx*B gene is part of a cluster of genes with homology to subunits of mitochondrial complex 1. The sequence motif CysXXCysXXXCysPro occurs twice in the 23-kDa polypeptide suggesting that the protein contains two 4Fe–4S iron–sulphur centers (Dupuis et al., 1991). Although this polypeptide is a component of the HP fragment, Dupuis et al. (1991) have indicated that its hydrophobicity profile suggests it is not an intrinsic membrane protein. It is therefore unlikely to contain center N-2 since this iron–sulphur center is considered to be associated with electron transfer to ubiquinone and would be expected to be located within the hydrophobic domain of the membrane (Ragan, 1987). An 18-kDa protein whose N-terminal sequence is homologous to FrxB has been purified from thylakoid membranes of *Chlamydomonas reinhardtii* (Wu et al., 1989). Wu et al. (1989) have shown that the EPR spectrum of purified FrxB resembled those of a two [4Fe–4S] center. The FrxB protein could be isolated from thylakoid membranes using a high salt wash in agreement with the view that it is not an integral membrane protein. Indeed, Wu et al. (1989) have reported that FrxB can bind to the chloroplast DNA origin of replication, although the significance of this observation is not known.

Another chloroplast gene, *psb*G, is located adjacent to *ndh*C in *Pisum sativum* (Nixon et al., 1989) and *ndh*J in *M. polymorpha* and *N. tabacum* (Pilkington et al., 1991b). The *psb*G gene was originally thought to be a component of photosystem II (PS II) (Steinmetz et al., 1986) but it now seems likely to be an *ndh* gene (Nixon et al., 1989). The PsbG protein has homology with the ORF7 protein of the formate hydrogen lyase complex of *E. coli* (Bohm et al., 1990), and to an ORF protein encoded in the mitochondrial DNA of *Paramecium* (Pritchard et al., 1989). The *psb*G genes have also been identified in *Synechocystis* 6803 (Mayes et al., 1990). One of these is adjacent to *ndh*C (Steinmuller et al., 1989). Recently a gene cluster has been identified in the filamentous cyanobacterium *Plectonema boryamum* which contains *ndh*1, *frx*B, *ndh*6, and *ndh*4-L (Takahashi et al., 1991). These genes are cotranscribed and occur in the same order as found in the chloroplast genomes. The body of data described above establish that the chloroplast and cyanobacterial genomes encode homologues of the subunits of mitochondrial complex 1. It seems most probable that they encode an NADH dehydrogenase, but the evidence for presence of an NADH dehydrogenase in oxygenic photosynthetic membranes will be discussed in Section III.D.

C. The Function of NADH Dehydrogenase in the Photosynthetic Membranes of Purple Bacteria

The majority of purple nonsulphur bacteria are facultative phototrophs, thus an obvious function of NADH dehydrogenase is in respiratory electron transfer. Along with the other respiratory complexes it contributes to the generation of Δp during aerobic growth in the dark. Marrs et al. (1972) have shown that an NADH dehydrogenase-deficient mutant of *R. capsulatus* is not able to grow chemoheterotrophically with oxygen as an electron acceptor. This mutant M-1 was still able to grow phototrophically, indicating that NADH dehydrogenase is not essential under phototrophic conditions. However, photosynthetic membranes do possess NADH dehydrogenase activity (Klemme, 1969) and this raises the question of the function of this enzyme. In a purple bacterium, such as *R. capsulatus*, photosynthetic electron transfer is cyclic and involves a photochemical reaction center which operates as a ferrocytochrome *c*-ubiquinone oxidoreductase and a cytochrome bc_1 complex which functions as a ubiquinol–ferricytochrome *c* oxidoreductase (Jackson, 1988). The electron transfer system shares components with respiratory chains of this bacterium (Figure 5). Thus, several primary dehydrogenases can reduce ubiquinone. Ubiquinol can be reoxidized via electron transfer to a variety of

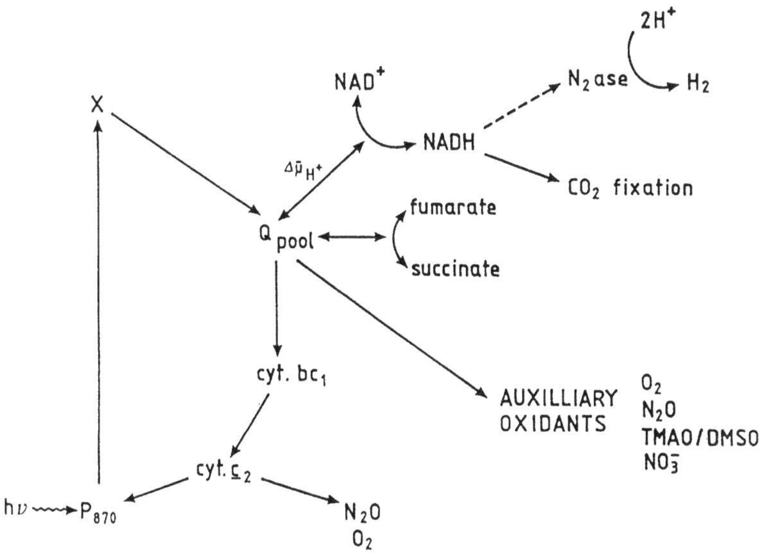

Figure 5. Scheme for photosynthetic and respiratory electron transfer in *R. capsulatus.* NADH dehydrogenase can act as an NADH-ubiquinone oxidoreductase and a generator of $\Delta\tilde{\mu}_{H}{}^{+}$ under chemoheterotrophic growth conditions, or as a ubiquinol-NAD^+ oxidoreductase and a consumer of $\Delta\tilde{\mu}_{H}{}^{+}$ under phototrophic growth conditions.

electron acceptors (Ferguson et al., 1987; Richardson et al., 1988). However, in the absence of external oxidants it has been known for many years that photophosphorylation in chromatophore membranes occurs within a narrow range of ambient redox potential (Loach, 1966). This probably reflects the optimal redox poising of the components of the cyclic electron transfer pathway. Presumably, the bacteria have mechanisms to allow adjustment of redox potential *in vivo*.

NADH dehydrogenase may have a special role in the maintenance of a balanced redox poise (Jones et al., 1990). Although no direct measurements are available for the concentration of free nucleotides in bacteria, it is widely assumed that the E_h for NAD^+/NADH is within 30 mV of its midpoint potential (Voordow et al., 1983). This is low enough to cause complete reduction of the ubiquinone pool if the reaction were allowed to come to equilibrium. NADH dehydrogenase is, however, coupled to Δp and thus the tendency of NADH to reduce ubiquinone in anaerobic, illuminated cells will be decreased ($2\,\Delta E = n\Delta p$ where n is the $H^+/2e$ ratio). Indeed, under phototrophic conditions NADH dehydrogenase may act as consumer of Δp (Figure 5). This is likely to be the case during growth on reductants such as succinate (Jones et al., 1990) or sulfide (Brune and Truper, 1986) which feed electrons directly into the ubiquinone pool. In this case the Δp generated during photosynthetic electron transfer will drive electron flow from succinate to NAD^+ (Figure 5). NADH can then be oxidized via a variety of metabolic pathways (Ferguson et al., 1987; Richardson et al., 1988).

D. The Function of NADH Dehydrogenase in Chloroplasts and Cyanobacteria

Respiratory electron transport is a well-established activity in cyanobacteria (Peschek, 1987), but during the last decade data have emerged which suggest that chloroplasts of green algae also catalyzed respiratory electron transfer to oxygen (Chlororespiration). Bennoun (1982) obtained evidence in *C. reinhardtii* and *Chlorella pyrenoidosa* for direct respiratory control of plastoquinone redox state. It has been suggested that chlororespiration could explain the Kok effect, usually attributed to mitochondrial respiration (Husic and Tolbert, 1987), as PS I inhibiting electron transfer to a terminal oxidase at light intensities above a certain threshold (Peltier and Sarrey, 1985). Direct evidence for the existence of an NADH dehydrogenase in chloroplasts of *C. reinhardtii* has come from Godde and Trebst (1980) who demonstrated PS I-dependent electron transfer from NADH to methylviologen (MV). This reaction was sensitive to retonone, although electron transfer from H_2O to PS I was unaffected. Godde (1982) went on to solubilize in detergent and partially purify a rotenone-sensitive NAD(P)H dehydrogenase. These observations, together with the identification of homologues of mitochondrial complex 1 in chloroplast and cyanobacterial genomes, raise the question of the function of NADH dehydrogenase in oxygenic photosynthetic membranes.

NADH as an Electron Donor to PS I in Cyanobacteria

In photosynthetically active vegetative cells of cyanobacteria it has been shown that light inhibits respiration (Jones and Myers, 1963; Scherer and Boger, 1982). However, provided DCMU is present, electron transfer from NADPH and NADH to PS I can be observed (Sturzl et al., 1984). A physiological situation in which the electron transfer from NADH to PS I may have importance is in nitrogen fixation, particularly in heterocysts. In heterocysts of *Anabaena* nitrogenase is maintained in an environment of low oxygen tension, a contributing factor being the lack of PS II (Tel-Orr and Stewart, 1977). It is assumed that ATP generation in heterocysts is produced via cyclic photophosphorylation (Almon and Bohme, 1982), but for the generation of reduced ferredoxin, the source of electrons for nitrogenase is unclear. Sugars, synthesized in vegetative cells, are transported to the heterocyst where they can be stored or degraded via glycolysis to yield NADH or via the oxidative pentose phosphate cycle to yield NADPH. Light-driven electron transfer from NAD(P)H via PS I to ferredoxin may generate the reductant for nitrogen fixation. An alternative source of reduced ferredoxin is via pyruvate–ferredoxin reductase, but in principle both pathways might operate (see Scherer et al., 1988 for discussion).

NADH as an Electron Donor to PS I in C. reinhardtii

In *C. reinhardtii*, reduction of CO_2 can be linked to oxidation of hydrogen. This light-dependent process, termed photoreduction (Gaffron, 1944), involves only PS I. Presumably, electrons can be fed from hydrogen to PS I via plastoquinone, but the pathway of electron transfer into and out of the plastoquinone pool is unclear. Maione and Gibbs (1986) have suggested that electron transfer to the plastoquinone pool could occur via an H_2-plastoquinone oxidoreductase, or via a ferredoxin–plastoquinone oxidoreductase. However, since photoreduction of CO_2 is inhibited by up to 75% by rotenone (Maione and Gibbs, 1986), this result is perhaps more consistent with the involvement of an H_2-NAD(P) oxidoreductase generating reduced pyridine nucleotide which is then oxidized via the rotenone-sensitive NADH dehydrogenase as described by Godde and Trebst (1980). During photoheterotrophic growth with acetate as a carbon source, an active NADH dehydrogenase would be expected to be functional (Maione and Gibbs, 1986).

Recently, Peltier and Schmidt (1991) have reported that chlororespiratory activity in *C. reinhardtii* was elevated in nitrogen-deficient cells. This was accompanied by increased amounts of NADH dehydrogenase in thylakoid membranes (measured as NADH-*p*-nitrobluetetrazolium chloride oxidoreductase activity) and the synthesis of cytochromes h_1 and h_2 which have recently been suggested to be involved in chlororespiration (Lemire et al., 1986). Ravenel and Peltier (1991) have also presented data which suggests that antimycin A and myxothiazol inhibit chlororespiration in *C. reinhardtii*. Since these inhibitors do not affect electron transfer via the cytochrome b_6/f complex it has been suggested that a "plastoquinol-oxidizing" complex resembling the cytochrome bc_1 complex may be function in chlororespiration (Peltier and Schmidt, 1991; Ravenel and Peltier, 1991). Since antimycin A is known to inhibit photoreduction of CO_2 in *C. reinhardtii* (Maione and Gibbs, 1986) the data suggest that a common pathway of electron transfer between NADH and PS I operates during photoreduction and chlororespiration (Figure 6). If this were the case, then myothiazol would be expected to inhibit photoreduction in *C. reinhardtii*, but the results of such an experiment have not been reported.

A Role for Homologues of Subunits of Mitochondrial Complex 1 in Chloroplast Electron Transfer?

The simplest interpretation of the existence of homologues of subunits of mitochondrial complex 1 in the chloroplast genome is that they are components of an NADH-plastoquinone oxidoreductase. Support for such a view is strongest in *C. reinhardtii* (see Section III.B), but in chloroplasts of higher plants NADH-plastoquinone oxidoreductase activity appears to be either very low or negli-

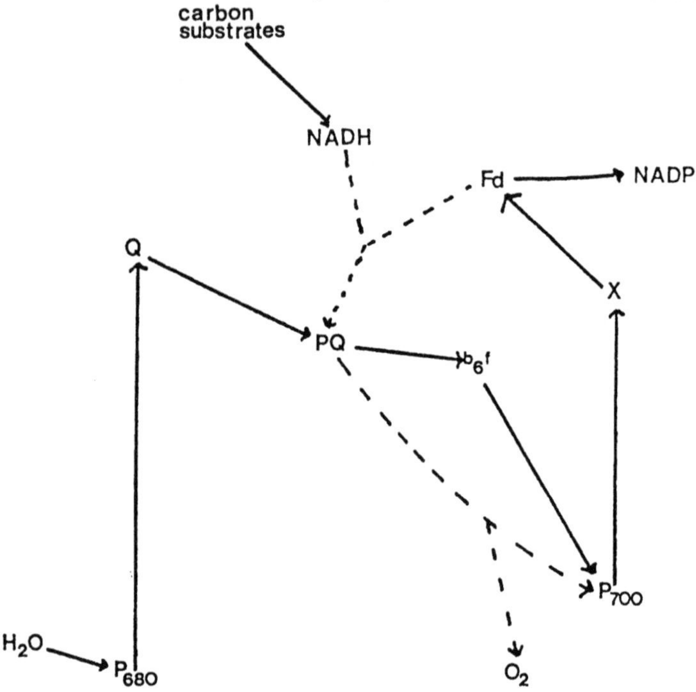

Figure 6. Scheme for electron transfer in chloroplasts and cyanobacteria. A typical Z-scheme is illustrated (——) with additional pathways which have been identified in *C. reinhardtii* chloroplasts (- - - - -). The NADH-plastoquinone oxidoreductase and antimycin-sensitive "plastoquinol–oxidizing" complex may participate in chlororespiration and in electron transfer to PS I.

gible. Godde and Trebst (1980) were unable to detect light-driven PS I-dependent electron transfer from NADH to MV in spinach chloroplasts under the conditions where this reaction occurred in chloroplasts of *C. reindardtii*. Rich (1991) has reported that the rate of reduction of the plastoquinone pool by NADH in mature pea thylakoids is very slow. Assuming a turnover number of $100s^{-1}$ for NADH dehydrogenase, it has been calculated that less than 1 NADH dehydrogenase is present in mature pea thylakoids per 100 photosynthetic electron transfer chains (Rich, 1991). The $t_{1/2}$ for plastoquinol oxidation in mature pea thylakoids is at least 10 min. (P.R. Rich, personal communication), suggesting that chlororespiratory activity in higher plant thylakoids may be negligible. In contrast, Garab et al. (1989) have presented evidence for chlororespiration in pea. However, the interpretation of these measurements is made difficult by the use of protoplasts and open cell preparations, but not isolated thylakoids.

Although NADH dehydrogenase may not be significant in mature chloroplasts, it is possible that it is active during chloroplast development from undifferentiated protoplasts or that it has a role in electron transfer in non-photosynthetic plastids (see Mullet, 1988 for a discussion of chloroplast development). Non-photosynthetic plastids include those which are specialized for the accumulation of carotenoids (chromoplasts), starch, and lipid (amyloplasts). Very little is known about the membrane bioenergetics of these nonphotosynthetic plastids and the question of whether they are capable of respiration needs to be addressed.

It is interesting that no homologue of the 51-kDa subunit of complex 1, which is considered to contain FMN and to be involved in NADH oxidation (Pilkington et al., 1991a), has been identified in the chloroplast genome. The simplest explanation is that this homologue is nuclear-coded and has not yet been identified. However, another possibility is that the chloroplast homologues in higher plant thylakoids catalyze the reduction of plastoquinone, but do not oxidize NADH. In view of the emerging picture of mitochondrial complex 1 as a chimera of electron transfer pathways of independent origins (Weiss et al., 1991), it is possible that an alternative electron donor is used in chloroplasts. A plausible electron donor would be reduced ferredoxin.

The existence of a ferredoxin–plastoquinone oxidoreductase has been considered for many years in order to explain cyclic electron transfer around PS I (Bendall, 1982). Cyclic electron transfer is sensitive to antimycin A (Hosler and Yochum, 1985; de Wolf et al., 1988), a property which is shared with the chlororespiratory and photoreduction pathways of *C. reinhardtii* (Maione and Gibbs, 1985). Figure 6 shows a scheme for electron transfer from reduced ferredoxin to PS I via a ferredoxin–plastoquinone reductase composed of homologues of mitochondrial complex 1 and the antimycin-sensitive "plastoquinol–oxidizing" complex identified by Ravenel and Peltier (1991). A similar situation may exist in cyanobacteria where the NADH dehydrogenase which has been identified is not of the mitochondrial-type (Alpes et al., 1989).

E. Concluding Remarks on NADH Dehydrogenases in Photosynthetic Membranes

Biochemical, spectroscopic, and molecular genetic evidence indicate that an NADH dehydrogenase, which is similar to mitochondrial complex 1, is present in purple photosynthetic bacteria. There has been dramatic progress in the understanding of the structure and genetic organization of mitochondrial complex 1. DNA sequence comparisons suggest that the mitochondrial (and probably the bacterial) NADH dehydrogenase may have evolved as a chimera of electron transfer proteins of independent origin. The surprising finding that homologues of subunits of mitochondrial complex I are encoded by the chloroplast and cyanobacterial genomes leads to the possibility that a mitochondrial-type NADH dehydrogenase is present in thylakoid membranes. However, the only strong evidence for a mitochondrial-type NADH dehydrogenase comes from work on chloroplasts of *C. reinhardtii* where the enzyme appears to function in chlororespiration and in transfer of electrons to PS I. The evidence for an active NADH dehydrogenase in mature higher plant chloroplasts is not strong, but it is possible that the enzyme might be active in non-photosynthetic plastids or developing chloroplasts. Alternatively, in both chloroplasts and cyanobacteria the homologues of subunits of mitochondrial NADH dehydrogenase might be components of a redox complex which catalyzes electron transfer to plastoquinone, but does not use NADH as an electron donor.

ACKNOWLEDGMENTS

Thanks to Ben Berks, Stuart Ferguson, Steve Hanlon, Peter Rich, and Ross Williams for helpful discussions.

AUTHORS' NOTE

The literature survey for this review was completed in April 1992.

ABBREVIATIONS

H^+-Thase:	proton translocating NAD(P)-transhydrogenase
MV:	methylviologen
DCMU:	dichlorophenyldimethylurea
Δp:	proton electrochemical gradient (volt)
$\Delta\tilde{\mu}_{H^+}$:	proton electrochemical gradient (joule mol^{-1})

REFERENCES

Almon, A. & Bohme, H. (1982). Phosphorylation in isolated heterocysts from the blue-green alga *Nostoc muscorum*. Biochim. Biophys. Acta 679, 279–286.

Alpes, I., Scherer, S., & Boger, P. (1989). The respiratory NADH dehydrogenase of the cyanobacterium *Anabaena variabilis*: Purification and characterization. Biochim. Biophys. Acta 973, 41–46.

Anderson, K.B. & Van Meyenburg, K. (1977). Charges of nicotinamide adenine dinucleotides and adenylate energy charge as regulatory parameters of the metabolism in *Escherichia coli*. J. Biol. Chem. 252, 4151–4156.

Beatty, J.T. & Gest, H. (1981). Biosynthetic and bioenergetic functions of citric acid cycle reactions in *Rhodopseudomonas capsulata*. J. Bact. 148, 584–593.

Bendall, D.S. (1982). Photosynthetic cytochromes of oxygenic organisms. Biochim. Biophys. Acta 683, 119–151.

Bennoun, P. (1982). Evidence for a respiratory chain in the chloroplast. Proc. Natl. Acad. Sci. USA 79, 4352–4350.

Berks, B.C. & Ferguson, S.J. (1991). Simplicity and complexity in electron transfer between NADH and *c*-type cytochromes in bacteria. Biochem. Soc. Trans. 19, 581–587.

Böhm, R., Sauter, M., & Bock, A. (1990). Nucleotide sequence and expression of an operon in *Escherichia coli* encoding formate hydrogen lyase components. Mol. Microbiol. 14, 231–243.

Bragg, P.D., Davies, P.L., & Hou, C. (1972). Function of energy-dependent transhydrogenase in *Escherichia coli*. Biochem. Biophys. Res. Commun. 47, 1248–1255.

Brune, D. & Truper, H.G. (1986). Non-cyclic electron transport in chromatophores from photolithotrophically-grown *Rhodobacter sulfidophilus*. Arch. Microbiol. 145, 295–301.

Carlenor, E., Persson, B., Glaser, E., Andersson, B., & Rydstrom, J. (1988). On the presence of a nicotinamide nucleotide transhydrogenase in mitochondria from potato tuber. Plant Physiol. 88, 303–308.

Chang, D.W.B., Hou, C., & Bragg, P.D. (1992). Anomalous effect of uncouplers on respiratory chain-linked transhydrogenation in *Escherichia coli* membranes: Evidence for a localized proton pathway. Arch. Biochem. Biophys. 293, 246–253.

Chen, S. & Guillory, R.J. (1981). Studies on the interaction of Arylazido-β-alanyl NAD$^+$ with the mitochondrial NADH dehydrogenase. J. Biol. Chem. 256, 8318–8323.

Chomyn, A., Mariottini, P., Cleeter, M.W.J., Ragan, C.I., Matsuno-Yagi, A., Hatefi, Y., Doolittle, R.F., & Attardi, G. (1985). Six unidentified reading frames of human mitochondrial DNA encode components of the respiratory chain NADH dehydrogenase. Nature 314, 5922–5927.

Chomyn, A., Cleeter, M.W.J., Ragan, C.I., Riley, M., Doolittle, R.F., & Attardi, G. (1986). URF6, last unidentified reading frame of Human mt DNA, codes for an NADH dehydrogenase subunit. Science 234, 614–618.

Clark, A.J., Cotton, N.P.J., & Jackson, J.B. (1983). The relation between membrane ionic current and ATP synthesis in chromatophores from *Rhodopseudomonas capsulata*. Biochim. Biophys. Acta 723, 440–453.

Clark, A.J. & Jackson, J.B. (1981). The measurement of membrane potential during photosynthesis and during respiration in intact cells of *Rhodopseudomonas capsulata* by both electrochromism and by permeant ion redistribution. Biochem. J. 200, 389–397.

Clarke, D.M. & Bragg, P.D. (1985). Purification and properties of reconstitutively active nicotinamide nucleotide transhydrogenase of *Escherichia coli*. Eur. J. Biochem. 149, 517–523.

Clarke, D.M., Loo, T.W., Gillam, S., & Bragg, P.D. (1986). Nucleotide sequence of the pntA and pntB genes encoding the pyridine nucleotide transhydrogenase of *Escherichia coli*. Eur. J. Biochem. 158, 647–653.

Cotton, N.P.J., Clark, A.J., & Jackson, J.B. (1981). The effect of venturicidin on light and oxygen-dependent electron transport, proton translocation, membrane potential development and ATP synthesis in intact cells of *Rhodopseudomonas capsulata*. Arch. Microbiol. 129, 94–99.

Cotton, N.P.J., Lever, T.M., Nore, B.F., Jones, M.R., & Jackson, J.B. (1989). The coupling between protonmotive force and the NAD(P) transhydrogenase in chromatophores from photosynthetic bacteria. Eur. J. Biochem. 182, 593–603.

Cotton, N.P.J., Myatt, J.F., & Jackson, J.B. (1987). The dependence of the rate of transhydrogenase on the value of the protonmotive force in chromatophores from photosynthetic bacteria. FEBS Lett. 219, 88–92.

Cotton, N.P.J., Myatt, J.F., & Jackson, J.B. (1987). The dependence of the rate of transhydrogenase on the value of the protonmotive force in chromatophores from photosynthetic bacteria. FEBS Lett. 219, 88–92.

Csonka, L.S., & Fraenkel, D.G. (1977). Pathways of NADPH formation in *Escherichia coli*. J. Bact. 252, 3382–3391.

Cunningham, I.J., Baker, J.A., & Jackson, J.B. (1992a). Reaction between the soluble and membrane-associated proteins of the transhydrogenase of *Rhodospirillum rubrum*. Biochim. Biophys. Acta, in press.

Cunningham, I.J., Williams, R., Palmer, T., Thomas, C.M., & Jackson, J.B. (1992b). The relation between the soluble factor associated with H$^+$-transhydrogenase of *Rhodospirillum rubrum* and the enzyme from mitochondria and *Escherichia coli*. Biochim. Biophys. Acta, in press.

Davidson, E. & Daldal, F. (1987). Primary structure of the bc$_1$ complex of *Rhodopseudomonas capsulta*. J. Mol. Biol. 195, 13–24.

Dawes, E.A. & Senior, P.J. (1973). The role and regulation of energy reserve polymers in microorganisms. Adv. Microbial Physiol. 10, 135–266.

de Wolf, F.A., Galmiche, J.M., & Krayenhof, R. (1988). The antimycin sensitivity of flash-induced ATP synthesis in photosystem I enriched subchloroplast vesicles. FEBS Lett. 235, 278–289.

Duine, J.A., Frank Jan, J., & van der Meer, R. (1982). Different forms of quinoprotein aldose-(glucose-) dehydrogenase in *Acinetobacter calcoaceticus*. Arch. Microbiol. 131, 27–31.

Dupuis, A., Skehel, J.K., & Walker, J.E. (1991). A homologue of a nuclear-coded iron-sulfur protein subunit of bovine mitochondrial complex I is encoded in chloroplast genomes. Biochemistry 30, 2954–2960.

Earle, S.R. & Fisher, R.R. (1980). Reconstitution of bovine heart mitochondrial transhydrogenase: a reversible proton pump. Biochemistry 19, 561–569.

Earley, F.G.P. & Ragan, C.I. (1984). Photoaffinity labelling of mitochondrial NADH dehydrogenase with arylazido-amorphigenin, an analogue of rotenone. Biochem. J. 224, 525–534.

Earley, F.G.P., Patel, S.D., Ragan, C.I., & Attardi, G. (1987). Photolabelling of a mitochondrially encoded subunit of NADH dehydrogenase with [^{3}H] dihydrorotenone. FEBS Lett. 219, 108–113.

Eidels, L. & Preiss, J. (1970). Carbohydrate metabolism in *Rhodopseudomonas capsulata*: Enzyme titers, glucose metabolism, and polyglucose polymer synthesis. Arch. Biochem. Biophys. 140, 75–89.

Enander, K. & Rydstrom, J. (1982). Energy-linked nicotinamide nucleotide transhydrogenase: Kinetics and regulation of purified and reconstituted transhydrogenase from beef heart mitochondria. J. Biol. Chem. 257, 14760–14766.

Eytan, G.D., Carlenor, E., & Rydstrom, J. (1990). Energy-linked transhydrogenase: Effects of valinomycin and nigericin on the ATP-driven transhydrogenase reaction catalyzed by reconstituted vesicles. J. Biol. Chem. 265, 12949–12954.

Eytan, G.D., Eytan, E., & Rydstrom, J. (1987a). Energy-linked nicotinamide nucleotide transhydrogenase: Light-driven transhydrogenase catalyzed by transhydrogenase from beef heart mitochondria reconstituted with bacteriorhodopsin. J. Biol. Chem. 262, 5015–5019.

Eytan, G.D., Persson, B., Ekebacke, A., & Rydstrom, J. (1987b). Energy-linked nicotinamide-nucleotide transhydrogenase: Characterization of reconstituted ATP-driven transhydrogenase from beef heart mitochondria. J. Biol. Chem. 262, 5008–5014.

Fearnley, I.M., Runswich, M.J., & Walker, J.E. (1989). A homologue of the nuclear coded 49 kDa subunit of bovine mitochondrial NADH-ubiquinone reductase is coded in chloroplase DNA. EMBO J. 8, 665–672.

Ferguson, S.J., Jackson, J.N., & McEwan, A.G. (1987). Anaerobic respiration in the *Rhodospirillaceae*: characterization of pathways and evaluation of roles in redox balancing during photosynthesis. FEMS Microbiol. Rev. 46, 117–143.

Fisher, K. & Kaplan, N.O. (1973). Studies on the mitochondrial energy-linked pyridine nucleotide transhydrogenase. Biochemistry 12, 1182–1188.

Fisher, R.R. & Earle, S.R. (1980). Reconstitution of bovine heart mitochondrial transhydrogenase: a reversible proton pump. Biochemistry 19, 561–569.

Fisher, R.R. & Earle, S.R. (1982). Membrane-bound pyridine dinucleotide transhydrogenases. In: The Pyridine Nucleotide Coenzymes (Everse, J., Anderson, B.M., & You, K.S., Eds.) pp. 279–324. Academic Press, New York.

Fisher, R.R. & Guillory, R.J. (1971). Resolution of enzymes catalyzing energy-linked transhydrogenation—Interaction of transhydrogenase factor with the *Rhodospirillum rubrum* chromatophore membrane. J. Biol. Chem. 246, 4679–4686.

Friedrich, T., Hofhaus, G., Ise, W., Nehls, U., Schmitz, B., & Weiss, H. (1989). A small isoform of NADH-ubiquinone oxidoreductase (complex 1) without mitochondrially encoded subunits is made in chloramphenicol-treated *Neurospora crassa*. Eur. J. Biochem. 180, 173–180.

Friedrich, T., Strohdeicher, M., Hofhaus, G., Preis, D., Sahm, H., & Weiss, H. (1990). The same domain motif for ubiquinone reduction in mitochondrial or chloroplast NADH dehydrogenase and bacterial glucose dehydrogenase. FEBS Lett. 265, 37–40.

Gaffron, H. (1942). Reduction of carbon dioxide coupled with the oxyhydrogen reaction in green algae. J. Gen. Physiol. 26, 241–267.

Galante, Y.M. & Hatefi, Y. (1979). Purification and molecular and enzymic properties of mitochondrial NADH dehydrogenase. Arch. Biochem. Biophys. 192, 559–568.

Gasab, G., Lajko, F., Mistardy, L., & Marton, L. (1989). Respiratory control over photosynthetic electron transport of higher-plant cells: Evidence for chlororespiration. Planta 179, 349–358.

George, C.L. & Ferguson, S.J. (1984). Immunochemical identification of a two-subunit NADH-ubiquinone oxidoreductase from *Paracoccus dentrificans*. Eur. J. Biochem. 143, 567–573.

George, C.L., Ferguson, S.J., Cleeter, M.W.J., & Ragan, C.I. (1986). Structural relationships between the NADH dehydrogenase of *Paracoccus denitrificans* and bovine heart mitochondria as revealed by immunological cross reactivities. FEBS Lett. 198, 135–149.

Gerolimatos, B. & Hanson, R.L. (1978). Repression of *Escherichia coli* pyridine nucleotide transhydrogenase by leucine. J. Bact. 134, 394–400.

Godde, D. & Trebst, A. (1980). NADH as electron donor for the photosynthetic membrane of *Chlamydomonas reinhardtii*. Arch. Microbiol. 127, 245–252.

Godde, D. (1982). Evidence for a membrane-bound NADH-plastoquinone oxidoreductase in *Chlamydomonas reinhardtii CW15*. Arch. Microbiol. 127, 197–202.

Golby, P., Carver, M., & Jackson, J.B. (1990). Membrane ionic currents in *Rhodobacter capsulatus*—Evidence for electrophoretic transport of K^+, Rb^+, and $NH_4{}^+$. Eur. J. Biochem. 187, 589–597.

Hanson, R.L. (1979). Kinetic mechanism of pyridine nucleotide transhydrogenase from *Escherichia coli*. J. Biol. Chem. 254, 888–893.

Hanson, R.L. & Rose, C. (1980). Effects of an insertion mutation in a locus affecting pyridine nucleotide transhydrogenase (pnt::Tn5) on the growth of *Escherichia coli*. J. Bact. 141, 401–404.

Hatefi, Y. (1985). The mitochondrial electron transport and oxidative phosphorylation system. Ann. Rev. Biochem. 543 1015–1069.

Hatefi, Y., Haavik, A.G., & Griffiths, D.E. (1962). Studies on the electron transfer system: XL. Preparation and properties of mitochondrial DPNH-coenzyme Q reductase. J. Biol. Chem. 257, 1676–1680.

Haywood, G.W., Anderson, A.J., & Dawes, E.A. (1988). The role of NADH- and NADPH-linked acetoacetyl-CoA reductases in the poly-hydroxybutyrate synthesizing organism *Alcaligines eutrophus*. FEMS Microbiol. Res. 52, 259–264.

Heron, C., Smith, S., & Ragan, C.I. (1979). An analysis of the polypeptide composition of bovine heart mitochondrial NADH-ubiquinone oxidoreductase by two-dimensional polyacrylamide gel electrophoresis. Biochem. J. 181, 435–443.

Homyk, M. & Bragg, P.D. (1979). Steady-state kinetics and the inactivation by 2,3-butandione of the energy independent transhydrogenase of *Escherichia coli* cell membranes. Biochim. Biophys. Acta 571, 201–217.

Hosler, J.P. & Yocum, C.F. (1985). Evidence for two cyclic photophosphorylation reactions concurrent with ferrodoxin-catalyzed non-cyclic electron transport. Biochim. Biophys. Acta. 808, 21–31.

Houghton, R.L., Fisher, R.J., & Sanadi, D.R. (1976). Control of $NAD(P)^+$ - transhydrogenase levels in *Escherichia coli*. Arch. Biochem. Biophys. 176, 747–752.

Hu, P.S., Persson, B., Carlenor, E., Hartog, A.F., Hoog, J.O., Jornvall, H., Berden, J.A., & Rydstrom, J. (1988). Properties and sequence of a nicotinamide nucleotide-binding site of mitochondrial nicotinamide transhydrogenase from beef heart. 5th European Bioenergetics Conference, Short Reports 5, 65–65.

Husic, D.W. & Tolbert, N.E. (1987). Inhibition of glycolate and D-lactate metabolism in a *Chlamydomonas reinhardtii* mutant deficient in mitochondrial respiration. Proc. Nat. Acad. Sci. USA 84, 1555–1559.

Ise, W., Haiker, H., & Weiss, H. (1985). Mitochondrial translation of subunits of the rotenone-sensitive NADH: ubiquinone reductase in *Neurospora crassa*. EMBO J. 1, 2075–2080.

Jackson, J.B. (1988). In: Bacterial Photosynthesis—Bacterial Energy Transduction (Anthony, C., Ed.) pp. 317–375. Academic Press, New York.

Jackson, J.B. (1991). The proton-translocating nicotinamide adenine dinucleotide transhydrogenase. J. Bioenergetics Biomembranes 23, 715–741.

Jaworowoski, A., Mayo, G., Shaw, D.C., Campbell, H.D., & Young, I.G. (1981). Characterization of the respiratory NADH dehydrogenase of *Escherichia coli* and reconstitution of NADH oxidase in *ndh* mutant membrane vesicles. Biochemistry 29, 3621–3628.

John, P. & Whatley, F.R. (1975). *Paracoccus denitrificans* and the evolutionary origin of the mitochondrion. Nature 254, 495–498.

Jones, L.W. & Myers, J. (1963). A common link between photosynthesis and respiration in blue-green algae. Nature 199, 670–672.

Jones, M.R., Richardson, D.J., McEwan, A.G., Ferguson, S.J., & Jackson, J.B. (1990). *In vivo* redox poising of the cyclic electron transport system of *Rhodobacter capsulatus* and the effects of the auxiliary oxidants, nitrate, nitrous oxide, and trimethylamine-N-oxide, as revealed by multiple short flash excitation. Biochim. Biophys. Acta. 1017, 209–216.

Keister, D.L. & Yike, N.J. (1966). Studies on an energy-linked pyridine nucleotide transhydrogenase in photosynthetic bacteria I—Demonstration of the reaction in *Rhodospirillum rubrum*. Biochem. Biophys. Res. Commun. 24, 519–525.

Klemme, J.H. (1969). Studies on the mechanism of NAD-photoreduction by chromatophores of the facultative phototroph *Rhodopseudomonas capsulata*. Z. Naturforsch. 246, 67–76.

Klemme, J. H. (1976). Unidirectional inhibition of phosphoenol pyruvate carboxykinase from *Rhodospirillum rubrum* by ATP. Arch. Microbiol 107, 198–192.

Konings, A.W.T., & Guillory, R.J. (1972). Specificity of the transhydrogenase factor for chromatophores of *Rhodopseudomonas spheroides* and *Rhodospirillum rubrum*. Biochim. Biophys. Acta 283, 334–338.

Kuroda, S., Tanizawa, K., Sakamoto, Y., Tanaka, H., & Soda, K. (1990). Alanine dehydrogenases from two Bacillus species with distinct thermostabilities: Molecular cloning, DNA and protein sequence determination and structural comparison with other NAD(P)-dependent dehydrogenases. Biochemistry 29, 1009–1015.

La Monica, R.F. & Marrs, B.L. (1976). The branched respiratory system of photosynthetically grown *Rhodopseudomonas capsulata*. Biochim. Biophys. Acta 423, 431–439.

Lascelles, J. (1960). The formation of ribulose 1:5-diphosphate carboxylase by growing cultures of *Athiorhodaceae*. J. Gen. Microbiol. 23, 499–510.

Lee, C. P., Simard-Duquesne, N., Ernster, L., & Hoberman, H.D. (1965). Stereochemistry of hydrogen transfer in the energy-linked pyridine nucleotide transhydrogenase and related reaction. Biochim. Biophys. Acta 105, 397–409.

Lemaire, C.P., Girard-Bascou, J., Wollman, F.A., & Bennoun, P. (1986). Studies on the cytochrome *b6/f* complex I characterization of the complex subunits in *Chlamydomonas reinhardtii*. Biochim. Biophys. Acta 851, 229–238.

Lever, T.M., Palmer, T., Cunningham, I.J., Cotton, N.P.J., & Jackson, J.B. (1991). Purification and properties of the H^+-nicotinamide nucleotide transhydrogenase from *Rhodobacter capsulatus*. Eur. J. Biochem. 197, 247–255.

Leyland, M.L. & Kelly, D.J. (1991). Purification and characterization of a monomeric isocitrate dehydrogenase with dual coenzyme specificity from the photosynthetic bacterium *Rhodomicrobium vannielii*. Eur. J. Biochem. 202, 85–93.

Liang, A. & Houghton, R.L. (1981). Coregulation of oxidized nicotinamide adenine dinucleotide (phosphate) transhydrogenase and glutamate dehydrogenase activities in enteric bacteria during nitrogen limitation. J. Bact. 146, 997–1002.

Loach, P. (1966). Primary oxidation-reduction changes during photosynthesis in *Rhodospirillum rubrum*. Biochemistry 5, 592–600.

Londesborough, J.C. & Dalziel, K. (1968). The equilibrium constant of the isocitrate dehydrogenase reaction. Biochem. J. 110, 217–222.

Lundquist, R. & Olivera, B.M. (1971). Pyridine nucleotide metabolism in *Escherichia coli*. J. Biol. Chem. 246, 1107–1116.

Maione, T.E. & Gibbs, M. (1986). Association of the chloroplastic respiratory and photosynthetic electron transport chains of *Chlamydomonas reinhardtii* with photoreduction and oxyhydrogen reaction. Plant. Physiol. 80, 364–368.

Marrs, P.L., Stahl, C.L., Lien, S., & Gest, H. (1972). Biochemical physiology of a respiration-deficient mutant of the photosynthetic bacterium *Rhodopseudomonas capsulta*. Proc. Natl. Acad. Sci. 169, 916–920.

Matsubayashi, T., Wakasugi, T., Shinozaki, K., Yamaguchi-Shinozakik, Zaita, N., Hidaka, T., Meng, B. Y., Ohto, C., Tanaka, M., Kato, A., Maruyama, T., & Sugiura, M. (1987). Six chloroplast genes (*ndh*A-F) homologous to human mitochondrial genes encoding components of the respiratory chain NADH dehydrogenase are actively expressed: Determination of the splice sites. Mol. Gen. Genet. 210, 385–393.

Matsushita, H., Ohnishi, T., & Kabeck, H.R. (1987). NADH-ubiquinone oxidoreductases of the *Escherichia coli* aerobic respiratory chain. Biochemistry 96, 7732–7737.

Mayes, S.R., Cook, K.M., & Barber, J. (1990). Nucleotide sequence of the second *psb* G gene in *Synechocystis* 6803: Possible implications for *psb* G function as a NAD(P)H dehydrogenase subunit gene. FEBS Lett. 262, 49–54.

Meijer, E.M., Schuitenmaker, M.G., Boogerd, F.C., & Stouthamer, A.H. (1978). Effects induced by rotenone during aerobic growth of *Paracoccus denitrificans* in continuous culture. Arch. Microbiol. 119, 119–127.

Meinhardt, S.W., Kula, T., Yagi, T., Lillich, T., & Ohnishi, T. (1987). EPR characterization of the iron-sulfur clusters in the NADH: Ubiquinone oxidoreductase segment of the respiratory chain in *Paracoccus denitrificans*. J. Biol. Chem. 262, 8702–8706.

Meinhardt, S.W., Matsushita, K., Kaback, H.R., & Ohnishi, T. (1989). EPR characterization of -2160 the iron-sulfur containing NADH-ubiquinone oxidoreductase of the *Escherichia coli* aerobic respiratory chain. Biochemistry 28, 2153.

Meinhardt, S. W., Matsushita, K., Kaback, H. R., & Ohnishi, T. (1989). EPR characterization of the iron-sulphur containing NADH-ubiquinone oxidoreductase of the *Escherichia coli* aerobic respiratory chain. Biochemistry 28, 2153–2160.

Mitchell, P. (1966). Chemiosmotic coupling in oxidative and photosynthetic phosphorylation. Biol. Rev. Cambridge Philos. Soc. 41, 445–502.

Mitchell, P. (1972). Chemiosmotic coupling in energy transduction: A logical development of biochemical knowledge. J. Bioenergetics 3, 5–24.

Mullet, J.E. (1988). Chloroplast development and gene expression. Ann. Rev. Plant Physiol. 39, 475–502.

Nixon, P.J., Gounaris, K., Coomber, S.A., Hunter, C.N., Dyer, T.A., & Barber, J. (1989). *psb* G is not a photosystem two gene but may be an *ndh* gene. J. Biol. Chem. 264, 14129–14135.

Ohyama, K., Fukuzawa, H., Kohchi, T., Shirai, H., Sano, T., Sano, S., Umersono, K., Shiki, Y., Takeechi, M., Chang, Z., Aota, S., Inokuchi, H., & Ozeki, H. (1986). Chloroplast gene organization deduced from complete sequence of liverwort *Marchantia polymorpha* chloroplast DNA. Nature 322, 572–574.

Ohyama, K., Fukuzawa, H., Kohchi, T., Sano, T., Sano, S., Shirai, H., Umesono, K., Shiki, Y., Takeuchi, M., Chang, Z., Aota, S., Inokuchi, H., & Ozeki, H. (1988). Structure and organization of *Marchantia polymorpha* chloroplast genome. J. Mol. Biol. 203, 281–298.

Ormerod, J.G. & Gest, H. (1962). Hydrogen photosynthesis and alternative metabolic pathways in photosynthetic bacteria. Bact. Revs. 26, 51–66.

Oshima, T. & Drews, S. (1981). Isolation and partial characterization of the membrane bound NADH dehydrogenase from the phototrophic bacterium *Rhodopseudomonas capsulatus*. Z. Naturforsch 36c, 400–406.

Palmer, T. & Jackson, J.B. (1990). A rapid burst preceding the steady-state rate of H^+-transhydrogenase during illumination of chromatophores of *Rhodobacter capsulatus*: Implications for the mechanism of interaction between protonmotive force and enzyme. FEBS Lett. 277, 45–48.

Palmer, T. & Jackson, J.B. (1992a). Nicotinamide nucleotide transhydrogenase from *Rhodobacter capsulatus*; the H^+/H^- ratio and the activation state of the enzyme during reduction of acetyl pyridine adenine dinucleotide. Biochim. Biophys. Acta 1099, 157–162.

Palmer, T. & Jackson, J.B. (1992b). H^+-transhydrogenase in chromatophores from *Rhodobacter capsulatus* after periods of continuous illumination and short flash illumination. Biochim. Biophys. Acta 1098, 21–26.

Palmer, T., Williams, R., Cotton, N.P.J., Thomas, C.M., & Jackson, J.B. (1992). Inhibition of H^+-transhydrogenase from photosynthetic bacteria by N,N^1-dicyclohexylcarbodiimide, in preparation.

Peltier, G. & Sarrey, F. (1985). The kok effect and the light-inhibition of chlororespiration in *Chlamydomonas reinhardtii*. FEBS Lett. 228, 259–262.

Pennington, R.M. & Fisher, R.R. (1981). Dicyclohexylcarbodiimide modification of bovine heart mitochondrial transhydrogenase. J. Biol. Chem. 256, 8963–8969.

Peltier, G. & Schmidt, G. W. (1991). Chlororespiration: an adaption to nitrogen deficiency in *Chlamydomonas reinhardtii*. Proc. Natl. Acad. Sci. USA 88, 4791–4795.

Persson, B., Berden, J.A., Rydstrom, J., & Van Dam, K. (1987). ATP-driven transhydrogenase provides an example of delocalized chemosmotic coupling in reconstituted vesicles and in submitochondrial particles. Biochim. Biophys. Acta 894, 239–251.

Peschek, G.A. (1987). In: Respiratory Electron Transport in the Cyanobacteria (Fay, P. & Van Baalen, G., Eds.), pp. 119–162. Elsevier, Amsterdam.

Petronilli, V., Persson, B., Zoratti, M., Rydstrom, J., & Azzone, G. F. (1991). Flow-force relationships during energy transfer between mitochondrial proton pumps. Biochim. Biophys. Acta 1058, 297–303.

Phelps, D.C. & Hatefi, Y. (1981). Inhibition of the mitochondrial nicotinamide nucleotide transhydrogenase by dicyclohexylcarbodiimide and diethylpyrocarbonate. J. Biol. Chem. 256, 8217–8221.

Phelps, D.C. & Hatefi, Y. (1984a). Interaction of purified nicotinamidenucleotide transhydrogenase with dicyclohexylcarbodiimide. Biochemistry 23, 4475–4480.

Phelps, D.C. & Hatefi, Y. (1984b). Effects of N,N'-dicyclohexylcarbodiimide and N-(ethoxycarbonyl)-2-ethoxy-1,2-dihyroquinoline on hydride ion transfer and proton translocation activities of mitochondrial nicotinamidenucleotide transhydrogenase. Biochemistry 23, 6340–6344.

Phelps, D.C. & Hatefi, Y. (1985). Mitochondrial nicotinamide nucleotide transhydrogenase: active site modification by 5′-(*p*-(fluorosulfonyl)benzoyl)adenosine. Biochemistry 24, 3503–3507.

Pilkington, S.J., Skehel, M., Gennis, R.B., & Walker, J.E. (1991a). Relationship between mitochondrial NADH-ubiquinone reductase and a bacteria NAD-reducing hydrogenase. Biochemistry 20, 2166–2175.

Pilkington, S.J., Skehel, J.M., & Walker, J.E. (1991b). The 30-kilodalton subunit of borine mitochondrial complex I is homologous to a protein coded in chloroplast DNA. Biochemistry 30, 1901–1908.

Pritchard, A.E., Venuti, S.E., Ghalambor, M.A., Sable, C.C., & Cummings, D.J. (1989). An unusual region of *Paramecium* mitochondrial DNA containing chloroplast-like genes. Gene 78, 121–134.

Ragan, C.I. (1987). Structure of NADH-ubiquinone reductase (complex I). Current. Top. Bioenerg. 15, 1–36.

Ragan, C.I., Galante, Y.M., Hatefi, Y., & Ohnishi, T. (1982a). Resolution of mitochondrial NADH dehydrogenase and isolation of two iron-sulfur proteins. Biochemistry 21, 590–594.

Ragan, C.I., Galante, Y.M., & Hatefi, Y. (1982b). Purification of three iron-sulfur proteins from the iron-protein fragment of mitochondrial NADH-ubiquinone oxidoreductase. Biochemistry 21, 2518–2524.

Ravanel, J. & Peltier, G. (1991). Inhibition of chlororespiration by myxothiazol and antimycin A in *Chlamydomonas reinhardtii*. Photosynth. Res. 28, 141–148.
Ragland, T.E., Kawasaki, T., & Lowenstein, J.M. (1966). Comparative aspects of some bacterial dehydrogenases and transhydrogenase. J. Bact. 91, 236–244.
Rich, P.R. (1991). Factors affecting turnover of the cytochrome b_6/f complex of their exploitation in the quantitation of lumend proton mobility and photorespiratory processes. 37th Harden Conference, Wye College, Kent, U.K.; Biochem. Soc. Publ., U.K., p. 37.
Richardson, D.K., King, G.F., Kelly, D.J., McEwan, A.G., Ferguson, S.J., & Jackson, J.B. (1988). The role of auxiliary oxidants in maintaining redox balance during phototrophic growth of *Rhodobacter capsultus* on propionate or butyrate. Arch. Microbiol. 150, 131–137.
Ritchie, G.A., Senior, P.J., & Dawes, E.A. (1971). The purification and characterization of acetoacetyl coenzyme A reductase from *Azotobacter beijerinckii*. Biochem. J. 112, 803–805.
Rydstrom, J. (1979). Energy linked nicotinamide nucleotide transhydrogenase: Properties of proton translocating and ATP driven transhydrogenase reconstituted from synthetic phospholipids and purified transhydrogenase from beef heart mitochondria. J. Biol. Chem. 254, 8611–8619.
Rydstrom, J. & Hoek, J. B. (1988). Physiological roles of nicotinamide nucleotide transhydrogenase. Biochem. J. 254, 1–10.
Rydstrom, J., Lee, C.P., & Ernster, L. (1981). Energy-linked nicotinamide nucleotide transhydrogenase. In: Chemosmotic Proton Circuits in Biological Membranes (Skulachev, V.P. & Hinkle, P.C., Eds.), pp. 483–503. Addison-Wesley, Reading.
Rydstrom, J., Persson, B., & Carlenor, E. (1987). Transhydrogenase linked to pyridine nucleotides. In: Pyridine Nucleotide Coenzymes: Chemical, Biochemical, and Medical Aspects (Dolphin, D., Poulson, R., & Avramovic, O., Eds.), Vol. 2B, pp. 433–460. John Wiley & Sons, New York.
Scharer, S. & Boger, P. (1982). Respiration of blue-green algae in the light. Arch. Microbiol. 132, 329–332.
Scharer, S., Almon, H., & Boger, P. (1988). Interaction of photosynthesis, respiration, and nitrogen fixation in cyanobacteria. Photosynth. Res. 15, 95–114.
Skulachev, V.P. (1970). Electric fields in coupling membranes. FEBS Lett. 11, 301–308.
Skulachev, V.P. (1974). Enzymic generators of membrane potential in mitochondria. Annals N. Y. Acad. Sci. 227, 188–202.
Steinmetz, A.A., Castronejo, M., Sayre, R.T., & Bogorad, L. (1986). Protein PSII-G: An additional component of photosystem II identified through its plastid gene in maize. J. Biol. Chem. 261, 2485–2488.
Steinmuller, K., Ley, A.A., Steinmatz, A.A., Sayre, R.T., & Bogorad, L. (1989). Characterization of the *ndhc-psb* G-ORF 157/159 operon of maize plastid DNA and of the cyanobacterium *Synechocystis* sp. pcc 6803, Mol. Gen. Genet. 216, 60–69.
Sturzl, E., Scharer, S., & Boger, P. (1984). Interaction of respiratory and photosynthetic electron transport, and evidence for membrane-bound pyridine-nucleotides dehydrogenases in *Anabaena variabilis*. Physiol. Plant 6, 479–483.
Takahashi, Y., Shonai, F., Fujita, Y., Kohchi, T., Ohyama, K., & Matsubara, H. (1991). Structure of a cotranscribed gene cluster, *ndh*1-*frx*B-*ndh*6-*ndh*4L, cloned from the filamentous cyanobacterium *Pletonema boryanum*. Plant Cell Physiol. 32, 969–981.
Tayeh, M.A. & Madigan, M.T. (1987). Malate dehydrogenase in phototrophic purple bacteria: Purification, molecular weight, and quaternary structure. J. Bact. 169, 4196–4202.
Taylor, M.A. & Jackson, J.B. (1985). Threshold dependence of bacterial growth on the protonmotive force. FEBS Lett. 192, 199–203.
Taylor, M.A. & Jackson, J.B. (1985). Threshold dependence of bacterial growth on the protonmotive force. FEBS Lett. 192, 199–203.
Tel-Or, E. & Stewart, W.D.P. (1977). Photosynthetic components and activities of nitrogen-fixing isolated heterocysts of *Anabaena cylindrica*. Proc. R. Soc. London B. 198, 61–86.

Tempest, D.W. & Neijssel, O.M. (1984). The status of Y_{ATP} and maintenance energy as biologically-interpretable phenomena. Ann. Rev. Microbiol 38, 459–486.

Veech, R.L. (1987). Pyridine nucleotides and control of metabolic processes. In: Pyridine nucleotide coenzymes, Part B (Dolphin, D., Poulson, R., & Avramovic, O, Eds.), pp. 79–104. John Wiley, New York.

Veech, R.L., Eggleston, L.V., & Krebbs, H.A. (1969). The redox state of free nicotinamide adenine dinucleotide phosphate in the cytoplasm of rat liver. Biochem. J. 115, 609–619.

Voordow, G., van der Vies, S.M., & Themmen, A.P.N. (1983). Why are two different types of pyridine nucleotide transhydrogenase found in living organisms? Eur. J. Biochem. 131, 527–533.

Wakabayashi, S. & Hatefi, Y. (1987a). Amino acid sequence of the NAD(H) binding region of the mitochondrial nicotinamide nucleotide transhydrogenase modified by N^1,N^1-dicyclohexylcarbodiimide. Biochem. Internat. 15, 667–675.

Wakabayashi, S. & Hatefi, Y. (1987b). Characterization of the substrate-binding sites of the mitochondrial nicotinamide nucleotide transhydrogenase. Biochem. Internat. 15, 915–924.

Wang, D-C., Meinhardt, S. W., Sackmannm, U., Weiss, H., & Ohnishi, T. (1991). The iron-sulfur clusters in the two related forms of mitochondrial NADH: ubiquinone oxoreductase made by *Neurospora crassa*. Eur. J. Biochem. 197, 257–264.

Weiss, H., Friedrich, T., Hofhaus, G., & Preis, D. (1991). The respiratory-chain NADH dehydrogenase (complex I) of mitochondria. Eur. J. Biochem. 197, 563–576.

Wu, M., Nie, Z.Q., & Yang, J. (1989). The 18-KD protein that binds to the chloroplast DNA replicative origin is an iron-sulfur protein related to a subunit of NADH dehydrogenase. Plant Cell 1, 551–557.

Xu, X., Matsuno-Yagi, A., & Yagi, T. (1991). The NADH-binding subunit of the energy-transducing NADH-ubiquinone oxidoreductase of *Paracaccus denitrificans*: Gene cloning and deduced primary structure. Biochemistry 30, 6422–6428.

Yagi, T. (1986). Purification and characterization of NADH dehydrogenase complex from *Paracoccus denitrificans*. Arch. Biochem. Biophys. 250, 302–311.

Yagi, T. & Dinh, T.R. (1990). Identification of the NADH-binding subunit of NADH-ubiquinone oxidoreductase of *Paracoccus denitrificans*. Biochemistry 29, 5515–5520.

Yagi, T., Xu, X., & Matsuno-yagi, A. (1991). Gene cluster of the energy-transducing NADH-ubiquinone oxidoreductase (NDH-1) of *Paracoccus denitrificans*. Biol. Chem. Hoppe-Seylar 372, 547–555.

Yamaguchi, M. & Hatefi, Y. (1989). Mitochondrial nicotinamide nucleotide transhydrogenase: NADPH binding increases and NADP binding decreases the acidity and susceptibility to modification of cysteine-893. Biochemistry 28, 6050–6056.

Yamaguchi, M., Hatefi, Y., Trach, K., & Hoch, J.A. (1988). The primary structure of the mitochondrial energy-linked nicotinamide nucleotide transhydrogenase deduced from the sequence of cDNA clones. J. Biol. Chem. 263, 2761–2767.

Yang, X. & Trumpower, S.L. (1986). Protonmotice Q cycle pathway of electron transfer and energy transduction in the three subunit ebiquinol-cytochrome c oxidoreductase complex of *Paracoccus denitrificans*. J. Biol. Chem. 263, 11962–11970.

Zannoni, D. & Ingledew, W.J. (1983). *Rhodopseudomonas capsulatus* respiratory dehydrogenase mutants: An electron paramagnetic resonance study. FEMS Microbiol. Lett. 17, 331–334.

Zimmermann, U. (1982). Electric field-mediated fusion and related electrical phenomena. Biochimica et Biophysica Acta 694, 227–277.

STRUCTURAL ELEMENTS INVOLVED IN THE ASSEMBLY AND MECHANISM OF ACTION OF RUBISCO

Steven Gutteridge and Tomas Lundqvist

Advances in Molecular and Cell Biology
Volume 10, pages 287–335.

ISBN: 1-55938-710-6

I. INTRODUCTION

Since the last reviews encompassing cellular and molecular aspects of Rubisco*-related research (Andrews and Lorimer, 1988; Gutteridge and Gatenby, 1988; Gutteridge, 1991; Hartman, 1992), various new insights can be reported. At the cellular level, there is now a better understanding of the role of protein factors required for assisting the assembly of the protein. Indeed the interactions between these factors (termed chaparonins) and Rubisco appear to be common for many other unfolded proteins. The process of protein folding in the cell apparently requires more than just the primary sequence, and as a result of recent systematic studies of Rubisco folding with the purified components, chaparonin function is better understood (Gatenby et al., 1992; Viitanen et al., 1992)

The interaction between the enzyme and stromal regulatory factors in chloroplasts is also known in greater detail; for example, the function of activase is clearer (Robinson and Portis, 1989; Portis, 1991; Lan et al., 1992). Similarly, further studies concerning the source and fate of the phosphorylated natural inhibitor, 2CA1P (Gutteridge and Julien, 1989; Seeman et al., 1990; Parry et al., 1992) is providing information on the regulation of CO_2 fixation and its relationship to the light reactions.

At the molecular level, the basis for enzyme infidelity is not completely unraveled, but some of the structural elements that contribute to the diversion of substrate into unwanted reactions have been identified, although others remain elusive. The combination of crystallographic analysis with *in vitro* mutagenesis is leading to a detailed understanding of the function of individual amino acids (see Schneider and Lindqvist, 1992).

One outcome of mutagenesis is that the results can be quite sobering. Very often, more is discovered about an amino acid that contributes nothing to function than might be desired. Two factors may be at play here. First, a mundane and frustrating reason is that our favorite amino acid simply is not part of the essential catalytic framework. Second, the partial reactions of a complex enzyme mechanism may require the combined intervention of more than one residue, and a single mutation may not be adequate to suppress all catalytic activity. From a more positive perspective, mutagenesis has provided a means of dramatically perturbing and interrupting the events required to achieve product formation and thus reveal details of the underlying chemistry. In some cases this might be exploited to redirect the reaction into new and novel products. The ability to introduce unnatural side chains into protein structures promises to widen the choice of group substitutions beyond the natural ones available.

This chapter does not attempt to cover all aspects of Rubisco research; rather it is meant to reemphasize the view that the reactions catalyzed by Rubisco are mutable and thus the partitioning of substrate is not a fixed constraint. The enzyme is still evolving, and it is tantalizing to predict that we may soon be able to understand this protein from synthesis through assembly to the function of individual amino acids in catalysis.

Retrospective

The basic functional unit of the enzyme is a dimer of large subunits (L_2), a form of Rubisco isolated from some photosynthetic prokaryotes. The most studied of this form II Rubisco is from the purple nonsulfur bacterium, *Rhodospirillum rubrum* (Schloss et al., 1979). The L-subunit is composed of 466 amino acids, giving a M_r of 51 kDa. The enzyme is no longer isolated from the authentic host; rather, much larger quantities are obtained from recombinant material expressed in *Escherichia coli* (Pierce and Gutteridge, 1986).

The most abundant species of Rubisco is form I which is composed of eight large and eight small subunits (L_8S_8). The L-subunits vary in length from 473 amino acids in cyanobacteria to 480 in some higher plant species. The S-subunits also range from 110 to 140 residues, depending on the source of the protein. This complicated aggregate is also the most efficient carboxylase and exists throughout all oxygenic photosynthetic species from bacteria to higher plants. A number of different form I genes have been expressed in heterologous hosts, but those most exploited are the genes coding for the enzyme from *Synechococcus* PCC6301. The protein assembles correctly into active holoenzyme irrespective of whether the individual subunits are cotranscribed or expressed separately from different vectors or hosts (Andrews, 1989; Tabita and Small, 1990; Gutteridge, 1991).

Both forms of Rubisco catalyze the same primary reactions of carboxylation and oxygenation of ribulose 1,5-bisphosphate. The former leads to two molecules of 3P-glycerate (i.e., net CO_2 fixation), whereas oxygenation produces 2P-glycolate plus 3P-glycerate. All species require activation through carbamylation with a molecule of CO_2 and binding of a Mg^{2+} ion. The major difference between species is the relative rates of the two catalytic reactions. Form II enzymes are less efficient at CO_2 fixation by almost a factor of 10 compared with the best hexadecameric species (Jordan and Ogren, 1981). The overwhelming problem yet to be solved is the determination of the structural differences in the enzyme that contribute to this variation. At present the expectation is that a comparison of the high-resolution structures of three species of Rubisco with distinct relative specificities will indicate which elements are important.

The enzyme is by far the major source of reduced carbon in the biosphere, responsible for some 10^{11} tons of atmospheric carbon fixed per year. At ambient concentrations of CO_2 and O_2 that exist at the present time, many major crop plants lose nearly one-third of the ribulose–P_2 substrate through oxygenation. In the absence of any overwhelming metabolic requirement for plant photorespiration other than to recycle the product of oxygenation, P-glycolate, this remains a major loss of reduced carbon for these plants. At present there is still no evidence to indicate that plants would not benefit if endowed with a superior version of the enzyme, whether identified from natural sources or designed rationally from structural and functional information. Both pursuits are actively in progress.

II. CELLULAR ASPECTS

A. Synthesis and Assembly of the Holoenzyme

Irrespective of the source of the structural genes that code for Rubisco, both form I and II enzymes can be expressed in heterologous hosts. A major obstacle is that only the genes from prokaryotic photosynthetic organisms provide active assembled enzyme, at least in bacterial hosts. Those expression systems based on eukaryotic genes may generate soluble protein, but the assembly mechanism of the bacterium appears to be incompatible with that operative in eukaryotes. Therefore, although 25% of the soluble cell contents might be functional Rubisco using the genes from either *R. rubrum* or *Synechococcus*, in the case of wheat or maize Rubisco, no intact or active hexadecameric enzyme has been isolated (see Gutteridge and Gatenby, 1988).

The structural genes for the L- and S-subunits of *Synechococcus* Rubisco comprise a dicistronic operon, with the S-subunit downstream of the L-subunit gene, separated by about 90 bases (Shinozaki and Sugiura, 1985). The intervening sequence harbors secondary structural elements that may be important for coordinate expression of the smaller subunit. Both subunits are translated from a single large message and the open reading frames preceded by recognizable ribosome binding sites. The same genes from *Anabena* are organized in a similar fashion, but in this case the separation between the subunits is larger (Gurevitz et al., 1985). The product of this construct does not produce a full complement of the smaller subunit, and the excess L-subunits not involved in aggregate formation are insoluble. This led the investigators to propose an unlikely mode of assembly of the hexadecamer.

The first studies of the assembly of Rubisco involved mild acid treatment of the hexadecameric cyanobacterial enzyme in high sulfate close to the pI of the protein (Andrews and Ballment, 1983). The differential solubilities of the subunits ensured that the L8 core precipitated leaving behind a significant amount of the S-subunits in solution. The resulting L8 core, unlike that of higher plant origin, was recoverable and reassociation of the S-subunits generated active holoenzyme. These studies suggested that if an L8 core assembled first, then association with S-subunits to holoenzyme requires no assistance from other protein factors. Furthermore, S-subunits obtained from higher plant sources are almost equally effective at reconstituting the holoenzyme (Andrews and Lorimer, 1985).

The ability of a bacterial heterologous host, such as *E.coli*, to synthesize active hexadecameric enzyme—albeit from prokaryotic sources—suggested that an assembly machinery was probably not required to generate an L8 core. However, pulse chase studies of the association of newly synthesized Rubisco subunits in chloroplast stroma with high molecular weight proteins, implicated a necessity for protein factors in assembly (Ellis, 1981). Although antibodies raised against the Rubisco "large subunit binding protein" (LSBP) of stroma were found to cross-react with proteins of many organisms, it was the similarity of the LSBP sequence and Gro EL from *E.coli* that indicated protein factors involved in folding nonnative polypeptides might be a general cellular mechanism (Hemmingsen, et al., 1988). From those systematic studies showing the requirement for protein chaparonins to mediate assembly of form II Rubisco L-subunits at normal physiological temperatures (Viitanen et al., 1992), in addition to the positive effect achieved by coexpression of chaparonins and Rubisco in bacterial hosts (Goloubinoff et al., 1989), it is now clearer at what stage in Rubisco assembly that chaparonins are required. During purification of an L8 core from bacterial extracts, there is evidence that the core can reversibly dissociate to L dimers. This suggests that involvement of chaparonins is required to orchestrate folding of monomeric L-subunits that, once released, associate to dimers.

From these recent studies of the folding of the simplest dimeric Rubisco, it is clear that chaparonins are not absolutely required for the protein to achieve its final folded form, at least *in vitro*. Once suitable conditions of protein concentration and temperature are achieved, significant amounts of active enzyme can be regenerated in their absence. However, rapid expression of protein from high-copy vectors at elevated temperatures *in vivo* producing large amounts of unfolded protein, requires the presence of chaparonins as essential cofactors. For those investigators attempting to identify elements of Rubisco structure critical for function, the absence of an efficient expression system to construct and assess mutations in higher plant enzyme is a frustrating obstacle. The identification and purification of protein factors required to mediate folding may be the prelude to designing a system compatible with higher plant subunits. However, isolated L-subunits from a hexadecameric enzyme have yet to be unfolded and refolded to generate an L8 core. Second, the factors that are required for assembling the higher plant enzyme in a heterologous host have yet to be identified. The inability to achieve formation of assembled L_8S_8 by coexpressing higher plant L- and S-subunits in bacteria suggests that bacterial assembly mechanisms may not be compatible with plant-type systems. Potentially, additional mediating factors may yet be identified that are required for this fascinating and fundamental process.

B. Expression of Recombinant Rubisco

The first plasmids that produced active recombinant Rubisco were obtained from a library of *Rhodospirillum rubrum* cDNA (Somerville and Somerville, 1984). These constructs not only provided the complete sequence of the L-subunit of a form II enzyme (Nargang et al., 1984), but also removed any further doubts that the oxygenase activity was a characteristic of Rubisco function. Although the first attempts to synthesize the protein in *E. coli* were fraught with inconsistencies, adequate precautions to retain the plasmid with appropriate selection pressure soon indicated that even a low-copy plasmid like pRR2119 could generate huge quantities of protein from modest growth volumes (Pierce and Gutteridge., 1985). As a result of this initial success, enough protein was purified to satisfy the crystallographers and provide the first high-resolution structure of the enzyme (Schneider et al., 1986; Lundqvist and Schneider., 1989). Most recently, plasmids have been designed around pUC vectors with appropriate unique restriction sites to aid *in vitro* mutagenesis strategies (Gutteridge et al., 1988; Lorimer et al., 1988; see also Larimer et al., 1986).

The first successful attempts to express the hexadecameric version of the enzyme was achieved using the dicistronic operon from *Synechococcus* PCC6301 (Gatenby et al., 1985). Initially, the genes were expressed behind a temperature-sensitive P_L promoter from a low-copy number plasmid. Although this resulted in production of active assembled enzyme, the levels were far from spectacular and not enough to achieve the same results that were obtained with form II Rubisco expressed from pRR2119. Construction of an expression vector based on pUC plasmids have largely overcome yield problems (Gutteridge et al., 1986; Kettleborough et al., 1987) and the recombinant hexadecameric enzyme can be obtained at yields approaching 25% of soluble contents in *E.coli*. Enough protein was purified to generate a high-resolution structural map of the *Synechococcus* enzyme (Newman and Gutteridge, 1990; Newman et al., unpublished)

C. Plant Expression Systems

Expression of the genes coding for the two subunits of Rubisco in eukaryotes requires the coordinated control of both cytosolic and chloroplastic synthetic processes. The S-subunit is synthesized in the cytosol as a precursor protein with a 40 to 60 amino acid extension to the N-terminus required for targeting the protein to the chloroplast stroma. The precursor passes through the double envelope of the chloroplast membrane in an ATP-dependent process (Soll and Waegamann, 1992; Waegamann and Soll, 1992) and appears in the stroma where a peptidase removes the target peptide (for review see Bascomb et al., 1992). Presumably the existence of preassembled L-subunit cores in the chloroplast sequester the newly arrived mature S-subunits to form the active hexadecameric aggregate.

The control of S-subunit synthesis is light and tissue-dependent in higher plants, and in many cases there is a heterogeneous population of the protein resulting from expression of multigene copies. Even in *Arabadopsis*, there are 4 versions of the S-subunit gene (Krebbers et al., 1988). None of the sequence differences are at a region of the protein involved in L- and S-subunit interfaces and thus unlikely to alter the activity of the enzyme significantly.

The L-subunit gene is localized on the plastid genome and synthesized on chloroplastic ribosomes. There is evidence that once synthesized the L-subunit is posttranslationally modified by the action of a peptidase that removes the first one (Met) or two (Met.Ser) amino acids (Houtz et al., 1989). The truncated subunit is then blocked by acetylation. Other posttranslational alterations have been identified. A Lys at position 14 of the subunit is methylated in some plant species by a specific N-methyl transferase enzyme (Houtz et al., 1991), although the significance of this modification remains obscure. The subunit is susceptible to proteolysis of the N-terminus once released from plant material, particularly plants starved of nitrate. Two Lys residues of the N-terminus at positions 8 and 14 are exposed to solution and readily cleaved by trypsin (Gutteridge et al., 1986). Loss of the N-terminal peptide up to Lys 8 is tolerated by the protein, but removal of the second peptide from 9 to 14 causes a dramatic loss of activity (Gutteridge et al., 1986) through weakened affinity for ribulose-P_2 binding (Phillips et al., 1987). Thus, this region plays some role in enzyme activity that might be modulated by methylation of Lys 14.

The inability of isolated higher plant L-subunits to refold *in vitro*, or apparently in heterologous expression systems, suggests that the chloroplast has its own unique set of assembly proteins. Unlike bacterial chaparonin 60 (cpn 60), cpn 60 of the chloroplast is isolated as two distinct subunits and has only 46% homology with the bacterial counterpart (Hemmingsen et al., 1988). The plastid has its own version of cpn 10, the protein that assists release of the folded subunit from cpn 60. Presumably these factors act to coordinate the release of a newly synthesized and folded L-subunit until a dimer or octamer can form in a sea of active hexadecamer. Conversely, it might also be considered that preexisting active proteins are protected from unwanted association with unfolded, itinerant polypeptides. Nevertheless, the modification of the L-subunits post-translationally may also be a component of the assembly mechanism of higher plant Rubisco (Roy, 1989).

The absence of a suitable expression system for higher plant Rubisco, apart from the natural host, has meant that alternative hexadecameric species have been exploited for mutagenesis. The genes for the cyanobacterial enzyme have proved particularly effective for producing large quantities of protein for detailed investigations of changes in structure and function. However, a means of determining the response that a photosynthetic organism might develop to such mutations has yet to be designed satisfactorily.

Ideally, a photosynthetic host is required that grows heterotrophically without the presence of Rubisco. It also has to be an organism that transforms stably with modified Rubisco genes. An attempt was made to engineer such a host using cyanbacteria, but this may not be the ideal choice. These organisms have a unique way of packaging the enzyme in carboxysomes, and they also possess a CO_2 concentrating mechanism (Badger, 1987; Price and Badger, 1991). Nevertheless, the organism can be transformed relatively efficiently and DNA integrated into host genome by homologous recombination. Unfortunately, all attempts to construct a mutant that lacked Rubisco genes failed even though the organism was grown nonphotosynthetically on a variety of reduced carbon compounds (Pierce et al., 1989). Nevertheless, replacement of the host Rubisco genes with only the L-subunit gene from *Rhodospirillum rubrum* produced a mutant organism that responded to the relative concentrations of CO_2 and O_2 consistent with a decrease in carboxylation efficiency associated with *R. rubrum* Rubisco. Interestingly, these organisms lacked the ability to form carboxysomes, suggesting that either S-subunits or DNA adjacent to the native Rubisco genes may be required for carboxysome assembly (Kaplan et al., 1990).

The location of the L-subunit gene in the chloroplast of photosynthetic eukaryotes is a major obstacle to investigating function through mutagenesis. Thus, most studies understandably have focused on the nuclear encoded S-subunit genes that are accessible for manipulation. The presence of multiple-gene copies (see e.g. Dean et al., 1985) has precluded mutagenesis studies, but significant insight into control of promoter function—for example, in response to light, tissue specificity, and time (Coruzzi et al., 1984)—have been forthcoming. Plants transformed with S-subunit antisense produce reduced quantities of the holoenzyme (Rodermel et al., 1988). The unpredictability of the response of transgenics expressing antisense genes provides plants with a wide variation in the amounts of active Rubisco. A question often raised is whether C3 plants need to invest as much nitrogen in synthesizing the prodigious quantities of Rubisco that naturally accumulates in chloroplasts. Initial interpretations of the response of the mutant plants to smaller amounts of Rubisco suggested that transformants could survive adequately with less than 60% of the enzyme (Quick et al., 1991). However, these results were acquired with plants grown in much less than natural illumination. When the transgenics are grown in light conditions usually encountered in the field, it is clear that all of the enzyme is fully employed fixing CO_2 (Hudson et al., 1991).

D. Interaction between Rubisco and Stromal Components

The confirmation that essentially all Rubisco synthesized in the plant is involved with assimilating CO_2 means that recent proposals about the regulation of enzyme activity through interaction with other stromal components (Gutteridge, 1991; Portis, 1991) must play a significant role in determining the productivity of photosynthesis. Two components that have been investigated in detail are 2′-carboxy arabinitol 1-phosphate (2CA1P), a potent naturally occurring inhibitor of the enzyme, and activase.

2CA1P

The diurnal change in Rubisco activity detected initially in young tomato seedlings and then in many plant species (Vu and Bowes, 1984; Servaites et al., 1986) is due to the binding of 2CA1P at the active site of activated Rubisco *in vivo* (Gutteridge et al., 1986; Berry et al., 1987). The inhibitor is a monophosphate that resembles the six-carbon intermediate of carboxylation, 2-carboxy 3-keto arabinitol 1,5-bisphosphate (3k2CABP), and the tight-binding synthetic inhibitor, 2′-carboxy arabinitol bisphosphate (2CABP). The monophosphate accumulates in plants in the dark, often to amounts exceeding Rubisco active site concentrations (5 mM) with an affinity for the activated form of the enzyme in the nanomolar range. Nevertheless, the inhibitor dissociates from the enzyme long enough for it to be degraded by a specific phosphatase (Gutteridge and Julien., 1989; Holbrook et al., 1989), relieving the inhibition of Rubisco that is already activated and thus primed to restart CO_2 fixation.

For inhibition to be truly diurnal, it might be expected that the phosphatase would be inactive in the dark and activated by light. Although some investigators have reported activation of the phosphatase by reducing equivalents likely to abound only in light conditions, this phenomenon has not been widely confirmed. For example, the phosphatase isolated from tobacco retains significant activity in the absence of thiol-reducing equivalents, suggesting that the enzyme may always be functional. However, the activity of the phosphatase isolated from potato or *Phaseolus* does respond to the presence of a number of stromal components, such as NADPH and fructose-P_2. Plants do not synthesize large quantities of the phosphatase, and it has yet to be purified to homogeneity: only then will its control and regulation be fully understood (see e.g. Parry et al., 1992).

Table 1. Substrate Specificity of Tobacco 2CA1P Phosphatase

Substrate	*Rel. Activity*
2CA1P	100
2CABP	250
2CRBP	60
4CABP	140
3P-(D)glycerate	3
3P-(L)glycerate	40
2P-glycolate	2
ribulose-P_2	1

Notes: 2CA1P: 2′-carboxy arabinitol 1-phosphate.
2CRBP: 2′-carboxy ribitol 1,5-bisphosphate.

Partially purified preparations of 2CA1P phosphatase have been characterized in terms of its specificity for substrates, and Table 1 indicates that it is relatively specific for phosphate esters that have the arabinitol configuration at the C2 carbon. Furthermore, the enzyme dephosphorylates molecules that have at least three C centers; that is, phosphoglycolate is not a substrate for this enzyme.

Identification of a specific phosphatase suggests that the first product of the breakdown of the inhibitor in plants must be 2′-carboxy arabinitol (2CA). The obvious fate of this rather unreactive molecule is to be recycled by a kinase back to inhibitor. A kinase has yet to be isolated with this activity, although the presence of a pool of 2CA in some plants (Seeman et al., 1990) and the appearance of label into 2CA1P from radioactive 2CA fed to plants indicates that this is the most likely source of the monophosphate. The origin of significant amounts of 2CA in plants has also to be unraveled.

Activase

One of the contenders (along with cpn 60) as the second most abundant protein in the stroma of C3 plants is a species termed activase. The protein was so-named as a result of its identification in *Arabidopsis*. Conditionally lethal mutants of the plant requiring high CO_2 for growth were found to contain normal Rubisco (Somerville et al., 1983). The enriched CO_2 requirement correlated with the absence of two related stromal proteins. Subsequent studies pointed to the involvement of these proteins in the activation mechanism of Rubisco, potentially assisting carbamylation of the inactive enzyme (see Portis, 1991 and refs. cited therein).

Further investigations have provided a more complete understanding of the function of activase in terms of the response of Rubisco to the presence of this protein (Robinson and Portis., 1989). The active site of Rubisco is designed to bind sugar bisphosphates with some affinity. Not only does the stroma contain ribulose-P_2 in quantity (5–20 mM may accumulate) but also molecules that closely resemble the substrate and compete effectively for the enzyme active site (e.g., 2CA1P, fructose-P_2). Some of these molecules have high affinity for the inactive form of the enzyme; others favor the activated ternary complex. In either case the enzyme is severely inhibited. Addition of activase to Rubisco inhibited by such substrate analogues removes the inhibition by altering the affinity of the enzyme for the inhibitor. Presumably the K_m for ribulose-P_2 is similarly affected, but the stromal concentrations of the substrate are high enough to offset any decline in turnover due to the loss of binding affinity.

The advantage of a protein with this function is evident during catalytic turnover of Rubisco. As ribulose-P_2 is consumed, there is a significant decline in Rubisco activity to about one-third of the initial rate of substrate consumption. The inactivation has been traced to the accumulation of different isomers of ribulose-P_2 produced at the active site of the enzyme as a consequence of normal turnover (Edmondson et al., 1991; Jensen et al., 1992; also see page 311). Activase has the ability to relieve the inhibition by these substrate analogues and the initial rates of turnover persist until the majority of the substrate is depleted. Clearly, this would suggest that activase interacts with Rubisco to alter the affinity of the enzyme for binding bisphosphates; however, that interaction has yet to be fully characterized.

The above describes two potential regulatory mechanisms that modulate the activity of Rubisco *in vivo*. From an extensive survey of many plant species, regulation due to 2CA1P and activase may be operative in the same plant.

III. MOLECULAR ASPECTS

A. Protein Structure

Within the past decade, there has been a major step forward in the accumulation of structural data on Rubisco. Since the first complete primary sequence of the *R. rubrum* enzyme reported by Somerville and colleagues (Nargang et al., 1984), there are now some 500 or more L-subunit sequences in data banks, and the three-dimensional structures of four species of the enzyme have been solved to high resolution. A number of site-specific mutants of *R. rubrum* (Schneider et al., 1992) and *Synechococcus* (Newman et al., unpublished) have also been crystallized for comparison with wild-type enzyme. The expectation is that this wealth of structural information will provide the explanation for the differences in activities between species of Rubisco.

B. Primary Sequences

Figure 1 shows the best alignment of the primary sequences of the L-subunits of those enzymes that have been crystallized and the structures solved to high resolution. Until the three-dimensional structure became available, only one amino acid had known function, although specific roles had been assigned to some others based on chemical modification and active site-directed inhibitor effects (see Hartman, 1992 for a recent review). Lysine occupies position 201 (191 in *R. rubrum* Rubisco) in the primary sequence of the spinach L-subunit, and is absolutely essential for enzyme activity. It is one of the few residues surrounded by a short stretch of conserved amino acids that occur in all Rubisco L-subunit sequences. A molecule of CO_2 reacts to form a carbamate at the ε-amino group of this residue. The formation of the carbamate to generate EC (see Figure 2) is the preamble to the activation of the enzyme by completing the Me ion binding site. The ternary complex between enzyme, CO_2, and metal (ECM) that emerges is the only active form of the enzyme, and to date any attempt to remove this amino group has produced an inactive enzyme. Many substrate and reaction intermediate analogues favor binding to this form of the enzyme [e.g., 2CABP and 2CA1P (EQC)], a feature that has been exploited to obtain the structure of the quaternary 2CABP complex

```
                                                50                                                    100
MSPQTETKAS VGFKAGVKDY KLTYYTPEYE TLDTDILAAF RVSPQPGVPP EEAGAAVAAE SSTGTWTTVW TDGLTNLDRY KGRCYHIEPV AGEENQYICY
.......... ........E. .........Q .K........ ..T....... .......... .......... .....S.... .....R..R. V..KD...A.
   MPK.QSA A.Y....... .......D.T PK...L.... .F.......A D.....I... .......... ..L..DM... ..K....... Q....S.FAF

             MDQSSRYV N.ALKEEDLI AGGEHV.C.Y IMK.KA.YG- YV.T..F... .....NVE.C .---.DDFTR GVDALVY.VD EAR.LTK.AYP
                                                             H


                                               150                                                    200
VAYPLDLFEE GSVTNMFTSI VGNVFGFKAL RALRLEDLRI PVAYVKTFQG PPHGIQVERD KLNKYGRPLL GCTIKPKLGL SAKNYGRAVY ECLRGGLDFT
.......... .......... .......... .......... .P........ .......... .......... .......... .......... ..........
I......... .....IL... .........I .S.....I.F ...L...... .......... L.......M. .......... .......... ..........

..LFDRNITD .KAMAS..LT M..NQ.MGDV EYAKMH.FYV .E..RAL.D. .SVN.SALW- ---.V...VV .TI....... RP.PFAE.CH AFWL.....I
             I   L                                               L   EVDGGL


                                               250                                                    300
KDDENVNSQP FMRWRDRFLF CAEALYKAQA ETGEIKGHYL NATAGTCEDM MKRAVFAREL GVPIVMHDYL TGGFTANTTL SHYCRDNGLL LHIHRAMHAV
.......... .......... .......... .......... ........E. I......... .......... ........S. A......... ..........
.....I.... .Q........ V.D.IH.S.. .......... .V..P...E. ....E..K.. .M..I...F. .A........ AKW.....V. ..........

.N..QG.-.. .APL..TI.V -AD.MRR..D ....A.LFSA .I..DDPFEI IA.GEYVL.T -FGENA.AL. VD.YV.GAAA ITTA.RRFPDNF..Y...G...
    P              A                                                  S V                                G
```

```
                                                   350                                                  400
IDRQKNHGMH FRVLAKALRL SGGDHIHSGT VVGKLEGERD ITLGFVDLLR DDYTEKDRSR GIYFTQSWVS TPGVLPVASG GIHVWHMPAL TEIFGDDSVL
........I. .........M .......... .......... .......... ..FV.Q.... ......D... L.....E... .......... ..........
....R...I. ......C... .....L.... .......DKA S.......M. E.HI.A.... .VF...D.A. M......... .......... V.........

TSP..R-.YT AF.HC.MA.. Q.ASG..T.. MG..M...SS ------.RAI AYMLTQ.EAQ .PFYR...GG MKACT.II.. .MNALR..GF F.NL.NAN..
    S                              F                                                                        I


                                                   450
QFGGGTLGHP WGNAPGAVAN RVALEACVQA RNEGRDLARE GNTIIREATK WSPELAAACE VWKEIKFEFP AMDTV           spinach
.......... .......... ........K. ........Q. ..E.....C. .......... .....V.N.A .V.VLDK         tobacco
.......... .......T.. .......... .......Y.. .GD.L...G. ........LD L........E T..KL           Synechococcus

TA...AF..I D.PVA..RSL .Q.WQ.WRDG VPVL-.Y... HKELA.-.FE -..GD.DQIY P..ALGV.DT RSALPA          Rhodospirillum
                                                         F          G R                             rubrum
```

Figure 1. The primary sequences of the L-subunits of three species of Rubisco with known three-dimensional structures. The hexadecameric species from plants, spinach and tobacco, and from cyanobacterium *Synechococcus* PCC6301, aligned without recourse to gaps except at the extreme ends of the sequence. In contrast, the best alignment of the sequence of the dimeric Rubisco of *Rhodospirillum rubrum* requires the presence of gaps and loop-outs. The former are indicated as dashes, the latter are indicated in bold type below the alignment and positioned between those residues they are adjacent to in the sequence.

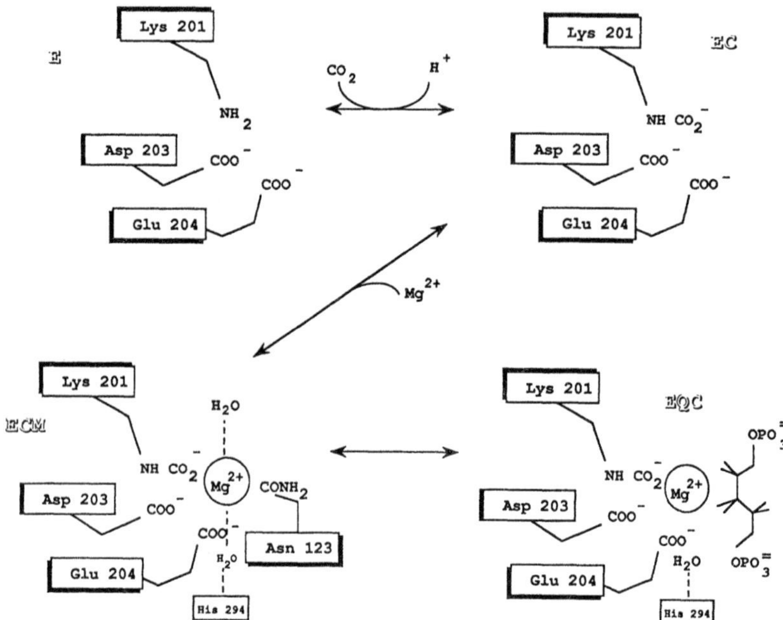

Figure 2. The reversible equilibria between inactive Rubisco and the cofactors CO_2 and Mg^{2+} producing activated enzyme. Those groups of the enzyme that are involved in stabilizing the carbamylated form of the active site Lys (201) and coordinating the metal are identified. "E" denotes the inactive enzyme and shows those acidic residues of loop 2 that may determine the pK_a of Lys 201. The deprotonated form of Lys is carbamylated relatively slowly by a molecule of CO_2 generating "EC" (also inactive), followed by rapid coordination of the essential metal to generate "ECM". This is the active form of Rubisco and those groups that occupy the primary positions around the Mg^{2+} are shown. The ECM ternary complex can bind bisphosphates such as ribulose-P_2 substrate or its inhibitory analogues to form a quaternary complex "EQC". Specific ligands of the metal are displaced on binding bisphosphates. The quaternary complex formed with the intermediate analogue, 2CABP and ECM, is essentially irreversible.

of the spinach, tobacco, and *Synechococcus* enzymes (Chapman et al., 1988; Andersson et al., 1989; Newman and Gutteridge, 1990).

The homology between the form II and form I primary sequences are strikingly low at about 28% comparing spinach and *R. rubrum*, whereas between form I species the similarities range from 80% (if cyanobacteria is included) to 90% between higher plant sequences. Prior to the availability of a three-dimensional structure, the role of a number of amino acids was inferred using active site-directed probes and the first applications of site specific mutagenesis. In this respect, the absence of homology was particularly helpful because it was suspected that those few amino acids conserved across all Rubisco species must have essential functions, either structurally or in catalysis. Hartman and colleagues used both approaches to particular advantage to attempt to map out the reactive groups that composed the active site and to obtain details of their function (Hartman, 1992).

C. Higher Levels of Structure

Only one form II enzyme structure has been determined to date, namely Rubisco from *R. rubrum* at 1.7 Å resolution. Three structures of the form I hexadecameric enzyme are now available that include *Synechococcus* Rubisco and the higher plant species, spinach, and tobacco are now complete to 2.2, 2.4 and 2.6 Å resolution, respectively. Indeed, the first three-dimensional structure of Rubisco, initially to only 2.9 Å resolution, was of the form II dimer of *Rhodospirillum rubrum*. The enzyme was a recombinant version of the authentic wild-type species expressed in *E. coli*. The first effective expression vector, pRR2119, had the L-subunit gene fused in-frame with β-galactosidase which provided a 24 amino acid extension to the N-terminus of the L-subunit (Larimer et al., 1985). Fortunately the excess sequence had no influence on enzyme activity, but may have proved useful for subsequent crystallization. The plasmid is unstable, but can be coaxed into synthesizing large quantities of active enzyme in the bacterium if suitable precautions are employed to ensure efficient selection (Pierce and Gutteridge, 1986). The resulting gram quantities of purified enzyme ultimately revealed the first complete crystallographic structure. The initial low-resolution maps indicated both the organization of the two L-subunits in the dimer and showed that each subunit was composed of two domains. One of the domains is an eight-stranded α/β barrel accounting for two-thirds of the primary sequence at the C-terminus of the L-subunit (Branden et al., 1984). The remaining one-third of the sequence is organized as antiparallel β-sheets with helical segments on both sides. These various elements are illustrated in Figure 3A.

The barrel structure is a motif found in many enzymes that has proved extremely adaptable for supporting various catalytic functions (Farber and Petsko, 1990). In the case of Rubisco, the bisphosphate substrate binds in a relatively extended conformation along the C-terminal surface of the barrel, with the phosphate and hydroxyl groups interacting with amino acids located in loops that extend above and over the rather elliptical barrel. One face of the substrate is thus directed toward the barrel surface and away from the solution, imparting the necessary stereochemical constraints on the catalytic events that follow. Those amino acids involved in binding to the metal ion and directing the catalysis are also located in these loop regions.

Very often enzymes composed of more than one domain catalyze reactions between two or more substrates, and Rubisco conforms to this organization. With increasing resolution of the structure of the dimeric enzyme, it was clear that each L-subunit could not function independently. Amino acids of the N-terminal domain also contribute to active site structure, not within the same subunit but rather with

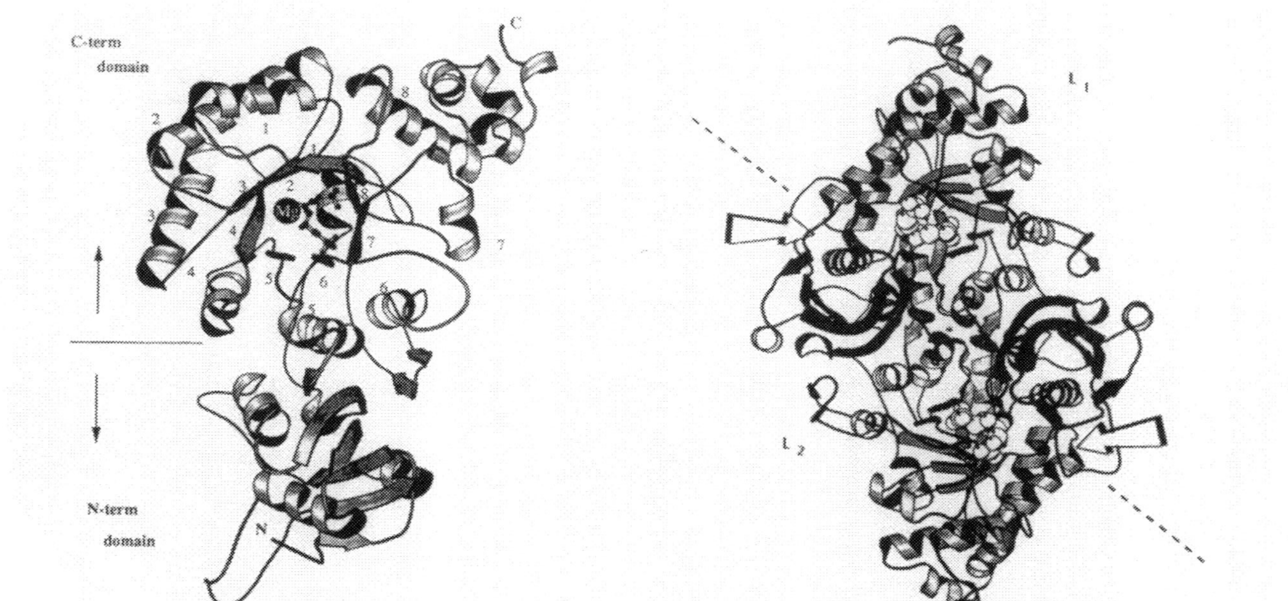

Figure 3. The domain structure of the L-subunit of Rubisco. (**A**) This Molscript display (Kraulis, 1991) of *R. rubrum* Rubisco clearly shows the alternating β-sheets and α-helices of the eight-stranded barrel C-terminal motif. The active site is situated at the C-terminal ends of the sheets and indicated by the location of the Mg^{2+} ion and bound substrate molecule. The N-terminal domain is composed of anti-parallel sheets with adjacent helical regions. (**B**) The minimum active unit of Rubisco is a dimer of L-subunits. The dashed line denotes the demarcation between each individual L-monomer. A twofold axis of symmetry (*) exists at the center of the dimer, perpendicular to the display. In this orientation the entrances to the active sites of the enzyme are shown by the arrows, although both are closed off by amino acids of the N-terminal domain of the partner subunit.

amino acids of a second L-subunit (see Figure 3B). Thus the active site is shared between two large subunits making the minimum functional entity a dimer (Larimer et al., 1987; Schneider et al., 1988).

Structure of the Spinach Enzyme

The structure of hexadecameric Rubisco from spinach is now refined to 2.4 Å resolution and is of the stable quaternary complex formed between activated enzyme and 2′-carboxy arabinitol bisphosphate (2CABP) (Andersson et al., 1989; Knight et al., 1990). The structure therefore closely resembles the organization of the active site immediately following the reaction between substrate CO_2 and the enediolate intermediate of the bisphosphate, generating the six-carbon intermediate, 3k2CABP.

Although there is only 28% homology between the primary sequences of spinach Rubisco L-subunits and the enzyme from *R. rubrum*, the structural elements composing the two domains are almost identical. Apart from the absence of S-subunits in the form II enzyme, the only other striking difference is the disposition of the C-terminal tail of the L-subunits and the positions of some of the loops around the barrel. Thus the active sites of the hexadecameric enzyme also require the interaction of amino acids of both domains of two L-subunits; subsequently the organization of the L8 core of the protein is best described as a tetramer of L-subunit dimers. Each L-subunit dimer has a twofold axis of symmetry that runs along the interface of the two subunits. In nearly all hexadecameric enzymes a Cys residue of each subunit forms an intradimeric bridge straddling this twofold axis (Newman and Gutteridge, 1990; Ranty et al., 1991)

The structure of the form II dimer does not have the tight binding inhibitor at the active site, and therefore cannot be used to infer the basis for catalytic differences of the two enzymes. However, there are two regions of the dimer that have ill-defined electron density associated with two loops of the L-subunit: one resides in the N-terminal domain between amino acids 60–70, and the other resides in the C-terminal domain composing loop 6. Both carry amino acids that are essential for catalytic activity. In the structure of the stable quaternary complex of form I enzyme, the loops are locked in a closed position over the active site and are well defined. It is therefore likely that movement of these two loops plays a significant part in the catalytic mechanism of the enzyme (see Branden et al., 1991).

The S-Subunit

Figure 4 shows the positions of the subunits of the hexadecameric enzyme, particularly the S-subunits relative to the L8 core. They are located around the fourfold axis of the molecule and situated between each L-subunit dimer (Chapman et al., 1988; Knight et al., 1990). Part of the primary sequence forms a loop that extends down into the prominent channel that runs down the fourfold axis of the

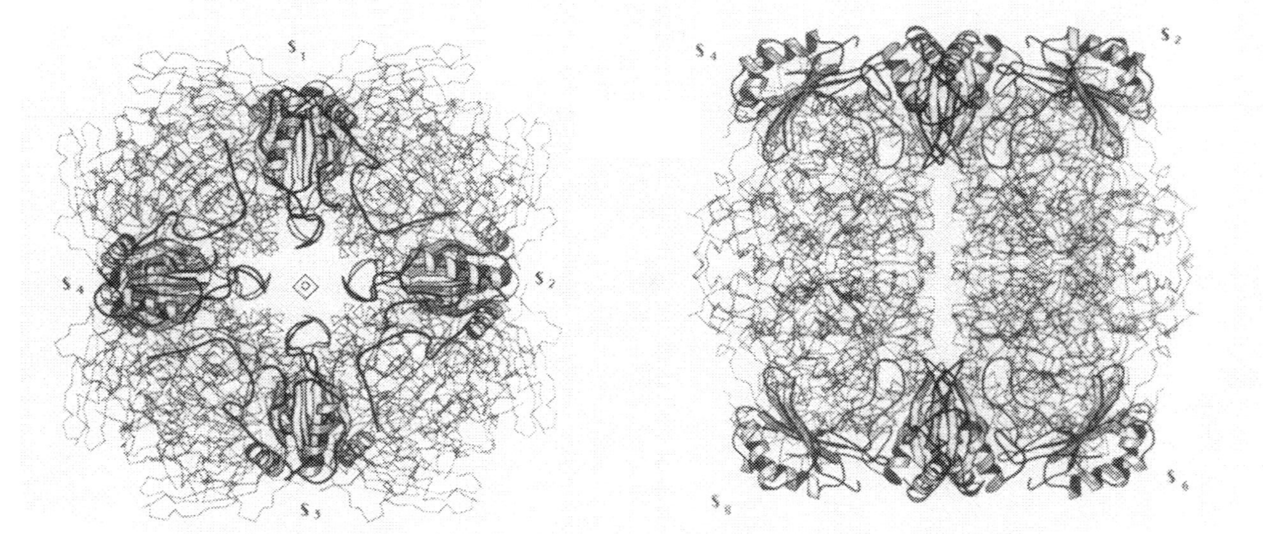

Figure 4. The organization of the subunits of the L_8S_8 Rubisco. In (**A**) the display is from above the fourfold axis of symmetry of the enzyme. The four L-subunit dimers are shown only as $C\alpha$ traces whereas the structural elements of the S-subunits have been displayed with Molscript. The four S-subunits shown, reside at the top of the molecule situated between each L-subunit dimer, with a loop extending into but not obscuring the prominent central channel of the L8 core. In (**B**) the other four S-subunits are found to occupy the same position at the bottom of the core.

molecule (Figure 4A). This loop is absent in the S-subunits of cyanobacterial Rubisco.

The major interactions between the S-subunit and the L-subunit occur with the C-terminal barrel, involving amino acids of the shorter loops at the N-terminus of the domain surface; that is, the bottom of the barrel. A comparison of the $C\alpha$ traces of the barrel domain of the form I enzyme with the form II enzyme shows that they are almost superimposable except for a segment around $\alpha 7$ and $\alpha 8$ (Schneider et al., 1990). The interaction with S-subunits involves loops at the C-terminal end of these two helices. Naturally, it has been speculated that the difference in catalytic specificities of the two enzymes might be due to this structural difference. However, an analysis of the partitioning of ribulose-P_2 between carboxylation and oxygenation, catalyzed by purified recombinant L8 core of *Synechococcus* Rubisco and compared with the reconstituted holoenzyme, indicates that there is little difference between the two species. This suggests that, at least in the case of the *Synecococcus* enzyme, the S-subunits do not influence the direction of carboxylation or oxygenation (Gutteridge, 1991). Similarly, reconstitution of purified L8 core with S-subunits from spinach Rubisco, generates a hybrid holoenzyme that still retains the specificity of cyanobacterial Rubisco (Andrews and Lorimer, 1987; however see Read and Tabita, 1992).

Unfortunately a comparison of the ternary complex of the dimeric enzyme with the quaternary structure of the hexadecamer may not be the most revealing for discerning the function of S-subunits. The hexadecamer has two of the flexible loops around the active site in more closed and rigid positions than the dimer, and it might be that there is similar movement of the loops connecting helices 7 and 8 with the barrel. A better comparison would be the structure of the quaternary complexes of the L8 core and the intact hexadecamer. However, only the latter structure has been solved to date, and the purified core is proving somewhat less amenable to crystallization. So far our insights concerning the function of S-subunits come from studies of the catalytic characteristics of the isolated L8 core before and after reconstitution with S-subunits (Andrews, 1989; Gutteridge, 1991).

D. Other Structures of Rubisco

Five other structures of various complexes of the dimeric enzyme have been solved. An activated ternary complex with CO_2 and Mg^{2+} ions was generated by soaking the crystals of the inactive enzyme with the cofactors (Lundqvist and Schneider, 1991a). A structure of the binary complexes between inactive enzyme and the product molecule, 3P-glycerate, and the tight-binding inhibitor, 2CABP, have also been solved (Lundqvist and Schneider, 1989, 1990). The latter reveals that the absence of a carbamate at Lys 191, and thus bound metal leaves the inhibitor free to reside at the active site "upside-down"; that is, the 1-phosphate occupies the position normally taken by the 5-phosphate group and vice versa.

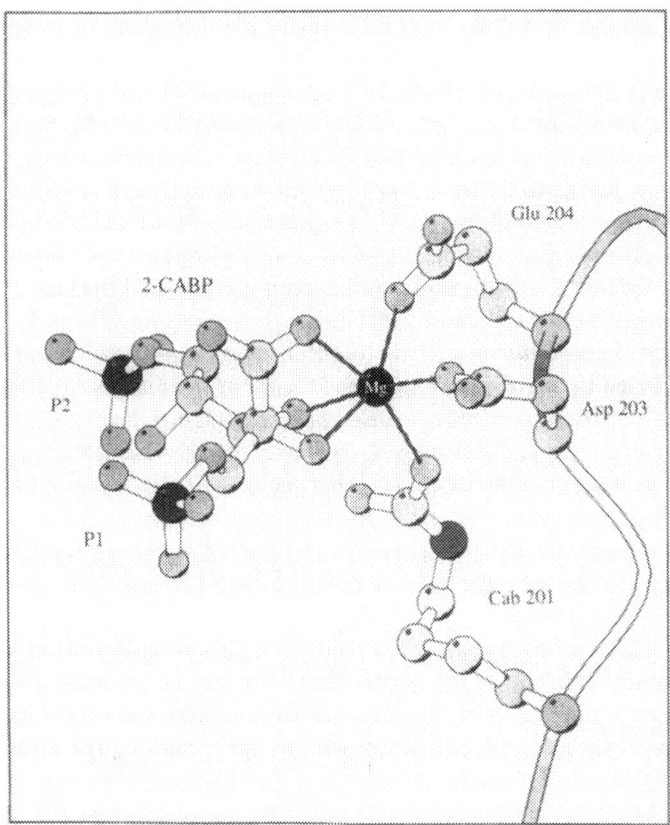

Figure 5. The identity of the six ligands to the essential Mg^{2+} ion at the active site of Rubisco. The structure is that of the spinach Rubisco quaternary complex (Knight et al., 1990) with those groups of the tight binding intermediate analogue, 2CABP, also shown. The residues of the enzyme that occupy three positions around the metal in this complex are the carbamate of Lys 201 and the carboxyl groups of Asp 203 and Glu 204; all located in loop 2 of the barrel. The groups of the inhibitor completing the rather distorted octahedral geometry include the C2 and C3 hydroxyls along with an "O" atom of the 2′-carboxyl.

Most revealing has been the structure of the activated ternary complex with the substrate ribulose-P_2 at the active site (Lundqvist and Schneider, 1991b). The enzyme is inactive in the crystal because one of the loop elements critical for activity is held away from the active site by crystal lattice interactions. Although the complex is unable to turn substrate over, the structure that emerges offers some tantalizing details that may be critical for describing catalytic events; details that have gained some support from mutagenesis studies (see page 318). Finally, a mutant of the dimeric enzyme D193N has also been solved (Schneider et al., 1992), and the details are discussed in the context of the effect of mutagenesis on metal coordination and function (see page 317).

With regard to the hexadecameric enzyme, two developments have been pursued. The first extended the resolution of the enzyme quaternary structure to 2.4 Å (Knight et al., 1990). With this detail, those amino acids around the active site that were involved in binding the metal and 2CABP reaction intermediate analogue were identified for the first time (Figure 5). In addition, the structure confirmed that one of the carbamino oxygen groups of Lys 201 is a ligand of the metal. Coordination to the enzyme is completed by the acidic side chains of Asp 203 and Glu 204 that reside in the same loop as Lys 201. Those groups on the inhibitor that coordinated to the metal were also resolved and were identified as the C2 and C3 hydroxyls, that is, *cis*-conformation (Knight, personal communication) along with one of the oxygen atoms of the 2′-carboxyl. The second development has involved solving the structure of the quaternary complex of *Synechococcus* Rubisco to 2.2 Å using molecular replacement techniques (Newman and Gutteridge, 1990). The cyanobacterial structure has been delayed by the size of the problem because the protein crystallizes with a complete L_8S_8 molecule in the asymmetric unit. Nevertheless, the model has confirmed that the orientation of the C2 and C3 hydroxyls of the inhibitor is *cis* and it will provide the first opportunity of comparing the same complex of two different species of Rubisco with quite distinct catalytic specificities.

E. Spectroscopic Analysis of the Active Site

Prior to the resolution of the three-dimensional structure of the Rubisco quaternary complex, some progress had been made in mapping the interaction and orientation of bisphosphates at the active site. These investigations involved EPR and NMR methods (Miziorko and Sealy, 1984; Styring and Branden, 1985) and exploited the ability of alternate metal ions to support activation and occupy the active site in place of Mg^{2+}. With suitable isotopic enrichment of the inhibitor 2CABP, it was determined that the hydroxyl at C2 of the molecule and the oxygen of the 2′ carboxyl group bind to the metal in the quaternary complex. Enrichment of the inhibitor or carbamate with ^{13}C provided an upper limit for the distances of these C centers from the metal (Pierce and Reddy, 1986). It was also established that the two phosphates of the molecule were at different distances from the metal. The advent of a structural solution for the quaternary complex from X-ray analysis not only confirmed most of these assignments but also indicated which of those protein groups are essential for stabilization of the complex.

One major contribution of spectroscopic investigations complemented the crystallographic data of the higher plant quaternary complex. At the initial resolutions that the structure was solved, it was unclear which orientation the 2CABP occupied at the active site, given that the "wrong" orientation was possible with inactive enzyme. Phosphate NMR studies with inhibitor selectively enriched with ^{17}O either in the 1-phosphate or 5-phosphate positions clearly placed the 1-phosphate closer to the metal than the 5-phosphate, confirming that the inhibitor is indeed oriented correctly (Lorimer et al., 1989). Subsequently, increased resolution of the X-ray data confirmed this orientation, and also showed that the C2 and C3 hydroxyls have *cis*-conformation. These studies have become particularly useful for analyzing the effects of mutations on the structure of the active site, especially in the case of mutant dimeric enzyme that cannot be crystallized as the quaternary complex.

A further application of NMR has unraveled some of the dynamics of Rubisco turnover, particularly determination of the order of interaction with substrate molecules. For the first time, it was established that the obligatory and initial step of the catalytic cycle involved the deprotonation of ribulose-P_2 to enediol, and that the gaseous second substrates, CO_2 or O_2, were not essential for the enzyme to catalyze this reaction (Gutteridge et al., 1984; Pierce et al., 1986). Combining the structural studies with those details determined spectroscopically and kinetically, the following sections provide a more complete description of the reaction chemistry mediated by the enzyme.

F. The Catalytic Cycle

The reactions catalyzed by Rubisco are unique, requiring only a second molecule of CO_2 and a Mg^{2+} ion as obligatory cofactors to render the enzyme catalytically competent. The three reversible equilibria that lead to formation of the activated enzyme shown in Figure 2 are also apparently unique to Rubisco. The only active form of the enzyme is the ternary complex formed as a result of carbamylation of the ε-amino group of an active site Lys residue and coordination of Mg^{2+} (Lorimer and Miziorko, 1980; Lorimer, 1981).

The active site carbamate is not the only one that forms between Lys groups and CO_2 on the enzyme. ^{13}C NMR indicates the existence of others. The significant feature of the active site carbamino group is the presence of carboxyl groups of adjacent acidic amino acids that occupy appropriate positions to offer a Mg^{2+} ion a suitable coordination site. The carbamate is thus stabilized by the presence of the Mg^{2+} ion, and the C nucleus exhibits quite a distinct shift in the NMR spectrum to a lower field (O'Leary et al., 1979).

A question remains as to the formation of the carbamate. The reaction is monitored by simply following the restoration of activity and thus reports on the appearance of the ternary complex ECM, not the conditions that must exist at the active site prior to EC formation. The Lys residue involved must first deprotonate for CO_2 to react, yet the amino group is in close proximity to adjacent acidic groups which should normally favor the protonated species. Two factors may contribute to reduce the pK_a of this Lys. First, the side chain is buried among hydrophobic side chains of the barrel to an extent that would discourage the existence of a positive charge on the amino group. Second, the presence of the Mg^{2+} ion may assist in carbamate formation by coordinating to the adjacent carboxyl groups of acidic amino acids. Some evidence supporting such a role for the metal is based on the close similarity of the K_d for CO_2 (Christeller and Laing, 1978) as a cofactor and the K_m for CO_2 as substrate in catalysis. The metal ion is a major determinant of the affinity of the enzyme for CO_2 in catalysis and may similarly affect the reactions required to activate the enzyme.

G. Chemistry of the Enzymic Reactions

A detailed consideration of the primary reactions catalyzed by Rubisco has recently been considered by Schloss (1990), therefore only a superficial treatment will be given here, emphasizing more what is known about the identity of those groups of the enzyme that contribute to the catalytic cycle. The processes that lead to carboxylation of ribulose-P_2 (see Figure 6) involve at least five partial reactions and the formation of three potentially unstable intermediates. Without the stabilizing influence of the enzyme, the intermediates would breakdown to various

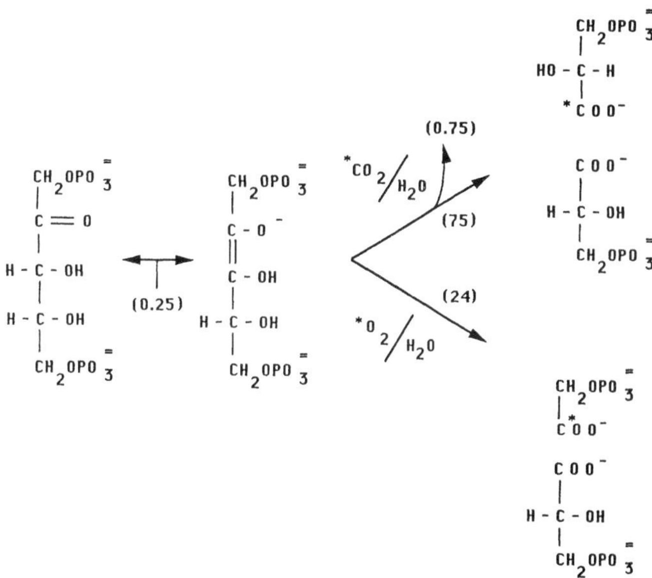

Figure 6. The primary catalytic activities of Rubisco. The products generated from ribulose-P_2 by the overall primary reactions of carboxylation and oxygenation are shown. The percentage distribution of ribulose-P_2 into various major and minor products during turnover by a higher plant enzyme in ambient concentrations of CO_2 and O_2 is also given. The first obligatory step of catalysis by activated Rubisco is the abstraction of the C3 proton of ribulose-P_2 to generate an enediolate ion of the bisphosphate.

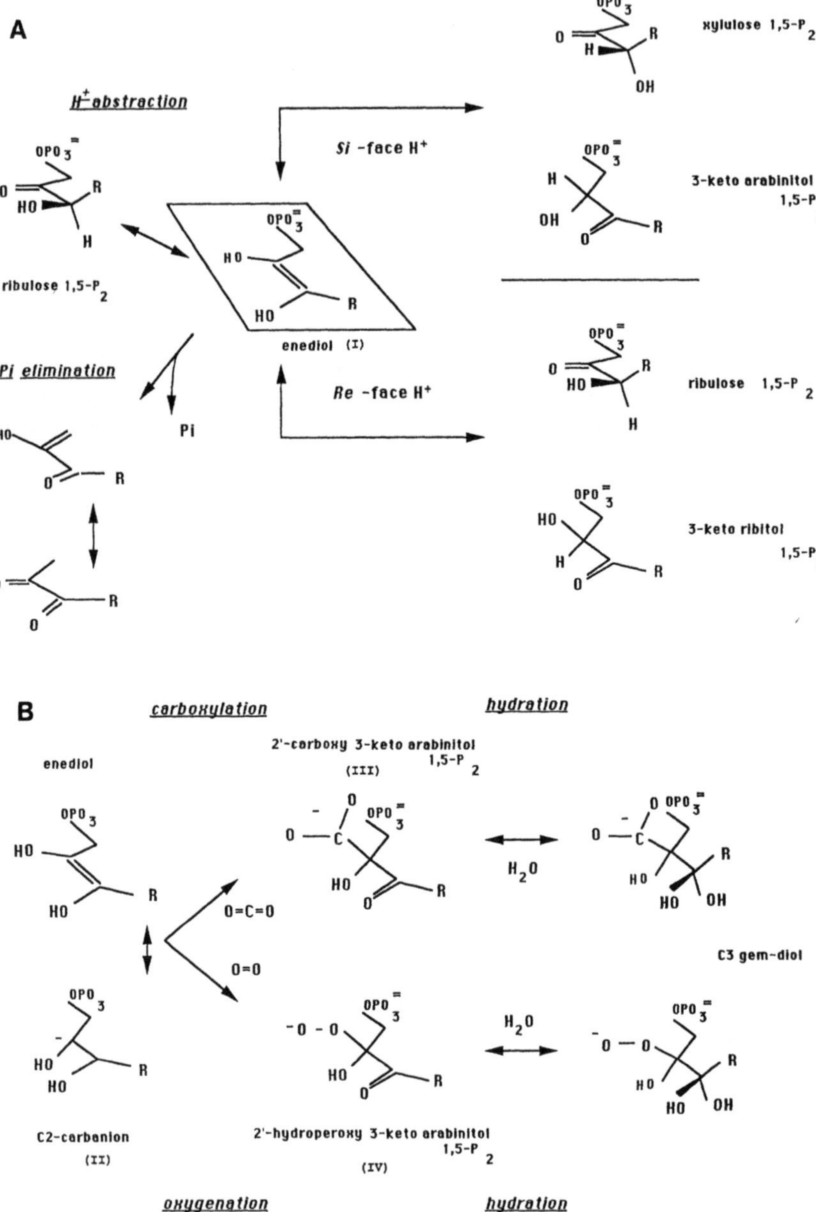

Figure 7. The chemistry of the reactions catalyzed by Rubisco. In (**A**) the fate of the enediol in terms of minor secondary reactions that occur with wild-type enzyme are shown. In (**B**) the primary reactions catalyzed by the enzyme proceed through carboxylation or oxygenation of the enediol generating 3k2CABP (III) or 2 hydroperoxy, 3-keto arabinitol-P_2 (IV), respectively. The interaction with water to hydrate the

(continued)

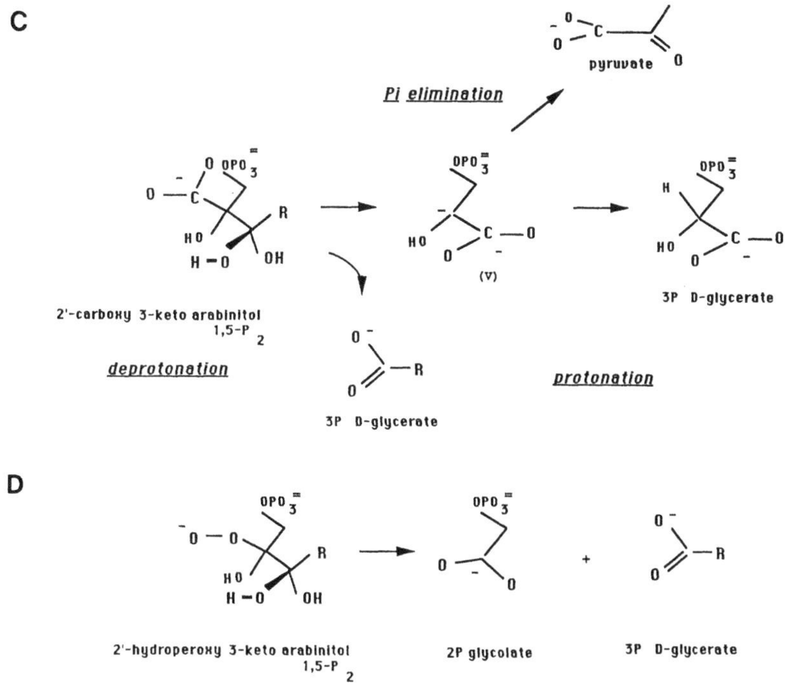

Figure 7. (continued) carbonyl at C3 is formally shown at this step (but see Figure 9) followed by deprotonation of the *gem*-diol molecule that severs the C2–C3 bond. (**C**) Carboxylation requires one more stereospecific protonation (v) step to release the second molecule of 3P-glycerate as the D-isomer. In (**D**) the products of oxygenation are generated directly. "R" is the lower two C centers, C4 and C5 of the substrate, i.e., -CHOH-CH_2-OPO_3.

unwanted products. Rubisco therefore has been constrained to evolve into more than just an efficient carboxylase, becoming a species that not only minimizes the partitioning of intermediates into oxygenation but also other secondary wasteful reactions.

These secondary reactions shown in Figure 7A, are evident from the small but detectable loss of substrate through partitioning of the enediol intermediate into inhibitory substrate analogues, xylulose 1,5 bisphosphate (Edmondson et al., 1990) and 3 keto-arabinitol bisphosphates (Zhu and Jensen, 1991). At the final step of product formation during carboxylation, the carbanion precursor of P-glycerate eliminates phosphate to form pyruvate (see Figure 7C). The appearance of inhibitory secondary products often complicates kinetic analysis of Rubisco causing underestimation of initial rates. Superimposed on these effects is excess substrate inhibition caused by high concentrations of HCO_3 that may be required to achieve optimal catalytic rates. Unlikely partitioning events are also observed with high concentrations of Rubisco due to the slow dehydration of HCO_3 to the substrate CO_2 (see e.g. Parry et al., 1989).

The first step in the catalytic process is the formation of the C2,C3-enediol (I) of ribulose-P_2 (Figure 7A). Investigation of the interaction of ribulose-P_2 with activated Rubisco indicates that the molecule is bound at the active site, with the C2 carbonyl oriented toward the metal ion and the *si*-face away from the protein surface of the barrel and toward the solution (Lorimer et al., 1988). Polarization of the carbonyl by the metal ion lowers the pK_a of the C3 group of the molecule making it susceptible to deprotonation. Abstraction of the proton from C3 is enhanced by some, as yet unidentified group acting as a base. From the orientation of the C2 and C3 hydroxyl groups of 2CABP at the active site suggests that the enediol most likely assumes a *cis*-configuration.

The next step of the reaction is less clear as its progress probably depends upon the localization of the electron density of the enediolate. Two conformations might be envisaged with the molecule either as an enediolate, or with density more toward the C2 center and thus stabilized as a carbanion (Figure 7B). Unfortunately, evidence is not available to discriminate between these conformations, although there might be species' differences here. Assuming that the more efficient carboxylases (i.e., the hexadecameric species) have evolved a more effective means of stabilizing the intermediate of carboxylation, 3k2CABP, then a tetrahedral arrangement of the C2 center might be favored in the approach to the transition state prior to carboxylation. A carbanion rather than enediolate would satisfy this conformation, and the identity of amino acids at the active site may preferentially stabilize this configuration. For example, the presence of Thr 173 in the hexadecameric species is close enough to the C2 center to H-bond to a hydroxyl that would exist at this position in a carbanion (see page 310). This Thr is not present in the dimeric enzyme since the position is occupied by Ile (164), and thus enediolate may be stabilized at this active site.

If the C2 center has indeed stabilized as a carbanion, then little reversal of the deprotonation step that formed enediol would be expected. There is tentative evidence to support this proposal based on NMR analysis of enediol formation. Exchange of the C3 proton from ribulose-P_2 by the hexadecameric enzyme with solvent deuterium in suboptimal concentrations of HCO_3 is catalyzed at rates that approximate the overall rate of carboxylation (Gutteridge et al., 1984). In contrast, dimeric Rubisco of *R. rubrum* readily catalyzes the exchange reaction in similar conditions. Thus an enediolate intermediate might be more favored than a carbanion at the active site of the dimer.

Three species compete for the newly formed enediol intermediate: CO_2, molecular O_2, and protons. The most efficient enzymes have evolved some means of discriminating between these various substrates. CO_2 reacts electrophilically with the C2 center to form the six-carbon intermediate, 3k2CABP (III). Hydration of the C3 carbonyl by water and subsequent deprotonation (Figure 7C) cleaves the intermediate between C2 and C3 generating two 3C compounds. The lower three carbons are the D-isomer of 3P-glycerate. The upper three carbons with the newly captured CO_2 form a carbanion (IV) of the second molecule of 3P-glycerate. The majority of this molecule is protonated stereospecifically, again from the *si*-face to generate the D-isomer. A very small percentage of the carbanion (0.75%) eliminates phosphate and forms pyruvate (Andrews and Kane, 1991).

The steps that involve reaction of molecular oxygen with the enediol intermediate are less well characterized and based on those operating for carboxylation. It is well known that carbanions are relatively reactive with molecular oxygen, although the mechanism of oxygen activation is not understood. The difficulty is obtaining evidence that a triplet state of the carbanion exists that would readily react with the triplet ground state oxygen molecule to form the putative 2-hydroperoxide 3-keto intermediate. Hydration, followed by deprotonation results in formation of 2P-glycolate and 3P-glycerate as the products of oxygenation (Figure 7D).

The enediol intermediate is also susceptible to reprotonation rather than react with CO_2 or O_2. Simple reversal of proton abstraction regenerates the substrate ribulose-P_2; however, this must proceed stereospecifically to regain the natural substrate. Once in about every 500 turnovers of the substrate reprotonation is not specific, and a number of isomeric forms of the bisphosphate have been identified including xylulose bisphosphate (protonation at C3; Edmondson et al., 1990) and 3-keto pentitol bisphosphate (protonation at C2; Zhu and Jensen, 1991).

One potential secondary reaction that has not been detected, at least with wild-type Rubisco, is the β-elimination of the 1-phosphate from the enediol intermediate. Since elimination occurs with the aci-acid precursor of 3P-glycerate, then it might also be expected with the carbanion of the bisphosphate. However, it is not yet clear whether the formation of pyruvate occurs on the enzyme, or through premature release into solution of the 3P-glycerate precursor where phosphate elimination would be more likely to occur than protonation.

H. The Role of Active Site Groups in Catalysis

Although we have extensive details of the chemistry involved in carboxylation, the advent of a three-dimensional structure for the enzyme constrains us to explain the partial reactions in terms of the essential groups that compose the active site. The presence of bound Mg^{2+}, in conjunction with various critical amino acids and water, provide the components that drive the catalytic events. There are at least five quite distinct steps that are required to generate product from ribulose-P_2, and all occur within the confines of the same site. It is now clear with the structures available that the enzyme achieves this chemistry by moving different groups in and out of the active site at discrete times during catalysis. A combination of changes to the groups in the coordination sphere of the Mg^{2+} ion and the movement of at least two flexible loop elements of the protein containing critical amino acids provide the choreography underpinning the chemistry of the reactions.

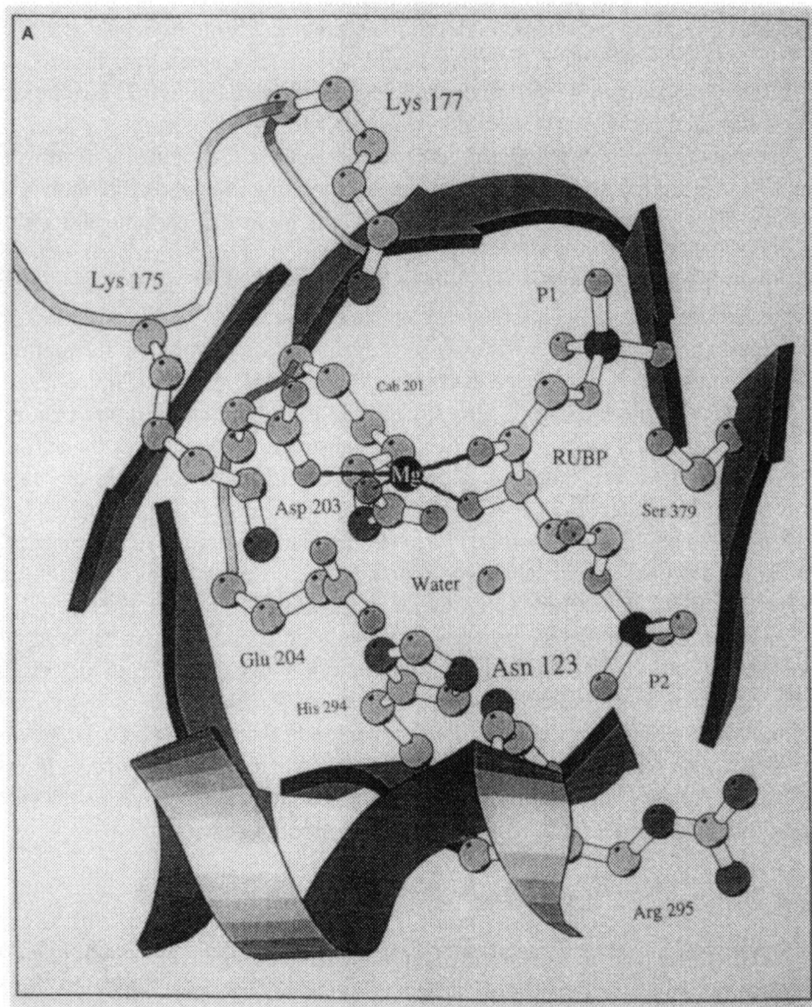

Figure 8. Two structures of quaternary complexes of Rubisco. (**A**) The complex between the activated ternary complex of *R. rubrum* Rubisco and ribulose-P_2 substrate. The display shows the organization of the critical residues around the substrate molecule and Mg^{2+} ion that must occur immediately after the activated enzyme encounters the substrate. Those groups that may assist the abstraction of the C3 proton include the carbamino "O", 1-phosphate (P1) or the water. (**B**) The second complex is a close approximation to the organization of the active site that exists at the time immediately after the fixation of CO_2 substrate by Rubisco. The ligands to the metal are those already identified in Figure 5. In addition, there are those involved in stabilizing the newly formed intermediate.

(continued)

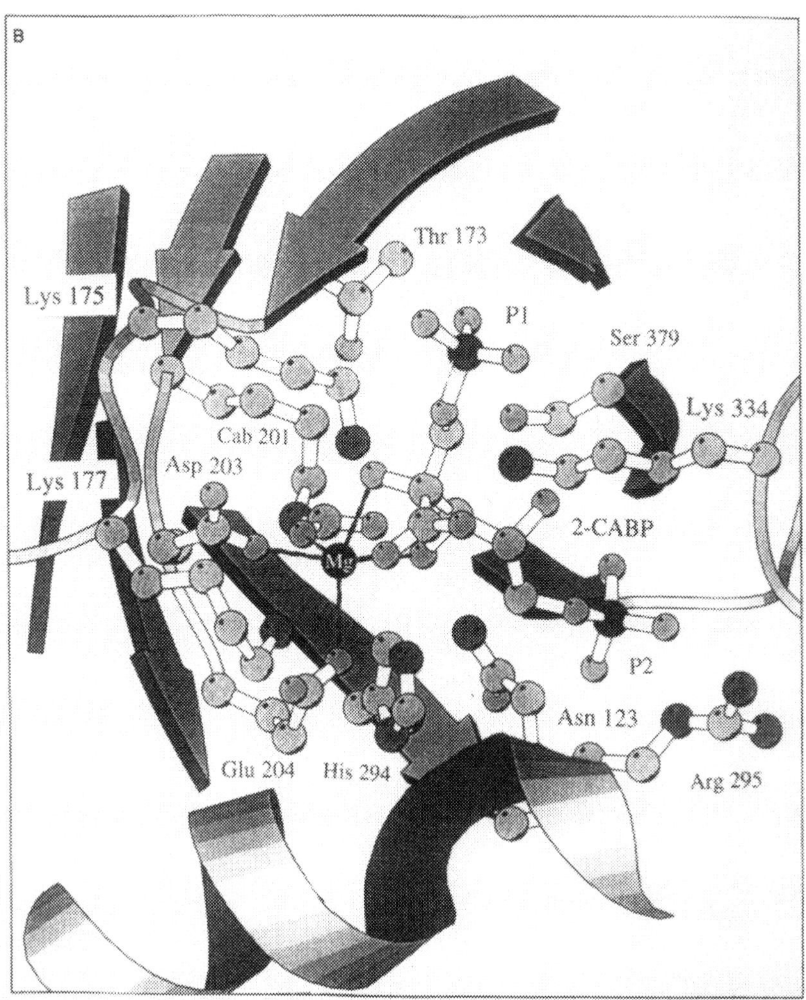

Figure 8. (continued) For example, Lys 334 interacts in conjunction with the metal to retain the 2′-carboxyl group emerging after CO_2 reacts. His 294 stabilizes the *gem*-diol due to the hydration of the C3 carbonyl. Glu 204 may be able to move into and away from the metal coordination sphere, depending on the charged state of His 294. Both of these residues may be involved in the deprotonation step that cleaves the C2–C3 bond of the intermediate The presence of basic groups of Lys l75 and 177 that can interact with the acidic residues 203 and 204 provide a means of modulating the acid–base potential of the active site. Glu 60 (not shown) and Asn123 of the N-terminal domain of the other L-subunit are in close proximity to these basic groups and the newly formed carboxyl, respectively.

In reaching some definition of the function of individual amino acids that catalyze the multiple steps of carboxylation and oxygenation, we have only two structures to exploit; and one of those is not ideal. The first of the structures in Figure 8A shows the complex between activated enzyme and the substrate ribulose-P_2. The view is from above the C-domain barrel with the side chains of residues included that are in close proximity to the metal or interact with groups on the substrate. This quaternary complex was achieved by soaking crystals of nonactivated *R. rubrum* Rubisco with HCO_3 and Mg^{2+} to form an activated ternary complex before soaking in the substrate (Lundqvist and Schneider, 1991b). The bisphosphate is not consumed because one element, loop 6, is unable to fold into the active site. The question remains, therefore, whether this structure is a true representation of the state of the active site with bound substrate.

Figure 8B is a more detailed view from above the barrel domain of the quaternary complex of spinach Rubisco with 2CABP, showing those groups of the N- and C-terminal domain that are in close proximity to the metal and the inhibitor. Two amino acids of the N-domain, Glu 60 and Thr 65, are not shown for clarity but are also essential for active site integrity (see also Figure 10, page 325). Residues of loop 2 and groups of the inhibitor that coordinate the metal, initially identified in Figure 6, are shown again; for example, amino acids involved in binding the inhibitor at the site such as Ser 379 that interacts with C4 hydroxyl of 2CABP, and residues of loop 5 such as Arg 295 that bind the 5-phosphate are also included. Some of these amino acids, along with the metal, are intimately involved in the chemistry; for example, Lys 334 located in loop 6 is essential for stabilizing reaction intermediates. The conformation of this complex closely resembles that adopted by the enzyme immediately following fixation of substrate CO_2 by enediol. The absence of a bound water molecule at the active site suggests that 2CABP mimics the conformation adopted by the hydrated *gem*-diol state of 3k2CABP.

Assuming that both structures provide a close approximation to the state of the active site at those two different stages of the catalytic cycle and, probably more importantly, that both the dimeric and hexadecameric Rubisco species utilize the same mechanism at each step, then the following series of illustrations are an attempt to reconcile the structural and functional data. Where possible, the results of mutagenesis studies are incorporated into the schemes. Nevertheless, because of those caveats just described, the assignments must be considered of "low resolution" until more enzyme complexes are solved and mutagenesis studies help define function less ambiguously.

Enediol Formation

Since the first demonstration that the relative specificity of Rubisco could be altered by replacement of Mg^{2+} by other metals (see e.g. Christeller, 1981), it has been speculated that most, if not all catalytic events require the particular properties of coordinated Mg^{2+} (see e.g. Gutteridge and Gatenby, 1988). Those amino acids

Table 2. The Partial Activities of Mutants of Dimeric Rubisco

L2	*Carboxylation*	*Rel. Specificity*	*H⁺ Abstraction*	*3k2CABP*	*QC-Stability*[a]
wt	100	16	100		100
D2O3N	–	–	–	+++h	25
E2O4Q	–	–	–	---dc	
K175G[b]	<0.01	–	–	+++h	
K201C	–	–	–	+++dc	
N123V	3.0	3.0	30.0		50
E60C[c]	0.05 (5.0)	(3.0)			

Notes: [a]Quaternary complex (QC) stability for the dimer relates to that quantity (expressed as %) of enzyme that forms a complex relative to wild type.
[b]Data taken from Hartman, 1992.
[c]The values in parentheses were those obtained from mutant after reactivation with iodoacetate (Smith et al., 1990).
+++h hydrolysis > decarboxylation in high Mg^{2+}; ---dc no decarboxylation detected in Mg.
+++dc decarboxylation, no hydrolysis detected.

that bind the metal must therefore also be implicated in directing the course of the reactions. The acidic amino acids Asp 203 (193) and Glu 204 (194), which are adjacent to the carbamylated Lys residue in loop 2 of the barrel domain, bind to the Mg^{2+} (see Figure 8). Mutagenesis of the two side chains to the corresponding amides destroys overall catalysis and enediol formation is not detected. It might be expected that replacement of a ligand of the metal would abolish the ability of the enzyme to activate, but a small percentage of the quaternary complex between Asn 203 or Gln 204 with 2CABP is formed and, in the case of Asn 203, some of the partial reactions are catalyzed (Lorimer et al., 1988; Gutteridge et al., 1989).

The replacement of Asp 203 and Glu 204 causes differential effects on the fate of 3k2CABP (see Table 2) suggesting that each is essential for different steps of catalysis. The Asn mutant exhibited altered partitioning of the intermediate; that is, between decarboxylation and hydrolysis, depending on the concentration of Mg^{2+} ion—the decarboxylation of 3k2CABP diminished significantly in favor of hydrolysis as Mg^{2+} was increased. In contrast, Gln 204 exhibited less decarboxylation, but no hydrolysis was detected with this mutation. Since decarboxylation is equated with binding of the intermediate with nonactivated enzyme, the effect of the metal suggests that increased Mg^{2+} concentrations has increased the proportion of the mutants in the form of the more productive ECM ternary complex. These data suggested that Asp 203 participates in enediol formation, but is not essential for hydrolysis, whereas Glu 204 is required for both reactions. Clearly an amide group is unable to participate directly in proton abstraction; rather these mutants may simply reflect the requirement for Mg^{2+} to be bound correctly at the site to polarize the carbonyl of the substrate. A structural analysis of the Asn mutant has indicated that some alterations to the positioning of side chains in the inactive enzyme accompanied this mutation (Soderlind et al., 1992).

The identity of the group that acts in conjunction with Mg^{2+} to catalyze the removal of the C3 proton from ribulose-P_2 is still obscure. Apart from Asp 203 and

Glu 204, Table 2 indicates that one other amino acid essential for this step is Lys at position 175 (166). The residue is in loop 1 of the C-terminal barrel and has the pK_a expected of the group involved in proton abstraction (Hartman et al., 1985), at least as measured in the ECM ternary complex. Replacement of this residue by a Gly completely abolishes the ability of the enzyme to catalyze enediol formation, yet hydrolysis of 3k2CABP is not impaired (Lorimer and Hartman, 1988). Based on these results, this amino acid was identified as the essential base in the first step of catalysis. However, from the crystallographic model of the quaternary complex, the amino group of this residue is too far from the C3 proton of the substrate to function in this capacity, assuming that it occupies a similar position relative to the C3 of 2CABP. One other amino acid that is close to the C3 proton is His 294 (287), but this residue is not absolutely essential for catalysis since replacement with Asn generates a mutant that still retains 1 % of normal catalytic activity (Lorimer et al., 1988).

What are suitable choices for the role of essential base? Figure 8A offers a tantalizing view of the organization of the active site of the enzyme with ribulose-P_2 bound and suggests three other possible candidates. A scrutiny of the sequence of amino acids in loops 7 and 8 indicate two possible sites for interaction of 1-phosphate. The first is that position filled by the 1-phosphate of 2CABP involving a short helical segment composed of triple Gly (403-405) in loop 8 of the C-terminal barrel (Gly. Gly. Ala in dimeric Rubisco and triose phosphate isomerase) and residues of loop 1. The second involves Gly 403 of loop 8, but in combination with Ser 379 (368) and Gly 380 (369) of loop 7 (again similar to the phosphate binding site of triose phosphate isomerase). The 1-phosphate of ribulose-P_2 in the substrate quaternary complex occupies this position. In this orientation, one of the oxygens of the phosphate is within 4.0 Å of the C3 proton of the substrate and thus could assist enediol formation as a base. Whether the phosphate can adopt a pK_a that would be consistent with this proposal has yet to be tested, but it is clear that this may be one more group that alters its position at the active site during turnover. In this complex, His 294 interacts with the remaining water molecule that is not displaced on substrate binding.

Although the data is only to 2.6 Å resolution, leaving some ambiguity as to the relative positions of, for example, C3-hydroxyl of the substrate, nevertheless the locations of critical amino acid side chains are interesting and offer some explanation of the results of mutagenesis studies. The structure, if applicable to the functioning enzyme shows the C2 carbonyl of the substrate interacting with both the Mg^{2+} ion and the ε-amino group of Lys175 (166). In this configuration enediol formation would involve the concerted effect of metal and Lys175 as acids in conjunction with phosphate as the base to abstract the C3 proton.

Assuming the conformation of the C3 hydroxyl of the substrate is identical to that of 2CABP, then one other group within 3.0 Å of the C3 proton is a carbamino oxygen of Lys 201 that may also assist in the abstraction. A protonated carbamino group would be uncharged and unlikely to remain tightly bound to the metal, unless the newly acquired proton were rapidly transferred to, for example, the polarized C2 oxygen of the enediolate. Consequently, the enediol generated by proton abstraction would also have *cis*-conformation with both C2 and C3 oxygens bound to the metal.

The carbamate is essential for Rubisco activity, and in an attempt to probe its function a Cys residue was introduced at position 201 (191) instead of Lys in *R. rubrum* Rubisco. The mutation generated an inactive enzyme that was unable to catalyze any partial reactions or perform overall catalysis, yet still formed a quaternary complex with CO_2, Mg^{2+}, and 2CABP more stable than wild type. Although the absence of the ε-amino group of Lys 191 precluded formation of an active site carbamate, other Lys residues are close to the active site; therefore it was feasible they might offer a suitable carbamylation site. NMR studies indicated that the CO_2 molecule was not bound in the form of a carbamate; rather it was trapped as a carbonate ion. The organization of the metal relative to the carbonate and inhibitor was similar to that found in the wild-type quaternary complex; that is, the inhibitor was oriented correctly (Gutteridge, Reddy, Madden, and Lorimer, unpublished).

This result suggested that if all the components for activity are present at the site and yet there is no catalysis, then the carbamino group plays an essential role in catalytic events. However, when the K201C mutant and ribulose-P_2 (instead of 2CABP) were combined in the presence of CO_2 and Mg^{2+}, the expected quaternary complex was not detected. It was also established that the time required to achieve complex formation with 2CABP took hours rather than a few seconds, indicating that the lifetime of a ternary complex between CO_2, Mg^{2+} and mutant is too shortlived or stable in the presence of ribulose-P_2 to support catalysis. Normally formation of the carbamate would retain the CO_2 long enough for Mg^{2+} to bind and stabilize the active ternary complex. In the case of 2CABP with the mutant, the affinity of the inhibitor for the small quantity of ternary complex that exists fleetingly is so great that this species is preferentially trapped, and the equilibrium between cofactors and mutant, like wild type, is toward complex formation. This result also suggests that using quaternary complex formation with 2CABP may be a misleading indicator of the integrity of the active site following mutagenesis.

These examples illustrate the difficulty and potential dangers of extrapolating from a less than ideal structure to try to obtain evidence to support a hypothetical model. The *R. rubrum* Rubisco–substrate quaternary complex is inactive, presumably because loop 6 is unable to participate in completing the productive topography. The most revealing experiment will be the solution of the structure of the hexadecameric enzyme with ribulose-P_2 or xylulose-P_2 bound at the active site rather than 2CABP. Indeed, the ability to scrutinize such a structure might also confirm the potential role that the third candidate plays in the process, namely the single molecule of water that interacts with active site His (294).

In the activated ternary complex there are at least two of these molecules bound to Mg^{2+}. One is displaced by binding the C2 carbonyl of the bisphosphate to the metal, but the other that interacts with His 294 after the substrate binds might be retained to drive the hydrolysis step of catalysis (see Figure 9D). Interestingly, if the negative charge of a carbanion is substantially localized at C2, then this would be effectively balanced by the emergence of carbonyl character at C3. The question that then arises is whether this carbonyl is substantially hydrated at this stage of the reaction, that is *before* reaction with gaseous substrates (see also Cleland, 1990). These assignments are therefore illustrated as dashed lines in Figure 9B.

Reaction with CO_2 or O_2

The formation of enediol at the active site of the enzyme may be responsible for some dramatic conformational changes that sets up the next stage of the reaction: CO_2 fixation, or oxygenation. It was discussed previously (see Section III.I) that there might be species differences in terms of the stabilization of the enediol at the active site; for example, the presence of Thr at 173 may stabilize a carbanion more effectively than Ile. Furthermore, the 1-phosphate may at this stage move from loop 7 to interact with groups of loop 8 (Figure 9B). In so doing, Thr 65 (53) of the flexible N-terminal domain loop and Lys 334 of the C-terminal flexible loop (loop 6) move in to interact with the 1-phosphate. Based on the structure of the quaternary complex, loop 6 of the C-terminal barrel must move to a position that allows Lys 334 (329) at the apex of the loop to assist the capture of a molecule of CO_2. Figure 9C shows the Mg^{2+} remaining center stage coordinated to C2 and C3 "O" of the *cis*-enediol intermediate. Asn 123 is no longer coordinated directly to the metal, but remains within 3.0 Å of Lys 334 and the newly formed carboxyl. The presence of electron density localized on the C2 center of the enediol, in conjunction with the polarizing effect of Mg^{2+} and the positively charged amino group of Lys 334 on the oxygens of the CO_2 molecule, make the C of the gas an effective electrophile. The resulting 2′-carboxyl group that emerges is stabilized by the Mg^{2+} and Lys until the next step of the reaction is complete.

The sequence of events at this stage are clearly quite speculative because of the tenuous marriage of the structures of the *R. rubrum*–substrate complex and the spinach 2CABP quaternary complex. Nevertheless, the quest for a structural answer to the basis of species variation in substrate partitioning and catalytic efficiency require that the events involving fixation of the gaseous substrates be known in detail. If indeed loop 6 has already moved to a closed position during enediol formation (and mutagenesis studies support this), then the juxtaposition of charge on the metal and Lys 334 might compose a weak binding site for substrate CO_2. The nature of the organization of enediol/carbanion and captured CO_2 relative to the primary active site groups determines the fate of emerging 2′-carboxyl of 3k2CABP. Studies of the differences in ^{13}C discrimination by Rubisco species (Rieske and O'Leary, 1985) suggest that at this step the organization of the transition intermediate at the active site of the dimer is more "product-like" (i.e.,

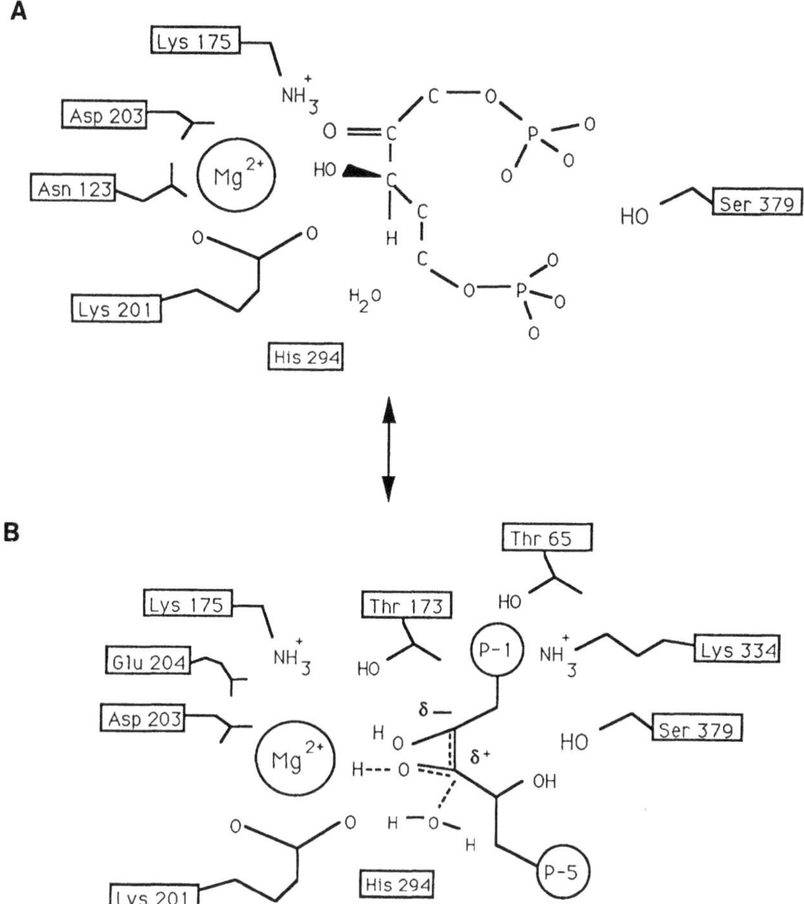

Figure 9. A hypothetical scheme for the role of active site groups during catalysis. The configuration of the active site and substrate, ribulose-P_2 shown in (**A**) is taken from the structure of the ribulose-P_2 quaternary enzyme complex of *R. rubrum* Rubisco (see the text). Assuming a *cis* orientation about C2 and C3, then three candidates are in close proximity to the C3 proton of the bisphosphate (*dashed lines*). The combination of the polarizing effects of metal and Lys l 75 and a base to accept the proton, i.e., carbamino "O", phosphate or water, generates the enediol shown in (**B**). (**B**) Reorganization of the desaturation of the bisphosphate due to loss of the C3 proton depends on the identity of those active site groups involved in stabilization. Negative charge may become localized at C2 to give a more carbanion type structure that, in turn, leads to carbonyl character at C3, stabilized by the interaction of Thr 173 of loop 1. The proximity of water to the C3 center may then hydrate this carbonyl. The persistence of a tetrahedral conformation around C2 may favor the next step of catalysis—CO_2 fixation, where C2 adopts sp3 character formally. At what stage of the reaction the two flexible loops of the enzyme adopt a closed conformation is unclear,

(continued)

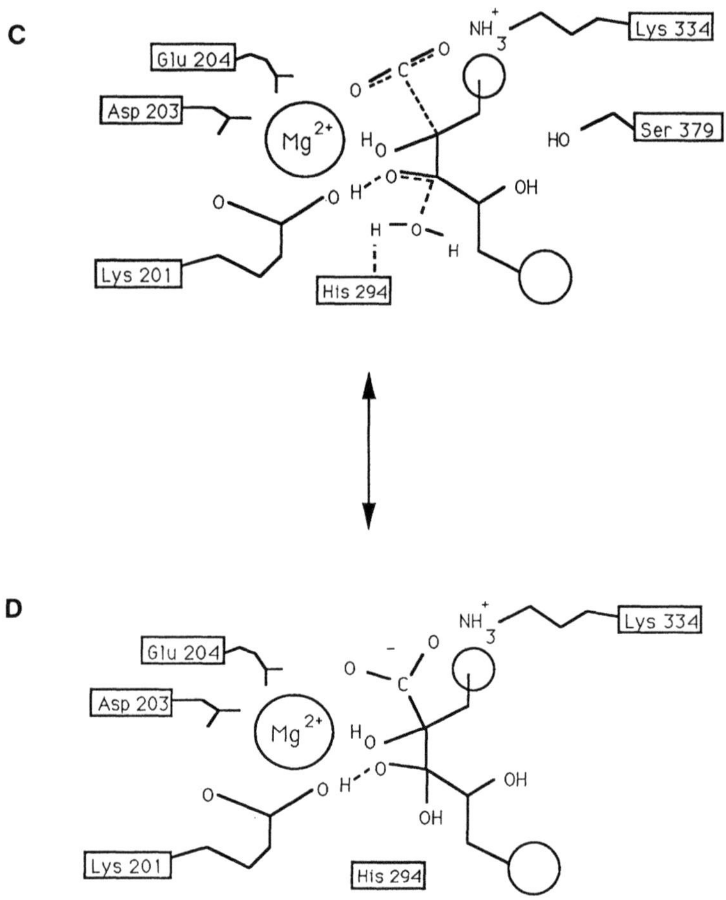

C

D

Figure 9. (continued) but the requirement for the 1-phosphate to move away from loop 7 and toward loop 8 requires the assistance of Thr 65. Likewise, the involvement of loop 6 residues in enediol formation suggests that this C-terminal loop has also closed over the site. (**C**) The CO_2 substrate is polarized by the combined effects of at least two positive charges; one on the metal, the other due to the ε-amino group of Lys 334. The electrophilic nature of the C of CO_2 that is induced close to the C2 anionic center of the enediol, leads to carboxylation. (**D**) Reorganization of the groups at C3 to produce hydrated *gem*-diol of 3k2CABP leads to an active site structure analagous to that observed in the 2CABP quaternary complex of the hexadecameric enzyme. Deprotonation of the hydrate cleaves the C2–C3 bond leaving the lower three C centers as 3P-glycerate (D-isomer) and the upper three as a carbanion. (**E**) The carbanion is protonated from the *si* face to produce the second molecule of 3P-glycerate, potentially a suitable role for Lys 175. The 1-phosphate may also adopt the same position as that of the substrate, i.e., at loop 7, to minimize the elimination of the phosphate group prior to protonation.

(continued)

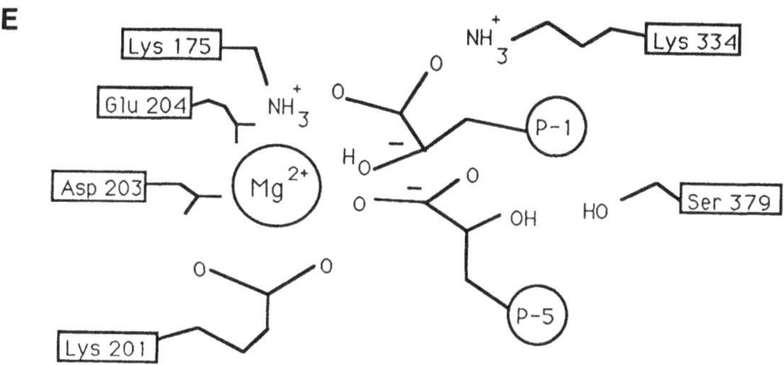

Figure 9. (continued)

resembling Figure 9D), and the intermediate more "substrate-like" (Figure 9C) with the hexadecamer.

The location of water has to be carefully controlled at this stage of the reaction. During the approach and polarization of CO_2, H_2O would effectively compete with the carbanion of the bisphosphate and thus must be excluded; yet water is essential for the next step. At this stage, the active site is closed off from the bulk solution as groups move in to stabilize the captured CO_2. Therefore, a molecule of water must still reside within the active site, but restrained some distance from the newly formed 2′-carboxyl. Formation of the carboxyl bond between enediol and CO_2 is balanced by the carbonyl group at C3 of the intermediate, thus completing 3k2CABP formation. Traditionally, the remaining bound water molecule in close proximity to the C3 carbonyl has been thought to hydrate the intermediate to the *gem*-diol at this stage of catalysis (see Figure 7B). Deprotonation of one of the C3 hydroxyls results in cleavage of the C2–C3 bond and formation of 3P-glycerate from the lower three carbons of the bisphosphate. The upper three carbons are retained as a carbanion prior to protonation to generate the second 3P-glycerate molecule.

Clearly, a residue such as Lys 334 (329) that is intimately involved in activating gaseous substrates and stabilizing the resulting 2′-carboxyl or hydroperoxy group of the intermediate might also be essential for determining the partitioning of that intermediate. Mutations of this critical Lys residue are functionally revealing. Alteration to a residue with nonbasic character would most likely destroy overall catalysis, since the intermediate can no longer respond to the stabilizing influence of an amino group, and indeed, replacement with Cys or Met has just this effect. The mutants are still capable of enediol formation as judged by their ability to catalyze the washout of ^3H from [^3H-C3] ribulose-P_2; however, even though the enediol form of ribulose-P_2 is unstable and susceptible to oxidation in free solution, these mutants are unable to catalyze oxygenation of this intermediate.

When Arg was introduced in *Synechococcus* Rubisco to retain the basic character of the side chain at this position, overall catalysis is preserved but is accompanied by a dramatic decline in relative specificity by some two orders of magnitude. In fact, the mutant is almost devoid of carboxylase activity. These results indicate that oxygenation does not arise simply through an opportunistic reaction between molecular oxygen and enediolate; rather, the intermediate so formed (presumably hydroperoxide) also requires stabilization by active site amino acids.

The first indication that a dynamic component is implicated in catalytic activity, is deduced from a comparison of the structures of the dimeric enzyme and hexadecameric quaternary complex. In the former, the region of the L-subunit composing loop 6 is ill-defined and distant from the active site, whereas the structure of the quaternary complex shows the loop closed over the active site with Lys 334 interacting with the 2′-carboxyl group of the inhibitor (see Fig. 8B). Thus, the amino acids composing loop 6, even distant from Lys 334, are critical to the catalytic events to ensure both flexibility and that the amino group of this residue achieves an optimal position at the active site. That this positioning is critical to partitioning of the bisphosphate substrate was observed after two Rubisco mutants were isolated in *Chlamydamonas rheinhardii* (Chen and Spreitzer, 1989).

A high CO_2-requiring mutant of the organism was localized to the L-subunit of Rubisco where Ala had changed to Val at position 331. A revertant of this mutation, viable in normal air concentrations of CO_2, still retained the V331A change, but had acquired a second mutation at position 342 where Thr was replaced by Ile. The second mutation had suppressed most of the loss of catalytic efficiency due to the first alteration. A number of changes were introduced into the recombinant *Synechococcus* enzyme (Table 3) to show that relative specificity indeed is sensitive to the identity of the residues in these positions, even though they are distant from the active site.

Table 3. The Relative Activities of Mutants of Hexadecameric Rubisco

L_8S_8	*Carboxylation*	*Rel Specificity*	*H^+-Abstraction*	*QC-Stability*[a]
wt	100	58	100	>100
V331A	5	30	0.6	1.0
V331G	0.5	20	0.3	1.5
V331L	5	50	0.2	10.0
T342A	0.2	20	0.2	1.0
T342L	40	56	40	5.0
T342M	6	56	8.0	1.3
K334R	0.5	0.3	12	<2.0
K334C	—	—	5	<0.5
K334M	—	—	3	<0.5

Note: [a]The value shown for the hexadecamer is the $t_{1/2}$ of exchange of 2CABP from the complex (in hours) compared to wild type.

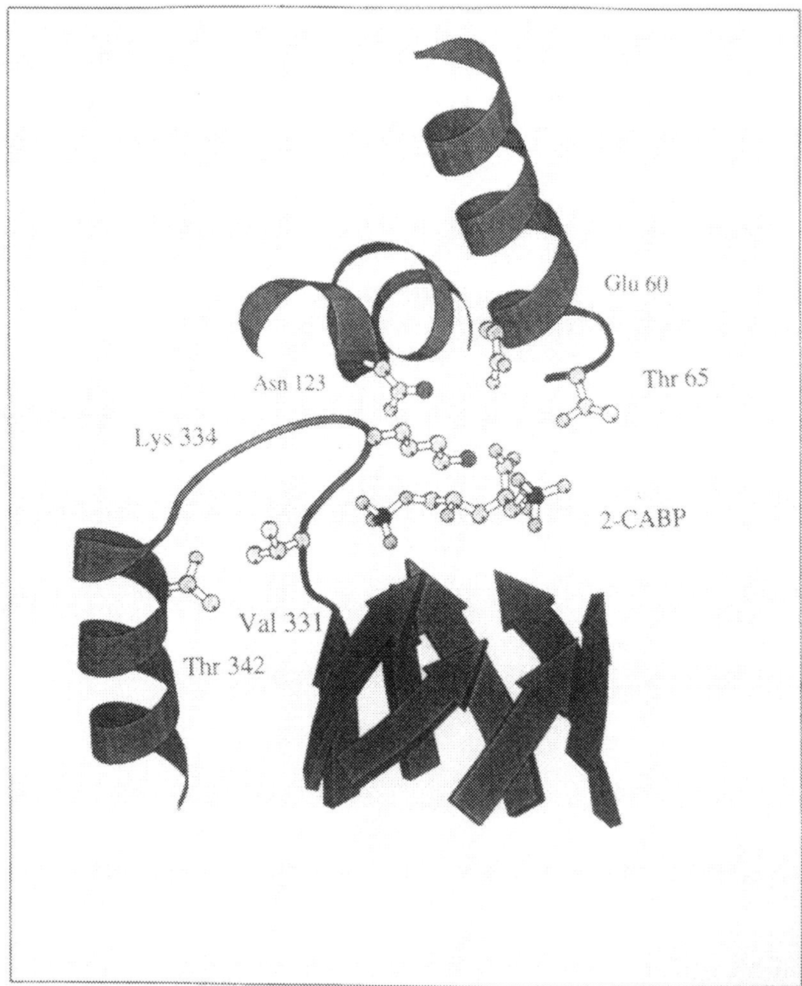

Figure 10. The location of some of the active site amino acids of Rubisco that influence relative specificity. The positions occupied by these residues in the spinach quaternary complex are visualized from behind loop I of the barrel. Glu 60, Thr 65, and Asn 123 reside in the N-terminal domain of the second L-subunit. Lys 334 at the apex of loop 6 is involved stabilizing the intermediate of carboxylation or oxygenation. Thr 342 and Val 33l form a hydrophobic pocket that determines the flexibility of loop 6.

The structure of the enzyme quaternary complex allows the effects of these mutations to be rationalized. The side chains of V331 and T342 form a hydrophobic pocket at the base of loop 6 (see Figure 10) and thus determine the geometry and presumably mobility of this loop. The introduction of a smaller hydrophobic side chain at 331 is compensated for by adjustment of the loop (and potentially other elements) to retain the packing in this region, with the result that K334 no longer occupies an optimal position at the active site. The 2′-carboxyl group is not stabilized as efficiently as in wild type, and both relative specificity and turnover are altered. The smaller size of the Ala side chain is almost completely suppressed by the larger hydrophobic side chain of Ile at position 342. From the site-specific mutations it is clear that 331 position is more sensitive to the identity of the amino acid than is position 342 and that a wide range of lower relative specificities are obtained. A major response from changes at 342 were only observed with the smallest and largest side chains.

One expectation from this series was that increasing the length of the hydrophobic side chain at either position might not only suppress a decline in specificity, as with the T3421 change, but actually induce an increase. Table 3 shows that this expectation was not realized and that relative specificity remained the same or declined. Finally, those mutations affecting loop 6 mobility also altered the rate of enediolate formation based on the rate of ^{3}H exchange from 3-^{3}H ribulose-P_2. This was unexpected if the function of loop 6 is to ensure the presence of a group to stabilize the 2′-carboxyl or hydroperoxide intermediates of catalysis. Clearly the amino acids composing this loop are required for events earlier than reaction with the second substrate; a situation implied from the inactivity of crystals of the *R. rubrum* Rubisco ternary complex with bound ribulose-P_2.

I. Other Amino Acids Influencing Relative Specificity

Other site-specific mutations that alter relative specificity have been identified. Two of them are located in loops of the N-terminal domain that compose that part of the active site contributed by the second L-subunit. A conserved Glu residue at position 48 (60) interacts with the amino group of Lys 329 (334) in the quaternary complex. Replacement of this residue in the L dimer of *R. rubrum* Rubisco destroys nearly all activity, testifying to its essential requirement for catalysis. However, a mutant that has Cys at this position can be reactivated 100-fold (see Table 2) by using iodoacetic acid to carboxymethylate the residue (Smith et al., 1990). This lengthens the side chain by some 1.5 Å with a sulfur interposed between the β- and γ-methylene groups, and also alters the geometry of the side chain. Nevertheless, the mutant is active and has reduced relative specificity. Clearly the other Cys residues of the L-subunit are not immune to iodoacetic acid, and thus the overall catalytic activity suffers because of these alterations. Nevertheless, this was one of the first demonstrations that the relative specificity of the enzyme could be changed through site-specific mutagenesis.

More recently, a second amino acid of the N-terminus has been identified that, when replaced, alters relative specificity. Asn 123 (111) of the L-subunit is also conserved and in the quaternary structure is within 4 Å of Lys 334 (329), Mg^{2+} ion, and 2′-carboxyl group of the inhibitor. Replacement of this residue with Val changes the partitioning of the bisphosphate substrate dramatically. Not only is the ratio of the products of carboxylation and oxygenation affected (i.e., P-glycerate to P-glycolate), other products are also detected. An analysis of the nature of these products indicates that protons become more effective second substrates and a variety of isomers of ribulose-P_2 are formed. It is also clear from ^{31}P NMR that phosphate elimination is detectable with this mutant.

The rationale for these effects, at least of this one change, is complicated. In the quaternary complex, Asn 123 not only interacts with the metal but also with the 2′-carboxyl of 2CABP. Without a structure for the mutant, it is difficult to explain the functional consequences of this one amino acid change. NMR studies of the metal-binding site suggest that the site might have been altered by the replacement of the amide with a more hydrophobic residue, which could explain the change in relative specificity. Interestingly, replacing Asn with acidic residues, Asp or Glu, results in a completely inactive species, whereas with no side chain such as Gly-like N123V, activity is retained but with much altered partitioning (Chene et al., 1992).

J. Final Protonation: The Second Product Molecule

Deprotonation of the *gem*-diol six-carbon intermediate leads to bond cleavage between C2 and C3. The amino acid responsible for this step has not been identified, but could be a His residue (287 in the dimer) that is at the active site of all Rubisco species. While the lower three carbons of the intermediate form one molecule of 3P-glycerate, the other three must be protonated with inversion of configuration around C2 before release as the upper, second molecule of 3P D-glycerate. Protonation from the *si*-face of the carbanion that remains after bond cleavage satisifies these requirements.

Stabilization of the carbanion is possible by delocalization of the charge at C2 over the carboxyl group to give a planar aci-acid conformation. However, if the C2 retains the negative charge, then the sp3-type configuration of C2 would resemble that of the putative carbanion form of enediolate, with the C2 "O" group interacting with Thr 173. Whether one conformation is more prevalent than the other may depend on the active site conformation adopted during formation of the enediolate. The amino acid completing the final stereospecific protonation is also unknown, although it might be speculated that if the carboxyl is stabilized in the same position at the active site as 3k2CABP (i.e., between Lys 334 and Mg^{2+}), then Lys 175 is a suitable candidate for this role. Although the fate of substrate protons have not been emphasized in the schemes described above, one notable outcome of NMR-exchange studies in D_2O indicates that the proton that eventually resides at the C2 center of the upper 3P-glycerate originates from solution, and is not the proton abstracted from the C3 position of ribulose-P_2 (Gutteridge et al., 1984). In Figure 9E, the positions of the various active site groups are shown based on these assignments. The phosphate group of the product precursor may move to the same position occupied by the 1-phosphate of the substrate (i.e., between loops 7 and 8) to maximize orbital overlap with the electrons of the carbanion and thereby minimize phosphate elimination (Andrews and Kane, 1991).

This final protonation is not required in the case of the oxygenated intermediate (IV) where hydrolytic cleavage leads directly to P-glycolate generation involving the C1 and C2 centers of the substrate (see Figure 6E). Based on mutatgenesis, the mechanism must now account for stabilization of the oxygenated intermediate by active site amino acids, and that it does not arise through an opportunistic reaction between molecular oxygen and enediol. One unusual feature of the reaction that may provide more insight into the mechanism of oxygenation is the emission of light detected when oxygenation is catalyzed with Mn^{2+} rather than Mg^{2+} ions. The chemiluminescence has been attributed to singlet oxygen formation (Mogel and McFadden, 1991) generated during bisphosphate turnover. This is unlikely (Gutteridge, Diner, and Chase, unpublished results) and probably arises as a result of the metal achieving an excited state during oxygen activation in the presence of enediol and then relaxing back to the ground state with subsequent emission of light.

K. Alternate Designs

In terms of relative specificity, the differences between the best Rubisco species and those less well-endowed examples is only a factor of 1 kcal/mol of energy (Lorimer, personal communication), or essentially the equivalent of the value expected of a relatively weak H-bond. If the three-dimensional structures of the dimeric enzyme could be compared directly to the quaternary complex of the hexadecamer, there are some significant differences to which this small value could be related—for example, elements 7 and 8 of the barrel domain (Schneider et al., 1990). Unfortunately, the two structures are not equivalent because the crystalline dimer is not an active species. A better comparison is the quaternary complex of the *Synechococcus* Rubisco with spinach. In this case we are attempting to relate structural divergence to only about 0.5 kcal, a factor of two variation between these two species. If that difference is smeared over the whole dimeric functional unit, then crystallographic analysis will be hard pressed to identify those essential components contributing to this.

Nevertheless, the functional deformities wrought by the present round of mutagenesis has been most useful in indicating that changes to amino acids within the primary sphere of the active site may not allow the fine tuning of the geometry of the site that will be necessary to alter partitioning of the reactions subtly. It may be that those residues more distant from the site, yet critical for the correct positioning of essential side chains (e.g., the amino acids affecting loop dynamics), are the ones to be targeted. Some evidence for this approach has been forthcoming from mutations in the C-terminal tail of the L-subunit of the *Synechococcus* enzyme. The last 10 amino acids of the tail occupy positions distant from the active site. They stretch over the top of loop 6 and interact with amino acids composing the loop. Replacement and or deletion in this region of the L-subunit affects relative specificity positively and negatively, presumably through modulating the movement of loop 6. There are now two further reports of mutations that have produced increased relative carboxylation efficiencies (Harpel and Hartman, 1992; Read and Tabita, 1992).

A comparison of the sequences of the L-subunits of higher plant Rubisco that are known to differ in catalytic efficiencies, such as spinach Rubisco and sunflower carboxylase, would support investigating residues distant from the active site. In these two cases, sunflower Rubisco is at least 16% more effective at fixing CO_2 than spinach, yet all the amino acids that compose the active site are identical. Clearly the basis for the differences must lie outside this primary sphere, and it is the small but significant alterations to the conformation of the active site induced by those less-conserved amino acids outside this region that define the kinetic properties of the enzyme. Such alterations might also involve residues quite distant from the active site including those of the S-subunit. Although no effect on relative specificity was detected from reconstituting the L8 core of *Synechococcus* Rubisco with spinach S-subunits, the change could be considered conservative. A hybrid species sporting the S-subunits of an aquatic alga that has a primary sequence so distinct from other form I enzymes that it may offer a different architecture produces a dramatic increase in specificity (Read and Tabita, 1992).

In conclusion, the attempts to engineer Rubisco with altered specificity have been successful, thus identifying regions and individual groups that contribute to determining partitioning. How these groups do this is still not understood and will require detailed structural analysis. A question remains as to whether the quaternary complex with 2CABP bound is the most revealing structure in this context. Even though some seven structures now exist of Rubisco, crystallographers must be encouraged to complete the determination to high resolution of other complexes (e.g., with bound ribulose-P_2 or xylulose-P_2), and also choose some of the better characterized specificity mutants that must harbor distinct structural changes. Those natural mutants of the higher plant enzymes that have been identified with superior relative specificities, namely wheat and sunflower Rubisco (Parry et al., 1990; Gutteridge, 1990), are phylogenetically rather distant relations. This would suggest that there is more than one structural means of achieving the same ends, but until a system of dissecting the structure of higher plant Rubisco becomes available, there is no opportunity to test any of these assignments.

ACKNOWLEDGMENTS

We wish to thank Stefan Knight for contributing structural information of the spinach enzyme, and Janet Newman for the *Synechococcus* Rubisco prior to publication. Per Kraulis kindly assisted in the production of the Molscript displays of the Rubisco molecule. We also acknowledge the continuing contribution of George Lorimer on enzyme mechanisms.

NOTES

*Rubisco: ribulose 1,5-bisphosphate carboxylase/oxygenase

#The numbering of the amino acids are based on the primary sequence of spinach Rubisco. Where it is necessary to specify the numbering for the same residue of the dimeric enzyme sequence, these are enclosed in parentheses; for example, Lys 201 (191).

REFERENCES

Anderson, K. & Caton, J. (1987). Sequence analysis of the Alcaligenes eutrophus chromosomally encoded ribulose bisphosphate carboxylase large and small subunit genes and their gene products. J. Bact. 169, 4547–4558.

Andersson, I., Knight, S., Schneider, G., Lindqvist, Y., Lundqvist, T., Branden, C-I., & Lorimer, G. (1989). X-ray structure of the active site of Rubisco. Nature (London) 337, 229–234.

Andrews, T.J. (1988). Catalysis by cyanobacterial ribulose bisphosphate carboxylase large subunits in the complete absence of small subunit. J. Biol. Chem. 263, 12213–12219.

Andrews, T.J. & Ballment, B. (1983). The function of the small subunits of ribulose bisphosphate carboxylase-oxygenase. J. Biol. Chem. 258, 7514–7518.

Andrews, T.J. & Kane, H. (1991). Pyruvate is a by-product of catalysis by ribulose bisphsophate carboxylase/oxygenase. J. Biol. Chem. 266, 9447–9452.

Andrews, T.J. & Lorimer, G.H. (1987). Rubisco: structure, mechanisms and prospects for improvement. In: The Biochemistry of Plants, Vol. 10, pp. 131–218. Academic Press, New York.

Andrews, T.J. & Lorimer, G.H. (1985). Catalytic properties of a hybrid between cyanobacterial large subunits and higher plant small subunits of ribulose bisphosphate carboxylase/oxygenase. J. Biol. Chem. 260, 4632–4636.

Badger, M.R. (1987). The CO_2 concentrating mechanism in aquatic photoautotrophs. In: The Biochemistry of Plants (Hatch, M.D. & Boardman, N.K., Eds.) Vol. 10, pp. 219–275. Academic Press, San Diego.

Bascomb, N.F., Gutteridge, S., & Lubben, T.H. (1992). Organellar targeting in plants. In: Plant Protein Engineering (Shewry, P. & Gutteridge, S., Eds.) pp. 142–163. Cambridge University Press, U.K.

Berry, J.A., Lorimer, G.H., Seemann, J.R., Meek, J., & Freas, S. (1987). Isolation, identification and synthesis of 2′ carboxy arabinitol 1-phosphate, a diurnal regulator of ribulose bisphosphate carboxylase activity. Proc. Natl. Acad. Sci. USA 84, 734–738.

Branden, C.-I., Lindqvist, Y., & Schneider, G. (1991). Protein engineering of Rubisco. Acta Crystallographia, in press.

Branden, C.I., Schneider, G., Lindqvist, Y., Andersson, I., Knight, S., & Lorimer, G.H. (1986). Phil. Trans. Roy. Soc. London, Ser. B. 313, 359–365.

Chapman, M.S., Suh, S.W., Cascio, D., Smith, W.W., & Eisenberg, D. (1988). Science 24, 71–74.

Chen, Z. & Spreitzer, R.J. (1989). Chloroplast intragenic suppression enhances the low CO_2/O_2 specificity of mutant ribulose bisphosphate carboxylase/oxygenase. J. Biol. Chem. 264, 3053–3055.

Chene, P., Day, A.G., & Fersht, A.R. (1992). Mutation of asparagine 111 of rubisco from *Rhodospirillum rubrum* alters carboxylase/oxygenase specificity. J. Mol. Biol. 225, 891–896.

Christeller, J.T. (1981). The effects of bivalent cations on ribulose bisphosphate carboxylase/oxygenase. Biochem. J. 193, 839–844.

Coruzzi, G., Broglie, R., Edwards, R., & Chua, N.-H. (1984). Tissue specific and light regulated expression of a pea nuclear gene encoding the small subunit of ribulose bisphosphate carboxylase. EMBO J. 3, 1671–1679.

Dean, C., van den Elzen, P., Tamaki, S., Dunsmuir, P., & Bedbrook, J. (1985). Linkage and homology analysis divides the eight genes for the small subunit of petunia ribulose bisphosphate carboxylase into three gene families. EMBO J. 4, 3055–3061.

Edmondson, D.L., Kane, H.J., & Andrews, T.J. (1990). Substrate isomerization inhibits ribulose bisphosphate carboxylase/oxygenase during catalysis. FEBS Lett. 260, 62–66.

Ellis, R.J. (1981). Chloroplast proteins: Synthesis, transport and assembly. Ann. Rev. Plant Physiol. 32, 111–137.

Ellis, R.J. & van der Vies, S.M. (1991). Molecular chaparones. Ann. Rev. Biochem. 60, 321–347.

Farber, G.K. & Petsko, G.A. (1990). The evolution of α/β barrel enzymes. Trends Biochem. Sci. 15, 228–234.

Gatenby, A.A., Van der Vies, S., & Bradley, D. (1985). Assembly in *E. coli* of a functional multi-subunit Rubisco from a blue-green alga. Nature 314, 617–620.

Gatenby, A.A., Viitanen, P.V., Speth, V., & Grimm, R. (1993). Identification, cellular localization and participation of chaparonins in protein folding. In: Molecular Processes of Photosynthesis 7 (Barber, J., Ed.), Vol. 7. Oxford University Press, London.

Goloubinoff, P., Gatenby, A.A., & Lorimer, G.H. (1989). GroE heat shock proteins promote assembly of foreign prokaryotic Rubisco oligomers in *E. coli*. Nature (London) 337, 44–47.

Gurevitz, M., Somerville, C.R., & Mcintosh, L. (1985). Pathway of assembly of ribulose bisphosphate carboxylase/oxygenase from Anaebena 7120 expressed in *E. coli*. Proc. Natl. Acad. Sci. USA 82, 6456–6550.

Gutteridge, S. (1990). Limitations of the primary events of CO_2 fixation in photosynthetic organisms: The structure and mechanism of Rubisco. Biochim. Biophys. Acta 1015, 1–14.

Gutteridge, S. (1991). The relative catalytic specificities of the large subunit core of *Synechococcus* ribulose bisphosphate carboxylase/oxygenase. J. Biol. Chem. 266, 7359–7362.

Gutteridge, S. & Gatenby, A.A. (1987). Molecular analysis of the assembly, structure, and function of Rubisco. In: Oxford Surveys of Plant Molecular and Cell Biology (Miflin, B.J., Ed.), Vol. 4, pp. 95–135. Oxford Press, London.

Gutteridge, S. & Julien, B. (1989). A phosphatase from chloroplast stroma of Nicotiana tabacum hydrolyses 2′-carboxy arabinitol 1-phosphate, the natural inhibitor of Rubisco to 2′-carboxy arabinitol. FEBS Lett. 254, 225–230.

Gutteridge, S., Millard, B., & Parry, M.A. (1986). Inactivation of ribulose bisphosphate carboxylase by limited proteolysis. FEBS Lett. 196, 263–268.

Gutteridge, S., Reddy, G.S., & Lorimer, G.H. (1989). The synthesis and purification of 2′-carboxy arabinitol 1-phosphate, a natural inhibitor of ribulose bisphosphate carboxylase, investigated by ^{31}P NMR. Biochem. J. 260, 711–716.

Gutteridge, S., Parry, M., Schmidt, C.N.G., & Feeney, J. (1984). An investigation of ribulose bisphosphate carboxylase activity by high-resolution 1H NMR. FEBS Lett. 170, 355–359.

Gutteridge, S., Phillips, A.L., Kettleborough, K., Parry, M.A.J., & Keys, A.J. (1986). Expression of bacterial Rubisco genes in *E. coli*. Phil. Trans. Roy. Soc. London B313, 433–445.

Gutteridge, S., Pierce, J., & Lorimer, G.H. (1988). Details of the reactions catalyzed by mutant forms of Rubisco. Plant Physiol. Biochem. 26, 675–682.

Gutteridge, S., Parry, M., Schmidt, G., & Lorimer, G.H. (1985). Nature of the activation and active site of ribulose bisphosphate carboxylase from the e.p.r. of transition state enzyme-manganese complexes. Biochem. Soc. Trans. 13, 629–631.

Gutteridge, S., Sigal, I., Thomas, B., Arentzen, R., Cordova, A., & Lorimer, G.H. (1984). A site specific mutation within the active site of ribulose bisphosphate carboxylase of *Rhodospirillum rubrum*. EMBO J. 3, 2737–2743.

Harpel, M.R. & Hartman, F.C. (1992). Enhanced CO_2/O_2 specificity of a site-directed mutant of ribulose bisphosphate carboxylase/oxygenase. J. Biol. Chem. 267, 6475–6478.

Hartman, F.C. (1992). Structure-function relationships of ribulose bisphosphate carboxylase/oxygenase as suggested by site-directed mutagenesis. In: Plant Protein Engineering (Shewry, P.R. & Gutteridge, S., Eds.), pp. 61–92. Cambridge University Press, Cambridge.

Hartman, F.C. & Lee, E. (1989). Examination of the function of active site lysine 329 of ribulose bisphosphate carboxylase/oxygenase as revealed by proton exchange reaction. J. Biol. Chem. 246, 11784–11789.

Hartman, F.C., Milanez, S., & Lee, E.H. (1985). Ionization constants of two active site lysyl e-amino groups of ribulose bisphosphate carboxylase/oxygenase. J. Biol. Chem. 260, 13968–13975.

Hemmingsen, S.M., Woolford, C., van der Vies, S.M., Tilly, K., Dennis, D.T., Georgopolous, C.P., Hendrix, R.W., & Ellis, R.J. (1988). Homologous plant and bacterial proteins chaperone oligomeric protein assembly. Nature 333, 330–334.

Holbrook, G.P., Bowes, G., & Salvucci, M.E. (1989). Degradation of 2-carboxy arabinitol 1-phosphate by a specific chloroplast phosphatase. Plant Physiol. 90, 673–678.

Houtz, R.L., Stults, J.T., Mulligan, R.M., & Tolbert, N.E. (1989). Post translational modifications in the large subunit of ribulose bisphosphate carboxylase/oxygenase. Proc Natl. Acad. Sci. USA 86, 1855–1859.

Hudson, G.S., Evans, J.R., von Caemmerer, S., Arvidsson, Y.B.C., & Andrews, T.J. (1992). Reduction of ribulose bisphosphate carboxylase/oxygenase content by antisense RNA reduces photosynthesis in transgenic tobacco plants. Plant Physiol. 98, 294–302.

Jordan, D.B. & Ogren, W.H. (1981). Species variation in the specificity of ribulose bisphosphate carboxylase. Nature 291, 513–515.

Kaplan, A. (1990). Analysis of high CO_2 requiring mutants indicates a central role for the 5′-flanking region of rbcL and for the carboxysomes in cyanobacterial photosynthesis. Can J. Bot. 68, 1303–1310.

Keegstra, K., Olsen, L.J., & Theg, S.M. (1989). Chloroplastic precursors and their transport across the envelope membranes. Ann. Rev. Plant Physiol. 40, 471–501.

Kettleborough, K., Parry, M.A.J., Burton, S., Gutteridge, S., Keys, A.J., & Phillips, A.L. (1987). The role of the N-terminus of the L-subunit of ribulose bisphosphate carboxylase investigated by construction and expression of chimaeric genes. Eur. J. Biochemistry 170, 335–342.

Kingston-Smith, A.H., Major, I., Parry, M.A. & Keys, A.J. (1992). Properties of a phosphatase in French bean leaves that hydrolyses 2′-carboxy arabinitol 1-phosphate. Biochem. J. 287, 821–825.

Knight, S., Andersson, I., & Branden, C.-I. (1990). Crystallographic analysis of ribulose 1,5-bisphosphate carboxylase from spinach at 2.4 Å resolution: subunit interactions and active site. J. Mol. Biol. 215, 113–160.

Kraulis, P. J. (1991). J. Appl. Crystallog. 24, 946–950.

Krebbers, E., Seurinck, J., Herdies, L., Cashmore, A.R., & Timko, M.P. (1988). Four genes in two diverged subfamilies encode the ribulose bisphosphate carboxylase small subunit polypeptides of *Arabidopsis thaliana*. Plant Mol. Biol. 11, 745–749.

Lan, Y., Woodrow, I.E., & Mott, K.A. (1992). Light dependent changes in ribulose bisphosphate carboxylase activase activity in leaves. Plant Physiol., in press.

Larimer, F.W., Machanoff, R., & Hartman, F.C. (1986). A reconstruction of the gene for ribulose bisphosphate carboxylase from *Rhodospirillum rubrum* that expresses the authentic enzyme in *Eschericia coli*. Gene, 41, 113–120.

Larimer, F.W., Lee, E.H., Mural, R.J., Soper, T.S., & Hartman, F.C. (1987). Intersubunit location of the active site of ribulose bisphosphate carboxylase/oxygenase as determined by *in vivo* hybridization of site-directed mutants. J. Biol. Chem. 262, 15327–15329.

Lorimer, G.H. (1981a). The carboxylation and oxygenation of ribulose 1,5-bisphosphate. The primary events of photosynthesis and photorespiration. Ann. Rev. Plant Physiol. 32, 349–383.

Lorimer, G.H. (1981b). Ribulose bisphosphate carboxylase—Amino acids sequence of a peptide bearing the activator carbon dioxide. Biochemistry 20, 1236–1240.
Lorimer, G.H. & Hartman, F.C. (1988). Evidence supporting lysine 166 of *Rhodospirillum rubrum* ribulose bisphosphate carboxylase as the essential base which initiates catalysis. J. Biol. Chem. 263, 6468–6471.
Lorimer, G.H. & Miziorko, H.M. (1980). Carbamate formation of the e-amino group of a lysyl residue as the basis for the activation of ribulose bisphosphate carboxylase by CO_2, and Mg^{2+}. Biochemistry 19, 5321–5328.
Lorimer, G.H., Gutteridge, S., & Reddy, G.S. (1989). The orientation of substrate and reaction intermediates in the active site of ribulose bisphosphate carboxylase. J. Biol. Chem. 264, 9873–9879.
Lorimer, G.H., Madden, M., & Gutteridge, S. (1988). Partial reactions of ribulose bisphosphate carboxylase: Their utility in the study of mutant enzyme. In: NATO Proceedings of Plant Molecular Biology (Von Wettstein and Chua, N.-H., Eds.), pp. 21–31. Plenum Press, New York.
Lundqvist, T. & Schneider, G. (1988). Crystal structure of the binary complex of ribulose bisphosphate carboxylase and its product 3-phospho-D-glycerate. J. Biol. Chem. 263, 3643–3646.
Lundqvist, T. & Schneider, G. (1989). Crystal structure of the complex of ribulose bisphosphate carboxylase and a transition state analogue, 2′-carboxy arabinitol 1,5-bisphosphate. J. Biol. Chem. 264, 7078–7083.
Lundqvist, T. & Schneider, G. (1991a). Crystal structure of the ternary complex of ribulose-1,5-bisphosphate carboxylase, magnesium (II) and activation carbon dioxide at 2.3 Å resolution. Biochemistry 30, 904–908.
Lundqvist, T. & Schneider, G. (1991b). Crystal structure of activated ribulose bisphosphate carboxylase complexed with its substrate ribulose bisphosphate. J. Biol. Chem. 266, 12604–12611.
Miziorko, H.M. & Sealy, R.C. (1984). Electron resonance studies of ribulose bisphosphate carboxylase: identification of activation cation ligands. Biochemistry 23, 479–485.
Mogel, S.N. & McFadden, B.A. (1990). Chemiluminescence of ribulose bisphosphate oxygenase reaction: Evidence for singlet oxygen production. Biochemistry 29, 8333–8337.
Nargang, F., Mcintosh, I., & Somerville, C.R. (1984). Nucleotide sequence of the ribulose bisphosphate carboxylase gene from *Rhodospirillum rubrum*. Mol. Gen. Genet. 193, 220–224.
Newman, J. & Gutteridge, S. (1990). The purification and preliminary X-ray diffraction studies of recombinant *Synechococcus* ribulose 1,5 bisphosphate carboxylase/oxygenase from *E. coli*. J. Biol. Chem. 265, 15154–15159.
O'Leary, M.H., Jaworski, R.J., & Hartman, F.C. (1979). ^{13}C nuclear magnetic resonance of the CO_2 activation of ribulose bisphosphate carboxylase from *Rhodospirillum rubrum*. Proc. Natl. Acad. Sci. USA 76, 673–675.
Parry, M.A., Keys, A.J., & Gutteridge, S. (1989). Variation in the specificity factor of C_3 higher plant Rubiscos determined by the total consumption of ribulose-P_2. J. Exp. Botany 40, 317–320.
Pierce, J. & Gutteridge, S. (1985). Large scale preparation of Rubisco from a recombinant system in *E. coli* characterized by extreme plasmid instability. Appl. Environ. Microbiol. 49, 1094–1100.
Pierce, J. & Reddy, G. (1986). The sites for catalysis and activation share a common domain. Arch. Biochem. Biophys. 245, 483–493.
Pierce, J., Andrews, T.J. & Lorimer, G.H. (1986). Reaction intermediate partitioning by ribulose bisphosphate carboxylase/oxygenase with different substrate specificities. J. Biol. Chem. 261, 10248–10256.
Pierce, J., Carlson, T.J., & Williams, J.G.K. (1989). A cyanobacterial mutant requiring the expression of ribulose bisphosphate carboxylase from a photosynthetic anaerobe. Proc. Natl. Acad. Sci. USA 86, 5753–5757.
Pierce, J., Lorimer, G.H., & Reddy, G. (1986). The kinetic mechanism of ribulose bisphosphate carboxylase: evidence for an ordered sequential reaction. Biochemistry 25, 1636–1644.
Portis, A.R. (1990). Rubisco activase. Biochim. Biophys. Acta 1015, 655–659.

Price, G.D. & Badger, M.R. (1991). Evidence of the role of carboxysomes on cyanobacterial CO_2-concentrating mechanisms. In: Proceedings of the Second International Symposium on Inorganic Carbon Uptake by Aquatic Photosynthetic Organisms (Colman, B., Ed.). Can. J. Botany 69, 963–973.

Quick, W.P., Schurr, U., Scheibe, R., Schulze, E.-D., Rodermel, S.R., Bogorad, L., & Stitt, M. (1991). Decreased ribulose bisphsophate carboxylase-oxygenase in transgenic tobacco transformed with antisense rbcS. I. Impact on photosynthesis in ambient growth conditions. Planta 183, 542–554.

Ranty, B., Lorimer, G.H., & Gutteridge, S. (1991). An intradimeric cross-link of large subunits of spinach ribulose-1,5-bisphosphate carboxylase/oxygenase is formed by oxidation of cysteine 247. Eur. J. Biochem. 200, 353–358.

Read, B. & Tabita, F.R. (1992). A hybrid ribulose bisphosphate carboxylase/oxygenase enzyme exhibiting a substantial increase in substrate specificity factor. Biochemistry 31, 5553–5560.

Robinson, S.P. & Portis, A.R. (1989). Rubisco activase prevents the *in vitro* decline in activity of Rubisco. Plant Physiol. 90, 968–971.

Rodermel, S.R., Abbott, M.S., & Bogorad, L. (1988). Nuclear organelle interactions: Nuclear antisense gene inhibits ribulose bisphosphate carboxylase enzyme levels in transformed tobacco plants. Cell 55, 673–681.

Roeske, C.A. & O'Leary, M.H. (1985). Carbon isotope effect on carboxylation of ribulose bisphosphate catalyzed by ribulose bisphosphate carboxylase from *Rhodospirilium rubrum*. Biochemistry 24, 1603–1607.

Roy, H.R. (1989). Rubisco assembly: A model system for studying the mechanism of chaperonin action. Plant Cell 1, 1035–1042.

Schloss, J.V. (1990).The kinetic properties of ribulose bisphosphate carboxylase. In: Enzymatic and Model Carboxylation and Reduction Reactions for Carbon Dioxide Utilization (Aresta, M. & Schloss, Eds.), pp. 321–345. J.V. Kluwer Academic Publishers, Netherlands.

Schloss, J.V., Phares, E.F., Long, M.V., Norton, I.L., Stringer, C.D., & Hartman, F.C. (1979). Isolation, characterization and crystallization of ribulose bisphosphate carboxylase from *Rhodospirillum rubrum*. J. Bacteriol. 137, 490–501.

Schneider, G. & Lindqvist, Y. (1992). Structural constraints on protein engineering. In: Plant Protein Engineering (Shewry, P.R. & Gutteridge, S., Eds.), pp. 44–58. Cambridge University Press, Cambridge.

Schneider, G., Lindqvist, Y., Branden, C.I., & Lorimer, G.H. (1986). The three dimensional structure of ribulose bisphosphate carboxylase from *Rhodospirillum rubrum* at 2.9 Å resolution. EMBO J. 5, 3409–3415.

Schneider, G., Knight, S., Andersson, I., Branden, C.I., Lindqvist, Y., & Lundqvist, T. (1990). Comparison of the crystal structures of L_2 and L_8S_8 Rubisco suggests a functional role for the small subunit. EMBO J. 9, 2045–2050.

Seemann, J.R., Kobza, J., & Moore, B.D. (1990). Metabolism of 2-carboxyarabinitol 1-phosphate and regulation of ribulose-1,5-bisphosphate carboxylase activity. Photosynthesis Res. 23, 119–130.

Servaites, J.C., Parry, M.A., Gutteridge, S., & Keys, A.J. (1986). Species variation in the predawn inhibition of ribulose bisphosphate carboxylase/oxygenase. Plant Physiol. 79, 1161–1163.

Shinozaki, K. & Sugiura, M. (1985). Genes for the large and small subunit of ribulose bisphosphate carboxylase/oxygenase constitute a single operon in a cyanobacterium *Anacystis nidulans*. Mol. Gen. Genet. 200, 27–32.

Smith, H.B., Larimer, F.W., & Hartman, F.C. (1990). An engineered change in substrate specificity of ribulose bisphosphate carboxylase/oxygenase. J. Biol. Chem. 265, 1243–1245.

Soll, J. & Waegamann, K. (1992). A functionally active protein import complex from chloroplasts. Plant J. 2, 253–256.

Somerville, C.R. & Somerville, S. (1984). Cloning and expression of the *Rhodospirilium rubrum* ribulose bisphosphate carboxylase gene in *E. coli*. Mol. Gen. Genet. 193, 214–219.

Somerville, C.R., Portis, A.R., & Ogren, W.L. (1982). A mutant of *Arabidopsis thaliana* which lacks activation of rubp-carboxylase *in vivo*. Plant Physiol. 70, 381–387.

Styring, S. & Branden, R. (1985). Identification of ligands to the metal ion in copper (II)-activated ribulose bisphosphate carboxylase/oxygenase by the use of EPR spectroscopy and ^{17}O-labeled ligands. Biochemistry 24, 6011–6019.

Sue, J.M. & Knowles, J.R. (1982). Ribulose-1,5-bisphosphate carboxylase: Fate of the tritium label in [3-^{3}H]ribulose 1,5-bisphosphate. Biochemistry 21, 5404–5410.

Tabita, F.R. & Small, C.L. (1985). Expression and assembly of cyanobacterial ribulose bisphosphate carboxylase/oxygenase in *E. coli* containing stoichiometric amounts of large and small subunits. Proc. Natl. Acad. Sci. USA 82, 6100–6103.

Timko, M.P., Kausch, A.P., Castrsana, C., Fassler, J., Herrera-Estrella, L., Van den Broeck, G., van Montagu, M., Schell, J., & Cashmore, A.R. (1985). Light regulation of plant gene expression by an upstream enhancer like element. Nature, 318, 579–582.

Viitanen, P.V., Gatenby, A.A., & Lorimer, G.H. (1993). Purified chaparonin 60 (groEL) interacts with non-native states of a multitude of *E. coli* proteins. Protein Sci., in press.

Vu, C.V., Allen, L.H., & Bowes, G. (1983). Effect of light and elevated atmospheric CO_2 on the ribulose biphosphate carboxylase activity and ribulose bisphosphate level of soybean leaves. Plant Physiol. 73, 729–734.

Waegamann, K. & Soll, J. (1991). Characterization of the protein import apparatus in outer envelopes of chloroplasts. Plant J. 1, 149–158.

Zhu, G. & Jensen, R.G. (1991). Fallover of ribulose bisphosphate carboxylase/oxygenase activity. Plant Physiol. 97, 1354–1358.

THE FERREDOXIN–THIOREDOXIN SYSTEM:

UPDATE ON ITS ROLE IN THE REGULATION OF OXYGENIC PHOTOSYNTHESIS

Bob B. Buchanan

Advances in Molecular and Cell Biology
Volume 10, pages 337–354.

ISBN: 1-55938-710-6

I. INTRODUCTION

Research during the past three decades has demonstrated that light provides specific regulatory agents that enable the carbon reactions of oxygenic photosynthesis to function effectively. The purpose of this chapter is to describe the regulatory function of the ferredoxin–thioredoxin system in this process. The article also updates our knowledge of the system and describes how that knowledge may be applied in the future.

II. REGULATION OF THE REDUCTIVE PENTOSE PHOSPHATE CYCLE

As is evident from Figure 1 that the principal and ultimate regulator of the carbon (historically the "dark") reactions is light. The first evidence that light plays a regulatory role in regulating the carbon reactions came from experiments showing that certain enzymes of the reductive pentose phosphate cycle were activated by illumination, namely NADP-glyceraldehyde 3-phosphate dehydrogenase (NADP-GAPDH) of intact leaves (Ziegler and Ziegler, 1965) and fructose 1,6-bisphosphatase (FBPase) of chloroplast preparations (Buchanan et al., 1967). Experiments on light/dark transition with intact green algae during this early period were interpreted as showing a light activation of certain enzymes of the cycle, including FBPase (Pedersen et al., 1966). Evidence obtained by a large number of laboratories since the 1960s has confirmed and extended these original findings. A summary of the early studies is given elsewhere (Buchanan, 1980, 1991).

In fulfilling its regulatory role, light, absorbed by chlorophyll and ultimately processed via ferredoxin, is converted to regulatory signals that modulate selected enzymes. Such regulation, which takes place after the target enzymes are synthe-

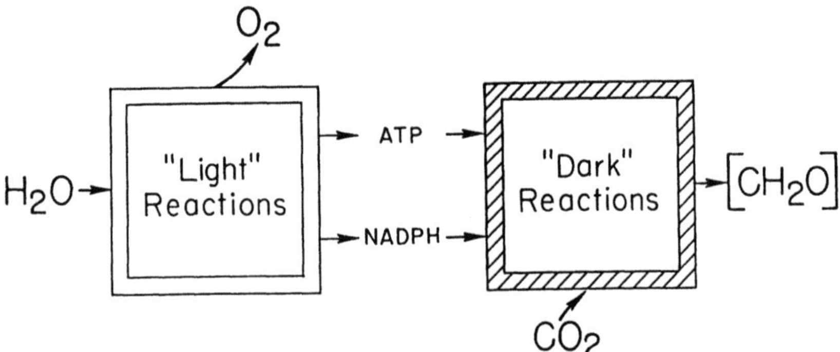

Figure 1. Contributions of the "light" to the "dark" reactions of photosynthesis in the assimilation of carbon dioxide.

sized and assembled, is essential because enzymes for degrading carbohydrates coexist in chloroplasts with enzymes of carbohydrate synthesis (Buchanan, 1980). Selected biosynthetic enzymes are light-activated, whereas degradative enzymes are light-deactivated. In this way, chloroplasts—and oxygenic prokaryotes—minimize the concurrent operation of reactions or pathways that operate in opposing directions ("futile cycling") and thereby maximize the efficiency of temporally disparate metabolic processes. The regulatory function of light thus maintains "enzyme order" by assuring that carbon dioxide assimilation takes place during the day, and carbohydrate degradation occurs primarily at night (Buchanan, 1980, 1991). Through the provision of triose phosphates (dihydroxyacetone phosphate) formed either from newly fixed carbon dioxide or from the breakdown of stored starch, chloroplasts are able to supply substrates for synthetic processes taking place in the cytosol. Primary among these processes is the synthesis of sucrose, a transport carbohydrate which in most plants satisfies day and night energy needs of non-photosynthetic (heterotrophic) tissues (Cséke and Buchanan, 1986).

III. ENZYMES REGULATED

The sensitivity of a metabolic pathway to regulation resides in a small number of the total steps in the pathway (ap Rees, 1980). Such regulatory steps typically have large, negative free energy changes and thus are essentially irreversible. The

Table 1. Enzymes of the Three Phases of the Reductive Pentose Phosphate Cycle

Phase of the Cycle	*Constituent Enzyme(s)*[a]
I. Carboxylation	
Carbon dioxide fixed into 3-phosphoglycerate	**Ribulose 1,5-bisphosphate carboxylase/oxygenase**
II. Reduction	
3-phosphoglycerate reduced to level of carbohydrate, net product formed	Phosphoglycerate kinase **Glyceraldehyde 3-phosphate dehydrogenase (NADP)**
III. Regeneration	
Initial carbon dioxide acceptor, ribulose 1,5-bisphosphate, regenerated	Triose phosphate isomerase Aldolase **Fructose 1,6-bisphosphatase** Transketolase **Sedoheptulose 1,7-bisphosphatase** Phosphopentoepimerase Phosphoriboisomerase **Phosphoribulokinase**

Note: [a]Regulatory enzymes are shown in bold type.

reactions that are substantially displaced from equilibrium in photosynthetic carbon assimilation—that is, those of the regeneration phase catalyzed by FBPase, sedoheptulose 1,7-bisphosphatase (SBPase), and phosphoribulokinase (PRK), and that of the carboxylation phase catalyzed by ribulose 1,5-bisphosphate carboxylase/oxygenase (Rubisco)—appear to be of greatest significance in controlling flux through the reductive pentose phosphate pathway (Table 1; see Bassham and Buchanan, 1982 for review). In addition, the enzyme catalyzing the reduction step of the cycle, NADP-glyceraldehyde 3-phosphate dehydrogenase (NADP-GAPDH), is under regulation (Table 1 and see below). The consequences of the deregulation of the regulatory enzymes of carbon metabolism, including those of the reductive pentose phosphate cycle, have been described elsewhere (Buchanan, 1991).

IV. THE FERREDOXIN–THIOREDOXIN SYSTEM: BACKGROUND

The ferredoxin–thioredoxin system consists of ferredoxin, FTR, and two different thioredoxins: thioredoxin *f* (named for its effectiveness in FBPase activation) and thioredoxin *m* (named for its effectiveness in NADP-malate dehydrogenase or NADP-MDH activation) (Buchanan et al., 1978; Jacquot et al., 1978; Wolosiuk et al., 1979). The system is now considered to function as a general mechanism of light-mediated enzyme regulation in all photosynthetic cells that evolve oxygen (Buchanan, 1980; 1991; Jacquot 1984; Edwards et al., 1985; Cséke and Buchanan, 1986; Knaff, 1989; Gilbert, 1990; Scheibe, 1990), including, as recently demonstrated, eukaryotic green algae (Huppe et al., 1990; Van Langendonckt and Vanden Driessche, 1992).

Thioredoxins are proteins, typically with a molecular mass of 12 kDa, that are widely if not universally distributed in the animal, plant, and bacterial kingdoms (Holmgren, 1985, 1989). Thioredoxins undergo reversible reduction and oxidation through changes in a disulfide group (S-S $\rightarrow$ 2 SH). In the ferredoxin–thioredoxin system, thioredoxin is reduced via an iron–sulfur enzyme, FTR, by ferredoxin, which itself is reduced by the electron transport system of illuminated chloroplast thylakoid membranes (Figure 2) [see Crawford et al. (1989) for references]. The enzymes so activated are oxidized and return to their inactive state in the dark (Figure 3). The evidence suggests that, as with other ferredoxin-linked enzymes, a noncovalent FTR-ferredoxin complex is the species active in promoting the reduction of thioredoxin by ferredoxin (Hirasawa et al., 1988).

As noted above, two different thioredoxins, designated *f* and *m*, are a part of the ferredoxin–thioredoxin system in chloroplasts (Buchanan, 1980, 1991; Jacquot, 1984). In the reduced state, thioredoxin *f* selectively activates enzymes of carbohydrate synthesis, including FBPase, SBPase, PRK, and NADP-GADPH. Thioredoxin *m* preferentially regulates (deactivates in the light) glucose 6-phosphate

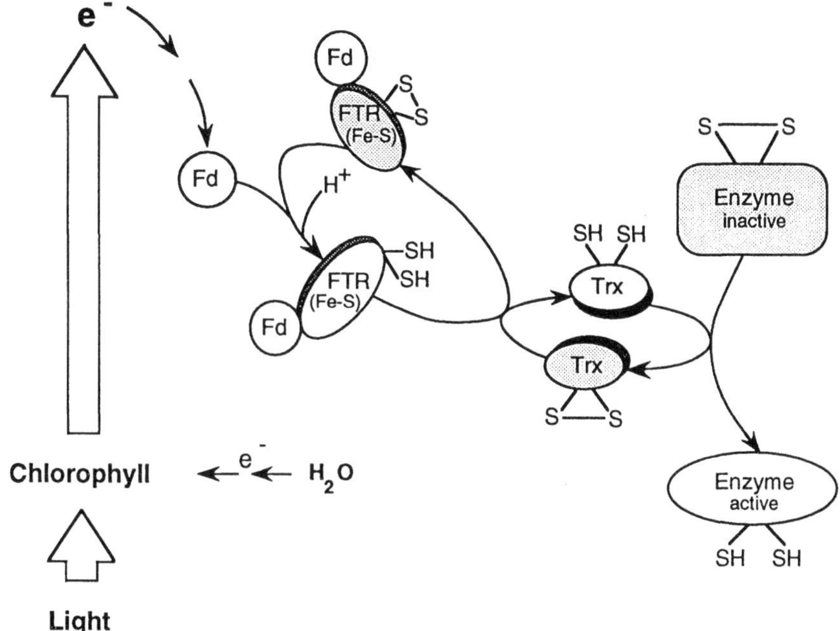

Figure 2. Light activation of biosynthetic enzymes by the ferredoxin–thioredoxin system. The reduction of FTR requires two protons from the medium and two electrons from ferredoxin, a one-electron carrier. As FTR appears to bind ferredoxin on an equimolar basis (Hirasawa, et al., 1988), it is believed that there is a sequential transfer of one electron from ferredoxin in the reduction of the enzyme. The role of the iron–sulfur cluster of FTR is an open question.

(i)	2 Enzyme (active) -SH HS-	+	2 Thioredoxin (oxidized) -S-S-	⟶	2 Enzyme (inactive) -S-S-	+	2 Thioredoxin (reduced) -SH HS-
(ii)	2 Thioredoxin (reduced) -SH HS-	+	O_2	⟶	2 Thioredoxin (oxidized) -S-S-	+	2 H_2O
Sum:	2 Enzyme (active) -SH HS-	+	O_2	⟶	2 Enzyme (inactive) -S-S-	+	2 H_2O

Figure 3. Dark deactivation of biosynthetic enzymes by the ferredoxin–thioredoxin system.

dehydrogenase, a regulatory enzyme of the oxidative pentose phosphate cycle. Thioredoxin *m* functions in chloroplasts in activating NADP-MDH, an enzyme that acts to concentrate carbon dioxide in C_4 photosynthesis. NADP-MDH also occurs in C_3 plants where it is believed to function as part of a light-dependent mechanism which transports excess reducing equivalents to the cytosol (Scheibe, 1990). Both thioredoxins are effective in activation of the "coupling factor" (CF1-ATPase), an enzyme functional in photophosphorylation. Evidence from several laboratories indicates that thioredoxin *m* preferentially activates algal glutamine synthetase (Tischner and Schmidt, 1982; Florencio et al., 1993).

Among the regulatory enzymes of photosynthetic carbon dioxide assimilation, Rubisco is the principal exception to the thioredoxin-linked mode of regulation (Buchanan, 1980, 1991). Evidence to date indicates that this enzyme is regulated by several light-effected mechanisms independently of sulfhydryl changes (see Chapter 7).

As noted above, the major groups of oxygenic photosynthetic organisms have been shown to utilize the ferredoxin–thioredoxin system in enzyme regulation. All eukaryotic oxygenic photosynthetic organisms examined contain both thioredoxins *f* and *m*. Cyanobacteria, on the other hand, contain thioredoxin *m*, but, based on current evidence, seem to lack a typical thioredoxin *f*—a situation that is in accord with the apparent difference in the evolutionary history of the two proteins (see below). It is significant that photosynthetic bacteria (anaerobic photosynthetic organisms that lack the ability to evolve oxygen) seemingly did not evolve a mechanism to regulate metabolic processes by thioredoxin, possibly owing to the absence of a suitable oxidant during growth. The reductive pentose phosphate cycle is apparently not light-regulated in anoxygenic photosynthetic organisms utilizing this pathway, a conclusion based initially on biochemical studies (Crawford et al., 1984) and later supported by sequence information on the PRK and FBPase from purple nonsulfur bacteria (Hallenbeck and Kaplan, 1987). It is noteworthy that certain anoxygenic photosynthetic bacteria, notably green sulfur species, do not contain the reductive pentose phosphate cycle and use an alternative, the reductive carboxylic acid cycle in which ferredoxin is used to drive the Krebs cycle in reverse (Bassham and Buchanan, 1982; Buchanan and Arnon, 1990).

The ferredoxin/thioredoxin system functions by changing the redox status of target enzymes. Biosynthetic enzymes are activated by a net transfer of reducing equivalents (hydrogen) from reduced thioredoxin to enzyme disulfide (S–S) groups (Figure 2), thereby yielding oxidized thioredoxin and reduced (SH) activated enzyme. Deactivation takes place through the oxidation (in the dark) of SH groups on reduced thioredoxin which in turn oxidizes the reduced (activated) enzyme (Figure 3). Enzymes of carbohydrate degradation regulated by this system show an opposite response; that is, a deactivation upon reduction and an activation on oxidation.

V. THE FERREDOXIN–THIOREDOXIN SYSTEM: NEW DEVELOPMENTS

In the past few years, impressive progress has been made on the ferredoxin–thioredoxin system:

1. FTR, thioredoxin *f* and the two isoforms of thioredoxin *m* have been purified and characterized at the protein level (Crawford et al., 1986; Maeda et al., 1986; Droux et al., 1987a,b,c; Tsugita et al., 1991);
2. genes have been cloned and characterized for thioredoxin *f* (Kamo et al., 1989; Lepiniec et al., 1992), thioredoxin *m* (Chang et al., 1986; Muller and Buchanan, 1989; Wedel et al., 1992), one subunit of an FTR (Szekeres et al., 1991), and several target enzymes (see below);
3. the thioredoxin *m* gene has been shown to be required for the growth of a cyanobacterium (Muller and Buchanan, 1989);
4. physiological evidence has been obtained for the functioning of the system in isolated intact chloroplasts (Crawford et al., 1989); and
5. studies on the crystal structure of a plant thioredoxin (*f*-type from spinach chloroplasts) have been initiated (Genovesiotaverne et al., 1991).

In a related important development, the regulatory sites of several thioredoxin-linked enzymes have been identified (Figure 4). FBPase (Marcus et al., 1988; Raines et al., 1988; Horsnell and Raines, 1991b; Kossmann et al., 1992) and NADP-MDH (Decottignies et al., 1988) show a regulatory site that contains a

Enzyme	Regulatory Site
FBPase	... Arg-Cys-Val-Val-Asn-Val-Cys-Gly... 174 179
NADP-MDH	... Glu-Cys-Phe-Gly-Val-Phe-Cys-Thr... 10 15
PRK	... Gly-Cys-Gly- .... Ile-Cys-Leu... 16 55
CF_1-ATPase	... Ile-Cys-Asp-Ile-Asn-Gly-Lys-Cys-Val... 199 205
Thioredoxin f	... Trp-Cys-Gly-Pro-Cys-Lys... 37 40

Figure 4. Regulatory sites of thioredoxin-linked chloroplast enzymes. References are given in the text.

cystine disulfide bridge near the center and amino terminus of the polypeptide chain, respectively. A situation similar to that of the FBPase holds for the regulatory (gamma) subunit of CF1-ATPase (Miki et al., 1988). In each of these three cases, the regulatory site is separate from the active site. In PRK, by contrast, the two sulfur groups of the regulatory site are a part of the active site (Porter et al., 1988; Roesler and Ogren, 1988; Raines et al., 1989; Horsnell and Raines, 1991a). PRK is also unique in that the regulatory cyst(e)ines are separated by a large number of residues (39). The positioning of the regulatory site relative to the active site of chloroplast FBPase, NADP-MDH, and PRK was confirmed in experiments with isolated intact chloroplasts (Crawford et al., 1989). In an interesting new development, two laboratories have come to the conclusion that, in addition to the thioredoxin-linked site identified in Figure 4, NADP-MDH contains a disulfide bond that must be reduced by a cellular thiol prior to enzyme activation (Hatch and Agnostico, 1992; Issakidis et al., 1992).

Pertinent to the sequences of the thioredoxins and target enzymes is the unfolding story on the structure of FTR. Amino acid sequence data from Schürmann's laboratory (P. Schürmann, personal communication) have built on earlier work (Szekeres et al., 1991) in showing that the variable (14 kDa) FTR subunit of spinach is structurally similar to its smaller (7 kDa) *Anacystis* counterpart. The results indicate that the cyanobacterial type protein served as a core for development of the larger higher plant subunit. The similar (16 kDa) FTR subunit, bearing the iron–sulfur chromophore and the catalytically active disulfide group, has to date been sequenced only from spinach. Nonetheless, based on biochemical data, it seems likely that the subunit will be conserved among different types of oxygenic organisms (Maeda et al., 1986; Droux et al., 1987a,b,c).

The nature of the regulatory sites of the target enzymes—FBPase, NADP-MDH, and CF1-ATPase versus PRK—suggests a difference in the evolutionary history of these two types of enzymes. In the future, it will be of interest to determine the phylogenetic relationship between these enzymes of chloroplasts and their counterparts from other sources; for example, chloroplast FBPase, NADP-MDH, and CF1-ATPase and the corresponding enzymes of animals and microorganisms that lack a thioredoxin regulation site.

In another recent development, phylogenetic studies have been initiated on the thioredoxins. The similarity between the *m*-type thioredoxins of chloroplasts and the thioredoxins from a variety of bacteria noted in a significant early study (Maeda et al., 1986) has been extended in a recent analysis (Hartman et al., 1990) in which 14 thioredoxin sequences were used to construct a minimal phylogenetic tree (Figure 5). When analyzed by a parsimony-based method, the bacterial thioredoxins clustered into three groups: one containing the photosynthetic purple bacteria (*Chromatium, Rhodospirillum, Rhodopseudomonas*), as well as the heterotrophs, *E. coli* and *Corynebacterium;* a second containing the photosynthetic green bacterium, *Chlorobium*; and a third containing cyanobacteria (*Anacystis* and *Anabaena*).

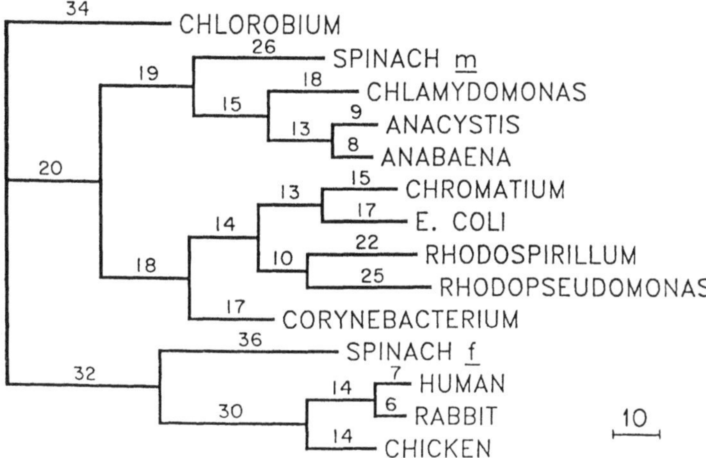

Figure 5. Phylogenetic tree for prokaryotic and eukaryotic thioredoxins based on amino acid sequences. The tree is parsimonious and unrooted. The bar indicates a distance of 10 amino acid replacements. The numbers on the lines indicate the number of amino acid replacements between the indicated thioredoxins. The vertical lines have no significance with regard to evolutionary distances.

These groupings are similar to those generated from earlier 16S ribosomal RNA analyses by Woese (1987). Animal thioredoxins formed a fourth group.

The two thioredoxins of chloroplasts (*f* and *m*) showed contrasting phylogenetic patterns. As predicted from prior studies, spinach chloroplast thioredoxin *m* grouped with its counterparts from cyanobacteria and eukaryotic algae, but thioredoxin *f* grouped with animal thioredoxins. The findings illustrate the potential of thioredoxin as a phylogenetic marker and suggest a relationship between the animal and *f*-type thioredoxins.

Finally, experiments to elucidate thermodynamic aspects of the ferredoxin–thioredoxin system are underway. The redox potentials of thioredoxin *m* and of two target enzymes, NADP-MDH and FBPase (Rebeille and Hatch, 1987a,b; Hutcheson and Ort, 1992), are consistent with the function of the system as depicted in Figures 2 and 3.

To sum up, the evidence is consistent with the view that the ferredoxin–thioredoxin system functions in diverse types of oxygenic photosynthetic organisms as a light-dependent mechanism for the regulation of both biosynthetic and degradatory enzymes. Aside from NADP-GAPDH, the enzymes of the reductive pentose phosphate cycle controlled by this system (FBPase, SBPase, and PRK) act to regenerate the carbon dioxide acceptor, ribulose-1,5-bisphosphate, from newly formed 3-phosphoglycerate. It seems likely that thioredoxin-linked enzymes limit this regeneration. In a recent development, the sites on several of the enzymes regulated by thioredoxin have been identified, with results of interest from the standpoint of protein structure as well as evolution. In the latter context, thioredoxins have emerged as a new evolutionary marker that is extending results obtained with other tools.

VI. INTERACTING SYSTEMS OF REGULATION

Biochemical processes are generally regulated not by one, but by several interacting systems of regulation. From early work, it was concluded that the ferredoxin–thioredoxin system acts jointly with other light-mediated systems; that is, light-driven shifts in pH, divalent cations, and in the concentration of metabolite effectors (Buchanan, 1980).

Since those studies, results from a number of laboratories support such a coordinate function of the different regulatory systems. Noteworthy among the metabolite effector studies are the demonstration of the inhibition of thioredoxin-linked NADP-MDH activation by NADP (Ashton and Hatch, 1983; Scheibe and Jacquot, 1983), the inhibition of activation of PRK by compounds such as 6-phosphogluconate (Gardermann et al., 1983), and the enhancement of thioredoxin-linked FBPase and SBPase activation by substrate (sugar bisphosphate) and divalent cations (Ca^{++} and Mg^{++}) (Hertig and Wolosiuk, 1983). Agents that alter hydrophobic interactions also enhance thioredoxin effects (Stein and Wolosiuk, 1988). In short, there seems little question that the ferredoxin–thioredoxin system functions jointly with mechanisms promoting light-dependent shifts in ions and metabolites in the regulation of a number of chloroplast enzymes.

VII. THE FUTURE

The elucidation of the role of thioredoxin in targeting specific sites on chloroplast enzymes opens the door to new technologies. By using protein engineering techniques, it is now possible to alter the thiol redox properties of proteins, including the capacity for regulation. Recent studies on the *in vitro* mutagenesis of *E. coli* thioredoxin is a case in point. Replacement of aspartate by aspargine at position 61 significantly increased the ability of this thioredoxin to activate FBPase when reduced either photochemically by ferredoxin and FTR or chemically by DTT (deLamotte-Guery et al., 1991). Similarly, a proline to histidine mutation in the active site increased the disulfide isomerase activity 10-fold (Lundstrom et al., 1992). Through site-directed mutagenesis, it was feasible to enhance the stability of PRK through modification of a thioredoxin-linked cysteine group (Milanez et al., 1991). In another study, it was found that substitution of cysteine for threonine at positions 21 and 142 of bacteriophage T4 lysozyme made possible the reversible redox regulation of the enzyme by thiol reagents (Matsumura and Matthews, 1989). These examples illustrate that the thiol redox properties of enzymes can be successfully modified by molecular genetic techniques, and raise the possibility that engineered alterations in regulation will find application in controlling enzyme activities, eventually in industrial processes.

Future applications of our knowledge of thioredoxin-linked enzyme regulation also exist with leaves. As the regulatory sites of chloroplast enzymes targeted by thioredoxin are not found in the corresponding proteins of animal or microbial cells, it may be feasible to design new thiol-specific herbicides. Such herbicides could target an enzyme equally important in all types of plants (e.g., chloroplast FBPase) or an enzyme more abundant in a particular plant group, such as NADP-MDH in C_4 species.

Finally, it is anticipated that new thioredoxin-linked enzymes will be identified in oxygenic photosynthetic cells. The ferredoxin–thioredoxin system has been reported to function in the regulation of enzymes outside the immediate arena of carbon dioxide assimilation; that is, in the metabolism of nitrogen (Tischner and Schmidt, 1982; Florencio et al., 1993), sulfur (Schwenn and Schriek, 1984), and glycerol, a lipid precursor (Gee et al., 1988). It is likely that future work will extend the function of thioredoxin to the regulation of other cell processes. Such processes may include ones of fermentative bacteria—organisms containing a novel thioredoxin system in which ferredoxin supplies reducing equivalents via a flavin enzyme (Hammel et al., 1983).

VIII. THE NADP–THIOREDOXIN SYSTEM

Thioredoxin is known to be a component of two types of enzyme systems in plants. Chloroplasts contain the above-described ferredoxin–thioredoxin system that links light to the regulation of enzymes of photosynthesis. The other system, which occurs outside plastids, is analogous to the one established for animals and most microorganisms in which thioredoxin (*h*-type) is reduced by NADPH and the flavin enzyme, NADP-thioredoxin reductase (NTR) (Eq. 1) (Suske et al., 1979; Johnson et al., 1987; Florencio et al., 1988).

$$\text{NADPH} + \text{H}^+ + \text{Thioredoxin } h_{\text{ox}} \xrightarrow{\text{NTR}} \text{NADP} + \text{Thioredoxin } h_{\text{red}} \quad (1)$$

Current evidence suggests that the NADP–thioredoxin system is widely distributed in plant tissues and is housed in the mitochondria, endoplasmic reticulum, and cytosol (Bodenstein-Lang et al., 1989; Marcus et al., 1991). Thioredoxin *h* sequences have been determined for the gene from two sources: tobacco (Marty and Meyer, 1991) and *Chlamydomonas* (Decottignies et al., 1991).

The seed is the only structure for which the NADP–thioredoxin system has been ascribed physiological activity in plants. Thioredoxin *h* reduces the intramolecular disulfide bonds of several types of seed proteins—thionins, α-amylase, and trypsin inhibitors (Johnson et al., 1987; Kobrehel et al., 1991)—and also reductively activates an enzyme of carbohydrate metabolism (pyrophosphate fructose-6-P, 1-phosphotransferase or PFP) (Kiss et al., 1991).

Recent evidence demonstrates that thioredoxin (*h*-type) reduces the intramolecular disulfide bonds of other seed proteins. Quantitatively, the most important group is comprised of storage proteins, which account for up to 80% of the total protein of seeds (Kasarda et al., 1976). In the case of cereals, these proteins are insoluble in aqueous solutions and are chemically inert until reduced. Our laboratory has recently found that representatives of the major wheat storage proteins—the gliadins and glutenins—are specifically reduced by thioredoxin (Kobrehel et al., 1992; Wong et al., 1993). The results provide evidence that the NADP–thioredoxin system functions in the reduction of the principal seed proteins (Figure 6), thereby increasing their proteolytic susceptibility and making amino acids (nitrogen and sulfur) available during germination. This conclusion is supported by related new findings showing that: (1) the Kunitz and Bowman-Birk trypsin inhibitors of

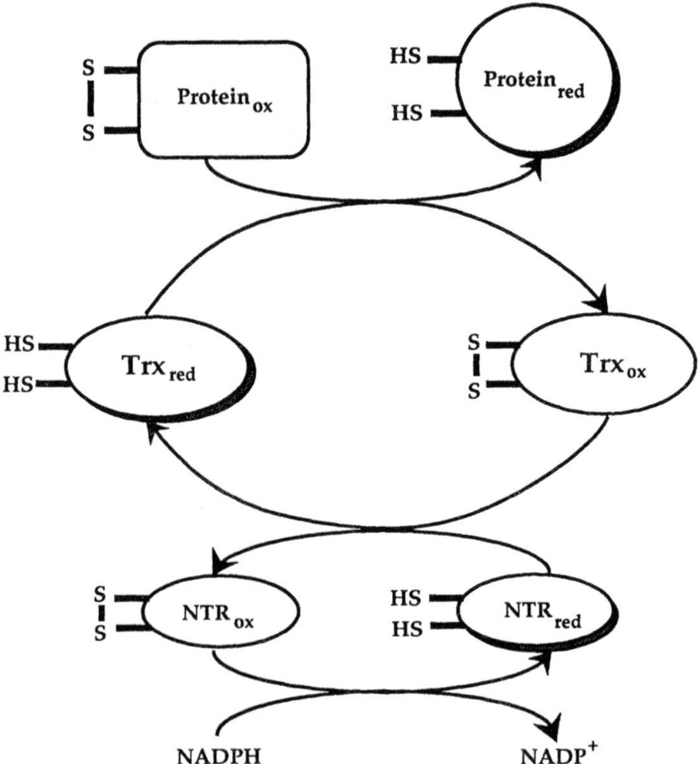

Figure 6. Role of the NADP–thioredoxin system in reducing intramolecular disulfide bonds of seed proteins.

soybean lose their ability to inhibit trypsin and become susceptible to proteolysis on reduction by thioredoxin (Jiao et al., 1992b); (2) the intramolecular disulfide bonds of the 2S protein from castor seed are specifically reduced by thioredoxin, whereas the intermolecular disulfides are not (Shin et al., 1993); and (3) the α-amylase/subtilisin inhibitor, a bifunctional disulfide protein of cereals that inhibits endogenous α-amylase and exogenous subtilisin, is specifically reductively inactivated by thioredoxin, suggesting that the NADP–thioredoxin system functions via this inhibitor protein in regulating α-amylase in germination (Jiao et al., 1992a). New questions raised by these findings are whether endosperm enzymes other than α-amylase are regulated during germination by a redox-linked deinhibition mechanism, and whether changes effected by the NADP–thioredoxin system are related to the breaking of dormancy (Mayer and Poljakoff-Mayber, 1989).

In a new line of research, thioredoxin is beginning to prove useful as a technological tool. Quite recent work indicates that thioredoxin, reduced by NADPH and NTR, may be of technological significance in improving the baking quality of flour as determined by laboratory (micro-farinograph) tests (Wong et al., 1993). Experiments to explore this and other potential applications of thioredoxin are currently in progress.

IX. CONCLUDING REMARKS

The mechanism of carbon dioxide assimilation by the reductive pentose phosphate cycle has been known for almost four decades. During this time, it has been established that light functions not only to fulfill ATP and NADPH requirements, but also to regulate selected enzymes of the cycle and of processes influencing its operation. In oxygen-evolving systems (chloroplasts and cyanobacteria), light absorbed by chlorophyll is converted to several different regulatory signals that alter the activities of selected enzymes—changes in pH, ion gradient, sulfhydryl groups, and concentrations of divalent cations and metabolite effectors. Collectively these signals 'inform' selected enzymes that the light is on so that the cell can direct available resources to increase growth and survival under a wide range of environmental conditions. In the case of sulfhydryl changes, the light signal is carried from chlorophyll-containing thylakoid membranes via ferredoxin to thioredoxins, which, through redox changes in their own sulfhydryl groups, bring about reversible changes in the sulfhydryl status of target enzymes. Such changes alter the activities of key enzymes and direct major biosynthetic and degradatory pathways in the appropriate direction. With certain enzymes, the light-produced alkalization of the stroma and the increase in the concentration of cations and selected metabolite effectors enhance the sulfhydryl changes. Current evidence suggests that it is possible to alter the response of enzymes of this type to thioredoxin through protein engineering.

In an outgrowth of the chloroplast research, thioredoxin has been found to function as a signal in seed germination. Here, thioredoxin acts by regulating (by reduction) the activity of a disulfide protein that inhibits α-amylase and thereby controls the breakdown (mobilization) of starch. Such a mechanism contrasts to that for the activation of chloroplast enzymes in which thioredoxin acts by reducing defined disulfide groups of the target enzymes independently of specific inhibitor proteins. The function of thioredoxin in enzyme regulation complements its role in reducing storage proteins, thereby leading to their mobilization and use by the developing seedling.

ABBREVIATIONS

NADP-GAPDH:	NADP-glyceraldehyde 3-phosphate dehydrogenase
FBPase:	fructose 1,6-bisphosphatase
FTR:	ferredoxin-thioredoxin reductase
Rubisco:	ribulose 1,5-bisphosphate carboxylase/oxygenase
SBPase:	sedoheptulose 1,7-bisphosphatase
PRK:	phosphoribulokinase
NADP-MDH:	NADP-malate dehydrogenase
CF1-ATPase:	chloroplast coupling factor.

REFERENCES

ap Rees, T. (1980). Integration of pathways of synthesis and degradation of hexose phosphates. In: The Biochemistry of Plants (Preiss, J., Ed.), Vol. 3, pp. 1–42. Academic Press, New York.

Ashton, A.R. & Hatch, M.D. (1983). Regulation of C_4 photosynthesis: physical and kinetic properties of active (dithiol) and inactive (disulfide) NADP-malate dehydrogenase from *Zea mays*. Arch. Biochem. Biophys. 227, 406–415.

Bassham, J.A. & Buchanan, B.B. (1982). Carbon dioxide fixation pathways in plants and bacteria. In: Photosynthesis (Govindjee, Ed.), pp. 141–189. Academic Press, New York.

Bodenstein-Lang, J., Buch, A., & Follmann, H. (1989). Animal and plant mitochondria contain specific thioredoxins. FEBS Lett. 258, 22–26.

Buchanan, B.B. (1980). Role of light in the regulation of chloroplast enzymes. Ann. Rev. Plant Physiol. 31, 341–374.

Buchanan, B.B. (1991). Regulation of CO_2 assimilation in oxygenic photosynthesis: the ferredoxin/thioredoxin system. Perspective on its discovery, present status and future development. Arch. Biochem. Biophys. 288, 1–9.

Buchanan, B.B. & Arnon, D.I. (1990). A reverse KREBS cycle in photosynthesis: Consensus at last. Photosyn. Res. 24, 47–53.

Buchanan, B.B., Kalberer, P.P., & Arnon, D.I. (1967). Ferredoxin-activated fructose diphosphatase in isolated chloroplasts. Biochem. Biophys. Res. Commun. 29, 74–79.

Buchanan, B.B., Wolosiuk, R.A., Crawford, N.A., & Yee, B.C. (1978). Evidence for three thioredoxins in leaves. Plant Physiol. 61, 38S.

Chang, J.L., Gleason, F.K., & Fuchs, J.A. (1986). Cloning, expression and characterization of the *Anabaena* thioredoxin gene in *Escherichia coli*. J. Bacteriol. 168, 1258–1264.

Crawford, N.A., Droux, M., Kosower, N.S., & Buchanan, B.B. (1989). Evidence for function of the ferredoxin/thioredoxin system in the reductive activation of target enzymes of isolated intact chloroplasts. Arch. Biochem. Biophys. 271, 223–239.

Crawford, N.A., Sutton, C.W., Yee, B.C., Johnson, T.C., Carlson, D.C., & Buchanan, B.B. (1984). Contrasting modes of photosynthetic enzyme regulation in oxygenic and anoxygenic prokaryotes. Arch. Microbiol. 139, 124–129.

Crawford, N.A., Yee, B.C., Hutcheson, S.W., Wolosiuk, R.A., & Buchanan, B.B. (1986). Enzyme regulation in C_4 photosynthesis: Purification, properties and activities of thioredoxin from C_4 and C_3 plants. Arch. Biochem. Biophys. 244, 1–15.

Cséke, C. & Buchanan, B.B. (1986). Regulation of the formation and utilization of photosynthetate in leaves. Biochim. Biophys. Acta 853, 43–64.

de Lamotte-Guery, F., Miginiac-Maslow, M., Decottignies, P., Stein, M., Minard, P., & Jacquot, J.P. (1991). Mutation of a negatively charged amino acid in thioredoxin modifies its reactivity with chloroplastic enzymes. Eur. J. Biochem. 196, 287–294.

Decottignies, P., Schmitter, J., Miginiac-Maslow, M., Lemarechal, P., Jacquot, J.P., & Gadal, P. (1988). Primary structure of the light-dependent regulatory site of corn NADP-malate dehydrogenase. J. Biol. Chem. 263, 11780–11785.

Decottignies, P., Schmitter, J.-M., Dutka, S., Jacquot, J.-P., & Miginiac-Maslow, M. (1991). Characterization and primary structure of a second thioredoxin from the green alga, *Chlamydomonas reinhardtii*. Eur. J. Biochem. 198, 505–512.

Droux M., Jacquot, J. -P., Miginiac-Maslow, M., Gadal, P., Huet, J.C., Crawford, N.A., Yee, B.C., & Buchanan, B.B. (1987a). Ferredoxin-thioredoxin reductase (FTR): An iron-sulfur enzyme linking light to enzyme regulation in oxygenic photosynthesis. Purification and properties of the enzyme from C_3, C_4 and cyanobacterial species. Arch. Biochem. Biophys. 252, 426–439.

Droux, M., Crawford, N.A., & Buchanan, B.B. (1987b). Mechanism of thioredoxin-linked activation of chloroplast fructose-1,6-bisphosphatase. C. R. Acad. Sci. Paris, III, 305, 335–341.

Droux, M., Miginiac-Maslow, M., Jacquot, J.-P., Gadal, P., Crawford, N.A., Kosower, N.S., & Buchanan, B.B. (1987c). Ferredoxin-thioredoxin reductase: A catalytically active dithiol group links photoreduced ferredoxin to thioredoxin functional in photosynthetic enzyme regulation. Arch. Biochem. Biophys. 256, 372–380.

Edwards, G.E., Nakamoto, H., Burnell, J.N., & Hatch, M.D. (1985). Pyruvate,Pi dikinase and NADP-malate dehydrogenase in C_4 photosynthesis: Properties and mechanisms of light/dark regulation. Ann. Rev. Plant Physiol. 36, 225–286.

Florencio, F.J., Gadal, P., & Buchanan, B.B. (1993). Thioredoxin-linked activation of the chloroplast and cytosolic forms of *Chlamydomonas reinhardtii* glutamine synthetase. Plant Physiol. Biochem. 31, 649–655.

Florencio, F.J., Yee, B.C., Johnson, T.C., & Buchanan, B.B. (1988). An NADP/thioredoxin system in leaves: Purification and characterization of NADP-thioredoxin reductase and thioredoxin *h* from spinach. Arch. Biochem. Biophys. 266, 496–507.

Gardemann, A., Stitt, M., & Heldt, H.W. (1983). Regulation of spinach ribulose-5-phosphate kinase by stromal metabolite levels. Biochim. Biophys. Acta 772, 51–60.

Gee, R.W., Byerrum, R.U., Gerber, D.W., & Tolbert, N.E. (1988). Dihydroxyacetone phosphate reductase in plants. Plant Physiol. 86, 98–103.

Genovesiotaverne, J.C., Jetzer, Y., Sauder, U., Hohenester, E., Hughet, C., Jansonium, J.N., Gardet-Salvi, L., & Schürmann (1991). Crystallization and preliminary X-ray diffraction studies of the spinach chloroplast thioredoxin *f*. J. Mol. Biol. 222, 459–461.

Gilbert, H.F. (1990). Molecular and cellular aspects of thiol-disulfide exchange. Adv. Enzymol. Rel. Areas Mol. Biol. 63, 69–172.

Hallenbeck, P., & Kaplan, S. (1987). Cloning of the gene for phosphoribulokinase activity from *Rhodobacter sphaeroides* and its expression in *E. coli*. J. Bacteriol. 169, 3669–3678.

Hammel, K.E., Cornwell, K.L., & Buchanan, B.B. (1983). Ferredoxin/flavoprotein linked pathway for the reduction of thioredoxin. Proc. Natl. Acad. Sci. USA 80, 3681–3685.

Hartman, H., Syvanen, M., & Buchanan, B.B. (1990). Contrasting evolutionary histories of chloroplast thioredoxins *f* and *m*. Mol. Biol. Evol. 7, 247–254.

Hatch, M.D. & Agostino, A. (1992). Bilevel disulfide group reduction in the activation of C4 leaf nicotinamide adenine dinucleotide phosphate-malate dehydrogenase. Plant Physiol. 100, 360–366.

Hertig, C.M. & Wolosiuk, R.A. (1983). Studies on the hysteretic properties of chloroplast fructose-l,6-bisphosphatase. J. Biol. Chem. 258, 984–989.

Hlirasawa, M., Droux, M., Gray, K.A., Boyer, J.M., Davis, D.J., Buchanan, B.B., & Knaff, D.B. (1988). Ferredoxin-thioredoxin reductase: properties of its complex with ferredoxin. Arch. Biochem. Biophys. 935, 1–8.

Holmgren, A. (1985). Thioredoxin. Ann. Rev. Biochem. 54, 237–271.

Holmgren, A. (1989). Thioredoxin and glutaredoxin systems. J. Biol. Chem. 264, 13963–13966.

Horsnell, P.R., & Raines, C.A. (1991a). Nucleotide sequence of spinach cDNA clone encoding chloroplast phosphoribulokinase from *Arabidopsis thaliana*. L. Plant Mol. Biol. 17, 183-184.

Horsnell, P.R. & Raines, C.A. (1991b). Nucleotide sequence of cDNA clone encoding chloroplast fructose-1,6-bisphosphatase from *Arabidopsis thaliana*. Plant Mol. Biol. 17, 185–186.

Huppe, H.C., de Lamotte-Guery, F., Jacquot, J.-P., & Buchanan, B.B. (1990). The ferredoxin-thioredoxin system of a green alga, *Chlamydomonas reinhardtii*. Identification and characterization of thioredoxins and ferredoxin reductase components. Planta 180, 341–351.

Hutcheson, R.S. & Ort, D.R. (1992). Investigation of the reductive activation of tomato stromal fructose bisphosphatase following chilling. Plant Phys. 99S, 61.

Issakidis, E., Miginiac-Maslow, M., Decottignies, P., Jacquot, J.P., Cretin, C., & Gadal, P. (1992). Site-directed mutagenesis reveals the involvement of an additional thioredoxin-dependent regulatory site in the activation of recombinant sorghum leaf NADP-malate dehydrogenase. J. Biol. Chem. 267, 21577–21583.

Jacquot, J.-P. (1984). Post-translational modification of proteins in higher plants chloroplasts: enzyme regulation by thiol-disulfide interchange. Physiol. Vég. 22, 487–507.

Jacquot, J.-P., Vidal, J., Gadal, P., & Schürmann, P. (1978). Evidence for the existence of several enzyme specific thioredoxins in plants. FEBS Lett. 96, 243–246.

Jiao, J., Yee, B.C., & Buchanan, B.B. (1992a) Thioredoxin-linked changes in properties of protease inhibitors of seeds. Plant Physiol. 99S, 57.

Jiao, J., Yee, B.C., Kobrehel, K., & Buchanan, B.B. (1992b). Effect of thioredoxin-linked reduction on the activity and stability of the Kunitz and Bowman-Birk soybean trypsin inhibitor proteins. J. Agric. Food Chem. 40, 2333–2336.

Johnson, T.C., Cao, R.Q., Kung, J.E., & Buchanan, B.B. (1987). Reduction of purothionin by the wheat seed thioredoxin system and potential function as a secondary thiol messenger in redox control. Planta 171, 321–331.

Johnson, T.C., Wada, K., Buchanan, B.B., & Holmgren, A. (1987). Reduction of purothionin by the wheat seed thioredoxin system and potential function as a secondary messenger in redox control. Plant Physiol. 85, 446–451.

Kamo, M., Tsugita, A., Wiessner, C., Wedel, N., Bartling, D., Hermann, RG., Aguilar, R., Gardet-Salvi, L., & Schürmann, P. (1989). Primary structure of spinach-chloroplast thioredoxin *f*. Protein sequencing and analysis of complete cDNA clones for spinach chloroplast thioredoxin *f*. Eur. J. Biochem. 182, 315–322.

Kasarda, D.D., Bernardin, J.E., & Nimmo, C.C. (1976). Wheat proteins. Adv. Cereal Sci. Technol. 1, 158–236.

Kiss, F., Wu, M.-X., Wong, J.H., Balogh, A., & Buchanan, B.B. (1991). Redox active sulfhydryls are required for fructose-2,6-bisphosphate activation of plant PFP (PPi fructose-6-phosphate 1-phosphotransferase). Arch. Biochem. Biophys. 287, 337–340.

Knaff, D. (1989). The regulatory role of thioredoxin in chloroplasts. Trends Biochem. Sci. 14, 433–434.

Kobrehel, K., Wong, J.H., Balogh, A., Kiss, F., Yee, B.C., & Buchanan, B.B. (1992). Specific reduction of wheat storage proteins by thioredoxin *h*. Plant Physiol. 99, 919–924.

Kobrehel, K., Yee, B.C., & Buchanan, B.B. (1991). Role of the NADP/thioredoxin system in the reduction of α-amylase and trypsin inhibitor proteins. J. Biol. Chem. 266, 16135–16140.

Kossmann, J., Mullerrober, B., Dyer, T.A., Raines, C.A., Sonnewald & Willmitzer, L. (1992) Cloning and expression analysis of the plastidic fructose-1,6-bisphosphatase coding sequence from potato: Circumstantial evidence for the import of hexoses into chloroplasts. Planta 188, 7–12.

Lepiniec, L., Hodges, M., Gadal, P., & Cretin, C. (1992). Isolation, characterization and nucleotide sequence of a full-length pea cDNA encoding thioredoxin *f*. Plant Mol. Biol. 18, 1023–1025.

Lundstrom, J., Krause, G., & Holmgren, A. (1992). A Pro to His Mutation in active site of thioredoxin increases its disulfide isomerase activity 10-fold: New refolding systems for reduced or randomly oxidized ribonuclease. J. Biol. Chem. 267, 9047–9052.

Maeda, K., Tsugita, A., Dalzoppo, D., Vilbois, F., & Schürmann, P. (1986). Further characterization and amino acid sequence of *m*-type thioredoxins from spinach chloroplasts. Eur. J. Biochem. 154, 197–203.

Marcus, F., Chamberlain, S.H., Chu, C., Masiarz, F.R., Shin, S., Yee, B.C., & Buchanan, B.B. (1991). Plant thioredoxin *h*: An animal-like thioredoxin occurring in multiple cell compartments. Arch. Biochem. Biophys. 287, 195–198.

Marcus, F., Moberly, L., & Latshaw, P. (1988). Comparative amino acid sequence of fructose 1,6-bisphosphatase: Identification of a region unique to the light regulated enzyme. Proc. Natl. Acad. Sci. USA 86, 5379–5383.

Marty, I. & Meyer, Y. (1991). Nucleotide sequence of a cDNA encoding a tobacco thioredoxin. Plant Mol. Biol. 17, 143–147.

Matsumura, M. & Matthews, B.W. (1989). Control of enzyme activity by an engineered disulfide bond. Science 243, 792–794.

Mayer, A.M. & Poljakoff-Mayber, A. (1989). The Germination of Seeds, 4th ed., p. 160. Pergamon Press, New York.

Miki, J., Maeda, J., Mukohata, Y., & Futai, M. (1988). The γ- subunit of ATP synthase from spinach chloroplasts. Primary structure deduced from the cloned cDNA sequence. FEBS Lett. 232, 221–226.

Milanez, S., Mural, R.J., & Hartman, F.C. (1991). Roles of cysteinyl residues of phosphoribulokinase as examined by site-directed mutagenesis. J. Biol. Chem. 266, 10694–10699.

Miziorko, H.M. & Lorimer, G.H. (1983). Ribulose-1,5-bisphosphate carboxylase/oxygenase. Annu. Rev. Biochem. 52, 5007–535.

Muller, E.G.D. & Buchanan, B.B. (1989). Thioredoxin is essential for photosynthetic growth. The thioredoxin *m* gene of *Anacystis nidulans*. J. Biol. Chem. 264, 4008–4014.

Pedersen, T.A., Kirk, M., & Bassham, J.A. (1966). Light-dark transients in levels of intermediate compounds during photosynthesis in air-adapted *Chlorealla*. Physiol. Plant. 19, 219–231.

Porter, M.A., Stringer, C.D., & Hartman, F.C. (1988). Nucleotide sequence of spinach cDNA clone encoding chloroplast phosphoribulokinase from *Arabidopsis thaliana*. J. Biol. Chem. 263, 123–129.

Raines, C.A., Lloyd, J.C., Longstaff, M., Bradley, D., & Dyer, T. A. (1988). Chloroplast fructose-1,6-bisphosphatase: The product of a mosaic gene. Nucleic Acids Res. 16, 7931–7942.

Raines, C.A., Longstaff, M., Lloyd, J.C. & Dyer, T.A. (1989). Complete coding sequence of wheat phosphoribulokinase: Developmental and light-dependent expression of the mRNA. Mol. Gen. Genetics 220, 43–48.

Rebeille, F. & Hatch, M.D. (1986a.). Regulation of NADP-malate dehydrogenase in C_4 plants: effect of varying NADPH to NADP ratios and thioredoxin redox state on enzyme activity in reconstituted systems. Arch. Biochem. Biophys. 249, 164–170.

Rebeille, F. & Hatch, M.D. (1986b.). Regulation of NADP-malate dehydrogenase in C_4 plants: relationship among enzyme activity, NADPH to NADP ratios and thioredoxin redox stats in intact maize mesophyll chloroplasts. Arch. Biochem. Biophys. 249, 171–179.

Roesler, K.R. & Ogren, W.L. (1988). Nucleotide sequence of spinach cDNA encoding phosphoribulokinase. Nucleic Acids Res. 16, 7192.

Scheibe, R. (1990). Light dark modulation—regulation of chloroplast metabolism in a new light. Botanica Acta 103, 327–334.

Scheibe, R. & Jacquot, J.P. (1983). NADP regulates the light activation of NADP-dependent malate dehydrogenase. Planta 157, 548–553.

Schwenn, J.D. & Schriek, U. (1984). A new role for thioredoxin in assimilatory sulfate reduction. Activation of the adenylsulfate kinase from the green alga, *Chlamydomonas reinhardtii* CW15. FEBS Lett. 170, 76–80.

Shin, S., Wong, J.H., Kobrehel, K., & Buchanan, B.B. (1993). Reduction of castor seed 2S albumin protein by thioredoxin. Planta 189, 557–560.

Stein, M. & Wolosiuk, R.A. (1988). The effect of chaotropic anions on the activation and the activity of spinach fructose-1,6-bisphosphatase. J. Biol. Chem. 262, 16171–16179.

Suske, G., Wagner, W., & Follmann, H. (1979). NADPH dependent thioredoxin reductase and a new thioredoxin from wheat. Z. Naturforsch. 34c, 214–221.

Szekeres, M., Droux, M., & Buchanan, B.B. (1991). The ferredoxin/thioredoxin reductase variable subunit gene from *Anacystis nidulans*. J. Bacteriol. 173, 1821–1823.

Tischner, R. & Schmidt, A. (1982). A thioredoxin-mediated activation of glutamine synthetase in synchronous *Chlorella sorokiniana*. Plant Physiol. 70, 113–116.

Tusgita, A., Yano, K., Gardet-Savli, L., & Schürmann, P. (1991). Characterization of spinach ferredoxin-thioredoxin reductase. Prot. Seq. Data Anal. 4, 9–13.

Van Langendonckt, A. & Vanden Driessche, T. (1992). Isolation, and characterization of different forms of thioredoxins from the green alga, *Acetabularia mediterranea*: Identification of an NADP/thioredoxin system in the extrachloroplastic fraction. Arch. Biochem. Biophys. 292, 156–164.

Wada, K. & Buchanan, B.B. (1981). Purothionin. A seed protein with thioredoxin activity. FEBS Lett. 124, 237–240.

Wedel, N., Clausmeyer, S., Hermann, R.G., Gardetsalvi, L., & Schürmann, P. (1992). Nucleotide sequence of cDNAs encoding the entire precursor polypeptide for thioredoxin *m* from spinach chloroplasts. Plant Mol. Biol. 18, 527–533.

Woese, C.R. (1987). Bacterial evolution. Microbiol. Rev. 51, 221–271.

Wolosiuk, R.A, Crawford, N.A., Yee, B.C., & Buchanan, B.B. (1979). Isolation of three thioredoxins from spinach leaves. J. Biol. Chem. 254, 1627–1632.

Wong, J.H., Kobrehel, K., Nimbona, C., Yee, B.C., Balogh, Kiss, F., & Buchanan, B.B. (1993). Thioredoxin and bread wheat. Cereal Chem. 70, 113–114.

Ziegler, H. & Ziegler, I. (1965). The influence of light on the $NADP^+$-dependent glyceraldehyde-3-phosphate dehydrogenase. Planta 65, 369–380.

IDENTIFICATION, CELLULAR LOCALIZATION, AND PARTICIPATION OF CHAPERONINS IN PROTEIN FOLDING

Anthony A. Gatenby, Paul V. Viitanen,

Volker Speth, and Rudolf Grimm

Advances in Molecular and Cell Biology
Volume 10, pages 355–388.

ISBN: 1-55938-710-6

ABSTRACT

Molecular chaperones are an abundant class of proteins that have apparently evolved to modulate the folding of a variety of other proteins in cells. The focus of this review is one group of molecular chaperones, the *chaperonins*, that comprise a sequence-related family of proteins, initially found in prokaryotes and in certain cellular organelles that derived from prokaryotes. These proteins possess an intriguing oligomeric molecular architecture, and mechanistically are perhaps the best understood of the molecular chaperones. The discovery, identification, and isolation of chaperonin proteins (and their co-chaperonins) from prokaryotic and eukaryotic cells is described. In addition, the role of chaperonins in facilitating protein folding and suppressing aggregation is discussed. In many microorganisms, chaperonins are also heat- and stress-induced proteins. Therefore, an understanding of the molecular details of their role(s) in protein folding could resolve the enigma of their cellular function during a physiological stress response.

I. INTRODUCTION

Cells invest large amounts of energy in the process of protein synthesis to achieve a correctly folded and functional protein. Many of the steps in this intricate pathway from template to product are understood in considerable detail, and yet the final stage, the folding of a polypeptide chain into its correct three-dimensional structure is still obscure. Not only must information in the primary amino acid sequence be converted rapidly into the correct structure, but this conversion must occur under conditions of high protein concentrations in the cell, in the presence of other rapidly folding species, and at physiological temperatures that are known to destabilize aggregation-prone folding intermediates. In addition, nascent polypeptides emerging from ribosomes in a vectorial fashion may initiate folding in the absence of the complete chain, and consequently not all of the information in the primary sequence is available for successful folding. Similar constraints would also apply to polypeptides that are translocated across the lipid bilayer of membranes prior to correct folding.

Earlier data, based on the successful spontaneous folding of a number of chemically denatured proteins *in vitro*, led to the view that protein folding in cells was a simple consequence of the completed chain being released from the ribosome. However, it is now appreciated that protein folding in cells is more complex than was initially considered. The folding process itself, like many other cellular processes, is subject to ordered regulation. In this case, the regulation is achieved by the interaction of incompletely folded polypeptides with a class of proteins known as molecular chaperones. This interaction with molecular chaperones partitions the polypeptides towards productive folding pathways by suppressing "off-pathway" reactions such as aggregation. The overall chaperonin-facilitated protein folding reaction appears to require the co-participation of two distinct chaperonin components, and for maximal efficiency, the hydrolysis of ATP.

Molecular chaperones are defined as proteins that influence the folding of other proteins, and yet are not components of the final structure (Hemmingsen et al., 1988). This is a very broad definition, with the result that many proteins are now considered as having a "molecular chaperone" function; for example, nucleoplasmin, hsp70, signal recognition particle, SecB, and several others (for recent reviews see Georgopoulos and Ang, 1990; Ang et al., 1991; Ellis, 1991; Ellis and van der Vies, 1991; Zeilstra-Ryalls et al., 1991). The subject of this review, the chaperonins, are a specialized family of molecular chaperones that are abundant, ubiquitous, stress-induced proteins which share characteristic similarities in their primary amino acid sequences and oligomeric molecular organization. Here we shall consider the mode of action of chaperonins, and their ancillary co-chaperonins, in the protein folding process.

II. GENERAL PROBLEMS INVOLVED IN PROTEIN FOLDING AND AGGREGATION

Several recent reviews give detailed descriptions of the molecular events that are important in successful protein folding (Jaenicke, 1987; Kuwajima, 1989; Creighton, 1990; Fischer and Schmid, 1990), and so in this review only an outline of the general principles will be presented in order to understand how chaperonins may function at the level of protein folding. The isomerization of proteins from the unfolded (U) state to the native (N) state involves the transient formation of folding intermediates (I). Although an increasing number of multistep folding pathways have been identified that can be populated by distinct partially folded states (Kuwajima, 1989), the folding of many proteins analyzed *in vitro* can be considered in its simplest form as a two step process.

$$U \rightarrow I \rightarrow N$$

The initial fast step involves the conversion of the unfolded (U) polypeptide to an intermediate state (I), often referred to as a "compact intermediate" or "molten globule". In the absence of denaturant, the U state will not be highly populated at any one time due to a rapid conversion to the thermodynamically more stable I state, a process that occurs on a millisecond time-scale. The I state is a collapsed, yet mobile structure that results from the rapid formation of specific secondary structural elements. Although the I state is compact, relative to the U state, it does not possess the close packing of the secondary structural elements typical of the N state. The amino acid side chains, therefore, exhibit greater fluctuation and consequently the core residues are accessible to solvent molecules. However, in the second rate-determining step these elements become organized into the specific tertiary structures associated with the N state. In contrast to the rapid and uncooperative transitions between the U and I states, those between the I and N states are both slow and cooperative. The I state presumably has a greater number of hydrophobic residues exposed on its surface than in the N state, since the I to N transition is associated with a large change in both enthalpy (ΔH) and in heat capacity (ΔC_p). Another characteristic of the I state is that it is less water-soluble than N, and therefore has a greater tendency to aggregate. This enhanced aggregation probably results from a larger exposure of hydrophobic surfaces that would otherwise be buried in the N state. It therefore follows that the "kinetic partitioning" of the I state between the N state or aggregated states (I_{agg}) is of considerable significance for the efficient folding of a protein under a given set of conditions.

$$
\begin{array}{c}
U \rightarrow I \rightarrow N \\
\downarrow \\
I_{agg}
\end{array}
$$

It is possible to experimentally control some of the factors that influence aggregation *in vitro* to obtain successful refolding of denatured proteins. For example, *in vitro* aggregation is frequently suppressed simply by lowering the concentration of the refolding protein, reducing the temperature of the reaction, or both. In contrast to the I to N transition, which is a first order isomerization reaction independent of the concentration of I, aggregation is an *n*th order reaction that is highly dependent on the concentration of I. Thus, partitioning to the aggregated state can be kinetically favored at higher protein concentrations. Consequently, a greater percentage of correctly folded protein molecules is often observed in dilute solutions where undesirable intermolecular interactions are minimized. Interactions between hydrophobic surfaces, such as those found on molten globules, can also be minimized by the simple expedient of reducing the temperature. For example, by suppressing hydrophobic interactions at low temperatures, conditions that lead to aggregation of ribulose bisphosphate carboxylase (Rubisco) can be minimized and partitioning to the native state favored (Viitanen et al., 1990; van der Vies et al., 1992).

In the complex cellular environment, it is not feasible to reduce protein concentrations to the levels that would prevent intermolecular aggregation events occurring between unfolded or partially folded molecules. Similarly, a reduction in temperature to reduce hydrophobic interactions is not practical for most living organisms. Because aggregation reactions are both temperature- and concentration-dependent, conditions within cells would seem unsuitable for protein folding reactions in the normal physiological range of temperatures and protein concentrations. Molecular chaperones have apparently evolved to help circumvent the aggregation problems *in vivo.*

III. CHAPERONIN MOLECULES

There are two general subfamilies of chaperonins, both of which are necessary for efficient chaperonin-assisted protein folding reactions. Viewed by electron microscopy (Hendrix, 1979; Pushkin et al., 1982; McMullin and Hallberg, 1988) the larger type usually contains fourteen identical subunits, each with a molecular mass of about 60 kDa that are arranged in two stacked rings of seven subunits each. Consequently, all members of this chaperonin subfamily exhibit a striking seven-fold rotational symmetry. Among the most intensively studied of these proteins are GroEL (of *Escherichia coli*), hsp60 (of yeast mitochondria), the Rubisco subunit binding protein (of plant chloroplasts), and the P1 protein (of mammalian mitochondria). They are all highly conserved at the level of amino acid sequence homology (Hemmingsen et al., 1988; Ellis and van der Vies, 1991; Zeilstra-Ryalls et al., 1991), and are collectively referred to as chaperonin 60 (cpn60) based on their characteristic subunit size. Recent evidence suggests that the double toroidal structure described above for most cpn60s may not necessarily represent a functional requirement. Thus, the purified mitochondrial cpn60 from chinese hamster ovary cells (P1 protein) is comprised of a single seven-membered ring, and this structure contains all of the information necessary for the recognition and binding of non-native protein substrates, the ATP-dependent formation of a stable complex with the co-chaperonin cpn10 (see below), and the chaperonin-facilitated refolding of chemically denatured Rubisco (Viitanen et al., 1992b). A single toroid structure has also been reported for mitochondrial cpn60 from moth sperm (Miller et al., 1990). In contrast to bacteria and mitochondria, which contain only a single type of cpn60 subunit (Hemmingsen et al., 1988; Reading et al., 1989), chloroplasts contain two distinct cpn60 polypeptides (α and β) that are present in roughly equal amounts (Hemmingsen and Ellis, 1986; Musgrove et al., 1987; Martel et al., 1990). It is not known whether the α- and β-subunits reside in the same or different cpn60 tetradecamers.

Members of the second chaperonin subfamily are also homooligomeric proteins; however, they are significantly smaller than cpn60 and usually contain subunits of about 10 kDa. This smaller chaperonin protein is known as GroES in *E. coli*, and is more generally referred to as chaperonin 10 (cpn10), again reflecting its subunit size. Homologs of cpn10 have now been identified in numerous prokaryotic species, and most recently in mammalian mitochondria (Lubben et al., 1990; Hartman et al., 1992) and higher plant chloroplasts (Bertsch et al., 1992). The available evidence suggests that, like cpn60, bacterial (Chandrasekhar et al., 1986) and mitochondrial (Hartman et al., 1992) cpn10 possess toroidal structures with sevenfold symmetry. Intuitively, this makes sense, since as we will see, cpn60 and cpn10 form a stable complex with each other in the presence of certain adenine nucleotides. The name cpn10 is actually somewhat misleading, since a functional chloroplast cpn10 homolog (Bertsch et al., 1992) has been shown to be comprised of identical subunits, each with a molecular mass of about 21 kDa (see below).

IV. EVIDENCE FROM CELLULAR AND SUBCELLULAR STUDIES THAT PROTEIN FOLDING IS MODULATED BY CHAPERONINS

A. Bacterial Chaperonins

The first chaperonins to be studied in detail were the GroES and GroEL proteins from *E. coli*. The *groE* genes were initially identified because mutations in them prevented the growth of several bacteriophages (reviewed by Georgopoulos and Ang, 1990; Ang et al., 1991; Ellis and van der Vies, 1991; Zeilstra-Ryalls et al., 1991). The *groE* chaperonins influence assembly of head or tail structures (depending on the particular phage), and the sites of these interactions were genetically defined. Subsequent studies revealed that the *groES* and *groEL* genes are essential for bacterial growth (Fayet et al., 1989), and that they constitute an operon whose expression is increased during heat shock. Following an appropriate stress, the cellular level of GroEL can be increased from about 2 to 10% of cell protein. Genetic evidence suggests that GroEL is involved in DNA replication, protein assembly and protein transport (Fayet et al., 1986; Jenkins et al., 1986; Goloubinoff et al., 1989a; Van Dyk et al., 1989; Phillips and Silhavy, 1990). Supporting biochemical data also indicates that GroEL functionally interacts with GroES (Chandrasekhar et al., 1986; Goloubinoff et al., 1989b; Viitanen et al., 1990; Baneyx and Gatenby, 1992). Proteins related to either GroES or GroEL have now been identified in numerous prokaryotic organisms, and they display a high degree of amino acid sequence homology (Zeilstra-Ryalls et al., 1991).

Significant amino acid sequence homology is also found between *E. coli* GroEL and the chloroplast cpn60 (Hemmingsen et al., 1988). This latter protein was implicated in the assembly of Rubisco in chloroplasts (reviewed by Gatenby and Ellis, 1990). Earlier experiments had shown that it was possible to express and assemble cyanobacterial Rubisco in *E. coli* (Gatenby et al., 1985), and so it became feasible to use bacterial molecular genetics to test if the *groE* gene products were involved in Rubisco assembly *in vivo*. This was achieved by either overproducing the chaperonins and looking for enhanced Rubisco assembly, or by using *groE* defective strains and anticipating a reduction in Rubisco assembly. Both approaches were successful, and verified a role for both GroES and GroEL in Rubisco assembly (Goloubinoff et al., 1989a). Recent studies have also demonstrated an involvement of GroE proteins in *nif* gene regulation and nitrogenase assembly (Govezensky et al., 1991), and the assembly of mammalian mitochondrial branched-chain α-keto acid decarboxylase (Wynn et al., 1992) and plant ferredoxin-$NADP^+$ oxidoreductase (Carrillo et al., 1992) in *E. coli.*

Although the GroE proteins assist in bacteriophage and Rubisco assembly, these are protein targets that are not usually present in *E. coli* cells. In attempts to define the normal role of chaperonins in bacterial cells, the technique of genetic suppression has been of value. Because successful polypeptide folding can be significantly influenced by temperature, it was suspected that some heat-sensitive mutations in bacteria could be folding mutants. In these types of mutants incubation at non-permissive temperatures might lead to destabilization of folding intermediates, with subsequent aggregation or proteolysis, resulting in the observed growth defects. Increased expression of the *groE* chaperonins in a range of heat-sensitive mutants grown at non-permissive temperatures might be able to correct growth defects by forcing either the folding or assembly of recalcitrant mutant polypeptides. Indeed, it was found that overexpression of the *groE* operon products resulted in suppression of heat-sensitive mutations in genes encoding biosynthetic enzymes, and secretory and structural proteins (Van Dyk et al., 1989). Similar suppression was also observed for the heat-sensitive mutant phenotype of several *dnaA* alleles (Fayet et al., 1986; Jenkins et al., 1986). It should be noted that this suppression requires overexpression of both *groES* and *groEL.* In addition, the levels of GroE proteins required are high—about 20–30% of total cell protein. This is significantly greater than the increased levels of GroE proteins that would be synthesized in cells containing only the chromosomal *groE* operon when plated at the higher non-permissive temperatures. Although high concentrations of GroE proteins are required to suppress these heat-sensitive mutations, perhaps by trapping unstable mutant proteins to allow partitioning towards correct folding, it is possible that normal levels of chaperonins in cells could help correct errors in protein folding arising from missense mutations, and thus alleviate some weak genetic folding defects. The conclusion from genetic data that many proteins in bacteria interact with chaperonins has recently been supported by the demonstration that there is a high-affinity binding of many cellular proteins to GroEL (Viitanen et al., 1992a).

B. Chloroplastic Chaperonins

The chloroplast cpn60 chaperonin was first encountered during studies on the biosynthesis of Rubisco in isolated chloroplasts. It was observed, following electrophoresis on non-denaturing polyacrylamide gels, that although Rubisco large subunits synthesized in chloroplasts could assemble into the holoenzyme, a significant proportion of the large subunits were stably associated with a large oligomeric protein (Barraclough and Ellis, 1980). This oligomeric protein, initially called the large subunit binding protein (known later as cpn60), was greater than 600 kDa in size, and contained subunits of 60 kDa (Barraclough and Ellis, 1980; Hemmingsen and Ellis, 1986). The results of time-course experiments during radiolabeling demonstrated that as radioactive large subunits became assembled into Rubisco holoenzyme, the radioactivity in the cpn60 oligomer declined. These observations raised the possibility that nascent Rubisco large subunits were specifically associated with cpn60 prior to assembly into holoenzyme, and that the cpn60 large subunit binary complex was an obligatory intermediate in the assembly of Rubisco (Barraclough and Ellis, 1980). Independent experiments also demonstrated that Rubisco large subunits synthesized *in vivo* or *in organello* can be recovered from intact chloroplasts in the form of two different sedimentation complexes of 7S and 29S on sucrose gradients (Roy et al., 1982). The 29S complex contains unassembled Rubisco large subunits associated with cpn60, and the 7S complex probably represents Rubisco dimers. When chloroplasts are incubated in light, it was observed that the newly synthesized large subunits present in both the 7S and 29S complexes disappear and are subsequently found in the assembled 18S Rubisco holoenzyme (Roy et al., 1982). This posttranslational assembly of Rubisco is accelerated in chloroplast extracts by the addition of ATP, but the 29S cpn60 oligomer remains intact (Bloom et al., 1983). In the presence of magnesium, however, ATP causes dissociation of the 29S cpn60 molecule, while a non-hydrolyzable analogue of ATP has no effect (Bloom et al., 1983; Musgrove et al., 1987). Although dissociation of the chloroplast cpn60 occurs at physiological concentrations of ATP, the low concentrations of chaperonins in these experiments would in themselves favor oligomer dissociation. As noted by the authors, the *in vivo* conditions might be quite different, and the higher concentration of chaperonins could permit oligomers to maintain their structure even in the presence of ATP (Musgrove et al., 1987). A complex set of reactions was proposed by Bloom et al. (1983) that requires nucleotides, magnesium, cpn60, and putative intermediates in the assembly of the Rubisco holoenzyme. More recently, purified chloroplast cpn60 has been used to successfully refold Rubisco *in vitro* using defined biochemical components. Under the conditions used, where chemically denatured Rubisco fails to revert spontaneously to its native state, the successful refolding of Rubisco can be obtained in the presence of chloroplast cpn60 and either bacterial (Goloubinoff et al., 1989b) or chloroplast cpn10 (P.V.V., unpublished) in a reaction that requires ATP hydrolysis.

As already noted, chloroplast cpn60 is composed of two types of subunits of 61 kDa and 60 kDa, known respectively as the α- and β-subunits (Musgrove et al., 1987). The two cpn60 subunits are highly divergent in their predicted amino acid sequences (Martel et al., 1990). Both subunits are encoded by nuclear genes and are imported into chloroplasts following synthesis of the precursor form by cytosolic ribosomes (Hemmingsen and Ellis, 1986). Historically, it was the isolation and analysis of genes for the α-subunit from plants that revealed a high degree of sequence homology to the *E. coli* GroEL protein, and ultimately led to the identification of the family of proteins that are now called chaperonins (Hemmingsen et al., 1988). Although the chloroplast cpn60 was originally demonstrated to be involved in Rubisco assembly (Barraclough and Ellis, 1980; Roy et al., 1982; Bloom et al., 1983), it undoubtedly plays a more general role in chloroplast biogenesis.

For example, many proteins imported into chloroplasts form a stable complex with the cpn60 oligomer (Gatenby et al., 1988; Lubben et al., 1989). As anticipated, the polypeptides that are captured by cpn60 following import can be released by the addition of MgATP, but not by a non-hydrolyzable analogue (Lubben et al., 1989). This suggests that proteins entering the chloroplast, presumably in non-native states, may need to interact with cpn60 to mediate their correct refolding, as has been observed for mitochondria (see later). An interesting relationship between the molecular mass of an imported polypeptide, and the degree to which it can form a stable binary complex with chloroplast cpn60 is shown in Figure 1. It is apparent from this graph, which was plotted using data from Lubben et al. (1989), that a greater proportion of larger imported polypeptides are associated with the chaperonin than are smaller polypeptides. This may simply indicate that binary complexes between chaperonins and larger polypeptides withstand better the rigors of the non-denaturing gel electrophoresis used in this analysis. A more interesting interpretation, however, is that following import into chloroplasts, the larger polypeptides have a longer occupancy time on the chaperonin prior to release and folding. This could result from a greater number of contact sites between larger polypeptides and chaperonins, leading to greater stability. Alternatively, during cycles of release and rebinding to cpn60, the larger polypeptides may refold at a slower rate and thus have exposed for longer periods the motifs that favor recognition by chaperonins. An analysis of the kinetics of import and association with cpn60 of the large and small subunits of Rubisco (Gatenby et al., 1988) also lends some support to the view that the degree of interaction with chaperonin is related to unit length. Nevertheless, additional studies are required to test this hypothesis more rigorously.

Another similarity with mitochondria and bacteria is that chloroplasts also contain a cpn10 homologue. This co-chaperonin was originally identified in pea chloroplasts by its ability to form a stable complex with bacterial cpn60 (GroEL), but only in the presence of ATP, and its ability to complement GroEL in the chaperonin-facilitated refolding of Rubisco (Lubben et al., 1990). More recently, a spinach chloroplast cpn10 was cloned from a cDNA library (Bertsch et al., 1992), and sequence analysis of the gene revealed some unexpected findings. The spinach protein is apparently synthesized as a higher M_r precursor and possesses a typical N-terminal chloroplast transit peptide. However, attached to the transit peptide is a single protein, comprised of two distinct prokaryotic-like cpn10 molecules linked in tandem. Remarkably, there are only 8 residues that are completely conserved in

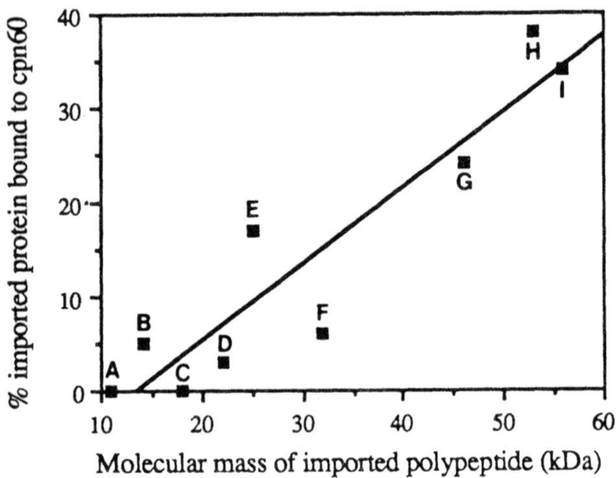

Figure 1. Relationship between the size of a number of polypeptides imported into isolated pea chloroplasts, and their ability to form a stable binary complex with the cpn60 oligomer. Polypeptides were synthesized using *in vitro* translation as radioactive precursor molecules and then imported. After 10 min. the chloroplasts were lysed, the proteins resolved by non-denaturing gel electrophoresis, and the amount of imported protein associated with the chaperonin quantitated. The precursor proteins are (**A**) ferredoxin, (**B**) small subunit of Rubisco, (**C**) superoxide dismutase, (**D**) light-harvesting chlorophyll *a/b* protein, (**E**) chloramphenicol acetyltransferase (**F**) pre-β-lactamase, (**G**) glutamine synthetase, (**H**) β-subunit of thylakoid ATP-synthase, (**I**) large subunit of Rubisco. The molecular mass of the imported protein is given after removal of the transit peptide. Data presented here is based on results obtained by Lubben et al. (1989).

the 16 bacterial cpn10 sequences currently known, and these same residues are also found in both halves of the "double" cpn10 molecule. This high degree of conservation of important amino acid residues suggests that both halves of the chloroplast cpn10 may be able to function independently, or perhaps, they perform different functions. Considering the unique and unexplained presence of two different cpn60 subunits in chloroplasts (α and β), the latter possibility remains particularly intriguing. By expressing in *E. coli* the full-length "double" cpn10 molecule, or its two halves independently, we have found that all three protein configurations are functional in bacteria. An *E. coli* strain was used that has a defective *groES* gene, thereby preventing bacteriophage morphogenesis (Fayet et al., 1986). When the two halves, or the full-length "double" cpn10, are expressed in this strain, growth of bacteriophage λ is restored (F. Baneyx, U. Bertsch, C. Kalbach, J. Soll, and A.A.G, unpublished). This indicates that the two halves of the chloroplast co-chaperonin polypeptide are, indeed, independently functional.

C. Mitochondrial Chaperonins

Both the cpn60 and cpn10 chaperonins are present in mitochondria. The cpn60 (or hsp60) was initially characterized as a protein that accumulated in the mitochondria of *Tetrahymena thermophila* during heat shock (McMullin and Hallberg, 1987). The purified protein contained subunits of 58 kDa, sedimented in sucrose gradients as a 20S to 25S complex, and was subsequently shown to be structurally related to the *E. coli* GroEL protein, and to proteins present in the mitochondria of fungi, plants, and animals (McMullin and Hallberg, 1988). Analysis of the cloned genes revealed a predicted amino acid sequence homology that was about 50% identical between the yeast protein and cpn60 from human mitochondria, bacteria, and chloroplasts (Reading et al., 1989).

In yeast, it is known that the mitochondrial cpn60 is encoded in the nucleus, is an essential gene product, and its accumulation is elevated following heat shock (Reading et al., 1989). Analysis of wild-type and mutant yeast strains shows that the cpn60 protein is required for the correct folding and assembly of imported proteins targeted to mitochondria (Cheng et al., 1989). When a strain with a temperature-sensitive mutation in the cpn60 gene (mif-4) is grown at the non-permissive temperature, the assembly of a number of imported proteins is defective, although synthesis and translocation is unimpaired (Cheng et al., 1989). This defect in assembly was traced to a disruption in the oligomeric state of cpn60 at the non-permissive temperature, which resulted in the pleiotropic non-assembly phenotype.

In subsequent studies (Ostermann et al., 1989), it was observed that a variety of proteins imported into isolated mitochondria in the absence of MgATP would become stably associated with cpn60, as also found in chloroplasts (Lubben et al., 1989). These bound proteins were protease-sensitive, suggesting a loose conformation, but on addition of MgATP the chaperonin·target polypeptide interaction was disrupted and the released protein adopted a more protease-resistant conformation. This indicates that the bound protein folds, at least partially, either on the surface of the cpn60 molecule or shortly after release. Recent data demonstrate that proteins imported into mitochondria may first have to interact with the hsp70 molecular chaperone, and that on release from hsp70 the target polypeptide sequentially interacts with cpn60 before complete folding and assembly is obtained (Manning-Krieg et al., 1991). The specific association between cpn60 and proteins synthesized within mitochondria also occurs. The α-subunit of F1-ATPase synthesized in maize mitochondria will bind to the endogenous cpn60 and is released by the addition of MgATP, demonstrating that the folding of nuclear- and mitochondrial-encoded proteins is mediated by the chaperonin (Prasad et al., 1990).

As described in more detail later, the efficiency of most chaperonin-facilitated protein folding reactions—mediated by purified bacterial cpn60 (GroEL)—is greatly enhanced when both MgATP and bacterial cpn10 (GroES) are also present. Indeed, under conditions where Rubisco (Goloubinoff et al., 1989b; Viitanen et al.,

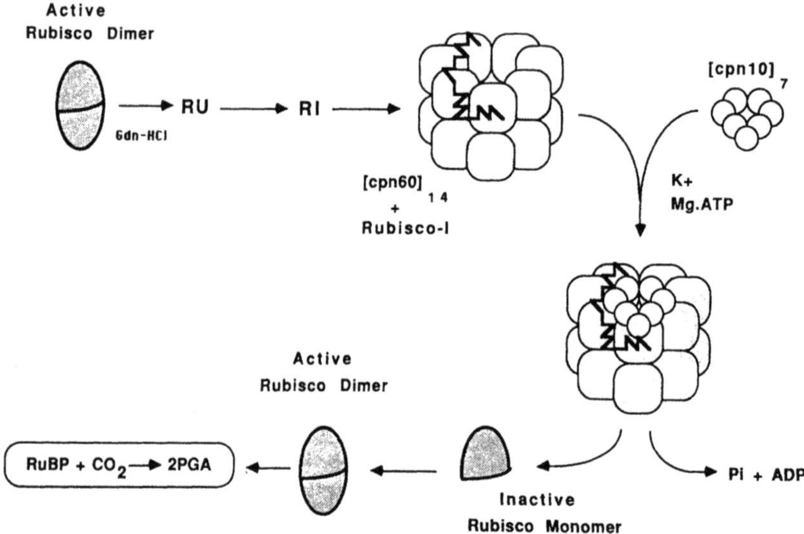

Figure 2. Diagram indicating the essential steps in an assay for the isolation of cpn10-like proteins from eukaryotic cells. The principal aim of the assay is to identify a protein species that will facilitate release of Rubisco from GroEL (cpn60), with subsequent folding and gain of catalytic activity. Functional dimeric Rubisco from *Rhodospirillum rubrum* is unfolded with guanidine hydrochloride to give RU. The RU sample is converted to folding intermediates (RI) by rapid dilution, and in the presence of the chaperonin 60 tetradecamer, the intermediates are trapped in a stable and nonfunctional binary complex. This complex is then used to assay column fractions for the presence of a cpn10-like protein, that in the presence of MgATP and potassium ions, will cause dissociation of Rubisco from cpn60. This discharge reaction can be monitored because the released Rubisco folds, dimerizes, and gives an active enzyme that can be assayed using a standard $[^{14}C]O_2$ incorporation reaction. This reaction has been successfully used to identify and isolate the cpn10 co-chaperonin proteins from mammalian mitochondria and higher plant chloroplasts, using *E. coli* GroEL as the chaperonin (Lubben et al., 1990).

1990, 1992a) and rhodanese (Martin et al., 1991) failed to refold to their native states spontaneously, the presence of both chaperonin components (cpn60 and cpn10) was absolutely required for successful refolding. These observations argue strongly that the co-chaperonin cpn10 plays a critical role in chaperonin-assisted protein folding reactions. It was therefore anticipated that cpn10 would also be present in mitochondria (or in any other cellular compartment that contains cpn60), despite the fact that it had only been found in prokaryotes. The strategy used to identify a mammalian mitochondrial cpn10 (from bovine and rat liver) was based on the premise that if such a protein were to exist, it might be functionally compatible with bacterial GroEL (Lubben et al., 1990). As noted above, this approach also enabled the identification of a unique "double" cpn10 homologue that is present in plant chloroplasts (Bertsch et al., 1992), and may be of future use in identifying co-chaperonins from other cellular compartments. Further details of this technique are outlined in Figure 2.

Aside from complementing GroEL in the chaperonin-facilitated refolding of Rubisco, both mitochondrial and chloroplast cpn10 form stable ATP-dependent complexes with GroEL. This important partial reaction (see later) also occurs between GroEL and GroES (Chandrasekhar et al., 1986; Viitanen et al., 1990), and was instrumental in the identification of the eukaryotic co-chaperonins (Lubben et al., 1990; Bertsch et al., 1992). The power of this approach lies in the fact that native GroEL (and its ATP-dependent complex with cpn10) is so large (>840 kDa) that it is well resolved from the majority of other proteins on an appropriate sizing column. One merely looks for a protein (a potential cpn10 homolog) that shifts in its retention time and co-elutes with GroEL—in an ATP-dependent fashion—during gel filtration chromatography.

The bovine mitochondrial cpn10 that was identified in this manner has been purified and subjected to partial amino acid sequence analysis (Bertsch et al., 1992). From the partial alignments shown in Figure 3, it is evident that it is highly homologous to the bacterial GroES protein (Chandrasekhar et al., 1986). A similar conclusion was reached for a functional mitochondrial cpn10 that was recently purified from rat liver (Hartman et al., 1992), and whose nearly complete sequence was determined by automated Edman degradation. Thus, the protein-folding machinery of mitochondria, like that of prokaryotes and chloroplasts, requires a co-chaperonin for full biological activity. Interestingly, the mammalian mitochondrial cpn60 appears to be more discriminating than its bacterial counterpart (GroEL) in its choice of a co-chaperonin since it is not functionally compatible with

```
     KFLPLFDPVLVE          GGIMLPEKSQGKVLQATVVAVGSGSKG
     .:  || |.|:|          |||:|.:...:|  .:.|:|||.|
  1  MNIRPLHDRVIVKRKEVETKSAGGIVLTGSAAAKSTRGEVLAVGNGRILE

               VGDKVLLPEXXGTK         VVLDDKDYFLF
               ||| |:: :::| |         :::.:.|.: :
 51  NGEVKPLDVKVGDIVIFNDGYGVKSEKIDNEEVLIMSENDILAIVEA*
```

Figure 3. Amino acid sequence homologies between the bovine mitochondrial cpn10 and *E. coli* GroES co-chaperonins. The lower, complete sequence in each pair of lines (numbered residues 1 and 51) is the predicted amino acid sequence deduced from the gene (Hemmingsen et al., 1988). The upper, broken line of each pair is the sequence of the mitochondrial cpn10 obtained by direct sequencing of peptide fragments (Bertsch et al., 1992).

bacterial GroES. The successful refolding of Rubisco, assisted by the single toroidal mammalian mitochondrial cpn60, was only observed in the presence of mammalian mitochondrial cpn10 in a reaction that also required ATP hydrolysis (Viitanen et al., 1992b); GroES could not substitute in this reaction.

D. Cytosolic Chaperonins

The significant role of chaperonins and co-chaperonins in facilitating protein folding in bacteria, plastids and mitochondria raises the question of whether these proteins are present in the cytosol of eukaryotic cells. Considering the basic problems of protein folding *in vivo* discussed earlier, and the indications that chaperonins have evolved to rectify these problems, it might be anticipated that wherever protein folding occurs in cells, protein-folding machinery should exist. This has certainly been observed for the hsp70 class of molecular chaperones, which are distributed throughout many of the compartments of eukaryotic cells (Ang et al., 1991; Ellis and van der Vies, 1991; Gething and Sambrook, 1992).

An important clue that the eukaryotic cytosol contained chaperonin-like proteins was the observation of limited amino acid sequence homology between chaperonins and the mouse cytosolic T-complex polypeptide 1 [TCP1] (Gupta, 1990). TCP1 is apparently an essential protein that is constitutively expressed in almost all cells, and exhibits enhanced synthesis during spermatogenesis. In a subsequent study (Trent et al., 1991), a heat-shock protein (TF55) from a thermophilic archaebacterium was found to possess 40% amino acid sequence identity to mouse TCP1, and formed a chaperonin-like double toroid. Although TF55 was able to bind unfolded proteins and exhibited ATPase activity, there was no direct evidence to show it was involved in protein folding. Direct evidence was provided, however, by the isolation of TCP1 or TCP1-related proteins from rabbit reticulocyte lysates that functioned as molecular chaperones during the *in vitro* folding of tubulin (Yaffe et al., 1992) or β-actin (Gao et al., 1992). This cytosolic chaperonin is organized as a multi-subunit toroid, requires MgATP for activity, and forms a binary complex with unfolded proteins. TCP1, therefore, possesses the structural and functional attributes of a molecular chaperone, although there are significant differences that indicate it is biochemically and structurally unique when compared to cpn60 (Lewis et al., 1992). For plant cells, there is also immunolocalization data indicating that chaperonin-related proteins are present in the cytosol (Grimm et al., 1991). These localization studies, using oat tissue, were initiated following the observation that a chaperonin-like protein copurified with the plant photoreceptor protein phytochrome (E. Mummert, C. Eckerskorn, A. A. G., R. G., V. S., and E. Schäfer, unpublished). Since phytochrome is a cytosolic protein, an association between the photoreceptor and chaperonins could indicate that the latter exist in the cytosol and facilitate folding of phytochrome. In support of this notion, it has recently been demonstrated that the refolding of chemically denatured phytochrome to a photoactive form is mediated by the GroEL chaperonin (Grimm et al., 1993). To identify

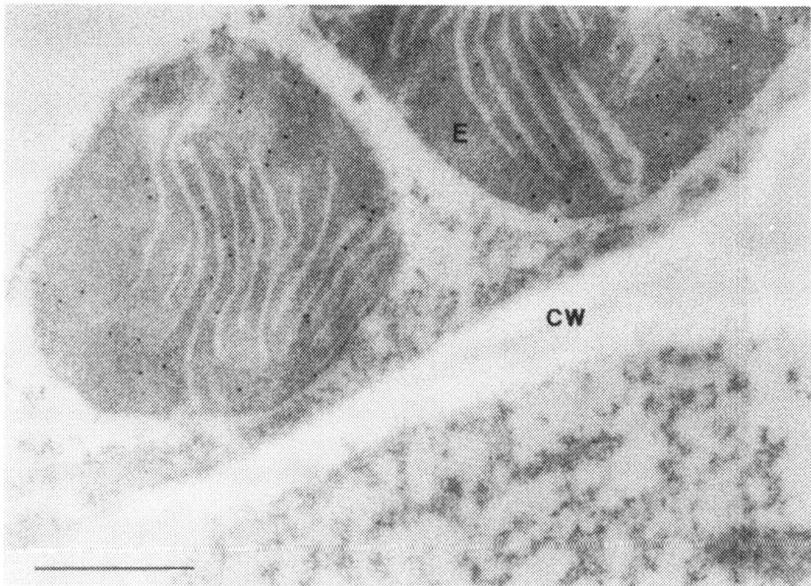

Figure 4. Localization of cpn60 in etioplasts. An electron micrograph of primary leaf cells from oat seedlings grown for 4 days in the dark. Sections were incubated with antisera raised against the chloroplast cpn60, followed by immunogold labeling (Grimm et al., 1991). Bar = 1 μm. E, etioplast; CW, cell wall.

cytosolic chaperonins, various antisera against plant and bacterial chaperonins were used in conjunction with electron microscopy and immunogold labeling.

With antisera raised against the pea chloroplast cpn60, immunoreactive material was selectively found in the stroma of etioplasts (Figure 4), maturing plastids, and chloroplasts, but was noticeably absent from the cytosol, mitochondria, and nuclei (Grimm et al., 1991). This indicated specificity of the antiserum for the chloroplast cpn60. Interestingly, when looking for homologues of chloroplast cpn60 in the plastid of the cryptomonad *Pyrenomonas salina*, no cross-reaction was observed with antisera against the chloroplast cpn60, but was observed with antisera against the *E. coli* GroEL protein (Figure 5). In this cryptomonad the plastoplasm is separated from the cytoplasm by four membranes, and the organisms are thought to represent an intermediate stage in the phyletic development of complex plastids from endosymbionts. The differential cross-reactivity with antisera suggests that the cryptomonad plastidic chaperonin is more related to bacterial GroEL than chloroplast cpn60; this is supported by the sequence of the corresponding cryptomonad gene (U. Maier, M. Merz, R.G., and A.A.G., unpublished).

In contrast to the highly specific staining of oat cells with antisera against the chloroplast cpn60, a more complex pattern is observed when using antisera against

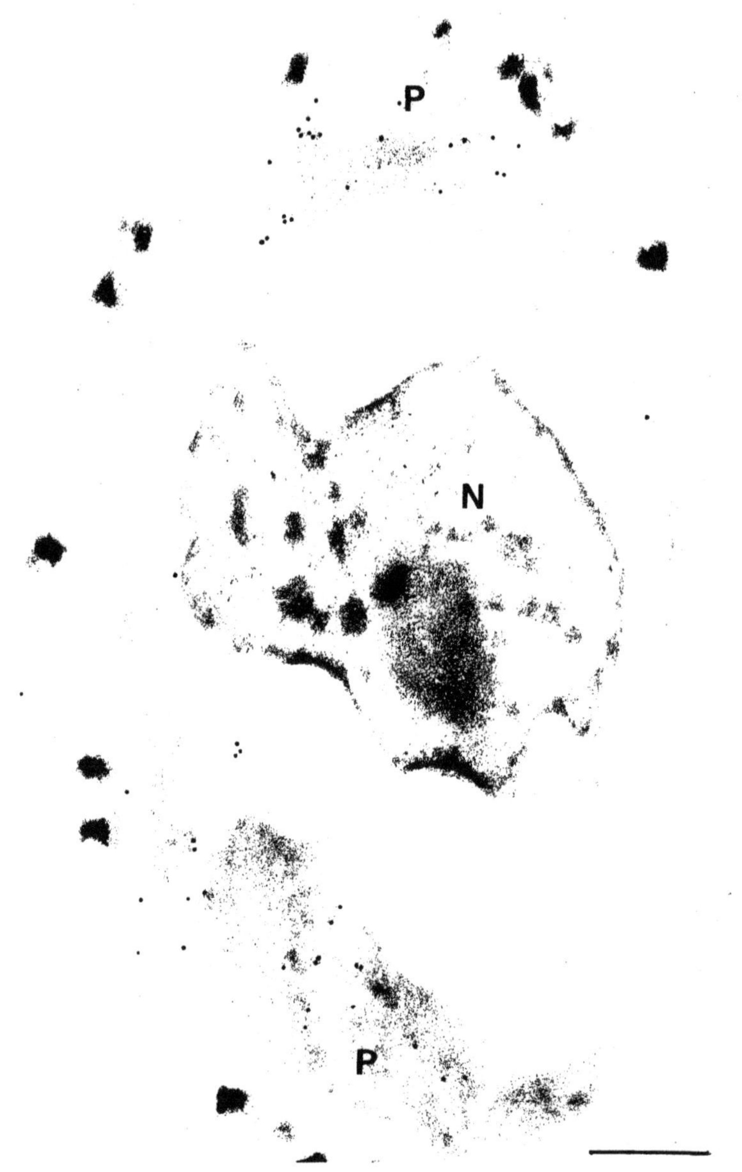

Figure 5. An electron micrograph of the cryptomonad *Pyrenomonas salina* incubated with antisera raised against the *E. coli* GroEL protein, followed by immunogold labeling. Bar = 1 μm. P, plastid; N, nucleus.

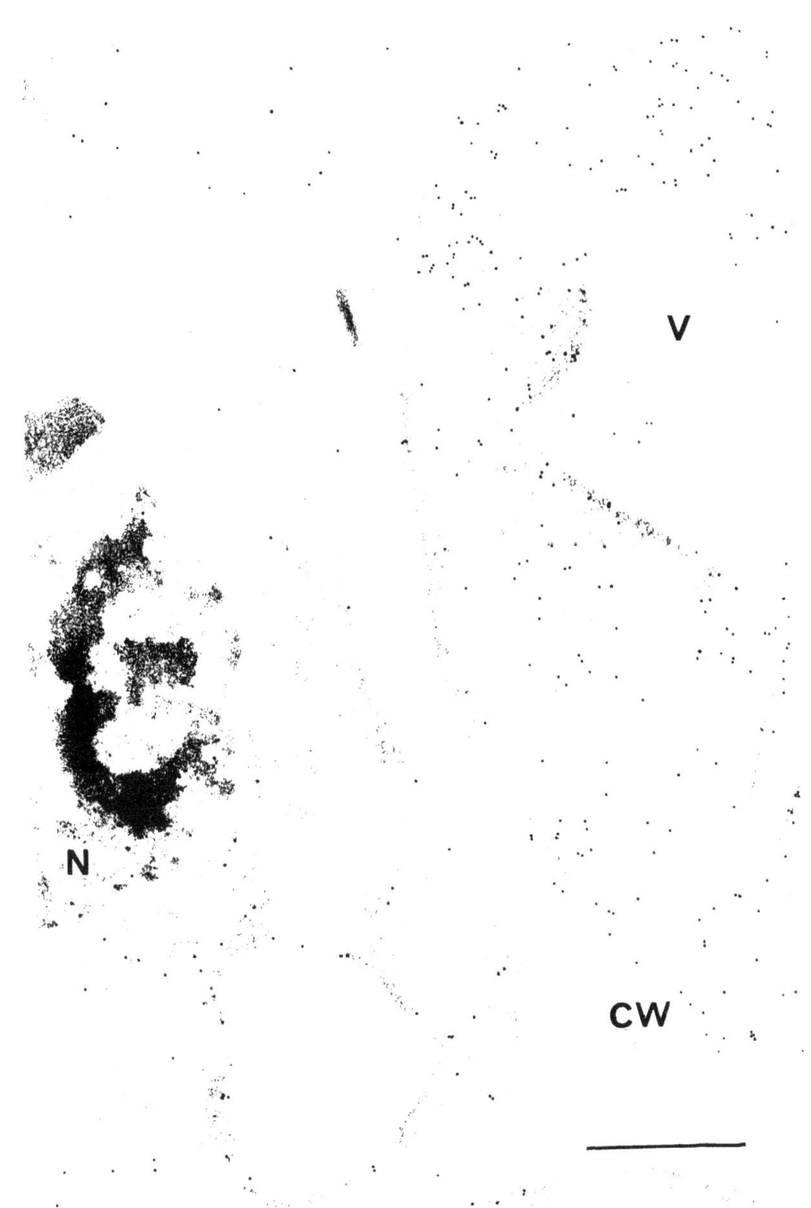

Figure 6. Cytosolic localization of GroEL-related antigen in plant cells. An electron micrograph of oat coleoptile cells grown in the dark for 4 days, followed by irradiation with red light (660 nm) for 2 hours. Cells were incubated with antiserum raised against *E. coli* GroEL protein, followed by immunogold-labeling. Bar = 1 μm. N, nucleus; CW, cell wall; V, vacuole.

the *E. coli* GroEL chaperonin. Immunoreactive material is distributed throughout the cytosol (Figure 6). The antigenically reactive proteins in the cytosol are concentrated in localized regions, but do not appear to be associated with either polysomes, rough endoplasmic reticulum, membranes, or with the cytoskeleton (Grimm et al., 1991). The distribution pattern is found in all tissues from seedlings grown in the dark, or during greening. Heat shock does not appear to change the staining pattern, or to alter the levels of the 60-kDa cross-reacting polypeptide. It is also evident that GroEL-related proteins are present in the nuclei of some plant cells, and are selectively attached to condensed chromatin structures. Analysis of the cpn60 isolated from purified barley nuclei indicates that it has a similar oligomer and subunit size as the GroEL protein, and cross-reacts with antisera raised against GroEL (A.A.G. and R.G., unpublished).

V. MOLECULAR INTERACTIONS BETWEEN CHAPERONINS AND TARGET POLYPEPTIDES DURING FOLDING *IN VITRO*

One of the major objectives in studying chaperonins is to understand the molecular mechanisms that enable them to assist in the folding, assembly, and transmembrane translocation of numerous other proteins. This requires analysis of both the purified chaperonin GroEL (cpn60) and the co-chaperonin GroES (cpn10) proteins individually, as well as their complex interactions with each other. This approach led to the first demonstration that chaperonins could influence the folding of a protein *in vitro*, and facilitate the formation of a correctly folded Rubisco enzyme (Goloubinoff et al., 1989b). The presence of the chaperonins during the refolding reaction prevented Rubisco aggregation, and thus favored correct folding by suppressing off-pathway reactions. A general two-step mechanism was proposed to account for this facilitated folding reaction and the results obtained from certain "order-of-addition experiments". The first step in this mechanism is the trapping of a labile Rubisco folding intermediate by GroEL before aggregation can commence. In its complex with GroEL, the bound Rubisco is prevented from aggregating, but at the same time is unable to progress spontaneously to its native state. In the second step, the sequestered Rubisco is released in a manner that permits it to attain its catalytically active native state. This second step is energy-dependent since it requires the hydrolysis of MgATP, and in the case of Rubisco the GroES co-chaperonin is also necessary.

Following these initial studies on Rubisco, the folding and interactions of several different purified proteins in the presence of GroE chaperonins have now been studied in some detail with some interesting mechanistic differences. Current examples include a more extensive analysis of Rubisco (Viitanen et al., 1990, 1992; Baneyx and Gatenby, 1992; van der Vies et al., 1992), pro-OmpA, and pre-phoA (Lecker et al., 1989), pre-β-lactamase (Laminet et al., 1990; Zahn and Plückthun,

1992), rhodanese (Martin et al., 1991; Mendoza et al., 1991), dihydrofolate reductase (Martin et al., 1991; Viitanen et al., 1991), citrate synthase (Buchner et al., 1991), α-glucosidase (Höll-Neugebauer et al., 1991), lactate dehydrogenase (Badcoe et al., 1991), phytochrome (Grimm et al., 1993), glutamine synthetase (Fisher, 1992) and several thermophilic enzymes (Taguchi et al., 1991). In addition, the binding and release of proteins to GroEL in cell extracts has been examined (Bochkareva et al., 1988; Viitanen et al., 1992a). From these studies some common themes have emerged. Perhaps the most striking observation is that the folding of many different proteins is influenced by chaperonins. These proteins bear little resemblance to each other with regard to size, shape, function, or cellular location. The interaction of GroE chaperonins with target proteins during folding is, therefore, a very general mechanism that enables cells to exert some control over the isomerization of a broad range of molecules. However, a single unifying mechanism to explain all of the results is lacking. Most discrepancies relate to the second part of the reaction—the discharge of the target protein from GroEL. Some target proteins require the presence of GroES for successful release from GroEL, while others do not. Even the requirement for ATP hydrolysis depends on the protein in question.

It is evident that during the chaperonin-facilitated folding of proteins, a number of partial reactions in the overall folding pathway can be identified and examined in more detail using purified components. These partial reactions include: (1) the interactions between GroEL and adenine nucleotides, (2) the ATP-dependent formation of a stable binary complex between GroEL and GroES, (3) the association of GroEL with non-native protein substrates to form a binary complex, and (4) the release of protein substrates from this GroEL binary complex. An understanding of these partial reactions should lead to a clearer understanding of how chaperonins participate in protein folding, and are discussed in more detail in the following sections.

A. The ATPase Activity of GroEL

In the absence of other protein components, purified GroEL can hydrolyze ATP to yield ADP and inorganic phosphate. This partial reaction has been referred to as the "uncoupled" ATPase activity of GroEL (Viitanen et al., 1990), to indicate an apparently wasteful hydrolysis of high energy phosphate, and to distinguish it from the "coupled" hydrolysis of ATP that is observed during chaperonin-assisted protein-folding reactions. Nevertheless, it represents a bonafide enzymatic activity and is a common property of prokaryotic (Ishihama 1976a,b; Hendrix, 1979; Chandrasekhar et al., 1986; Viitanen et al., 1990; Terlesky and Tabita, 1991) and eukaryotic cpn60s (Pushkin et al., 1982; Hemmingsen et al., 1988; Picketts et al., 1989). In general, the turnover numbers for ATP hydrolysis by GroEL are low in comparison to other known ATPases. Most values for the *E. coli* protein range from

0.06 to 0.21 molecules of ATP per second per protomer, although a larger value (~4.4 s^{-1}) has also been reported (Gray and Fersht, 1991).

Surprisingly little is known about the uncoupled ATPase reaction, with regard to its substrates, effectors, kinetics, or mechanism. Most of our knowledge is derived from studies on the purified *E. coli* GroEL. An early report (Ishihama et al., 1976b) suggested that the reaction catalyzed by GroEL is rather specific for ATP; CTP and UTP were not substrates, while the rate with GTP was only 10% that of ATP. In the same study it was shown that the uncoupled ATPase requires divalent cations. Magnesium ions were necessary for full catalytic activity; however, other divalent cations were partially effective ($Ca^{2+} > Mn^{2+} > Zn^{2+}$). More recently, it was shown that the ability of GroEL to hydrolyze ATP also depends on certain monovalent cations (Viitanen et al., 1990), in particular K^+ ions. Under conditions where Mg·ATP is saturating (300 μM), half-maximal activation of the ATPase occurs at a low concentration of about 80 μM K^+. This is about an order of magnitude lower than most other proteins that are activated by K^+ (Suelter et al., 1970). When ATP is subsaturating, however, the observed K^+-requirement increases by several orders of magnitude (M. Todd, G.H. Lorimer, and P.V. V., unpublished). In common with other K^+-requiring enzymes (Suelter et al., 1970), maximal activation of ATP hydrolysis is also observed with low concentrations of Rb^+ and NH_4^+, but not with Na^+, Li^+, or Cs^+. These results have obvious mechanistic implications, since parallel experiments have revealed similar K^+ requirements for the ATP-dependent, chaperonin-assisted reconstitution of Rubisco (Viitanen et al., 1990) and rhodanese (Mendoza et al., 1991). Potassium ions are also required for the ATP-dependent dissociation of a binary complex formed between GroEL and heat inactivated α-glucosidase (Höll-Neugebauer et al., 1991).

Thus, K^+ ions are required for both the "coupled" and "uncoupled" hydrolysis of ATP by GroEL. Indeed, these two activities are likely mediated through a common K^+-binding site(s) that exists somewhere on the chaperonin. Whether K^+ ions exert their effect at the level of ATP binding or hydrolysis remains to be determined. However, this monovalent cation requirement has been retained through evolution from bacteria to mammals. Recent studies with a purified chinese hamster mitochondrial cpn60 (Viitanen et al., 1992b) have shown that low concentrations of K^+ ions are also indispensable for its interactions with ATP. It is likely that the cpn60 of chloroplasts also requires K^+, although this has not yet been demonstrated.

With regard to kinetics, it was recently reported that the hydrolysis of ATP by GroEL is cooperative with respect to ATP (Gray et al., 1991). The data fit equally well to several different models for cooperativity, including that of Monod, Wyman, and Changeux. It was further observed that the extent of cooperativity is increased in the presence of GroES, as reflected by a change in the Hill coefficient from ~2 to ~3. The mechanistic conclusion from this kinetic analysis is that GroEL possesses multiple interactive binding sites for ATP. In light of its seven-fold rotational symmetry, it is likely that the native GroEL tetradecamer can bind either 7 or 14 molecules of adenine nucleotide as suggested by the authors. Whether all of these sites can hydrolyze ATP, or whether some only function allosterically, remains to be determined.

The uncoupled ATPase of GroEL is inhibited by its co-chaperonin GroES (Chandrasekhar et al., 1986; Viitanen et al., 1990; Martin et al., 1991), and by a functional homolog of GroES that has been purified from bovine and rat liver mitochondria (Lubben et al., 1991, Hartman et al., 1992). While the exact mechanism of such inhibition has not been determined, it likely results from the ATP-dependent formation of a stable complex between GroEL and GroES (see below). Thus, complete inhibition is only observed when there is sufficient GroES present to drive all of the GroEL into the inhibitory complex. Even when this condition is met, however, maximal inhibition is not immediately manifest, but gradually develops over a period during which GroEL catalyzes several rounds of ATP hydrolysis (Viitanen et al., 1990; Martin et al., 1991). Consequently, measurements of initial rates of GroEL ATPase activity underestimate the inhibitory effect of GroES (Gray et al., 1991).

The lag period preceding full inhibition could reflect the time required for the formation of the ATP-dependent complex between the two chaperonin components. A lag would also result if the inhibition by GroES required, as a prerequisite, the buildup of a coinhibitor such as ADP, or a slow conformational change in GroEL that results from complex formation. Regardless of the precise explanation, in the presence of an excess of GroES the inhibition steadily progresses to a maximum, and depending on the experimental conditions, the subsequent rate of ATP hydrolysis by GroEL can approach zero (Viitanen et al., 1990; Martin et al., 1991).

The observation that GroES inhibits the uncoupled ATPase of GroEL, and yet is essential for the ATP hydrolysis-dependent functions of the chaperonins, suggests that it might serve as a "coupling factor" (Viitanen et al., 1990). If this view were correct, then it would be expected that the properly coupled chaperonin system would only hydrolyze ATP in the presence of a suitable non-native target protein substrate. In support of this notion is the observation that non-native rhodanese stimulates ATP hydrolysis by the inhibited GroEL·GroES complex, with kinetics similar to that of the chaperonin-facilitated refolding of this protein (Martin et al., 1991). Upon completion of the *in vitro* refolding reaction, ATP hydrolysis ceased, but not before 130 molecules of ATP were hydrolyzed for each molecule of rhodanese refolded. The authors suggested that this was a rather trivial expenditure of cellular energy, relative to the overall cost of synthesizing a protein *in vivo*.

Other interactions between purified cpn60 and adenine nucleotides have also been reported. Most of these relate to the influence of ATP on the oligomeric state of the chaperonin. Using 4 M urea, Girshovich and co-workers (Lissin et al., 1990) dissociated GroEL into its 14 identical subunits and isolated a stable population of folded monomers. These monomers were unable to spontaneously reassemble into native GroEL particles unless certain adenine nucleotides were also provided. MgATP was the most effective in stimulating reassembly, although ADP and a non-hydrolyzable ATP analog (AMP-PNP) were partially active. Interestingly, both GroES and intact GroEL tetradecamers ($cpn60_{14}$) were able to potentiate the maximum effect of Mg·ATP during the *in vitro* reconstitution reaction. In contrast, the yeast mitochondrial cpn60 may not be capable of ATP-dependent self-assembly. Genetic experiments indicate that the *in vivo* assembly of this chaperonin into native oligomers requires pre-existing functional mitochondrial $cpn60_{14}$ (Cheng et al., 1990). It was suggested that the *de novo* assembly of the yeast chaperonin strictly requires chaperoning. However, it could not be excluded that the requirement for a functional chaperonin occurred at some step prior to the actual oligomerization event (e.g., perhaps at the level of monomer folding). Experiments analogous to those performed with the purified GroEL monomers (Lissin et al., 1990) have not yet been conducted with the yeast chaperonin.

The oligomeric state of the chloroplast cpn60 is also influenced by adenine nucleotides (Bloom et al., 1983; Hemmingsen and Ellis, 1986; Musgrove et al., 1987). In this case, however, Mg·ATP shifts the equilibrium towards dissociation. This effect is most pronounced at low temperatures and appears to be fully reversible. The bacterial GroEL tetradecamer also dissociates in the presence of MgATP, when experiments are conducted in apolar metrizamide (Hemmingsen et al., 1988). Although dissociation of chloroplast cpn60 occurs at physiological concentrations of ATP, these *in vitro* experiments were performed with dilute chaperonin solutions, a condition that in itself would favor dissociation. As acknowledged by the authors, the *in vivo* situation is probably entirely different (Musgrove et al., 1987).

It is clear that the particular effect of ATP on the chaperonin oligomer varies, depending on the source of the protein and the experimental conditions. However, taken together, the above results suggest that the binding and/or hydrolysis of ATP by $cpn60_{14}$ is accompanied by significant conformational changes. This is not surprising, considering that cpn60 possesses a molecular architecture that is ideally suited to propagate conformational changes arising from any of its monomeric subunits. Within the double toroidal structure, each of the 14 identical subunits are in direct contact with at least three (and perhaps four) of its nearest neighbors. This high degree of quaternary structural organization allows ample opportunity for communication between the monomers of a given ring, and also permits cross-talk between the two stacked rings of the "double donut".

In summary, it almost seems certain that ATP-induced conformational changes in $cpn60_{14}$ are central to the mechanism of chaperonin-facilitated protein folding. Perhaps the most obvious manifestation of these structural perturbations is the observed decrease in the affinity of $cpn60_{14}$ for its non-native protein substrates (Bochkareva et al., 1988; Laminet et al., 1990; Badcoe et al., 1991; Martin et al., 1991; Viitanen et al., 1991; Fisher, 1992; Viitanen et al., 1992a; Grimm et al., 1993). Structural changes in GroEL following the addition of MgATP have also been detected by enhanced susceptibility of the chaperonin to proteolysis (Baneyx and Gatenby, 1992). This enhanced susceptibility to proteolysis was also observed in the presence of ADP and non-hydrolyzable ATP analogues, suggesting that ATP hydrolysis is not required to achieve certain topological rearrangements in GroEL. ATP-dependent conformational changes have also been invoked in the mechanism of other molecular chaperones. For example, both the DnaK (Liberek et al., 1991) and BiP (Kassenbrock and Kelly, 1989) proteins show altered protease digestion patterns in the presence of adenine nucleotides.

B. ATP-Dependent Association of GroES with GroEL

A specific interaction between GroES and GroEL was first reported by Chandrasekhar et al. (1986), who found that (1) the two proteins would cosediment on glycerol gradients in the presence of MgATP, (2) GroES would bind to an immobilized GroEL affinity column in the presence of MgATP, and (3) the ATPase activity of GroEL could be inhibited by GroES. Thus, biochemical data supported the earlier genetic observations that the two chaperonin proteins functionally interact *in vivo*. Since stable GroEL·GroES complexes were not observed in the presence of a non-hydrolyzable ATP analog, it was initially suggested that formation of the chaperonin complex required the hydrolysis of ATP (Chandrasekhar et al. 1986). However, it was not shown that this analogue could interact with GroEL, precluding a definitive conclusion.

Subsequently, it was reported that GroEL·GroES complexes could form in the absence of added K^+ ions (Viitanen et al., 1990). Since K^+ is required for the GroEL ATPase activity, it was suggested that the interaction between the two chaperonin proteins might not require ATP hydrolysis after all. More direct evidence that this is indeed the case comes from the observation that GroEL·GroES complexes can also form in the presence of ADP (Lissin et al., 1990; Bertsch et al., 1992; Bochkareva et al., 1992). Moreover, rapid gel filtration experiments have shown that the adenine nucleotide that is stably associated with GroEL·GroES complexes—formed in the presence of ATP—is actually ADP (Bochkareva et al., 1992). In that experimental system, neither nucleotide bound stably to GroEL in the absence of GroES. Thus, formation of the GroEL·GroES complex not only requires adenine nucleotides, but actually stabilizes their binding to GroEL. The authors concluded that the native tetradecameric GroEL can accommodate 14 molecules of ATP (ADP) in the presence of GroES, and that this binding is cooperative.

The requirement for adenine nucleotides in the formation of the binary complex between chaperonin and co-chaperonin is rather unique. As described in detail later, most protein substrates that interact with GroEL only do so in their non-native states. Furthermore, such interactions occur spontaneously, and are actually destabilized by ATP or other adenine nucleotides. It is possible that GroES and GroEL always exist as binary complexes *in vivo* since the levels of ATP or ADP that are required for complex formation (submillimolar) are well within the physiological range. Dilution *in vitro*, for example during purification, would be expected to favor complex dissociation. Even so, a stable chaperonin complex has been purified from a thermophilic bacterium (Taguchi et al., 1991) and shown to assist in the *in vitro* refolding of several proteins. N-terminal amino acid sequence analysis indicates that this complex contains about an equal number of cpn60 and cpn10 protomers. This observation reveals that certain cpn60·cpn10 complexes are extremely stable—even in the absence of added adenine nucleotides—and supports the notion that this is the usual form of the two chaperonin proteins *in vivo*. In this regard, it is apparent that a mutation in GroEL that prevents the assembly of phage or Rubisco in *E. coli* results in a suboptimal interaction between the mutant chaperonin and GroES (Baneyx and Gatenby, 1992). This results in an inefficient discharge reaction that presumably inteferes with the kinetics of successful folding for some proteins *in vivo*, and probably accounts for the slower growth rates of strains with this mutation.

The location on cpn60 for the binding of cpn10 has not been determined. Considering that both structures are toroidal and possess sevenfold rotational symmetry, it seems intuitive that both faces of the double ring $cpn60_{14}$ should be able to bind a ring of $cpn10_7$. However, electron micrographs of the ATP-dependent complexes formed from the purified *E. coli* chaperonins suggest that $cpn10_7$ binds to $cpn60_{14}$ in an asymmetric manner (Saibil et al., 1991). Side views of these complexes are "bullet shaped", and exhibit a gross distortion that is thought to exist in one ring of the cpn60 oligomer. Similar observations have been made for the stable chaperonin complex of *T. thermophilus*. The simplest interpretation of these images is that they represent chaperonin complexes that consist of one molecule of $cpn10_7$ bound to one molecule of $cpn60_{14}$. This would agree with the stoichiometries reported for GroEL·GroES complexes isolated by sucrose density gradient centrifugation (Bockareva et al., 1992), and for the chaperonin-assisted refolding of dihydrofolate reductase (Martin et al., 1991).

To try and account for the unexpected stoichiometry, Creighton (1991) has suggested that GroEL is a nonsymmetric dimer of two 7-mers, only one of which can bind GroES. In theory, conversion of the apparently symmetrical $cpn60_{14}$ to a nonsymmetric state could result through its interaction with ATP, cpn10, or both. For example, the random association of $cpn10_7$ to either one of the initially identical rings of $cpn60_{14}$ might result in a conformational change in the other ring, such that it can no longer bind to $cpn10_7$ with high affinity or perhaps at all. Clearly, the former situation could profoundly affect the types of images that are observed during electron microscopic analysis of cpn60·cpn10 complexes, depending upon whether or not they were isolated under dissociating conditions.

In any event, the available evidence suggests that the two identical rings of bacterial $cpn60_{14}$ are not necessarily structurally or functionally equivalent within the "double donut". While this may be true, the results obtained with the mitochondrial chaperonins (Viitanen et al., 1992b) imply that all of the information necessary for the recognition and binding of non-native protein substrates and the ATP-dependent formation of a stable complex with cpn10 resides within a single heptameric ring of cpn60.

C. Binding of Polypeptides to Chaperonins

A pivotal partial reaction in chaperonin-facilitated protein folding is the ability to form stable binary complexes between GroEL and a wide range of proteins that are structurally unrelated in their native states. Clearly, the question of specificity of the interactions between chaperonins and target proteins in this partial reaction becomes paramount. Evidence obtained to date indicates that once proteins have folded to their N states there is little tendency to interact with chaperonins, suggesting that the basis for these interactions is a structural element or motif that is only accessible in proteins that are incompletely folded. Since exposed hydrophobic residues are a characteristic feature of many folding intermediates, perhaps they are responsible for the chaperonin·I state recognition event. Based on the observation that chaperonins can substitute for nondenaturing detergents to obtain successful folding of rhodanese, it was proposed that the interactions of hydrophobic surfaces that lead to aggregation can be prevented by the binding of GroEL to partly folded intermediates (Mendoza et al., 1991). Studies on the chaperonin-dependent folding of the monomeric enzymes, dihydrofolate reductase and rhodanese, indicate that GroEL stabilizes these proteins in a structure that resembles the molten globule state (Martin et al., 1991). The fluorescence properties of α-glucosidase bound to GroEL also suggests a molten globule state for the target protein (Höll-Neugebauer et al., 1991). In contrast, from measurements on the interaction of lactate dehydrogenase with chaperonins, it was concluded that GroEL binds to the unfolded and first transient intermediate in the folding pathway, and not to other later structures such as molten globules (Badcoe et al., 1991). It has also been suggested that improperly folded proteins are recognized by excessive stretches of solvent-exposed main-chain polar groups rather than binding to hydrophobic patches (Hubbard and Sander, 1991). The interaction of GroEL with two synthetic peptides has been studied using 2-D NMR and the analysis of transferred nuclear Overhauser effects (Landry and Gierasch, 1991; Landry et al., 1992). The peptides were stabilized as α-helices when bound to the chaperonin, although it is not clear if this is the structure recognized by GroEL, or merely the conformation that is assumed by the unstructured peptide as a consequence of binding. In any event, the proposal for amphipathic α-helix interaction with chaperonins could account for the diversity of proteins recognized, since such structures are formed very early in the folding pathway of many proteins. Side-chain hydrophobicity is also apparently important for peptide binding to GroEL (Landry et al., 1992). In a recent study (Schmidt and Buchner, 1992), it was also suggested that β-sheet structural elements are also recognized by GroEL, with the conclusion that the interaction of non-native states with GroEL depends primarily on the nature of early-folding intermediates, rather than specific elements of secondary structure.

The binding of proteins in non-native states to GroEL does not require the presence of MgATP or the co-chaperonin GroES. The binary complexes formed between GroEL and a target polypeptide are very stable, and in some instances can be isolated intact by size exclusion chromatography (Viitanen et al., 1991, 1992a; Baneyx and Gatenby 1992; Grimm et al., 1993). While sequestered on the chaperonin the target proteins are stabilized in non-native states and are unable to progress to the N state spontaneously. This has been demonstrated by the failure to detect catalytic activity when various enzymes are sequestered on GroEL, or by dramatically enhanced sensitivity to proteolysis of the immobilized target protein (Goloubinoff et al., 1989b; Laminet et al., 1990; Badcoe et al., 1991; Buchner et al., 1991; Höll-Neugebauer et al., 1991; Martin et al., 1991; Viitanen et al., 1991; Fisher, 1992). An apparent exception to the view that only proteins in non-native states will bind to GroEL are observations that native mouse dihydrofolate reductase and bacterial pre-β-lactamase are subject to a net unfolding when incubated with GroEL (Laminet et al., 1990; Viitanen et al., 1991). It is known, however, that the "native" states of these two proteins exist in slow conformational equilibria with a mixture of folded and unfolded structures. It is the latter species that are probably recognized by GroEL. As a result, by mass action, most of the native enzyme can eventually be sequestered on the chaperonin in an inactive form. The stoichiometry of binding to chaperonins appears to be one or two target polypeptides bound to each GroEL tetradecamer, with most studies favoring the lower number. In several documented examples, the formation of these binary complexes between GroEL and the non-native states of proteins inhibits the development of aggregates. Under appropriate conditions, rhodanese (Martin et al., 1991; Mendoza et al., 1991), Rubisco (Goloubinoff et al., 1989b; Viitanen et al., 1990), citrate synthase (Buchner et al., 1991), α-glucosidase (Höll-Neugebauer et al., 1991), and phytochrome (Grimm et al., 1993) will aggregate following dilution from a solution containing a chaotrope. If GroEL is present during dilution, aggregates are not formed because the partially folded states of these enzymes are trapped by the chaperonin. This stabilizes the I state in a form which not only prevents it from aggregation, but also stops it from proceeding to the native state.

$$\mathrm{U} \rightarrow \mathrm{I} + \mathrm{GroEL} \rightarrow \mathrm{GroEL{\cdot}I}$$

As noted previously, the formation of aggregates during refolding depends both on the concentration of the protein and the temperature at which the experiment is performed. Above a threshold concentration, defined as the critical aggregation concentration (van der Vies et al., 1992), the aggregation of Rubisco will occur until the concentration of I is reduced to a value at which it is no longer susceptible to aggregation. Four of the enzymes described earlier (Rubisco, rhodanese, citrate synthase, and α-glucosidase) which aggregate unless GroEL is present, can also refold in the absence of chaperonins at lower temperatures or protein concentrations. However, if GroEL is also present during these spontaneous folding reactions the folding is inhibited. This inhibition of successful folding is also observed for pre-β-lactamase (Laminet et al., 1990), dihydrofolate reductase (Martin et al., 1991; Viitanen et al., 1991), isocitrate dehydrogenase (Taguchi et al., 1991), lactate dehydrogenase (Badcoe et al., 1991; Taguchi et al., 1991), and glutamine synthetase (Fisher, 1992). The chaperonins, therefore, not only suppress aggregation, but also inhibit legitimate protein folding. The extent of inhibition observed should be a function of (1) the concentrations of GroEL and target protein, (2) the dissociation constant for the binary complex, and (3) the overall rate constant leading to the native state or other states that are unable to rebind to the chaperonin. The common step in these two apparently distinct mechanisms of suppressing aggregation and inhibiting folding is in the interaction of GroEL with unstable folding intermediates which interferes with the partitioning process. When unfolded proteins collapse to the I state, and the critical aggregation concentration is exceeded, they can rapidly aggregate. At lower concentrations they are given the opportunity to refold. If GroEL is present before the kinetic partitioning between the alternative productive (correctly folded) or nonproductive (misfolded) pathways is followed, then the I state is physically trapped by the chaperonin and neither pathway can be pursued. For binary complex formation to occur, GroEL must successfully encounter the refolding protein before it progresses to states that are no longer recognized by the chaperonin, as initially observed by Goloubinoff et al. (1989b). The latter includes both the native state and certain misfolded aggregated states.

D. Release of Polypeptides from Chaperonins

The dissociation, or discharge, of polypeptides bound to GroEL is effected by adenine nucleotides and the co-chaperonin GroES in the presence of potassium ions. Proteins bound to GroEL behave differently in their requirements for nucleotides and GroES in the dissociation reaction. Many proteins, such as dihydrofolate reductase (Martin et al., 1991; Viitanen et al., 1991), pre-β-lactamase (Laminet et al., 1990), lactate dehydrogenase (Badcoe et al., 1991), α-glucosidase (Höll-Neugebauer et al., 1991), glutamine synthetase (Fisher, 1992), and phytochrome (Grimm et al., 1993) can be released and fold to an active form by the addition of MgATP alone. However, it should be appreciated, that in most of these examples the presence of the co-chaperonin GroES potentiates this ATP-dependent discharge. GroES is therefore not necessarily required for the release process *in vitro*, but instead acts to increase its overall efficiency. These GroES-enhanced rates of release, while not essential to obtain the desired product *in vitro*, may be significant to the physiology of cells and account for the simultaneous requirement of both GroEL and GroES for cell viability (Fayet et al., 1989). This may explain why mutations that reduce the interaction between GroEL and GroES result in slower cell growth rates (Baneyx and Gatenby, 1992).

The hydrolysis of ATP is also not essential for the release of certain polypeptides *in vitro*. For example, lactate dehydrogenase (Badcoe et al., 1991), dihydrofolate reductase (Viitanen et al., 1991), and glutamine synthetase (Fisher, 1992) can be dissociated from GroEL in an active form by the addition of the non-hydrolyzable analogue 5′-adenylyl imidodiphosphate (AMP-PNP). It is also noteworthy that adenosine 5′-*O*-(3-thiotriphosphate) (ATPγS) is as effective as ATP in releasing catalytically active dihydrofolate reductase from GroEL (Viitanen et al., 1991), but not bound Rubisco. These observations suggest that *in vitro*, discharge of some proteins that are complexed to the chaperonin is mediated in part through the binding of adenine nucleotides to GroEL. In the presence of ATP, or a non-hydrolyzable analogue capable of producing a similar conformational change, there is a significant reduction in the affinity between GroEL and some target proteins. This shifts the equilibrium towards free enzyme, and spontaneous folding resumes. Interestingly, ATPγS and AMP-PNP were also partially effective in the ATP-dependent self-assembly of the GroEL tetradecamer from its monomeric state (Lissin et al., 1990). The nonspecific affinity of chaperonins for so many different target proteins probably requires a dissociation mechanism that results from a gross conformational rearrangement of GroEL when MgATP is present. Such rearrangements of GroEL have recently been detected by measuring changes in protease sensitivity upon adenine nucleotide addition (Baneyx and Gatenby, 1992).

There are several examples in which GroES is essential during the dissociation step for the successful recovery of a biologically active proteins. During the chaperonin-dependent refolding of rhodanese (Martin et al., 1991; Mendoza et al., 1991) and Rubisco (Goloubinoff et al., 1989b; Viitanen et al., 1990) the complete folding reaction must contain GroEL, GroES, MgATP, and potassium ions. Here it is important to distinguish between a requirement for GroES for the *release* of the polypeptide from GroEL, from a requirement for efficient recovery of the *biologically active* protein. For example, if a binary complex is prepared between GroEL and radioactive folding intermediates of Rubisco, the complex is stable and can be resolved by size-exclusion chromatography (Baneyx and Gatenby, 1992; Viitanen et al., 1992a). The addition of GroES and MgATP results in a substantial dissociation (85–90%) of the complex, and the appearance on the column of two new peaks corresponding to the active Rubisco dimer and a small amount of inactive folded monomer. In contrast, when MgATP is added in the absence of GroES, a significant proportion (50–75%) of Rubisco is discharged from the complex, but it is not resolved on the column and is not catalytically active. Thus, MgATP alone causes a conformational change in GroEL that weakens its affinity for the bound Rubisco, but the species released does not successfully progress to the native state, at least under conditions where spontaneous refolding is not possible. This indicates that the degree of foldedness of the discharged Rubisco differs depending on whether GroES is present or absent. In the presence of GroES, the bound Rubisco is able to progress to a state where it is not susceptible to aggregation upon release. A similar conclusion was previously reached for the chaperonin-dependent folding of rhodanese (Martin et al., 1991). In the absence of GroES, the rhodanese that was released from GroEL by MgATP alone was not active and instead formed aggregates. The authors suggested that in the absence of GroES there are repeated cycles of release and rebinding to the chaperonin which do not permit rhodanese to successfully advance to the N state. In contrast, in the presence of GroES the released rhodanese was found to be active.

VI. CONCLUDING REMARKS

As demonstrated in several examples of chaperonin-facilitated protein folding described in this review, the presence of cpn60 (GroEL) does not significantly increase the rate of refolding; rather the chaperonins inhibit non-productive reactions such as aggregation. Chaperonins differ in this respect from enzymes such as protein disulphide isomerase and peptidyl prolyl *cis-trans* isomerase that catalyze these slow rate-limiting steps, leading to an acceleration of protein folding (Fischer and Schmid, 1990). Chaperonins do not appear to actively direct correct folding by supplying steric information, but by binding to the non-native states of proteins and suppressing off-pathway aggregation reactions, correct folding is facilitated. GroES may prevent a premature release of the target protein from GroEL until it has advanced to a point where it is no longer susceptible to aggregation and is committed to the native state, a state that is not recognized by GroEL. Recent data suggest that the chaperonins represent only part of the protein-folding pathway in cells. Langer et al (1992) have observed the successive action of the DnaK, DnaJ, and GroEL molecular chaperones in a sequential fashion during chaperone-mediated protein folding. These aspects may have potential applications in the biotechnology industry where it is often important to obtain the synthesis and correct folding of foreign proteins produced in microorganisms (Gatenby et al., 1990). Overexpression of molecular chaperones can, in some cases, dramatically improve the production of foreign proteins in an active form (Goloubinoff et al., 1989a; Carrillo et al., 1992; Wynn et al., 1992).

Stress conditions, such as high temperature, that would be anticipated to destabilize folding intermediates and promote aggregation, often result in higher concentrations of chaperonins in cells. The physiology of organisms can readily be modified to adapt to harsh conditions that could interfere with protein folding by ensuring a compensatory synthesis of various molecular chaperones. The heat shock or stress response may, in part, be directed towards stabilizing protein-folding intermediates that would otherwise partition towards aggregation.

ACKNOWLEDGMENTS

We are indebted to our many colleagues who have worked with us on various aspects of molecular chaperones, and we would especially like to acknowledge the contributions of François Baneyx, Cathy Kalbach, Gail Donaldson, Sue Erickson-Viitanen, Pierre Goloubinoff, Bob LaRossa, George Lorimer, Tom Lubben, Uwe Maier, Martina Merz, Ram Seetharam, Eberhard Schäfer, Matthew Todd, Tina Van Dyk, and Saskia van der Vies. V.S. thanks the Deutsche Forschungsgemeinschaft for financial support.

REFERENCES

Ang, D., Liberek, K., Skowyra, D., Zylicz, M., & Georgopoulos, C. (1991). Biological role and regulation of the universally conserved heat shock proteins. J. Biol. Chem. 266, 24233–24236.

Badcoe, I.G., Smith, C.J., Wood, S., Halsall, D.J., Holbrook, J.J., Lund, P., & Clarke, A.R. (1991). Binding of a chaperonin to the folding intermediates of lactate dehydrogenase. Biochemistry 30, 9195–9200.

Baneyx, F. & Gatenby, A.A. (1992). A mutation in GroEL interferes with protein folding by reducing the rate of discharge of sequestered polypeptides. J. Biol. Chem. 267, 11637–11644.

Barraclough, R. & Ellis, R.J. (1980). Protein synthesis in chloroplasts. IX. Assembly of newly synthesized large subunits into ribulose bisphosphate carboxylase in isolated pea chloroplasts. Biochim. Biophys. Acta 608, 19–31.

Bertsch, U., Soll, J., Seetharam, R., & Viitanen, P.V. (1992). Identification, characterization, and DNA sequence of a functional "double" groES-like chaperonin from chloroplasts of higher plants. Proc. Natl. Acad. Sci. USA 89, 8696–8700.

Bloom, M.V., Milos, P., & Roy, H (1983). Light-dependent assembly of ribulose 1,5-bisphosphate carboxylase. Proc. Natl. Acad. Sci. USA 80, 1013–1017.

Bochkareva, E. S., Lissin, N.M., Flynn, G.C., Rothman, J.E., & Girshovich, A.S. (1992). Positive cooperativity in the functioning of molecular chaperone GroEL. J. Biol. Chem. 267, 6796–6800.

Bochkareva, E.S., Lissin, N.M., & Girshovich, A.S. (1988). Transient association of newly synthesized unfolded proteins with the heat-shock GroEL protein. Nature 336, 254–257.

Buchner, J,. Schmidt, M., Fuchs, M., Jaenicke, R., Rudolph, R., Schmid, F.X., & Kiefhaber, T. (1991). GroE facilitates refolding of citrate synthase by suppressing aggregation. Biochemistry 30, 1586–1591.

Carrillo, N., Ceccarelli, E.A., Krapp, A.R., Boggio, S., Ferreyra, R.G., & Viale, A.M. (1992). Assembly of plant ferredoxin-$NADP^+$ oxidoreductase in *Escherichia coli* requires GroE molecular chaperones. J. Biol. Chem. 267, 15537–15541.

Chandrasekhar, G.N., Tilly, K., Woolford, C., Hendrix, R., & Georgopoulos, C. (1986). Purification and properties of the groES morphogenetic protein of *Escherichia coli.* J .Biol. Chem. 261, 12414–12419.

Cheng, M., Y., Hartl, F.-U., & Horwich, A. L. (1990). The mitochondrial chaperonin hsp60 is required for its own assembly. Nature 348, 455–458.

Cheng, M.Y., Hartl, F.-U., Martin, J., Pollock, R.A., Kalousek, F., Neupert, W., Hallberg, E.M., Hallberg, R.L., & Horwich, A.L. (1989). Mitochondrial heat-shock protein hsp60 is essential for assembly of proteins imported into yeast mitochondria. Nature 337, 620–625.

Creighton, T.E. (1990). Protein folding. Biochem. J. 270, 1–16.

Creighton, T. E. (1991). Unfolding protein folding. Nature. 352, 17–18.

Ellis, R.J. (1991). Chaperone function: Cracking the second half of the genetic code. Plant J. 1: 9–13.

Ellis, R.J. & van der Vies, S.M. (1991). Molecular chaperones. Ann. Rev. Biochem. 60, 321–347.

Fayet, O., Louarn, J.-M., & Georgopoulos, C. (1986). Suppression of the *Escherichia coli dnaA46* mutation by amplification of the *groES* and *groEL* genes. Mol. Gen. Genet. 202, 435–445.

Fayet, O., Ziegelhoffer, T., & Georgopoulos, C. (1989). The *groES* and *groEL* heat shock gene products of *Escherichia coli* are essential for bacterial growth at all temperatures. J. Bact .171, 1379–1385.

Fischer, G. & Schmid, F.X. (1990). The mechanism of protein folding. Implications of *in vitro* refolding models for *de novo* protein folding and translocation in the cell. Biochemistry 29, 2205–2212.

Fisher, M.T. (1992). Promotion of the *in vitro* renaturation of dodecameric glutamine synthetase from *Escherichia coli* in the presence of GroEL (chaperonin-60) and ATP. Biochemistry 31, 3955–3963.

Gao, Y., Thomas, J.O., Chow, R.L., Lee, G.-H., & Cowan, N.J. (1992). A cytoplasmic chaperonin that catalyzes β-actin folding. Cell 69, 1043–1050.

Gatenby, A.A. & Ellis, R.J. (1990). Chaperone function: The assembly of ribulose bisphosphate carboxylase-oxygenase. Ann. Rev. Cell Biol. 6, 125–149.

Gatenby, A.A., Lubben, T.H., Ahlquist, P., & Keegstra, K. (1988). Imported large subunits of ribulose bisphosphate carboxylase/oxygenase, but not imported β-ATP synthase subunits, are assembled into holoenzyme in isolated chloroplasts. EMBO J. 7, 1307–1314.

Gatenby, A.A., van der Vies, S.M., & Bradley, D. (1985). Assembly in *E. coli* of a functional multi-subunit ribulose bisphosphate carboxylase from a blue-green alga. Nature 314, 617–620.

Gatenby, A.A., Viitanen, P.V., & Lorimer, G.H. (1990). Chaperonin assisted polypeptide folding and assembly: Implications for the production of functional proteins in bacteria. Trends Biotechnol. 6, 95–101.

Georgopoulos, C. & Ang, D. (1990). The *Escherichia coli groE* chaperonins. Semin. Cell Biol. 1, 19–25.

Gething, M.-J. & Sambrook, J. (1992). Protein folding in the cell. Nature 355, 33–45.

Goloubinoff, P., Gatenby, A.A., & Lorimer, G.H. (1989a). GroE heat-shock proteins promote assembly of foreign prokaryotic ribulose bisphosphate carboxylase oligomers in *Escherichia coli*. Nature 337, 44–47.

Goloubinoff, P., Christeller, J.T., Gatenby, A.A., & Lorimer, G.H. (1989b). Reconstitution of active dimeric ribulose bisphosphate carboxylase from an unfolded state depends on two chaperonin proteins and Mg-ATP. Nature 342, 884–889.

Govezensky, D., Greener, T., Segal, G., & Zamir, A (1991). Involvement of GroEL in *nif* gene regulation and nitrogenase assembly. J. Bact. 173, 6339–6346.

Gray, T. E. & Fersht, A. R. (1991). Cooperativity in ATP hydrolysis by GroEL is increased by GroES. FEBS Lett. 292, 254–258.

Grimm, R., Donaldson, G.K., van der Vies, S.M., Schäfer, E., & Gatenby, A.A. (1993). Chaperonin-mediated reconstitution of the phytochrome photoreceptor. J. Biol. Chem. 268, 5220–5226.

Grimm, R., Speth, V., Gatenby, A.A., & Schäfer, E. (1991). GroEL-related molecular chaperones are present in the cytosol of oat cells. FEBS Lett. 286, 155–158.

Gupta, R.S. (1990). Sequence and structural homology between a mouse T-complex protein TCP-1 and the "chaperonin" family of bacterial (GroEL, 60–65 kDa heat shock antigen) and eukaryotic proteins. Biochem. Int. 20, 833–841.

Hartman, D.J., Hoogenraad, N.J., Condron, R., & Hoj, P.B. (1992). Identification of a mammalian 10-kDa heat-shock protein, a mitochondrial chaperonin 10 homologue essential for assisted folding of trimeric ornithine transcarbamoylase *in vitro*. Proc. Natl. Acad. Sci USA 89, 3394–3398.

Hemmingsen, S.M. & Ellis, R.J. (1986). Purification and properties of ribulose bisphosphate carboxylase large subunit binding protein. Plant Physiol. 80, 269–276.

Hemmingsen, S.M., Woolford, C., van der Vies, S.M., Tilly, K., Dennis, D.T., Georgopoulos, C.P., Hendrix, R.W., & Ellis, R.J. (1988). Homologous plant and bacterial proteins chaperone oligomeric protein assembly. Nature 333, 330–334.

Hendrix, R.W. (1979). Purification and properties of groE, a host protein involved in bacteriophage assembly. J. Mol. Biol. 129, 375–392.

Höll-Neugebauer, B., Rudolph, R., Schmidt, M., & Buchner, J. (1991). Reconstitution of a heat shock effect *in vitro*: Influence of GroE on the thermal aggregation of α-glucosidase from yeast. Biochemistry 30, 11609–11614.

Hubbard, T.J.P. & Sander, C. (1991). The role of heat-shock and chaperone proteins in protein folding: Possible molecular mechanisms. Protein Eng. 4, 711–717.

Ishihama, A., Ikeuchi, T., & Yura, T. (1976a). A novel adenosine triphosphatase isolated from RNA polymerase preparations of *Escherichia coli*; I. Copurification and separation. J. Biochem. (Tokyo) 79, 917–925.

Ishihama, A., Ikeuchi, T., Matsumoto, A., & Yamamoto, S. (1976b). A novel adenosine triphosphatase isolated from RNA polymerase preparations of *Escherichia coli*; II. Enzymatic properties and molecular structure. J. Biochem. (Tokyo) 79, 927–936.

Jaenicke, R. (1987). Folding and association of proteins. Prog. Biophys. Mol. Biol. 49, 117–237.

Jaenicke, R. (1991). Protein folding: Local structures, domains, subunits and assemblies. Biochemistry 30, 3147–3161.

Jenkins, A.J., March, J.B., Oliver, I.R., & Masters, M. (1986). A DNA fragment containing the *groE* genes can suppress mutations in the *Escherichia coli dnaA* gene. Mol. Gen. Genet. 202, 446–454.

Kassenbrock, C.K. & Kelly, R.B. (1989). Interaction of heavy-chain binding protein (BiP/GRP78) with adenine nucleotides. EMBO J. 8, 1461–1467.

Kuwajima, K. (1989). The molten globule state as a clue for understanding the folding and cooperativity of globular-protein structure. Proteins 6, 87–103.

Laminet, A.A., Ziegelhoffer, T., Georgopoulos, C., & Plückthun, A. (1990). The *Escherichia coli* heat shock proteins GroEL and GroES modulate the folding of the β-lactamase precursor. EMBO J. 9, 2315–2319.

Landry, S.J. & Gierasch, L.M. (1991). The chaperonin GroEL binds a polypeptide in an α-helical conformation. Biochemistry 30, 7359–7362.

Landry, S.J., Jordan, R., McMacken, R., & Gierasch, L.M. (1992). Different conformations for the same polypeptide bound to chaperones DnaK and GroEL. Nature 355, 455–457.

Langer, T., Lu, C., Echols, H., Flanagan, J., Hayer, M.K., & Hartl, F.U. (1992). Successive action of DnaK, DnaJ and GroEL along the pathway of chaperone-mediated protein folding. Nature 356, 683–689.

Lecker, S., Lill, R., Ziegelhoffer, T., Georgopoulos, C., Bassford, P.J., Kumamoto, C.A., & Wickner, W. (1989). Three pure chaperone proteins of *Escherichia coli*—SecB, trigger factor, and GroEL—form soluble complexes with precursor proteins *in vitro*. EMBO J. 8, 2703–2709.

Lewis, V.A., Hynes, G.M., Zheng, D., Saibil, H., & Willison, K. (1992). T-complex polypeptide-1 is a subunit of a heteromeric particle in the eukaryotic cytosol. Nature 358, 249–252.

Lubben, T.H., Donaldson, G.K., Viitanen, P.V., & Gatenby, A.A. (1989). Several proteins imported into chloroplasts form stable complexes with the GroEL-related chloroplast molecular chaperone. Plant Cell 1, 1223–1230.

Lubben, T.H., Gatenby, A.A,, Donaldson, G.K., Lorimer, G.H., & Viitanen, P.V. (1990). Identification of a groES-like chaperonin in mitochondria that facilitates protein folding. Proc. Natl. Acad. Sci. USA 87, 7683–7687.

Liberak, K., Skowyra, D., Zylicz, M., Johnson, C., & Georgopoulos, C. (1991). The *Escherichia coli* DnaK chaperone, the 70-kDa heat shock protein eukaryotic equivalent, changes conformation upon ATP hydrolysis, thus triggering its dissociation from a bound target protein. J. Biol. Chem. 266, 14491–14496.

Lissin, N.M., Venyaminov, S.Y., & Girshovich, A.S. (1990). (Mg-ATP)-dependent self-assembly of molecular chaperone GroEL. Nature 348, 339–342.

Manning-Krieg, U.C., Scherer, P.E., & Schatz, G. (1991). Sequential action of mitochondrial chaperones in protein import into the matrix. EMBO J. 10, 3273–3280.

Martel, R., Cloney, L.P., Pelcher, L.E., Hemmingsen, S.M. (1990). Unique composition of plastid chaperonin-60: α- and β-polypeptide-encoding genes are highly divergent. Gene 94, 181–187.

Martin, J,. Langer, T., Boteva, R., Schramel, A., Horwich, A.L., & Hartl, F.-U. (1991). Chaperonin-mediated protein folding at the surface of groEL through a "molten globule"-like intermediate. Nature 352, 36–42.

McMullin, T.W., & Hallberg, R.L. (1987). A normal mitochondrial protein is selectively synthesized and accumulated during heat shock in *Tetrahymena thermophila*. Mol. Cell. Biol. 7, 4414–4423.

McMullin, T.W. & Hallberg, R.L. (1988). A highly evolutionarily conserved mitochondrial protein is structurally related to the protein encoded by the *Escherichia coli groEL* gene. Mol. Cell. Biol. 8, 371–380.

Mendoza, J.A., Rogers, E., Lorimer, G.H., & Horowitz, P.M. (1991). Chaperonins facilitate the *in vitro* folding of monomeric mitochondrial rhodanese. J. Biol. Chem. 266, 13044–13049.

Musgrove, J.E., Johnson, R.A., & Ellis, R.J. (1987). Dissociation of the ribulosebisphosphate-carboxylase large-subunit binding protein into dissimilar subunits. Eur. J. Biochem. 163, 529–534.

Miller, S.G., Leclerc, R.F., & Erdos, G.W. (1990). Identification and characterization of a testis-specific isoform of a chaperonin in a moth, *Heliothis virescens*. J. Mol. Biol. 214, 407–422.

Ostermann, J., Horwich, A.L., Neupert, W., & Hartl, F.-U. (1989). Protein folding in mitochondria requires complex formation with hsp60 and ATP hydrolysis. Nature 341, 125–130.

Phillips, G.J. & Silhavy, T.J. (1990). Heat-shock proteins DnaK and GroEL facilitate export of LacZ hybrid proteins in *E. coli*. Nature 344, 882–884.

Picketts, D. J., Mayanil, C. S. K., & Gupta, R. S. (1989). Molecular cloning of a chinese hamster mitochondrial protein related to the "chaperonin" family of bacterial and plant proteins. J. Biol. Chem. 264, 12001–12008.

Prasad, T.K., Hack, E., & Hallberg, R.L. (1990). Function of the maize mitochondrial chaperonin hsp60: Specific association between hsp60 and newly synthesized F1-ATPase alpha subunits. Mol. Cell. Biol. 10, 3979–3986.

Pushkin, A.V., Tsuprun, V.L., Solovjeva, N.A., Shubin, V.V., Evstigneeva, Z.G., & Kretovich, W.L. (1982). High molecular weight pea leaf protein similar to the groE protein of *Escherichia coli*. Biochim. Biophys. Acta 704, 379–384.

Reading, D.S., Hallberg, R.L., & Myers, A.M. (1989). Characterization of the yeast HSP60 gene coding for a mitochondrial assembly factor. Nature 337, 655–659.

Roy, H., Bloom, M., Milos, P., & Monroe, M. (1982). Studies on the assembly of large subunits of ribulose bisphosphate carboxylase in isolated pea chloroplasts. J. Cell Biol. 94, 20–27.

Saibil, H., Dong, Z., Wood, S., & Auf der Mauer, A. (1991). Binding of chaperonins. Nature. 353, 25–26.

Schmidt, M. & Buchner, J. (1992). Interaction of GroE with an all-β-protein. J. Biol. Chem. 267, 16829–16833.

Suelter, C. H. (1970). Enzymes activated by monovalent cations. Science 168, 789–795.

Taguchi, H., Konishi, J., Ishii, N., & Yoshida, M. (1991). A chaperonin from a thermophilic bacterium, *Thermus thermophilus*, that controls refoldings of several thermophilic enzymes. J. Biol. Chem. 266, 22411–22418.

Terlesky, K.C. & Tabita, F.R. (1991). Purification and properties of the chaperonin 10 and chaperonin 60 proteins from *Rhodobacter sphaeroides*. Biochemistry 30, 8181–8186.

Trent, J.D., Nimmesgern, E., Wall, J.S., Hartl, F.-H., & Horwich, A.L. (1991). A molecular chaperone from a thermophilic archaebacterium is related to the eukaryotic protein t-complex polypeptide-1. Nature 354, 490–493.

van der Vies, S.M., Viitanen, P.V., Gatenby, A.A., Lorimer, G.H., & Jaenicke, R. (1992). Conformational states of ribulose bisphosphate carboxylase and their interaction with chaperonin 60 (GroEL). Biochemistry 31, 3635–3644.

Van Dyk, T.K., Gatenby, A.A., & LaRossa, R.A. (1989). Demonstration by genetic suppression of interaction of GroE products with many proteins. Nature 343, 451–453.

Viitanen, P.V., Donaldson, G.K., Lorimer, G.H., Lubben, T.H., & Gatenby, A.A. (1991). Complex interactions between the chaperonin 60 molecular chaperone and dihydrofolate reductase. Biochemistry 30, 9716–9723.

Viitanen, P.V., Gatenby, A.A., & Lorimer, G.H. (1992a). Purified chaperonin 60 (groEL) interacts with the nonnative states of a multitude of *Escherichia coli* proteins. Protein Sci. 1, 363–369.

Viitanen, P.V., Lorimer, G.H., Seetharam, R., Gupta, R.S., Oppenheim, J., Thomas, J.O., & Cowan, N.J. (1992b). Mammalian mitochondrial chaperonin 60 functions as a single toroidal ring. J. Biol. Chem. 267, 695–698.

Viitanen, P.V., Lubben, T.H., Reed, J., Goloubinoff, P., O'Keefe, D.P., & Lorimer, G.H. (1990). Chaperonin-facilitated refolding of ribulose bisphosphate carboxylase and ATP hydrolysis by chaperonin 60 (groEL) are K^+ dependent. Biochemistry 29, 5665–5671.

Wynn, R.M., Davie, J.R., Cox, R.P., & Chuang, D.T. (1992). Chaperonins GroEL and GroES promote assembly of heterotetramers ($\alpha 2\beta 2$) of mammalian mitochondrial branched-chain α-keto acid decarboxylase in *Escherichia coli*. J. Biol. Chem. 267, 12400–12403.

Yaffe, M.B., Farr, G.W., Miklos, D., Horwich, A.L., Sternlicht, M.L., & Sternlicht, H. (1992). TCP1 complex is a molecular chaperone in tubulin biogenesis. Nature 358, 245–248.

Zahn, R. & Plückthun, A. (1992). GroE prevents the accumulation of early folding intermediates of pre-β-lactamase without changing the folding pathway. Biochemistry 31, 3249–3255.
Zeilstra-Ryalls, J., Fayet, O., & Georgopoulos, C. (1991). The universally conserved GroE (Hsp60) chaperonins. Ann. Rev. Microbiol. 45, 301–325.

TRANSLOCATION OF PROTEINS ACROSS CHLOROPLAST MEMBRANES

Barry D. Bruce and Kenneth Keegstra

Advances in Molecular and Cell Biology
Volume 10, pages 389–430.

ISBN: 1-55938-710-6

I. INTRODUCTION

Although chloroplasts contain their own genetic material, they are dependent upon the nuclear genome for the majority of their proteins. These nuclear-encoded proteins are synthesized on free cytosolic ribosomes and enter chloroplasts in a posttranslational manner (Figure 1). Since the early work of Dobberstein et al. (1977), it has become clear that almost all nuclear-encoded chloroplast proteins are synthesized as higher molecular weight precursors containing a short amino acid extension on the amino-terminus. This N-terminal extension is referred to as a transit peptide to distinguish it from other sorting signals such as the signal sequence involved in the secretion of proteins in prokaryotes.

A transit peptide is both necessary for transport into chloroplasts and sufficient to transport foreign proteins into chloroplasts. The basic outlines of the transport process are well described, and recent reviews can be consulted for details (Keegstra et al., 1989; De Boer and Weisbeek, 1991).

In addition to the recent work concerning protein transport into chloroplasts, great progress has been made in describing the molecular processes involved in protein targeting to the mitochondria (Geli and Glick, 1990; Neupert et al., 1990). Although this process shares many aspects of protein import into chloroplasts (Figure 1), it is the differences between the two organelles which we will emphasize.

In this chapter we will focus on recent attempts to identify components of the transport apparatus and understand the mechanism of the transport process. One of the important themes in recent work has been the emphasis on conformational changes in the transported protein during the translocation process.

II. PRECURSOR CONFORMATION

A. Role of Transit Peptide

Structural Features of Transit Peptides

In the past few years the amino acid sequences of over 50 different nuclear-encoded chloroplast proteins and their transit peptides have been determined.

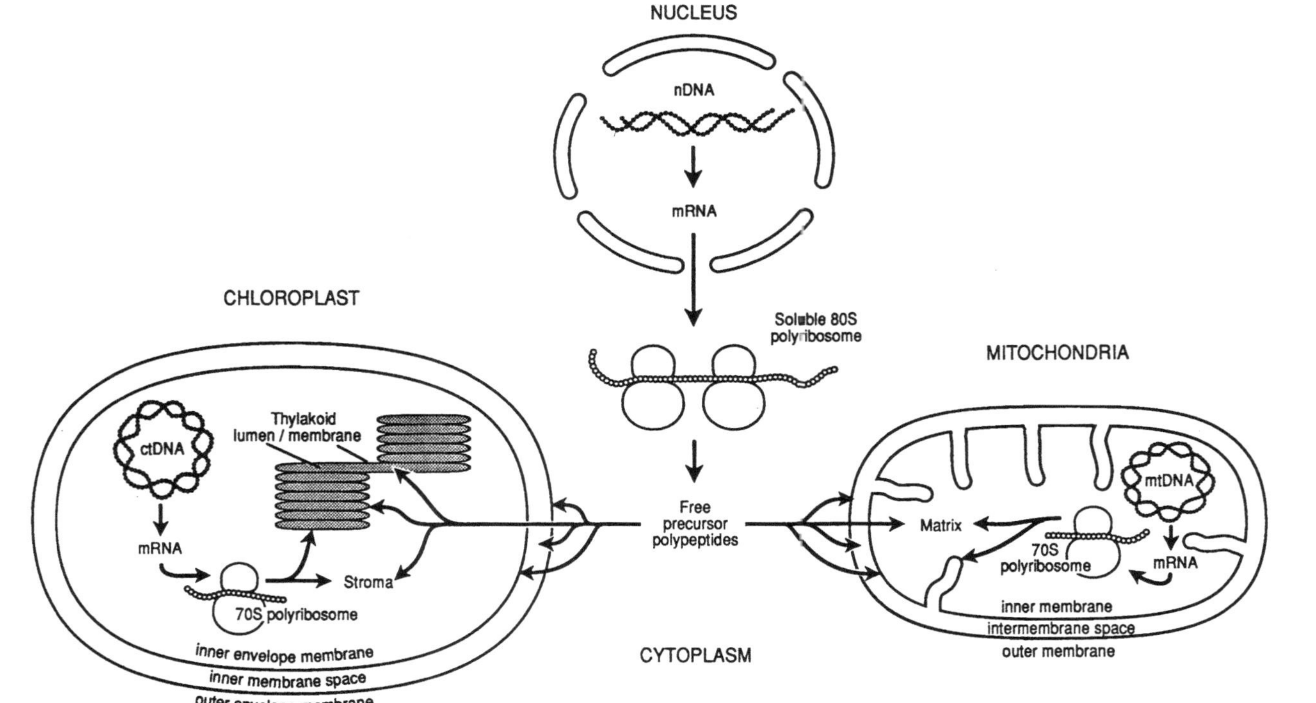

Figure 1. A schematic diagram comparing the synthesis and import of nuclear-encoded chloroplast and mitochondrial precursor proteins. Note the six different destinations available to chloroplast proteins as compared to four available to mitochondrial proteins.

These sequences have come from organisms representing diverse evolutionary groups including the green algae, gymnosperms, monocots, and dicots. Since many of the genes for chloroplast proteins exist in multiple copies, as members of small gene families, there are now sequences for ~260 different chloroplast transit peptides. To facilitate the analysis of these sequences, an international data base known as CHLPEP has been established (von Heijne et al., 1991).

From this data base it is clear that there is great variability in the length of transit peptides, which range from 28 to 139 amino acids, as well as an absence of any strong regions of similarity between transit peptides from different proteins. There is, however, similarity in length and sequence between transit peptides for the same protein from different organisms, as illustrated for the transit peptide of the precursor to the small subunit of Rubisco (pre-SSU) in Figure 2.

This apparent lack of similarity at the primary sequence level between transit peptides of different proteins has been interpreted in two ways (Keegstra, 1989b). First, it is still formally possible that there is a myriad of different receptors on the chloroplast surface that are specific for individual or small classes of precursors. However, concurring with existing models for the interaction of signal peptides with the endoplasmic reticulum (ER) (von Heijne, 1988) and for presequences with the mitochondrial import apparatus (Roise et al., 1988), the essential features shared by transit peptides could be found, not in their primary amino acid sequence, but

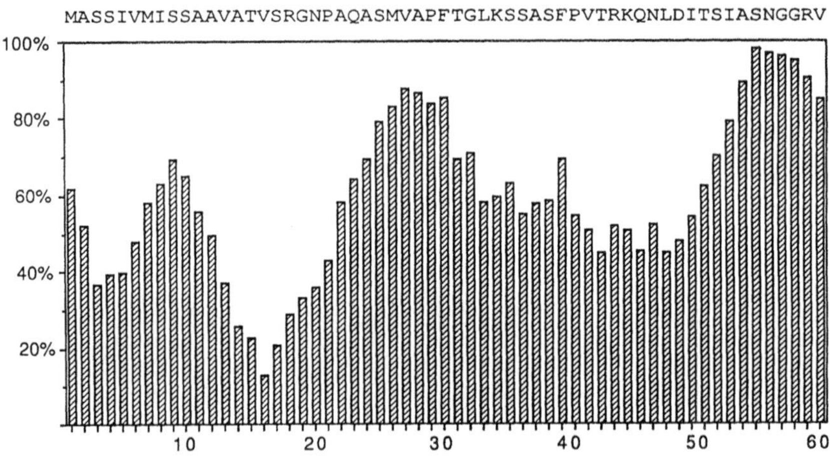

Figure 2. A homology plot for the transit peptides of 51 different small subunit precursors from 18 different plant species. After alignment with the consensus sequence, a score was assigned to each position representing the number of proteins which are identical at that position. No penalty was assigned for an inserted space. After a score was assigned to each position, a sliding window of the average of six positions was calculated. This average, expressed as a percent of identity, is plotted as a function of position.

in some property of their secondary structure. Indeed, through the analysis of many chloroplast transit peptides, some common secondary elements have been observed. For example, it has been noted that transit peptides are enriched in hydroxylated amino acids like serine and threonine and are depleted of tyrosine and negatively charged amino acids. Von Heijne has shown that chloroplast transit peptides contain three distinct domains: an uncharged amino-terminal domain of 10 to 12 amino acids; a central region which is nonamphiphilic containing hydroxylated and charged amino acids and which accounts for the variable length of the transit peptide; and a C-terminal domain of 10 to 12 amino acids which lacks leucine and lysine and is predicted to form an amphiphilic β-strand (von Heijne et al., 1989; Gavel and von Heijne, 1990).

In a more recent study, von Heijne and Nishikawa (1991) have used five different structure-predicting algorithms to show that, out of the 150 chloroplast transit peptide sequences analyzed, over 85% of the amino acids are found in a random coil conformation. This is compared to only 55% of the amino acids in a large control sample of globular proteins. They concluded that chloroplast transit peptides have evolved to be rather flexible peptides which have minimized their content of regular secondary structures such as α-helixes or β-sheets. These peptides form a random coil and are relatively insensitive to variations in length by insertions and deletions to the central regions of the coil as observed by Smeekens (1989).

Proteins destined for the thylakoid lumen, such as plastocyanin and the members of the oxygen-evolving complex (OEC) contain, in addition to the three domains described above, a fourth carboxyl-terminal domain known as the thylakoid transfer domain. This additional domain contains a central hydrophobic region and has all of the properties of a bacterial signal sequence. Lumenal proteins are first translocated across the chloroplast envelope by the stromal targeting domain, consisting of the first three regions, of the transit peptide. In the stroma, the stromal targeting domain is removed by a specific endopeptidase before the intermediate form is subsequently translocated across the thylakoid membrane where the remaining thylakoid transfer domain is removed by thylakoid peptidase (Hageman et al., 1986; Smeekens et al., 1986; Kirwin et al., 1989).

Interruption or Retardation of the Folding Pathway

Several different functions have been proposed for the presequence or signal peptide of precursors destined for mitochondria or secretion, respectively. It has also been proposed in the secretion of proteins in *E. coli* that one role of a leader sequence is to retard folding of precursor proteins. Interaction of a bacterial signal peptide with the mature passenger protein has been shown for several secreted proteins in *E. coli* (Collier et al., 1988; Liu et al., 1988). When these precursors are denatured by treatment with chaotropic agents, the refolding of the mature protein is significantly retarded by the signal peptide. In a set of papers put forward by her laboratory, L. Randall argues that the rate of this folding is what distinguishes proteins that are destined to be exported from those that are to remain in the cytoplasm. The rate of folding acts to kinetically partition proteins between: (1) export bound precursors, which are slow folding allowing them time to become associated with the molecular chaperone, SecB, and eventually with the export machinery, and (2) those which are very rapidly folded, thus escaping SecB and the export pathway and therefore remaining in the cell (Liu et al., 1989; Randall and Hardy, 1989; Hardy and Randall, 1991).

Protein import into chloroplasts and mitochondria occurs posttranslationally, allowing the precursors an opportunity to initiate folding before they are translocated across the organelle membranes. Yet, it has been reported that precursor proteins must be in an unfolded state before they can be translocated into mitochondria or chloroplasts (Eilers and Schatz, 1986; della-Cioppa and Kishore, 1988). Therefore, one possible function of a transit peptide or presequence may be to retard or inhibit the formation of the folded, tertiary structure of a mature protein.

Although the transit peptide may slow folding, as has been proposed, evidence suggests that it does not prevent folding per se. There are several examples where a precursor protein folds to form an active conformation, and there are even examples of precursors assuming active multimeric forms. It was recently shown that the presequence of purified pre-ornithine transcarbamylase (pre-OTC) and mitochondrial aspartate aminotransferase does not prevent the folding of these proteins when they are overexpressed in *E. coli.* Both of these precursors are able to assemble into enzymatically active forms that consist of trimeric and dimeric forms, respectively (Alteiri et al., 1989; Murakami et al., 1990). There has been at least one report of a chloroplast precursor which is able to fold into a conformation with enough similarity to the mature protein that it also has enzymatic activity. The precursor of 5-enolpyruvylshikimate-3-phosphate (EPSP synthase) maintains a high level of import competence while being as catalytically competent as the mature form (della-Cioppa et al., 1986).

Although the transit peptide allows the folding of some proteins, it probably does not become part of the folded structure itself; instead, it remains as an external domain of the protein. Using several different partially purified endopeptidases from the stroma of pea chloroplasts, Musgrove and co-workers observed that each of these peptidases removes only a small part of the transit peptide from pre-plastocyanin (pre-PC). They concluded that the mature part of the protein must exist in a tightly folded, protease-resistant conformation, whereas the transit peptide is "protruding into the medium" making it more accessible to the peptidases (Musgrove et al., 1989).

Perturbation of Lipid Bilayer Structure

Another possible activity associated with transit peptides and presequences is to temporarily disrupt lipid organization and, in so doing, provide an import pathway for the precursor. Although the nature of this rearrangement is unknown, different types of lipid reorganizations have been proposed (Reitveld and de Kruijff, 1986; Archer and Keegstra, 1990). Several different small peptides, including transit peptides and presequences, have been shown to have lipid disrupting properties (Batenburgh et al., 1987; Roise et al., 1988; Tamm, 1991).

A recent study on the conformation and membrane-binding properties of the signal sequence of the *E. coli* outer membrane protein, LamB, suggests that this peptide is able to spontaneously insert into the acyl chain region of the lipid bilayer (McKnight et al., 1991). The ability of the signal peptide to interact with the lipid bilayer is not affected by the mature portion of the protein; however, the mature domain alone does not interact with the bilayer. McKnight et al. (1991) propose that the N-terminal regions of the mature translocated proteins have evolved to be noninteractive with lipid bilayers.

Killian and co-workers have shown that when model membranes, with a lipid composition similar to *E. coli* membranes, are treated with synthetic peptides corresponding to the signal sequence of the *E. coli* outer membrane protein PhoE, they undergo a transition from a bilayer organization to a hexagonal type II (H_{II}) non-bilayer structure (Killian et al., 1990). This membrane disruption tends to be closely correlated to the translocation efficiency of the signal peptide. Killian et al. (1990) propose that a critical property of these signal sequences is to induce local changes in the lipid structure that are somehow involved in the translocation of proteins across the membrane.

This may also be the case with protein transport across the chloroplast envelope. A recent paper investigating the membrane activity of 20 amino acid fragments of the pre-SSU transit peptide have indicated that these peptides are also surface active and insert into lipid monolayers (van't Hof et al., 1991). These peptides interact with artificial bilayers formed from chloroplast envelope lipid extracts. The peptide 21-40 (corresponding to residues 21–40 of the transit peptide of pea pre-SSU) interacts with lipid extracts from the outer envelope. Peptide 41-60 interacts the strongest of all three peptides and is slightly more interactive with the inner envelope lipids. It was also observed that peptide 41-60 interacts with both galactolipids and sulfolipids. This may have increased significance since these lipids have been shown to form hexagonal H_{II} structures. These observations could explain the results of Perry et al. (1991) which indicated that peptide 21-40 provided the strongest inhibition of pre-SSU binding to the outer envelope and that peptide 41-60 was the most effective at inhibiting translocation possibly by interacting with some part of the transport process beyond the initial binding (Perry et al., 1991).

All of these experiments extend the possibility that one function of transit peptides is to alter the lipid organization in some way that permits proteins to translocate across the membrane. This alternative method of translocation is especially attractive for mitochondria and chloroplasts which are rich in non-bilayer-forming lipids. One specific model of lipid rearrangement has been proposed by Keegstra (1989a), but many other models are also possible. This involvement of a lipid bilayer reorganization could help to explain how many different foreign passenger proteins can be carried into chloroplasts and mitochondria as chimeric proteins with a range of transit peptides (Keegstra, 1989b).

Interaction with Cytosolic Factors

Two possible roles for cytosolic factors in protein transport are: (1) as molecular chaperones in which they function to provide an import competent conformation and/or act to prevent illegitimate folding or aggregation (discussed in more detail below); and (2) as participants in the targeting of precursors by directly interacting with the presequence or transit peptide.

Although a great deal is known about the role of cytosolic factors, such as the signal recognition particle (SRP) in protein transport to the ER and its possible homologue, SecB (Watanabe and Blobel, 1989), in protein export in *E. coli,* very little is known about the role of cytosolic factors during protein transport into chloroplasts and mitochondria. A few cytosolic factors involved in targeting to mitochondria have been partially characterized, including a hsp70 homologue (Deshaies et al., 1988), a presequence binding factor (Murakami and Mori, 1990; Murakami et al., 1990), and a 28-kDa targeting factor (Ono and Tuboi, 1988).

The one report on the involvement of a cytosolic hsp70 in stimulating protein import into isolated chloroplasts provides no evidence supporting direct interaction with the transit peptide (Waegemann et al., 1990). It has, however, been recently hypothesized by von Heijne and Nishikawa (1991) that the role of the chloroplast transit peptide is to bind to the molecular chaperone, hsp70. This interaction would be structurally possible since the transit peptide exists as an exposed part of the protein (Musgrove et al., 1989). Flynn and co-workers have already shown that two members of the hsp70 family, BiP, an ER-localized hsp70, and hsc70, a cytosolic homologue, have the ability to bind short synthetic peptides (Flynn et al., 1989). Von Heijne observed that of the four peptides having the highest affinity for BiP and hsc70 all but one completely lack any acidic amino acids, a property shared with most chloroplast transit peptides. Unfortunately, it is still not clear whether this peptide-binding property of BiP and hsc70 really possesses any sequence specificity (Flynn et al., 1989, 1991). An equally plausible model has hsp70 simply recognizing regions of a protein which have their peptide backbone exposed as would be the case in a unfolded protein.

In protein export in *E. coli,* the molecular chaperone, SecB, has also been suggested to interact directly with the signal sequence (Watanabe and Blobel, 1989). However, in more recent work, it has been clearly shown that there is no specific interaction or recognition between the leader peptide and SecB (Randall et al., 1990). Thus, eukaryotic presequences and transit peptides may be different than the signal sequences of prokaryotes in their ability to interact with soluble factors.

Interaction with Membrane Components

One of the roles of the presequence or transit peptide may be to specifically interact with some component of the transport apparatus. If this is the case, then a transit peptide or some part of it could interact with the transport apparatus in a way that is independent of the mature protein. By mimicking this interaction, binding and import could be inhibited in a competitive way by the addition of synthetic peptides corresponding to some or all of the transit peptide. Several studies have recently taken advantage of this approach to study protein import.

Mitochondrial Transit Peptides. An inhibitory effect of synthetic peptides on the import of a precursor protein was first reported by Gillespie and co-workers when they showed that a peptide corresponding to the presequence of pre-OTC could inhibit the import of authentic pre-OTC and other unrelated precursor proteins into mitochondria (Gillespie et al., 1985). This inhibition could be reversed by increasing the concentration of pre-OTC. Other studies have extended this observation to a synthetic peptide corresponding to the presequence of malate dehydrogenase. Both of these synthetic peptides work at the 5 to 10 μM level and are not simply a nonspecific interference since inhibition of import was not found when an equal concentration of a similarly charged peptide corresponding to the transit peptide of the pre-SSU from soybean chloroplast or pre-CF_1 from tobacco mitochondria were tested.

Gillespie went on to use the synthetic peptide in chemical cross-linking studies (Gillespie, 1987). This work showed that the peptide was able to be specifically cross-linked to a 30-kDa protein localized in the mitochondrial outer membrane. Gillespie concluded that the interaction of the synthetic peptide with this membrane component had all of the properties characteristic of a ligand binding to a receptor such as saturability, reversibility, and specificity.

However, not all peptides corresponding to presequences inhibit import. The 20-amino acid amino-terminus for the presequence of $F_1\beta$ precursor fails to significantly inhibit import of either $F_1\alpha$ or $F_1\beta$ precursor into mitochondria (Hoyt et al., 1991). However, when this group used peptides corresponding to longer regions of the $F_1\beta$ presequence, they found a much greater inhibition of import (Cyr and Douglas, 1991). They concluded that both structure and some minimum length is required to compete with full-length precursors for the import apparatus.

Apparently, the mechanism of this import inhibition must be conserved since the amino-terminus of the presequence to subunit IV of cytochrome *c* oxidase (COX) blocked not only its own import into yeast mitochondria but also the import of precursors for COX V_a, $F_1\beta$, malate dehydrogenase, ATP/ADP carrier, and even human OTC (Glaser and Cumsky, 1990a,b). These workers found that although these peptides form amphiphilic α-helical structures, which are known to interact with lipid membranes, they also appear to interact with some trypsin-sensitive factor on the mitochondrial surface. These peptides may interact sufficiently well with the transport apparatus to undergo partial translocation. It was found that one peptide was imported into the mitochondria in a $\Delta\Psi$-dependent fashion (Glaser and Cumsky, 1990c). This peptide was associated with the inner membrane and could not be removed with salts, chaotropic agents, or high pH.

Chloroplast Transit Peptides. Similar studies have also been initiated to characterize the interaction of transit peptides with the chloroplast translocation apparatus. Using a synthetic peptide, pL(1-20), corresponding to the amino-terminus of pre-LHCP from *Arabidopsis*, Buvinger et al. (1989) showed that this peptide inhibited the import of several pea and *Arabidopsis* precursors into isolated pea chloroplasts. They concluded that complete inhibition of import by this peptide at 160 μM occurred at some step after the initial binding of precursors and before the proteolytic cleavage event. Their results are not due to any lytic activity of the peptide nor can they be reversed by the addition of ATP. They concluded that the inhibition is a result of some specific interaction between the peptide and import machinery. The specific involvement of this region is strengthened by the observation of Karlin-Neuman and Tobin (1986) that this 20-amino acid region contains two stretches of amino acids that are very highly conserved throughout the LHCP family of precursor proteins.

In a follow-up study, synthetic peptides corresponding to 20-amino acid long regions of the transit peptide to pre-SSU were used to test the inhibition of binding and translocation of chloroplast precursor proteins (Perry et al., 1991). Each of these peptides inhibited the import of not only pre-SSU but also the import of precursors to plastocyanin, ferrodoxin, and LHCP. Although all of the peptides tested—p1-20, p21-40, p41-60, and p31-50—inhibited import to some extent, p21-40 substantially inhibited the binding of these precursors to chloroplasts. Peptide p41-60, however, was the most effective in the inhibition of translocation of the precursors. The results presented in this paper suggest, at least for pre-SSU, that it is the central region of the transit peptide which mediates binding to the chloroplast envelope, whereas the carboxyl-end is more important for translocation across the envelope. This paper also suggests that these four precursors use at least some of the same components in the translocation apparatus despite differing destinations. So concurring with the mitochondrial field, Perry proposes that the transit peptide contains multiple structural domains that interact independently at various points along the import pathway (Chu et al., 1989; Perry et al., 1991).

Blobel's group used synthetic peptides corresponding to 30 amino acids of the transit peptide of pre-Fd and pre-SSU (Schnell et al., 1991). Like the peptides used by Perry and co-workers, these synthetic peptides were able to inhibit the import of their full length precursors. However, these peptides were able to inhibit import at a much lower concentration (1–2 μM). This block in the import was not due to any disruption of the chloroplast integrity but was the result of the peptide interfering with precursor binding to the envelope. From their results, both Perry and Schnell concluded that the carboxy terminus of the transit peptide is acting to inhibit the import of the precursor. The peptides used by Schnell are longer than those used by Perry and contain both the central- and carboxy-terminal regions of the transit peptide; this difference could explain why the Schnell's peptides are active at a lower concentration and also are able to inhibit binding (Schnell et al., 1991).

B. Effect of Folding Conformation on Precursor Competence

It was proposed several years ago that polypeptide chain unfolding is required for a protein to be competent for membrane translocation. This has been shown in protein translocation in mitochondria (Eilers and Schatz, 1986; Chen and Douglas, 1987), in bacterial secretion (Randall and Hardy, 1986), in microsomes (Wiech et al., 1987), and in chloroplasts (della-Cioppa and Kishore, 1988).

The early work of Eilers and Schatz showed that when DHFR fusion proteins were allowed to bind methotrexate, a DHFR substrate analog, they became protease resistant, suggesting stabilization of a more tightly folded conformation (Eilers and Schatz, 1986). Methotrexate treatment also blocks the import of the DHFR fusion protein. These observations were interpreted to mean that protein unfolding was required for transport across the mitochondrial membranes.

It is still an open question whether protein conformation affects only the translocation competence of a precursor or whether it can also interfere with the initial membrane recognition or binding process. It has been shown by Verner and Lemire (1989) that for the import of a chimeric protein with a truncated presequence into isolated yeast mitochondria the import has to occur either with unfolded, nascent polypeptide or after the mature passenger protein moiety has been disrupted with urea treatment or by destabilizing point mutations. They concluded that the dependence of protein import on the precursor's conformation may not only be a result of the translocation apparatus's inability to translocate tightly folded precursors but also to the membrane or membrane receptors inability to successfully interact with the folded precursor (Verner and Lemire, 1989).

Whether complete unfolding is required for import competence is still not known. It is possible that for a precursor to become translocationally competent only a small subdomain needs to be unfolded at any given point in the translocation process. This was recently suggested when translocationally incompetent chimeric proteins (blocked by methotrexate) were found to still contain a membrane spanning domain of about 50 amino acids (Rassow et al., 1990). It has been calculated that if this stretch of amino acids assumed a fully extended conformation, it would be able to span the minimum distance of 15 to 20 nm, which is the measured distance across two membranes at a contact site (Neupert et al., 1990). Recent studies by Skerjanc have shown that an unfolded form of the precursor, pre-OTC, was found associated with a mitochondrial membrane fraction, intermediate in density relative to the inner and outer mitochondrial membrane (Skerjanc et al., 1990). They propose that this unfolded form represents a translocation intermediate that is associated with some region of the mitochondrial membrane, possibly a contact site.

However, the role of protein conformation and translocational competence is more uncertain in protein transport into chloroplasts. In one case, the interaction of a precursor, pre-EPSP, with the herbicide, glyophosphate, formed a complex that was imported much more slowly (della-Cioppa and Kishore, 1988). This was interpreted as evidence for precursor unfolding during import, similar to the methotrexate inhibition of DHFR import into mitochondria (Eilers and Schatz, 1986).

However, when chimeric precursors containing the plastocyanin transit peptide and DHFR were imported into chloroplasts, methotrexate did not inhibit the import of the fusion protein but did slow the import (America et al., 1992). The authors propose several explanations for the differences between chloroplasts and mitochondria, one of which is that chloroplast protein import does not require protein unfolding in the same way that is required for mitochondria. However, they favor the hypothesis that before or during translocation, the PC-DHFR fusion protein undergoes a substantial conformation change allowing methotrexate to be released. This explanation would suggest that the translocation mechanism in chloroplasts is different and possibly more "powerful" than the analogous process in mitochondria.

C. Role of Soluble Factors and Molecular Chaperones

Cytosolic factors may be involved in conveying import competence to posttranslationally synthesized precursors. As mentioned above, a molecular chaperone may unfold tertiary structures formed by precursors and/or prevent aggregation of unfolded, import competent precursors.

With the availability of purified precursors produced by overexpression in *E. coli*, it should become clear what factors are required for protein transport into chloroplasts and mitochondria. The first example of import of a purified precursor into isolated mitochondria did not require any additional cytosolic proteins, yet did require a pretreatment by chaotropic denaturation (Eilers and Schatz, 1986). In chloroplasts, similar results were obtained using overexpressed pre-ferredoxin. When this protein was denatured with 6 M urea and rapidly diluted into buffer, it was able to be imported into isolated chloroplasts without the addition of any cytosolic factors (Pilon et al., 1990b). However, in both of these studies, it cannot be ruled out that cytosolic factors could have an effect on either the efficiency or rate of precursor import.

In contrast to these reports there are several examples where soluble factors either are required for transport or act to stimulate the efficiency of translocation. Some of the better studied examples are discussed below.

Cytosolic Hsp70

In the mas3 mutant in yeast, Smith and Yaffe have shown that hsp70 is needed for posttranslational import of $F_1\beta$ and citrate synthase into mitochondria. It is significant that in the absence of hsp70, these precursors exist as two different pools, one which is rapidly imported and one which is imported very slowly. It is therefore possible that the rapidly imported pool of precursors is already in an import competent form, whereas the slowly imported pool represents precursors which must change to an import competent form (Smith and Yaffe, 1991).

The hybrid protein, pO-DHFR, overexpressed in *E. coli*, requires both a NEM sensitive factor and a cytosolic hsp70 from rabbit reticulocyte lysate to gain import competence (Sheffield et al., 1990). Hsp70 alone, although able to prevent aggregation and protein folding in a ATP-dependent manner, was unable to confer import competence. From this work, it would appear that import competence requires a multi-subunit protein complex with a molecular mass of 250 kDa, one component of which is a cytosolic hsp70. Successful import requires that the precursor dissociate from the complex and this step requires a reactive thiol group in the cytosolic complex.

In chloroplasts, it has also been suggested that pre-LHCP synthesized *in vitro* requires a cytosolic factor, possibly hsp70, to be import competent into isolated chloroplasts (Waegemann et al., 1990; Waegmann and Soll, 1991).

Mitochondrial Hsp70 (Ssc1)

Some of the strongest evidence for the involvement of hsp70 in protein translocation across organelle membranes comes from the laboratories of E. Craig and N. Pfanner (Kang et al., 1990). In yeast, one of the eight hsp70 genes, Sscl, encodes a soluble protein located in the mitochondrial matrix. When a temperature-sensitive mutant of this gene is grown at the non-permissive temperature, cells accumulate several precursors including hsp60, Sscl, and $F_1\beta$. *In vitro* studies with mitochondria from this mutant indicate that these precursors associate with the mitochondrial membrane but are unable to be translocated to a protease-protected location. It is possible that they remain stuck in contact sites. However, this block in translocation can be overcome if the precursor is first unfolded in 8 M urea. They conclude that the function of Sscl in mitochondria is to bind the precursor as it is threaded through a contact site. This interaction with the precursor acts to "pull" it through the membrane and may even facilitate its unfolding on the cytosolic side.

Shortly after the appearance of this paper, the Schatz group confirmed the involvement of a mitochondrial hsp70 in the translocation of precursors into yeast mitochondria (Scherer et al., 1990). This group took a biochemical approach to identify a 70-kDa matrix protein which was chemically cross-linked to a chimeric protein unable to be completely translocated. After purification of this protein, partial sequencing indicated that it is identical to the SSCl gene product described by Craig (Craig et al., 1987).

Chloroplast Hsp70

A similar mechanism may occur in chloroplasts, which have been shown to contain three different hsp70 homologues (Marshall et al., 1989). Von Heijne and Nishikawa have proposed that chloroplast precursors sequentially bind different hsp70 homologues and by being passed from one hsp70 member to the next, follow a "molecular bucket brigade" across the chloroplast envelope in a set of distinct chaperoned steps (von Heijne and Nishikawa, 1991). In chloroplasts there may be an additional inter-envelope transfer step since this organelle contains not only soluble hsp70 homologues in the stroma but also a hsp70 member associated with the envelope, possibly in the outer envelope (Marshall et al., 1989; Ko et al., 1992).

At least one stromal form of the chloroplasts hsp70 homologues has been shown to interact with newly imported proteins in the chloroplast stroma. The mature form of the Rieske Fe-S protein and ferredoxin-$NADP^+$ reductase have both been shown to interact with a stromal hsp70 upon import (Madueño et al., 1993; Tsugeki and Nishimura, 1993). In addition, Nechushtai and co-workers demonstrated that successful integration of LHCP into the thylakoid membrane requires a direct physical interaction between pLHCP and a stromal hsp70 (Yalovsky et al., 1992). However, recently this result has been challenged by Cline and co-workers (Payan and Cline, 1991; Yuan et al., 1993). They assert that a different stromal factor is required for protein integration into the thylakoid membrane and that this factor cannot be replaced by either hsp70 or by unfolding with chemical denaturants.

GroEL/GroES or Hsp60

GroEL or hsp60 homolgues have been found in a wide range of organisms and share a high degree of amino acid similarity. They have been found in many prokaryotes and in both mitochondria and chloroplasts (Hemmingsen et al., 1988; McMullin and Hallberg, 1988). There is even a report that a groEL related protein may be in the cytosol of plant cells (Grimm et al., 1991). These proteins also have a very broad range of specificity regarding the proteins that they assist in folding and assembly. The chloroplast groEL homologue has been shown to form high molecular weight complexes with a range of different nuclear- and chloroplast-encoded proteins, including the Rieske Fe-S protein (Madueño et al., 1993), ferredoxin $NADP^+$ reductase (Tsugeki and Nishimura, 1993), small subunit of Rubisco, LHCP, and CF_1-β (Lubben et al., 1989). However, certain imported proteins such as ferredoxin and superoxide dismutase fail to form a stable association with the chloroplast GroEL homologue, suggesting the association is not universal or obligatory in protein import and folding (Lubben et al., 1989). It is generally believed that the groEL homologues direct the assembly and folding of proteins after they have been successfully imported into the chloroplast and mitochondria (Cheng et al., 1989).

Recent work by Schatz and co-workers have shown that precursors interact only transiently with the mitochondrial hsp70 and that the extent of this interaction correlates with the rate of import. Interaction of these imported precursors with hsp60 can only occur after the precursors have been released from hsp70 in an ATP-dependent manner. They conclude, therefore, that there is a sequential action of mitochondrial chaperones in the import of precursors into the mitochondrial matrix (Manning-Krieg et al., 1991). It appears that precursors follow a series of chaperoned transport steps that begin with the interaction of newly synthesized proteins with a cytosolic hsp70. The protein then interacts with the organellar hsp70 during translocation, and finally upon import the processed precursor is associated with hsp60/cpn60 where it undergoes folding and/or assembly. Based on this prevailing model in the mitochondrial literature for the role of cytosolic hsp70 and mitochondrial hsp70 and hsp60, we have extended this model to chloroplast and have schematically represented it in Figure 3. Although the model is highly speculative and lacks solid supporting evidence, it is widely used as a working model for designing and interpreting experiments on protein import.

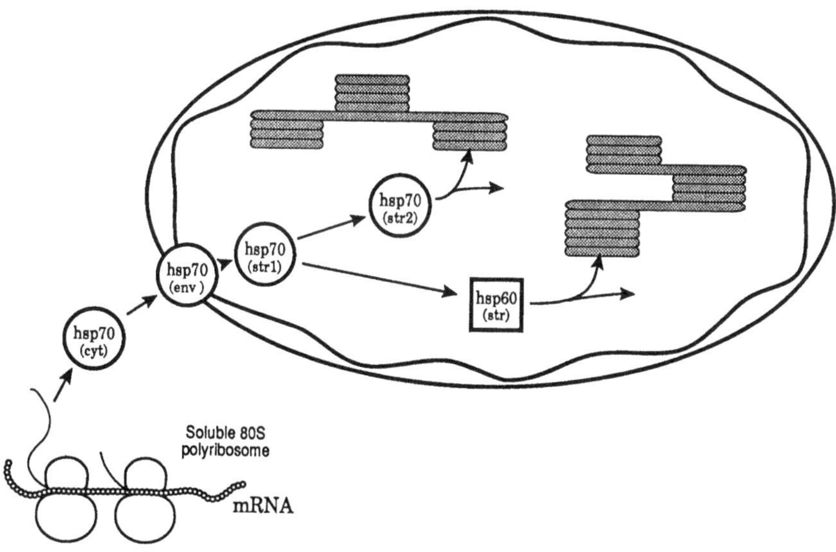

Figure 3. A model illustrating the possible involvement of cytosolic and chloroplast hsp70 and hsp60 homologues in the import and targeting of chloroplast precursors. Note the three different chloroplast hsp70 homologues and the sequential involvement with hsp60.

III. PRECURSOR INTERACTION WITH PUTATIVE RECEPTORS

A. Energy Requirements for Binding

Although protein translocation into chloroplasts has been clearly shown to require energy in the form of internal ATP (Theg et al., 1989), it was generally thought that binding would be energy-independent. However, after a more detailed investigation, it is now believed that binding also requires ATP, although at a much lower level than required to support translocation (Olsen et al., 1989). Binding displays a preference for ATP but is also supported by a broad range of nucleotides. This binding requires nucleotide hydrolysis since only hydrolyzable ATP analogs could support binding. The location of this ATP was originally shown to be internal to the outer envelope (Olsen et al., 1989). More recently the question of NTP localization has been refined and it has now been concluded that the site of NTP utilization for precursor binding to chloroplasts is in the intermembrane space between the two envelope membranes (Olsen and Keegstra, 1992).

B. Saturable Number of Binding Sites

The first step of protein import into organelles is the binding of the precursor to the surface of the organelle. If this precursor binding is receptor mediated, then there should be a finite number of sites which are protease sensitive, saturable, and specific for the precursor. Friedman and Keegstra developed a receptor–ligand binding assay for pre-SSU binding to the surface of pea chloroplasts. They calculated that there are approximately 3000 binding sites per chloroplast (Friedman and Keegstra, 1989). By assuming the average surface area of a chloroplast to be 75 μm^2, then the density of binding sites would be about $40/\mu m^2$. This low density will make these sites very difficult to detect by immunoelectron microscopy and will also be an obstacle to any biochemical characterization of the receptor.

To quantitate the number of translocation sites in mitochondria, Rassow and co-workers used methotrexate to inhibit the import of a DHFR chimeric protein. This protein accumulates as a translocation intermediate which they used to quantify the number of translocation sites per mitochondria as approximately 4200/mitochondria (Rassow et al., 1989). Although the total number of translocation sites on the mitochondria is comparable to the number of receptors found on chloroplasts, the density of these proteins on the organelle surface will undoubtedly be different. It should be noted, however, that the stoichiometry of precursor binding sites to translocation sites does not necessarily need to be 1:1.

It is interesting to note that, in mitochondria at least, there appears to be a protein import pathway which may not be mediated by receptors or involve the translocation apparatus. The protein apocytochrome *c* is believed to follow a very different import pathway into the mitochondria. This pathway is characterized by not requiring a presequence, ATP, a membrane potential, or a protease-sensitive receptor. It has been shown that apocytochrome *c* has special bilayer-penetrating properties that enable it to directly enter mitochondria (Jordi et al., 1992). This special pathway could be the explanation for earlier work which indicated that there is a 15-fold increase in the number of import sites for apocytochrome *c* and that these "binding sites" have a much lower dissociation constant (Henning et al., 1983). This sort of receptor and ATP-independent pathway may also be involved in the insertion of proteins into the chloroplast outer envelope (Salomon et al., 1990; Li et al., 1991) which will be discussed in more detail below.

C. Competition between Different Precursors

If binding to the chloroplast surface is mediated by finite population of proteinaceous receptors, then the question of receptor specificity arises. Do all precursors enter the chloroplast via the same binding and translocation apparatus, or are there different components for different precursors? The fact that synthetic peptides corresponding to the transit peptide of pre-SSU are able to competitively inhibit the binding and import of many different chloroplast precursors suggests that these precursors do indeed follow at least part of the same import pathway (Perry et al., 1991). This was also found to be the case with a synthetic peptide corresponding to part of the transit peptide of LHCP (Buvinger et al., 1989). Similar peptide competition experiments in mitochondria also suggest that many precursors share some steps in their import pathway in this organelle as well (Gillespie et al., 1985; Rassow et al., 1989; Glaser and Cumsky, 1990b; Cyr and Douglas, 1991).

D. Strategy for Identification of Possible Receptors

The identification of import receptors has been largely based on biochemical methods such as chemical cross-linking, immunoprecipitation, antibody inhibition, and the use of anti-idiotypic antibodies. In yeast, genetic evidence has been used to support the conclusion derived from the biochemical approaches. By the combination of these different approaches a fairly detailed model has been developed to explain the binding of precursors in mitochondria (Söllner et al., 1992). Unfortunately, it is premature to speculate how the chloroplast binding and translocation apparatus compares to this mitochondrial model. However, it is still instructive to examine the various strategies used with mitochondria and consider how they may be applied to chloroplasts.

Chemical Cross-Linking Studies

Early attempts to identify components of the yeast mitochondrial import apparatus involved photochemical cross-linking of a precursor protein to a 42-kDa outer membrane protein (Vestwebber et al., 1989). Another report using chemical cross-linking identified a different protein that was cross-linked to a synthetic peptide corresponding to a presequence (Gillespie, 1987). This work showed that the peptide was able to be specifically cross-linked to a 30-kDa protein localized in the mitochondrial outer membrane. The interaction of the synthetic peptide with this membrane component had the properties saturability, reversibility, and specificity, which are characteristic of a ligand binding to a receptor.

Immunoprecipitation and Antibody Inhibition

After raising antibodies against the major proteins of the mitochondrial outer membranes, Neupert's laboratory began testing these antibodies for their ability to inhibit protein import (Söllner et al., 1988). One of the proteins of the mitochondrial import apparatus has been identified in *Neurospora* by inhibiting import with antibodies against a 19-kDa mitochondrial outer membrane protein, MOM19 (Söllner et al., 1988). The involvement of this protein is confirmed by immunoprecipitation studies, where antibodies against this protein coprecipitate translocation intermediates in a complex containing not only MOM19 but also three other outer membrane protein. This 19-kDa protein has been shown to act as a receptor involved in the import of most proteins into the mitochondria (Kiebler et al., 1990b).

Another import receptor was also identified in *Neurospora* using monospecific antibodies against a 72-kDa mitochondrial outer membrane protein (MOM72) which was able to inhibit the import of the ADP/ATP carrier (AAC) (Söllner et al., 1988). These antibodies will also immunoprecipitate a complex between AAC and MOM72 after detergent solubilization. When the gene for this MOM72 was interrupted, the import of ACC was specifically impaired (Steger et al., 1990). This work was confirmed in another study that implicated a 70-kDa protein which accelerates the import of many precursor proteins. When this protein is either blocked with an anti-MAS70 antibody or removed by proteolysis, these treatments inhibit the import of many precursors. Hines and co-workers conclude that this protein functions as a receptor-like component which is involved in the early steps of import (Hines et al., 1990).

In similar experiments in yeast, antibodies against a 42-kDa outer membrane protein inhibited protein import into isolated yeast mitochondria (Vestwebber et al., 1989). Yeast cells which have had this gene depleted accumulate unprocessed precursors and eventually die. This protein is considered to be essential for protein import and growth (Baker et al., 1990; Baker and Schatz, 1991).

Membrane Fractionation

The general import receptor for most proteins in mitochondria, MOM19, was also identified by mild detergent solubilization of mitochondria and gel filtration. This protein fractionated as a complex with a molecular mass of 400 to 600 kDa. This detergent-solubilized complex contains not only MOM19 but also three other outer membrane proteins, MOM72, MOM22, and MOM38 (Kiebler et al., 1990a; Söllner et al., 1992).

Anti-Idiotypic Antibodies

Anti-idiotypic antibodies, raised against an antibody preparation directed against the presequence of COX IV, identified a 32-kDa protein on Western blots. This protein was shown to be an integral mitochondrial protein, and the anti-idiotypic antibodies recognizing it also inhibit import into isolated mitochondria. Polyclonal antibodies raised against this 32-kDa protein will also immunoprecipitate a precursor containing complex following a mild detergent solubilization of mitochondrial membranes. Immuno-electron microscopy, using these polyclonal antibodies, indicates that this protein is localized at contact sites where protein transport is thought to occur (Pain et al., 1990). However, this result is controversial because another group has concluded that this same protein functions as a phosphate translocator (Flügge et al., 1991).

E. Possible Chloroplast Import Receptors

Some of the same techniques described above for mitochondria have also been used to identify proteins involved in targeting precursors to chloroplasts.

The use of chemical cross-linking reagents has been employed by several groups. The first used a photoactivatable cross-linker attached to the pre-SSU (Cornwell and Keegstra, 1987). The modified precursor was bound to intact pea chloroplasts before activating the cross-linker with light. This method identified a cross-linked adduct that migrated with an apparent molecular mass of ~66 kDa. However, since the cross-linker used was not cleavable, the exact molecular mass of the envelope protein is not known and could even represent a dimer of two ~30-kDa proteins. The second group used a synthetic transit peptide as the cross-linking species (Kaderbhai et al., 1988). They identified two proteins, a 30-kDa component and a 52-kDa component. The 52-kDa protein most likely was the large subunit of Rubisco which is often present in envelope preparations. Alternatively, it could be the same as the 51-kDa protein identified below by Flügge as a member of the import apparatus (Hinz and Flügge, 1988). The authors concluded that the 30-kDa protein was the phosphate translocator found in the inner envelope membrane, the same protein Schnell and co-workers have identified with anti-idiotypic antibodies below (Schnell et al., 1990). The most recent effort to use chemical cross-linking was by Perry and Keegstra (1993). They used a label transfer cross-linking strategy that permitted the precise determination of the molecular weight of the cross-linked protein(s). This cross-linking approach used an *E. coli* overexpressed form of pre-SSU that was blocked at an early stage of transport. This approach identified two outer membrane proteins: an 86-kDa protein and a 75-kDa protein. Based on their observations, Perry and Keegstra conclude that pre-SSU first binds to the

86-kDa protein in an ATP-independent fashion and then is subsequently transferred to the 75-kDa protein in an ATP-requiring step. Membrane fractionation studies indicate that pre-SSU cross-linked to the 86-kDa protein is found only in the outer membrane. However, upon association with the 75-kDa protein pre-SSU fractionates in a more dense mixed envelope fraction. This suggests a two-step process where the precursor first associates with a receptor (86-kDa) in an ATP independent manner, and then is transferred to the translocation complex which forms a contact site between the inner and outer membrane of the chloroplastic envelope.

Using detergent solubilization, Soll and co-workers isolated a pre-SSU containing membrane complex that also contained the 86-kDa protein. They concluded that the 86-kDa protein is an integral part of the chloroplast translocation apparatus (Waegemann and Soll, 1991; Soll and Waegemann, 1992; Soll and Alefsen, 1993).

Using a very different approach to identify members of the import apparatus, Hinz and Flügge took advantage of the ATP dependence of import and were able to show that the presence of a precursor stimulated phosphorylation of a 51-kDa protein (Hinz and Flügge, 1988). This protein is found in the outer envelope and has been shown to only undergo phosphorylation under conditions which support precursor import. They speculate that this protein may be a putative receptor or intimately associated with the transport apparatus. However, there is still no direct evidence to support the involvement of this protein in protein import into chloroplasts.

The strategy that has yielded the most information but also the most controversial results has been the use of anti-idiotypic antibodies. A polyclonal anti-idiotypic antibody was prepared and shown to mimic the transit peptide of pre-SSU as determined by the ability of the antibody to block import of pre-SSU. The antibody reacted with two chloroplast proteins, a 52-kDa and a 30-kDa protein as determined by immunoblotting (Pain et al., 1988). The reaction against the 52-kDa protein was dismissed because this protein was identified as the large subunit of Rubisco. The 30-kDa protein was shown to be an integral membrane protein which was localized at the contact sites of the two envelopes. Schnell et al. later purified this protein and isolated a cDNA clone encoding the putative receptor (Schnell et al., 1990). Sequence comparisons between their clone and other sequences in the data base led to the conclusion that they had identified the same protein studied by Flügge as the phosphate translocator (Flügge et al., 1989; Willey et al., 1991). It is clear that the cDNA clones isolated by the two groups are very similar with only minor changes, probably due to species or varietal differences. A controversy remains regarding the function of the protein encoded by this gene. Schnell et al. (1990) argue that it functions as an import receptor. Flügge et al. (1991) argue that the protein functions as a phosphate translocator. Although it is possible to develop hypotheses which accommodate both results, these arguments seem unlikely and it is clear that further work will be needed to resolve this controversy. However, it is interesting to note that studies with anti-idiotypic antibodies attempting to identify import receptors for both chloroplasts (Schnell et al., 1990) and mitochondria (Pain et al., 1990) have identified proteins that have been identified by other groups as phosphate translocators (Flügge et al., 1991; Phelps et al., 1991).

IV. INTERACTION OF PRECURSORS WITH THE TRANSLOCATION APPARATUS

Transport of proteins into chloroplasts and mitochondria requires the translocation of proteins across two distinct lipid bilayer membranes. To simplify the complexity of this problem, it has been postulated that translocation occurs at contact sites where the membranes are held in close proximity and may even fuse transiently. Despite some evidence to support this attractive hypothesis many questions remain regarding the structure of contact sites and their role in protein transport. Some of the evidence and remaining problems are briefly summarized below.

A. Contact Sites

Both mitochondria and chloroplast are enclosed by two envelope membranes. In both organelles contact sites have been observed by both TEM of thin sections and freeze-fracture electron microscopy (Decker and Greenawalt, 1977; Cline et al., 1985). It has been proposed that the contact sites are important in the regulation of metabolite transport, and more recently they have been implicated in the import of cytosolically synthesized proteins. Currently, there are three different types of contact sites recognized in organelles and these different contact sites have been the subject of a recent review (Brdiczka, 1991).

Energy Transfer Contact Sites

These sites are generated by specific interactions between an energy-conserving kinase, such as creatine kinase or hexokinase, with both porin in the outer membrane and the adenine nucleotide translocator in the inner membrane. Rojo and co-workers have shown that purified mitochondrial creatine kinase is capable of producing an *in vitro* intermembrane contact between inner membrane vesicles and outer membrane sheets (Rojo et al., 1991). This kinase, which exists as an octomer, acts to draw the two membranes together forming a contact site.

Protein Transfer Contact Sites

When mitochondrial precursors were exposed to specific antibodies, their import was inhibited. When these precursors were immunolocalized, they were concentrated in regions where the two membranes were in close contact (Schwaiger et al., 1987). A similar pattern of immunolocalization was observed in chloroplasts using an anti-idiotypic antibody raised against the putative receptor for pre-SSU (Pain et al., 1988).

Skerjanc and co-workers, looking at the extent of folding of a fusion protein pre-OCAT, found that a unfolded form of the precursor was detected which specifically associated with a mitochondrial membrane fraction intermediate in density relative to the inner and outer mitochondrial membrane (Skerjanc et al., 1990). They speculated that this form of pre-OCAT represents an import intermediate that is associated with a putative contact site. It is interesting that this import intermediate still contains an intact presequence, suggesting that unfolding occurs prior to the presequence crossing the membrane and gaining access to the matrix localized protease.

In recent work, Schnell and Blobel constructed a fusion protein (pS/protA) consisting of the transit peptide and amino-terminus of the small subunit of Rubisco fused to the IgG-binding domain of protein A (Schnell and Blobel, 1993). With this construct they identified a bound form of pS/protA that is simultaneously accessible to the stromal processing protease and exogenously added thermolysin. They concluded that this form of pS/protA represents a transport intermediate that is spanning both the inner and outer envelope membranes. Using both biochemical and immunological methods they confirmed that this intermediate is localized to regions where the outer and inner membrane are closely appressed. This could represent the same membrane fraction that Perry and Keegstra identified as Fraction II that contained both the 86- and 75-kDa cross-linked proteins (Perry and Keegstra, 1994).

Other workers (Ardail et al., 1990) studying mitochondrial contact sites have proposed that a possible role for the increased amounts of negatively charged lipids such as cardiolipin may be to help neutralize the largely positively charged presequences of the mitochondrial precursors. This electrostatic interaction could be involved in facilitating the translocation of proteins through both membranes.

Lipid Transfer Contact Sites

Simbeni isolated yeast mitochondrial contact sites which were enriched with cardiolipin and phosphotidylethanolamine (PE)-lipids synthesized by enzymes on the inner membrane (Simbeni et al., 1991). This same membrane fraction was depleted in phosphatidylinositol and phosphatidylcholine relative to either the inner or outer membrane alone. Enzymatic assays of this fraction indicate an enrichment of both phosphatidylserine synthase and the PE-forming enzyme, phosphatidylserine decarboxylase. These results suggest that these contact sites function both as a site of lipid movement from the exterior where the primary synthesis occurs to the inner membrane and as the site where interconversion to cardiolipin and PE occurs.

Non-Bilayer Lipid Structures

It has been shown in many systems that compounds which interact with lipids can have a strong effect on the translocation of proteins across membranes. NMR studies by Jordi et al. (1990) with apocytochrome *c* and the well-studied drug phenethyl alcohol suggest that when the acyl chain region of a bilayer becomes disordered and therefore more fluid, it also becomes more active in protein translocation. It is not clear whether this drug-induced stimulation of translocation is via the normal route or simply an artifactual "bypass" route. Also, it is not evident from this work whether this is a feature of all import pathways or is unique to apocytochrome *c* which does not cross both membranes and is only able to reach the first aqueous compartment, the intermembrane space. It is interesting, however, that treatments which tend to increase the order of lipid bilayers such as cholesterol treatment and low temperature result in a reduced activity of protein translocation.

In yeast mitochondria, contact sites are associated with regions of the membrane that have increased amounts of both cardiolipin and PE. The lipids associated with these submitochondrial membranes fractions may exist in the hexagonal phase (H_{II}) since both of these lipids are able to form this phospholipid structure (Burger and Verkleij, 1990).

Hydrophilic Channels

When translocation intermediates of malate dehydrogenase and OTC are generated by a low-temperature treatment of isolated rat liver mitochondria, the precursors are processed to an intermediate size which is still largely exposed to the cytosolic face. When these intermediates are subjected to phase partitioning with the nonionic detergent, Triton X-114, they behave as hydrophilic molecules. Sztul et al. (1989) concluded that these proteins are residing in a microenvironment that is aqueous in nature, yet still mediated by integral membrane proteins.

More recent work by Simon and Blobel (1991) has implicated a protein-conducting channel which could possibly be the hydrophilic microenvironment that allows precursors access to both the matrix and the cytosol. This protein conducting channel would still involve integral membrane proteins, yet would not assume the direct interaction of precursors with the lipid bilayer.

B. Energy Requirements for Translocation

Several recent reviews have provided extensive coverage of the energy requirements for both binding and translocation of proteins into chloroplasts and mitochondria (Keegstra et al., 1989; Flügge, 1990; De Boer and Weisbeek, 1991). Consequently, we will provide only a brief summary and consider some explanations for the way in which energy is used.

Protein translocation across biological membranes is an energy-requiring process. This energy can be in the form of an ion gradient, a membrane potential, or the high energy phosphate bonds of ATP. The use of ATP may be to provide energy directly, or it may be via an indirect process such as the phosphorylation of a participating protein or through the formation of a $\Delta\Psi$ or ΔpH. Protein transport into chloroplasts is unique in that light may substitute for the addition of ATP. However, this utilization of light is equivalent to an internal addition of ATP, since these organelles retain their photosynthetic capacity *in vitro* and their ability to perform photophosphorylation.

Protein translocation across the envelope membranes of chloroplasts is fundamentally different from the analogous process in mitochondria or from bacterial protein secretion in that it does not require an electrochemical gradient. Only ATP is needed for protein movement across the chloroplast envelope. Although all studies have agreed on this point, the location of ATP utilization has been a point of controversy. It was initially reported that the ATP needed to support translocation was utilized in a site external to the inner membrane; more recently, several groups have convincingly shown that the ATP used for transport across the envelope is needed in the stroma (Pain and Blobel, 1987; Theg et al., 1989).

Although the role of ATP in protein translocation across membranes is still not understood, the following nonexclusive roles have been proposed:

1. ATP may be used in a cotranslational or posttranslational way to provide an energy source for some cytosolic factor(s) which act to maintain the precursor in a translocation competent form. This could be through the action of a putative "unfoldase". Several different cytosolic factors have been identified and discussed above that may act as molecular chaperones in this process. It should be noted that the *in vitro* protein synthesis systems used by most researchers already have an ATP level approaching 1 mM. It is therefore possible that the precursors produced by these methods may have already been rendered import-competent via an ATP-dependent process. The fact that purified pre-Fd and pre-PC overexpressed in *E. coli* do not require any additional factors for import casts some uncertainty on this role for ATP involvement (Pilon et al., 1990a; de Boer et al., 1991). However, the fact that pre-Fd and pre-PC are competent for import may be the result of their small molecular weight and hydrophilic nature. Overexpressed LHCP, on the other hand, has three hydrophobic transmembrane domains and requires a cytosolic factor for import competence (Waegemann et al., 1990; Waegemann and Soll, 1991).
2. The translocation apparatus itself may contain a subunit with an ATPase activity. A preliminary characterization of a chloroplast ATPase which is proposed to be involved in protein import has been reported (Pain and Blobel, 1987). Although the identity of this ATP-dependent "translocase" is not known in chloroplast or mitochondria, there is evidence that SecA, a member

of the protein translocation apparatus in *E. coli,* functions as an ATPase (Lill et al., 1989).

3. ATP could be used by a specific kinase that phosphorylates a component of the translocation apparatus. Protein phosphorylation has been shown to be the key event regulating the activity of many enzymes, often by causing conformational changes in the protein. In a similar fashion, phosphorylation of an envelope protein may induce some form of conformational change which induces the movement of the protein across the membrane. A 51-kDa protein of the envelope is phosphorylated during protein transport (Hinz and Flügge, 1988) and may explain some of the ATP requirement for translocation. However, this phosphorylation requires an ATP level more comparable to what supports binding than the level needed for translocation (Flügge and Hinz, 1986).

In mitochondria, the first model described above has also been proposed. Where artificially unfolded proteins did not require ATP for protein import, however, import of the authentic precursors requires ATP (Ostermann et al., 1990; Pfanner et al., 1990). More recent work by Pfanner and co-workers (1990) have refined the role of ATP on the import of precursors. They propose that ATP is required for two independent roles in import and assembly of mitochondrial proteins: (1) for the maintenance or generation of translocationally competent conformation of precursors in the cytosol, and (2) the refolding and targeting of precursors in the matrix through a cascade of interactions with chaperone-like components. Chloroplasts also probably require ATP for these pre- and posttranslocation events, yet they are distinct in that they require NTP for precursor binding and ATP for membrane translocation. The fact that a proton motive force cannot substitute for ATP suggests that some important differences exist in the translocation mechanisms between these two organelles.

V. PROCESSING OF PRECURSORS

Nuclear-encoded precursors entering either the chloroplast stroma or the mitochondrial matrix undergo specific processing by peptidases either during translocation or immediately afterward. One group of proteases involved are the soluble, metal-dependent peptidases known as stromal processing peptidase (SPP) and matrix processing peptidase (MPP). These are distinct from a second group of membrane-bound proteases which are found in bacterial transport or ER transport systems, known as signal peptidases. Chloroplasts and mitochondria also contain members of this second group known as thylakoidal processing peptidase (TPP) and inner membrane protease I, respectively. A third protease, mitochondria intermediate peptidase, has recently been found in the mitochondrial system and has no known homologue in chloroplasts.

A. Mitochondrial Endopeptidases

Matrix Processing Protease (MPP)

In mitochondria, the amino-terminal presequence is proteolytically cleaved by a matrix-localized metalloprotease which was first identified and characterized in yeast, rat liver, and maize (Bohni et al., 1980). Purification to homogeneity from *Neurospora* revealed two related subunits (50% similarity) of approximately 50 kDa in size which were named protease-enhancing protease (PEP) and matrix processing protease (MPP) (Hawlitschek et al., 1988). In yeast, these proteins are the products of the nuclear genes, *MAS1* and *MAS2*, respectively (Yang et al., 1988). Although they function together as a complex, it is believed that MPP contains the peptidase activity (Jensen and Yaffe, 1988). PEP alone has no proteolytic activity, but has a large stimulatory effect on MPP. PEP may also function in translocation of precursor proteins across the mitochondrial membranes (Pollock et al., 1988).

Intermediate Peptidase (MIP)

The presequences of most precursors are removed in a single step by MPP/PEP; however, several precursors undergo two sequential proteolytic cleavages. In this case, proteins such as pre-OTC are initially processed to an intermediate form by MPP. A second distinct matrix peptidase, mitochondrial intermediate peptidase (MIP), then cleaves the remaining eight amino acids from the amino-terminus of the intermediate yielding the mature protein. Recent work showed that the mature amino-terminus of twice-cleaved precursors is structurally incompatible with the peptidase, MPP (Isaya et al., 1991). These precursors have therefore evolved an additional peptide sequence, at the carboxy-terminus of their presequence, which is specifically recognized and cleaved by the second peptidase, MIP. It is not yet known what structural domains interfere with MPP activity or how this additional sequence overcomes this interference.

Inner Membrane Protease I

Another two-step processing pathway has been observed in mitochondrial for proteins destined for the intermembrane space (IMS) (Hartl et al., 1986, 1987). In this case, processing involves both MPP and an IMS peptidase. Schneider and co-workers have developed an *in vitro* assay for this protease and have named it inner membrane protease I (IMP I). It is an integral protein of the inner membrane, requires divalent cations and acidic phospholipids, and has a 21-kDa subunit (Schneider et al., 1991). This protease has many similarities to the leader peptidase in *E. coli*. It is also similar to the thylakoid processing protease, TPP, of chloroplasts (discussed below).

B. Chloroplast Endopeptidases

Stromal Processing Protease (SPP)

A major question regarding the import of chloroplast precursors is whether the same enzyme is involved in processing precursors destined for the various internal locations shown in Figure 1. Precursors for stromal proteins contain a transit peptide with a single domain. Precursors targeted for the thylakoid lumen, however, have a transit peptide composed of two parts: the amino-terminus, which acts as a envelope transfer domain; and an additional carboxy-terminal region, which functions as a thylakoid transfer domain. Both stromal and lumenal precursors are processed by the SPP. Action of SPP on lumenal precursors produces an intermediate form containing the thylakoid transfer domain.

SPP has been partially purified from pea chloroplasts and has been shown to have a molecular mass of 180 kDa. The fact that it is inhibited by metal chelators suggests that it is a metalloprotease (Robinson and Ellis, 1984a,b). It has been shown to catalyze the removal of the transit peptide from several stromal proteins including pre-SSU, pre-ACP, pre-Fd, and pre-LHCP, and to catalyze the processing of pre-PC to an intermediate form (Lamppa and Abad, 1987; Abad and Oblong, 1988; Abad et al., 1991). Several reports have shown that SPP processes precursors in two distinct steps, with the smaller product being equal to the form observed in intact chloroplasts (Robinson and Ellis, 1984b; Hageman et al., 1986). This observation raised the possibility that the partially purified SPP preparations were contaminated with a second endopeptidase. In fact, several endopeptidases have been purified from pea chloroplasts. One of these, EP-2, processes pre-PC *in vitro* to a product with the same apparent molecular weight as the intermediate seen *in vivo* (Musgrove et al., 1989). However, when the SPP preparation from the earlier work (Robinson and Ellis, 1984a) was further purified by phenyl Sepharose chromatography to give a fraction free of EP-2 , the *in vitro* processing of pre-PC still gave two cleavage products (Musgrove et al., 1989). This work suggests that SPP produces an initially larger processing intermediate before the final form. This is the case whether it is processing a stromal localized protein such as pre-SSU or a lumenal protein like pre-PC which has an additional thylakoid transfer domain. This two-step processing could suggest that the 180-kDa complex that has partially been purified may contain chloroplast homologues to both MPP and MIP, which function sequentially *in vivo* as has been described above for mitochondria.

Thylakoidal Processing Peptidase (TPP)

Cytoplasmically synthesized proteins destined for the thylakoid lumen follow a complex import pathway, crossing three different membranes, that is unique to chloroplasts. This pathway is really the sum of two sequential targeting events; first, the full length precursor is transported across the envelope where it is processed to

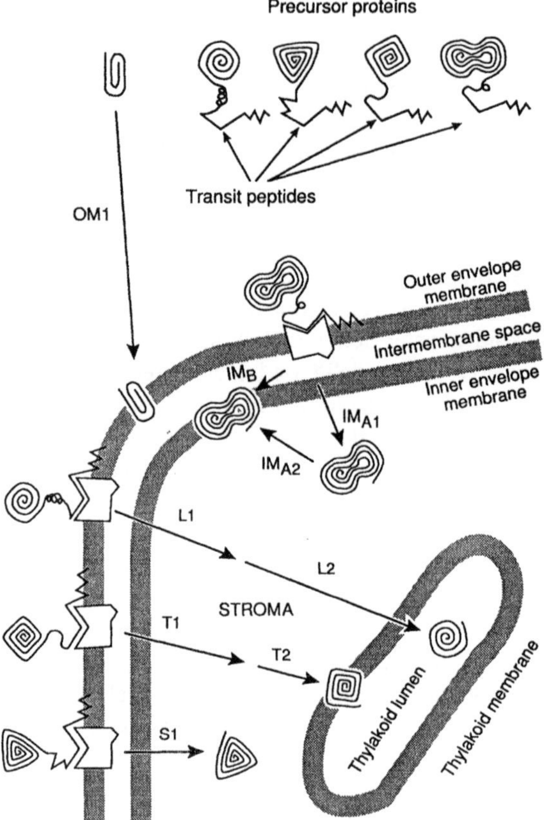

Figure 4. A diagram illustrating five of the six different chloroplast domains and their possible import pathways. S1 is the import pathway into the stroma. T1 and T2 are sequential steps in the import path into the thylakoid membrane. L1 and L2 are sequential steps in the import and targeting to the thylakoid lumen. OM1 is the path to the outer envelope membrane. IM_A and IM_B represent alternative pathways to the inner envelope membrane. The IM_A pathway has two steps and is referred to as the "conservative sorting pathway".

an intermediate form by SPP. Then in a second distinct step, this intermediate is transported across the thylakoid and processed to its mature form by a thylakoidal peptidase. This complex, two-step mechanism has been shown to be involved in the import of pre-PC and the 33 and 23-kDa subunits of the OEC (Smeekens et al., 1986; James et al., 1989; Kirwin et al., 1989).

TPP has been partially purified by the laboratory of C. Robinson (Kirwin et al., 1987). This peptidase is able to process both the intermediate form of PC and pre-PC to their mature forms. It shows no detectable activity against non-chloroplast proteins or against stromal localized proteins. Using this partially purified peptidase, an *in vitro* reconstituted system has been developed by Kirwin and co-workers to study the processing of thylakoid proteins (Kirwin et al., 1989). The reaction specificity of the thylakoid processing peptidase has been shown to be identical to that of *E. coli* leader peptidase and the ER signal peptidase. Both the thylakoid and *E. coli* enzyme correctly process the 23-kDa subunit of the OEC. TPP is also capable of correctly processing proteins which are normally transported across the bacterial plasma membrane or the ER membrane (Halpin et al., 1989). Both the TPP and leader peptidase function as endopeptidases, yet they are not inhibited by PMSF, EDTA, EGTA or any other standard protease inhibitor (Musgrove et al., 1989). This suggests that neither is a serine peptidase nor a metallo-peptidase, leaving the exact class of these proteases unknown. The TPP, like the IMP I from mitochondria, shares several critical features with signal peptidase (Schneider et al., 1991). This similarity with the signal-type peptidases was further confirmed in a study that indicated the residues at the -3 and -1 position have a very important role in defining the cleavage site of both peptidases (Fikes et al., 1990; Shackleton and Robinson, 1991).

VI. TARGETING OF PROTEINS WITHIN PLASTIDS

With both chloroplasts and mitochondria, many imported proteins need to be directed to locations other than the stromal or matrix compartment. As described below, certain proteins in both organelles need to be targeted to the two membranes surrounding the organelle or to the soluble space between these two membranes. However, the presence of the internal thylakoid membrane system in plastids makes targeting of proteins into chloroplasts more complex than into mitochondria. Because thylakoids are unique to chloroplasts and because of their importance for photosynthesis, the targeting of cytoplasmically synthesized proteins to the thylakoid membrane system has been extensively studied (see De Boer and Weisbeek, 1991 for a review). At least some of the integral and peripheral proteins of the thylakoid membrane as well as most of the soluble proteins of the thylakoid lumen are synthesized in the cytoplasm. In each case, the precursor proteins contain a transit peptide that directs the precursor across the two membranes of the envelope in a manner similar to that described above for stromal proteins (Figure 4). A second distinct step is then responsible for directing imported proteins to the thylakoid membrane system (Figure 4). Details of the intraorganellar targeting of both classes of thylakoid components are briefly described below.

A. Targeting to the Thylakoid Membrane

Although many different thylakoid membrane proteins are imported from the cytoplasm, the chlorophyll *a/b* binding (C*ab*) protein of the light harvesting complex has been most extensively studied. In this particular case, the transit peptide of the precursor functions only to direct the protein to the stromal space (Step T1 in Figure 4); the C*ab* transit peptide can be replaced with the transit peptide from the pre-SSU (Lamppa, 1988; Hand et al., 1989). The information directing the insertion of C*ab* into the thylakoid membrane (Step T2 in Figure 4) is contained within the mature sequence of C*ab* (Kohorn and Tobin, 1989).

The insertion process requires energy and the assistance of one or more unidentified stromal proteins (Cline, 1986; Chitnis et al., 1987). Many of the details regarding insertion of imported proteins into the thylakoid membrane are derived from studies where insertion has been reconstituted with isolated thylakoid membranes. For C*ab*, either the precursor form or the processed mature form can serve as the substrate for insertion (Cline et al., 1989). Insertion of C*ab* into thylakoids requires ATP and is enhanced by a proton motive force (Cline et al., 1989, 1992). The requirement for a stromal protein in the insertion process (Fulson and Cline, 1988) is consistent with the concept of chaperonin molecules being involved in protein targeting and assembly, but the identity of chaperonins involved has not yet been determined. Attractive candidates include the hsp 60 chaperones involved in Rubisco assembly or the hsp 70 molecules known to be present in the stromal space. Further work is needed to determine whether these or other unidentified chaperonins are the factors required for C*ab* insertion.

B. Transport to the Thylakoid Lumen

The first support for a two-step pathway for targeting to the thylakoid lumen came from the observation that the transit peptide of a lumenal precursor contained a complex two-domain transit peptide (Smeekens et al., 1985). Much stronger evidence derived from a kinetic analysis of the appearance and disappearance of intermediate-sized forms during the transport of plastocyanin precursor into intact chloroplasts (Smeekens et al., 1986). The authors postulated that the first step consisted of transport of the precursor to the stroma (Step L1 in Figure 4) where part of the transit peptide was removed to yield an intermediate that was transported across the thylakoid membrane during the second step (Step L2 in Figure 4) (Smeekens et al., 1986). However, it was not possible to exclude a second hypothesis involving transport to the intermembrane space of the chloroplast envelope during the first step followed by sequestration within a vesicle moving from the inner envelope membrane to the thylakoid membrane during the second step (Keegstra and Bauerle, 1988). Despite the considerable evidence that supports the notion of vesicular traffic from the inner envelope membrane to the thylakoids, there is no evidence that these vesicles include any cytoplasmically synthesized thyalkoid proteins. The vesicle pathway seems highly unlikely for lumenal proteins in view of the recent success in reconstituting the transport of lumenal proteins across isolated thylakoid membranes (Kirwin et al., 1989; Bauerle and Keegstra, 1991). It is difficult to imagine that isolated thylakoid membranes would possess the ability to transport lumenal precursors *in vitro* if the lumenal proteins arrived via vesicle traffic *in vivo*.

Although the two-step pathway of lumenal transport seems well established, several details of the second step remain unresolved. For example, transport across the thylakoid membrane is efficiently reconstituted with some proteins, whereas the process is very inefficient with others (Kirwin et al., 1989; Bauerle and Keegstra, 1991; Cline et al., 1992). The reasons for these differences are not clear. Additionally, it is not clear whether the precursor or the intermediate-sized form is the substrate for the translocation step. With plastocyanin only the precursor can be transported *in vitro* (Bauerle and Keegstra, 1991), whereas with OEC23 both the precursor and the intermediate-sized form serve as substrates (Kirwin et al., 1989). The energy requirements for transport across thylakoids are not simple. Indeed it currently seems that the energy requirements depend upon the precursor studied (Cline et al., 1992). It is unclear whether stromal factors are needed for translocation; again it may depend upon the protein under investigation. Thus, many important details of the second step in the transport process require additional work.

The second step of transport to the thylakoid lumen shares many features with bacterial transport. First, is the observation that the second domain of the transit peptide that directs transport across the thylakoid membrane is similar in structure to a bacterial signal peptide (Smeekens et al., 1985). Additionally, many features of the thylakoid transport apparatus are similar to those found in bacteria; for example, the lumenal peptidase that removes the last portion of the transit peptide (Halpin et al., 1989). Thus, targeting of proteins to the thylakoid lumen fits well with the "conservative sorting" hypothesis (Hartl and Neupert, 1990) which posits that the second step is derived from the bacterial ancestor that gave rise to chloroplasts (Weisbeek, 1989).

C. Multiple Transport Pathways into and Across the Thylakoid

Much of the conflicting literature concerning the targeting and transport of proteins to and across the thylakoid membrane may be resolved if multiple pathways coexist. These pathways may differ in their energetics, their involvement of soluble stromal factors, and the topogenic sequences that are recognized by their translocation apparatus. Recently, several manuscripts have identified and partially characterized different transport pathways in the thylakoid membrane (Cline et al., 1993; Henry et al., 1994; Madueño et al., 1994; Michl et al., 1994; Robinson et al., 1994; Yuan and Cline, 1994). The results of these manuscripts are summarized in Table 1. Although much work remains in identifying and characterizing their individual components, it is clear from Table 1 that the various pathways are distinct

Table 1. Multiple Import Pathways for Nuclear-Encoded Thylakoid Proteins

Characteristics	*Pathway I*	*Pathway II*	*Pathway III*	*Pathway IV*
ATP/NTP required	yes	no	yes	no
Thylakoid ΔpH dependent	yes	yes	yes	no
Stromal factor(s) involved	yes	no	yes	no
Presequence contains a bipartite lumen-targeting domain	yes	yes	no	yes
Mature protein contains thylakoid-targeting domain	no	no	yes	no
Protein destination	lumen	lumen	thylakoid	thylakoid
Precursors transported	PC, OE33	OE23, OE17	LHCP, Rieske Fe-S?	CFoII
References	Cline et al., 1993; Yuan & Cline, 1994; Robinson et al., 1994; Henry et al., 1994	Cline et al., 1993; Robinson et al., 1994; Henry et al., 1994	Cline et al., 1993; Madueno et al., 1994	Michl et al., 1994

in their requirements for ATP, a proton gradient, stromal factors, the topogenic sequences recognized, and in the precursors transported.

D. Targeting to the Envelope

The envelope of chloroplasts consists of two lipid bilayer membranes separating a soluble space between them. This structural motif is shared by mitochondria and gram negative bacteria. Although virtually all of the proteins of the chloroplast envelope are thought to be derived from the cytoplasm, relatively little is known regarding the targeting of these proteins to the envelope of chloroplasts. However, this topic has been extensively studied with mitochondria and these results can serve as a model for studies with chloroplasts.

Targeting to the Outer Envelope Membrane

Targeting of integral membrane proteins to the outer envelope membrane has been studied with two different proteins (Salomon et al., 1990; Li et al., 1991). In both cases, the precursor proteins are the same size as the mature protein and lack a cleavable transit peptide. Also, in both cases, the precursors are inserted into the outer membrane via a process that does not require energy (Salomon et al., 1990; Li et al., 1991). The insertion process does not utilize the receptors or transport apparatus that is used for targeting of proteins into chloroplasts (Li et al., 1991), although it is not possible to conclude that no receptors or envelope membrane proteins are involved in the insertion process. These results are similar to those observed with mitochondria. With both organelles it appears that evolution has developed unique pathways for targeting proteins to the outer envelope membrane of each organelle.

Targeting to the Inner Envelope Membrane

Targeting of proteins to the inner envelope membrane is very different from insertion of proteins in the outer envelope membrane, but has many similarities to the import of internal proteins. Although relatively few examples have been studied, the pattern that emerges is that precursors for inner membrane proteins have a cleavable transit peptide. Targeting to the inner envelope membrane requires energy and the same import apparatus that is used for the import of internal proteins (Flügge et al., 1989; Li et al., 1992). It was postulated that the information for targeting to the inner envelope membrane is contained within the transit peptide in a manner similar to thylakoid lumenal precursors (Dreses-Werringloer et al.,1991). However, studies with a different precursor have recently established that the transit peptide serves as a stromal targeting sequence and that the information directing the protein to the inner envelope membrane is contained within the mature protein (Li et al., 1992).

Regardless of whether the targeting information specifying the inner envelope membrane is present in the transit peptide or in the mature protein, the pathway of transport to the inner envelope membrane is an interesting unresolved problem. One possibility is that the precursor contains a "stop transfer" signal that halts translocation across the inner envelope membrane and causes the protein to become a resident of this membrane. In this case, it will be interesting to identify such a "stop transfer" sequence. It seems unlikely that it can simply be a membrane spanning domain, since several integral thylakoid proteins that contain membrane spanning domains are transported across the inner envelope membrane before insertion into the thylakoid membrane. The other possibility is that the precursor is transported completely into the stromal space and then is inserted into the inner envelope membrane as predicted by the conservative sorting hypothesis (Hartl and Neupert, 1990). If this pathway is used for targeting to the chloroplast envelope, as it appears to be in mitochondria (Hartl and Neupert, 1990), then it will be interesting to identify the signals that determine the specificity of membrane insertion. Some proteins will need to be inserted into the thylakoid membrane, while others are inserted into the inner envelope membrane. Whichever mechanism operates, and it is possible that both do, it will be interesting to determine how envelope membrane proteins are distinguished from thylakoid membrane proteins.

ACKNOWLEDGMENTS

Work in the authors' laboratory was supported by a grant from the Office of Basic Energy Science at the U.S. Department of Energy (K.K.), N.S.F. (K.K.), and by an NSF Postdoctoral Fellowship in Plant Biology (B.D.B.).

REFERENCES

Abad, M.S. & Oblong. (1988). Properties of a chloroplast enzyme that cleaves the chlorophyll *a/b* binding protein precursor: Optimization of an organelle-free reaction. Plant Physiol. 90, 117–124.

Abad, M.S., Pblong, J.E., & Lamppa, G. K. (1991). Soluble chloroplast enzyme cleaves pre-LHCP made in E. coli to a mature form lacking a basic N-terminal domain. Plant Physiol. 96, 1220–1227.

Alteiri, F., Mattingly, J.M., Rodriques-Berrocal, F.J., Youssef, J., Iriarte, T., Wu, T., & Martines-Carrion, M. (1989). Isolation and properties of a liver mitochondrial precursor protein to aspartate aminotransferase expressed in *E. coli*. J. Biol. Chem. 264, 4782–4786.

America, T., Hageman, J., Archer, K., Keegstra, K., & Weisbeck, P. (1992). Methotrexate does not block import of a DHFR fusion protein into chloroplasts. Submitted for publication.

Archer, E.K. & Keegstra, K. (1990). Current views on chloroplast protein import and hypotheses on the origin of the transport mechanism. J. Bioenerg. Biomembrane 22, 789–810.

Ardail, D., Privat, J. P., Egret-Charlier, M., Levrat, C., Lerme, F., & Louisot, P. (1990). Mitochondrial contact sites: Lipid composition and dynamics. J. Biol. Chem. 265, 18797–18802.

Baker, K.P., Schaniel, A., Vestweber, D., & Schatz, G. (1990). A yeast mitochondrial outer membrane protein essential for protein import and cell viability. Nature 348, 605–609.

Baker, K.P. & Schatz, G. (1991). Mitochondrial proteins essential for viability mediate protein import into yeast mitochondria. Nature 349, 205–208.

Batenburgh, A.M., van Esch, J.H., Leunissen-Bijvelt, J., Verkleij, A.J., & de Kruijff, B. (1987). Interaction of meliten with negatively charged phospholipids: Consequences for lipid reorganization. FEBS Lett. 223, 148–154.

Bauerle, C. & Keegstra, K. (1991). Full-length plastocyanin precursor is translocated across isolated thylakoid membranes. J. Biol. Chem. 266, 5876–5883.

Bohni, P.C., Gasser, S., Leaver, C., & Schatz, G. (1980). *The Organization and Expression of the Mitochondrial Genome*. Elsevier, Amsterdam.

Brdiczka, D. (1991). Contact sites between mitochondrial envelope membranes. Structure and function in energy and protein transfer. Biophys. Biochem. Acta, Reviews on Biomem.

Burger, K.N.J. & Verkleij, A.J. (1990). Membrane fusion. Experientia 46, 631–644.

Buvinger, W.E., Michel, H., & Bennett, J. (1989). A truncated analog of a pre-light-harvesting chlorophyll *a/b* protein II transit peptide inhibits protein import into chloroplasts. J. Biol. Chem. 264, 1195–1202.

Chen, W.J. & Douglas, M.G. (1987). The role of protein structure in the mitochondrial import pathway. J. Biol. Chem. 262, 15598–15604.

Cheng, M.Y., Hartl, F.U., Martin, J., Pollock, R.A., Kalousek, F., Neupert, W., Hallberg, E.M., Hallberg, R.L., & Horwich, A.L. (1989). Mitochondrial heat-shock protein hsp60 is essential for assembly of proteins imported into yeast mitochondria. Nature 337, 620–625.

Chitnis, P.R., Nechushtai, R., & Thornber, J.P. (1987). Insertion of the precursor of the light-harvesting chlorophyll *a/b* protein into the thylakoids requires the presence of a developmentally regulated stromal factor. Plant Mol. Biol. 10, 3–10.

Chu, T.W., Eftime, R., Sztul, E., & Strauss, A.W. (1989). Synthetic transit peptides inhibit import and processing of mitochondrial precursor proteins.

Cline, K. (1986). Import of proteins into chloroplasts. Membrane insertion of a thylakoid precursor protein reconstituted in chloroplast lysates. J. Biol. Chem. 261, 14804–14810.

Cline, K., Keegstra, K., & Staehelin. (1985). Freeze-fracture electron microscopic analysis of ultrarapidly frozen envelope membranes on intact chloroplasts and after purification. Protoplasma 125, 111–123.

Cline, K., Ettinger, W.F., & Theg, S.M. (1992). Protein-specific energy requirements for protein transport across or into thylakoid membranes: Two lumenal proteins are transported in the absence of ATP. J. Biol. Chem. 267, 2688–2696.

Cline, K., Fulsom, D.R., & Viitanen, P.V. (1989). An imported thylakoid protein accumulates in the stroma when insertion into the thylakoid is inhibited. J. Biol. Chem. 264, 14255–14232.

Cline, K., Henry, R., Li, C., & Yuan, J. (1993). Multiple pathways for protein transport into or across the thylakoid membrane. EMBO J. 12, 4105–4114.

Cline, K., Werner-Washburne, M., Lubben, T., & Keegstra, K. (1985). Precursors to two nuclear-encoded chloroplast proteins bind to the outer envelope before being imported into chloroplasts. J. Biol. Chem. 260, 3691–3696.

Collier, D.N., Bankaitis, V.A., Weiss, J.B., & Bassford, P.J. (1988). The antifolding activity of SecB promotes the export of the *E. coli* maltose binding protein. Cell. 53, 273–283.

Cornwell, K. & Keegstra, K. (1987). Evidence that a chloroplast surface protein is associated with a specific binding site for the precursor to the small subunit of ribulose-1,5-biscarboxylase. Plant Physiol. 85, 780–785.

Craig, E.A., Kramer, J., & Kosic-Smithers, J. (1987). SSC1, a member of the 70-kDa heat-shock protein multigene family of *S. cerevisae*, is essential for growth. Proc. Natl. Acad. Sci. USA 84, 4156–4160.

Cyr, D.M. & Douglas, M.G. (1991). Early events in the transport of proteins into mitochondria. Import competition by a mitochondrial presequence. J. Biol. Chem. 266, 21700–21708.

De Boer, A.D. & Weisbeek, P.J. (1991). Chloroplast protein topogenesis: Import, sorting and assembly. Biophys. Biochem. Acta, Reviews on Biomembranes 1071, 221–253.

de Boer, D., Pilon, R., Broeders, L., Oudshoorn, T., Lever, A., & Weisbeek, P. (1991). Import competition studied using purified plastocyanin precursor. In preparation.

Decker, G.L. & Greenawalt, J.W. (1977). Ultrastructural and biochemical studies of mitoplasts and outer membranes derived from French-pressed mitochondria. J. Ultrastruct. Res. 59, 44–56.

della-Cioppa, G., Bauer, S.C., Klein, B.K., Shah, D.M., Fraley, R.T., & Kishore, G.M. (1986). Translocation of the precursor of 5-enol-pyruvylshikimate-3-phosphate synthase into chloroplasts of higher plants *in vivo*. Proc. Natl. Acad. Sci. USA 83, 6873–6877.

della-Cioppa, G. & Kishore, G.M. (1988). Import of a precursor protein is inhibited by the herbicide glyphosphate. EMBO J. 7, 1299–1305.

Deshaies, R.J., Koch, B.D., Werner-Washburne, M., Craig, E.A., & Schekman, R. (1988). A subfamily of stress proteins facilitates translocation of secretory and mitochondrial precursor polypeptides. Nature 322, 800–805.

Dobberstein, B., Blobel, G., & Chua, N.H. (1977). *In vitro* synthesis and processing of a putative precursor for the small subunit of ribulose-1,5-bisphosphate carboxylase of *Chlamydomonas reinhardtii*. Proc. Natl. Acad. Sci. USA 74, 1082–1085.

Drese-Werringloer, U., Fischer, K., Wachter, E., Link, T.A., & Flügge, U.-I. (1991). cDNA sequence and deduced amino acid sequence of the precursor of the 37-kDa inner envelope polypeptide from spinach chloroplasts—its transit peptide contains an amphilphilic α-helix as the only detectable structural element. Eur. J. Biochem. 195, 361–363.

Eilers, M. & Schatz, G. (1986). Binding of a specific ligand inhibits import of a purified precursor protein into mitochondria. Nature 322, 228–232.

Fikes, J.D., Barkocy-Gallagher, G.A., Klapper, D.G., & Bassford, P.J. (1990). Sequence requirements for efficient processing and demonstration of an alternative cleavage site. J. Biol. Chem. 265, 3417–3423.

Flügge, U.-I. (1990). Import of proteins into chloroplasts. J. Cell Sci. 96, 351–354.
Flügge, U.-I., Fischer, K., Gross, A., Sebald, W., Lottspeich, F., & Eckerskorn, C. (1989). The triose phosphate-3-phosphoglycerate-phosphate translocator from spinach chloroplasts: Nucleotide sequence of a full length cDNA clone and import of the *in vitro* synthesized precursor protein into chloroplasts. EMBO J. 8, 39–46.
Flügge, U.-I., Weber, A., Fischer, K., Lottspeich, F., Eckerskorn, E., Waegemann, K., & Soll, J. (1991). The major chloroplast envelope polypeptide is the phosphate translocator and not the protein import receptor. Nature 353, 364–367.
Flügge, U.I. & Hinz, G. (1986). Energy dependence of protein translocation into chloroplasts. Eur. J. Biochem. 160, 563–570.
Flynn, G.C., Chappell, T.G., & Rothman, J.E. (1989). Peptide binding and release by proteins implicated as catalysts of protein assembly. Science 245, 385–390.
Flynn, G.C., Pohl, J., Flocco, M.T., & Rothman, J.E. (1991). Peptide-binding specificity of the molecular chaperone BiP. Nature 353, 726–730.
Friedman, A.L. & Keegstra, K. (1989). Chloroplast protein import. Quantitative analysis of precursor binding. Plant Physiol. 89, 993–999.
Fulson, D.R. & Cline, K. (1988). A soluble factor is required *in vitro* for membrane insertion of the thylakoid precursor protein, pLHCP. Plant Physiol. 88, 1146–1153.
Gavel, Y. & von Heijne, G. (1990). A conserved cleavage site motif in chloroplast transit peptides. FEBS Lett. 261, 455–458.
Geli, V. & Glick, B. (1990). Mitochondrial protein import. J. Bioenerg. Biomem. 22, 725–751.
Gillespie, L.L. (1987). Identification of an outer mitochondrial membrane that interacts with a synthetic signal peptide. J. Biol. Chem. 262, 7939–7942.
Gillespie, L.L., Argan, C., Taneja, A.T., Hodges, R.S., Freeman, K.B., & Shore, G. (1985). A synthetic signal peptide blocks import of precursor proteins destined for the mitochondrial inner membrane or matrix. J. Biol. Chem. 260, 16045–16048.
Glaser, S.M. & Cumsky, M.G. (1990a). Localization of a synthetic presequence that blocks protein import into mitochondria. J. Biol. Chem. 265, 8817–8822.
Glaser, S.M. & Cumsky, M.J. (1990b). A synthetic presequence reversibly inhibits protein import into yeast mitochondria. J. Biol. Chem. 265, 8808–8816.
Grimm, R., Speth, V., Gatenby, A.A., & Schafer, E. (1991). GroEL-related molecular chaperones are present in the cytosol of oat cells. FEBS Lett. 286, 155–158.
Grossman, A., Bartlett, S., & Chua, N.H. (1980). Energy dependent uptake of cytoplasmically synthesized polypeptides by chloroplasts. Nature 285, 625–628.
Hageman, J., Robinson, C., Smeekens, S., & Weisbeck, P. (1986). A thylakoid processing protease is required for complete maturation of the lumen protein plastocyanin. Nature 324, 567–569.
Halpin, C., Elderfield, P.D., James, H.E., Zimmermann, R., Dunbar, B., & Robinson, C. (1989). The reaction specificities of the thylakoidal processing peptidase and *E. coli* leader peptidase are identical. EMBO J. 8, 3917–3921.
Hand, J.M., Szabo, L.J., Vasconcelos, A.C., & Cashmore, A.R. (1989). The transit peptide of a chloroplast thylakoid membrane is functionally equivalent to a stromal targeting sequence. EMBO J. 8, 3195–3206.
Hardy, S.J.S. & Randall, L.L. (1991). A kinetic partitioning model of selective binding of non-native proteins by the bacterial chaperone SecB. Science 251.
Hartl, F.-U. & Neupert, W. (1990). Protein sorting to mitochondria: Evolutionary conservations of folding and assembly. Science 247, 930–938.
Hartl, F.-U., Ostermann, J., Guiard, B., & Neupert, W. (1987). Successive translocation in and out of the mitochondrial matrix: Targeting of proteins to the intermembrane space by a bipartite signal peptide. Cell 51, 1027–1037.

Hartl, F.-U., Schmidt, B., Wachter, E., Weiss, H., & Neupert, W. (1986). Transport into mitochondria and intramitochondrial sorting of the Fe/S protein of the ubiquinol-cytochrome *c* reductase. Cell 47, 939–951.

Hawlitschek, G., Schneider, H., Schmidt, B., Tropschug, M., Hartl, F.-U., & Neupert. (1988). Mitochondrial protein import: Identification of processing peptidase and of PEP, a processing enhancing protein. Cell 53, 785–806.

Hemmingsen, S.M., Woolford, C., van der Vies, S.M., Tilly, K., Dennis, D.T., Georgopoulos, C.P., Hendrix, R.W., & Ellis, J.R. (1988). Homologous plant and bacterial proteins chaperone oligomeric protein assembly. Nature 333, 330–334.

Henning, B., Koehler, H., & Neupert, W. (1983). Receptor sites involved in posttranslational transport of apocytochrome *c* into mitochondria: Specificity, affinity and number of sites. Proc. Natl. Acad. Sci. USA 80, 4963–4967.

Henry, R., Kapazoglou, A., McCaffery, M., & Cline, K. (1994). Differences between lumen targeting domains of chloroplast transit peptides determine pathway specificity for thylakoid transport. J. Biol. Chem. 269, 10189–10192.

Hines, V., Brandt, A., Griffiths, G., Horstmann, H., Brutsch, H., & Schatz, G. (1990). Protein import into yeast mitochondria is accelerated by the outer membrane protein MAS70. EMBO J. 9, 3191–3200.

Hinz, G. & Flügge, U.-I. (1988). Phosphorylation of a 51-kDa envelope membrane polypeptide involved in protein translocation into chloroplasts. Eur. J. Biochem. 175, 649–659.

Hoyt, D.W., Cyr, D.M., Gierasch, L.M., & Douglas, M.G. (1991). Interaction of peptides corresponding to mitochondrial presequences with membranes. J. Biol. Chem. 266, 21693–21699.

Isaya, G., Kalousek, F., Fenton, W.A., & Rosenberg, L.E. (1991). Cleavage of precursors by the mitochondrial processing peptidase requires a compatible mature protein or an intermediate octapeptide. J. Cell Biol. 113, 65–76.

James, H.E., Bartling, D., Musgrove, J.E., Kirwin, P.M., & Herrmann, R.R., C. (1989). Transport of proteins into chloroplasts: Import and maturation of precursors to the 33-, 23-, & 16-kDa proteins of the photosynthetic oxygen-evolving complex. J. Biol. Chem. 264, 19573–19576.

Jensen, R.E. & Yaffe, M.P. (1988). Import of proteins into yeast mitochondria: The nuclear MAS2 gene encodes a component of the processing protease that is homologous to the MAS1 encoded subunit. EMBO J. 7, 3863–3871.

Jordi, W., Hergersberg, C., & de Kruiff, B. (1992). Bilayer-penetrating properties enable apocytochrome *c* to follow a special import pathway into mitochondria. Eur. J. Biochem. 204, 841–846.

Kaderbhai, M.A., Pickering, T., Austen, B.M., & Kaderbhai, N. (1988). A photoactivatable synthetic transit peptide labels 30-kDa and 52-kDa polypeptides of the chloroplast inner envelope membrane. FEBS Lett. 232, 313–316.

Kang, P.-J., Ostermann, J., Shilling, J., Neupert, W., Craig, E.A., & Pfanner, N. (1990). Requirement for hsp 70 in the mitochondrial matrix for translocation and folding of precursor proteins. Nature 348, 137–143.

Karlin-Neumann, G.A. & Tobin, E.M. (1986). Transit peptides of nuclear-encoded chloroplast proteins share a common amino acid framework. EMBO J. 5, 9–13.

Keegstra, K. (1989a). A new hypothesis for the mechanism of protein translocation into chloroplasts.

Keegstra, K. (1989b). Transport and routing of proteins into chloroplasts. Cell. 56, 247–253.

Keegstra, K. & Bauerle, C. (1988). Targeting of proteins into chloroplasts. Bioessays 9, 15–19.

Keegstra, K., Olsen, L.J., & Theg, S.M. (1989). Chloroplastic precursors and their transport across the envelope membranes. Annu. Rev. Plant Physiol. 40, 471–501.

Kiebler, M., Pfaller, R., Söllner, T., Griffiths, G., Horstmann, H., Pfanner, N., & Neupert, W. (1990a). Identification of a mitochondrial receptor complex required for recognition and membrane insertion of precursor proteins. Nature 348, 610–616.

Killian, J.A., de Jong, A.M.P., Bijvelt, J., Verkleij, A.J., & Kruijff, B. (1990). Induction of non-bilayer lipid structures by functional signal peptides. EMBO J. 9, 815–819.

Kirwin, P.M., Elderfield, P.D., & Robinson, C. (1987). Transport of proteins into chloroplasts: Partial purification of a thylakoidal processing peptidase involved in plastcyanin biogenesis. J. Biol. Chem. 262, 16386–16390.

Kirwin, P.M., Meadows, J.W., Shackleton, J.B., Musgrove, J.E., Elderfield, P.D., Mould, R., Hay, N.A., & Robinson, C. (1989). ATP-dependent import of a lumenal protein by isolated thylakoid vesicles. EMBO J. 8, 2251–2255.

Ko, K., Bornemisza, O., Kourtz, L., Ko, Z., Plaxton, W.C., & Cashmore, A.R. (1992). Isolation and characterization of a cDNA clone encoding a cognate 70-kDa heat-shock protein of the chloroplast envelope. J. Biol. Chem. 267, 2986–2993.

Kohorn, B. & Tobin, E. (1989). A hydrophobic, carboxy-proximal region of a light-harvesting chlorophyll *a/b* protein. Plant Cell 1, 159–166.

Lamppa, G.K. (1988). The chlorophyll *a/b* binding protein inserts into the thylakoids independent of its cognate transit peptide. J. Biol. Chem. 263, 14996–14999.

Lamppa, G.K. & Abad, M.S. (1987). Processing of a wheat light-harvesting chlorophyll *a/b* protein precursor by a soluble enzyme from higher plant chloroplasts. J. Cell Biol. 105, 2641–2648.

Li, H.-m., Moore, T., & Keegstra, K. (1991). Targeting of proteins to the outer envelope membrane uses a different pathway than transport into chloroplasts. Plant Cell 3, 709–717.

Li, H.-m., Sullivan, T., & Keegstra, K. (1992). Submitted for publication.

Lill, R., Cunningham, K., Brundage, L.A., Ito, K., Oliver, D., & Wickner, W. (1989). SecA protein hydrolyzes ATP and is an essential component of the protein translocation ATPase of *E. coli*. EMBO J. 8, 961–966.

Liu, G., Topping, T.B., Cover, W.H., & Randall, L.L. (1988). Retardation of folding as a possible means of suppression of a mutation in the leader sequence of an exported protein. J. Biol. Chem. 263, 14790–14793.

Liu, G., Topping, T.B., & Randall, L.L. (1989). Physiological role during export for the retardation of folding by the leader peptide of maltose-binding protein. Proc. Natl. Acad. Sci. USA 86, 9213–9217.

Lubben, T.H., Donaldson, G.K., Viitanen, P.V., & Gatenby, A.A. (1989). Several proteins imported into chloroplasts form stable complexes with the GroEL-related chloroplast molecular chaperone. Plant Cell 1, 1223–1230.

Madueño, F., Bradshaw, S.A., & Gray, J.C. (1994). The thylakoid-targeting domain of the chloroplast Rieske Fe-S protein is located in the N-terminal hydrophobic region of the mature protein. J. Biol. Chem. 269, 17458–17463.

Madueño, F., Napier, J.A., & Gray, J. C. (1993). Newly imported Rieske iron-sulfer protein associates with both cpn60 and hsp70 in the chloroplast stroma. Plant Cell 5, 1865–1876.

Manning-Krieg, U.C., Scherer, P.E., & Schatz, G. (1991). Sequential action of mitochondrial chaperones in protein import into the matrix. EMBO J. 10, 3273–3280.

Marcote, M.J., Gonzalez-Bosch, C., Miralles, V.J., Hernandez-Yago, J., & Grisolia, S. (1989). Polyamines are sufficient to drive the transport of ornithine carbamoyltransferase into rat liver mitochondria: Possible effect on the mitochondrial membranes. Biochem. Biophys. Res. Comm. 158, 287–293.

Marshall, J.S., DeRocher, A.E., Keegstra, K., & Vierling, E. (1989). Identification of heat-shock protein hsp70 homologues in chloroplasts. Proc. Natl. Acad. Sci. USA 87, 374–378.

McKnight, C.J., Stradley, S.J., Jones, J.D., & Gierasch, L.M. (1991). Conformational and membrane-binding properties of a signal sequence are largely unaltered by its adjacent mature region. J. Biol. Chem. 88, 5799–5803.

McMullin, T.W. & Hallberg, R.L. (1988). A highly evolutionarily conserved mitochondrial protein is structurally related to the protein encoded by *E. coli* groEL gene. Mol. Cell Biol. 8, 371–380.

Michl, D., Robinson, C., Shackleton, J. B., Herrmann, R.G., & Klösgen, R.B. (1994). Targeting of proteins to the thylakoids by bipartite presequences-CF_0II is imported by a novel, 3rd pathway. EMBO J. 13, 1310–1317.

Murakami, K. & Mori, M. (1990). Purified presequence binding factor (PBF) forms an import-competent complex with a purified mitochondrial precursor protein. EMBO J. 9, 3201–3208.

Murakami, K., Tokunaga, F., Iwanaga, S., & Mori, M. (1990). Presequence does not prevent folding of a purified mitochondrial precursor protein and is essential for association with a reticulocyte cytosolic factor(s). J. Biochem. 108, 207–214.

Musgrove, J.E., Elderfield, P.D., & Robinson, C. (1989). Endopeptidases in the stroma and thylakoids of pea chloroplasts. Plant Physiol. 90, 1616–1621.

Neupert, W., Hartl, F.-U., Craig, E.A., & Pfanner, N. (1990). How do polypeptides cross the mitochondrial membranes? Cell 63, 447–450.

Olsen, L., Theg, S.M., Selman, B.R., & Keegstra, K. (1989). ATP is required for the binding of precursor proteins to chloroplasts.. J. Biol. Chem. 264, 6724–6729.

Olsen, L.J. & Keegstra, K. (1992). The binding of precursor proteins to chloroplasts requires nucleoside triphosphates in the intermembrane space. J. Biol. Chem. 267, 433–439.

Ono, H. & Tuboi, S. (1988). The cytosolic factor required for import of precursors of mitochondrial proteins into mitochondria. J. Biol. Chem. 263, 3188–3193.

Ostermann, J., Voos, W., Kang, P.J., Craig, E.A., Neupert, W., & Pfanner, N. (1990). Precursor proteins in transit through mitochondrial contact sites interact with hsp70 in the matrix. FEBS Lett. 277, 281–284.

Pain, D. & Blobel, G. (1987). Protein import into chloroplasts requires a chloroplast ATPase. Proc. Natl. Acad. Sci. USA 84, 3288–3292.

Pain, D., Kanwar, Y.S., & Blobel, G. (1988). Identification of a receptor for protein import into chloroplasts and its localization to envelope contact zones. Nature 331, 232–236.

Pain, D., Murakami, H., & Blobel, G. (1990). Identification of a receptor for protein import into mitochondria. Nature 347, 444–449.

Payan, L.A. & Cline, K. (1991). A stromal protein factor maintains the solubility and insertion competence of an imported membrane protein. J. Cell Biol. 112, 603–613.

Perry, S.E., Buvinger, W.E., Bennett, J., & Keegstra, K. (1991). Synthetic analogues of a transit peptide inhibit binding or translocation of chloroplastic precursor proteins. J. Biol. Chem. 266, 11882–11889.

Perry, S.E. & Keegstra, K. (1994). Envelope membrane proteins that interact with chloroplastic precursor proteins. Plant Cell 6, 93–105.

Pfanner, N., Rassow, J., Guiard, B., Söllner, T., Hartl, F.U., & Neupert, W. (1990). Energy requirements for unfolding and membrane translocation of precursor proteins during import into mitochondria. J. Biol. Chem. 265, 16324–16329.

Phelps, A., Schobert, C.T., & Wohlrab, H. (1991). Cloning and characterization of the mitochondrial phosphate transport protein gene from the yeast, *Saccharomyces cerevisiae*. Biochem. 30, 248–252.

Pilon, M., de Boer, A.D., Knols, S.L., Koppelmann, M., van der Graaf, R., deKrujiff, B., & Weisbeek, P.J. (1990a). Expression in *E. coli* and purification of a translocation competent precursor of the chloroplast protein ferrodoxin. J. Biol. Chem. 265, 3358–3361.

Pilon, M., Douwe de Boer, A., Luc Knols, S., Koppelman, M., van der Graaf, R. M., de Kruiff, B., & Weisbeck, P. (1990b). Expression in *E. coli* and purification of a translocation-competent precursor of the chloroplast protein ferredoxin. J. Biol. Chem. 265, 3358–3361.

Pollock, R.A., Hartl, F.-U., Cheng, J., Ostermann, A., Horwich, A., & Neupert, W. (1988). The processing protease of yeast mitochondria: The two cooperating components MPP and PEP are structurally related. EMBO. 7, 3493–3500.

Randall, L.L. & Hardy, S.J.S. (1986). Correlation of competence for export with lack of tertiary structure of the mature species: A study *in vivo* of maltose-binding protein in *E. coli*. Cell. 46, 921–928.

Randall, L.L., & Hardy, S.J.S. (1989). Unity in function in the absence of consensus in sequence: Role of leader peptides in export. Science 243, 1156–1159.

Randall, L.L., Toppings, T.B., & Hardy, S.J.S. (1990). No specific recognition of leader peptide by SecB, a chaperone involved in protein export. Science 248, 860–863.

Rassow, J., Guiard, B., Wienhaus, U., Herzog, V., Hartl, F.U., & Neupert, W. (1989). Translocation arrest by reversible folding of a precursor protein imported into mitochondria. A means to quantitate translocation sites. J. Cell Biol. 109, 1421–1428.

Rassow, J., Hartl, F.-U., Guiard, B., Pfanner, N., & Neupert, W. (1990). Polypeptides traverse the mitochondrial envelope in an extended state. FEBS Lett. 275, 190–194.

Reitveld, A., & de Kruijff, B. (1986). Phospholipids as a possible instrument for translocation of nascent proteins across biological membranes. Biosci. Rep. 6, 775–782.

Robinson, C. & Ellis, R.J. (1984a). Transport of proteins into chloroplast: Partial purification of a chloroplast protease involved in the processing of imported precursor polypeptide. Eur. J. Biochem. 142, 337–342.

Robinson, C. & Ellis, R.J. (1984b). Transport of proteins into chloroplasts: The precursor of ribulose bisphosphate carboxylase small subunit is processed in two steps. Eur. J. Biochem. 142, 343–346.

Robinson, C., Cai, D., Hulford, A., Brock, I.W., Michl, D., Hazell, L., Schmidt, I., Herrmann, R.G., & Klösgen, R.B. (1994). The presequence of a chimeric construct dictates which of two mechanisms are utilized for translocation across the thylakoid membrane: evidence for the existence of two distinct translocation systems. EMBO J. 13, 279–285.

Roise, D., Theiler, F., Horwath, S.J., Tomich, J.M., Richards, J.H., Allison, D.S., & Schatz, G. (1988). Amphilicity is essential for mitochondrial presequence function. EMBO J. 7, 649–653.

Rojo, M., Hovius, R., Demel, R.A., Nicolay, K., & Walliman, T. (1991). Mitochondrial creatine kinase mediates contact formation between mitochondrial formation. J. Biol. Chem. 266, 20290–20295.

Salomon, M., Fischer, K., Flügge, U.-I., & Soll, J. (1990). Sequence analysis and protein import studies of an outer envelope polypeptide. Proc. Natl. Acad. Sci. USA 87, 5778–5782.

Scherer, P.E., Krieg, U.C., Hwang, S.T., Vestweber, D., & Schatz, G. (1990). A precursor protein partly translocated into yeast mitochondria is bound to a 70-kDa mitochondrial stress protein. EMBO J. 9, 4315–4322.

Schneider, A., Behrens, M., Scherer, P., Pratje, E., Michaelis, G., & Schatz, G. (1991). Inner membrane protease I, an enzyme mediating intramitochondrial protein sorting in yeast. EMBO J. 10, 247–254.

Schnell, D.J. & Blobel, G. (1993). Identification of intermediates in the pathway of protein import into chloroplasts and their localization to envelope contact sites. J. Cell Biol. 120, 103–115.

Schnell, D.J., Blobel, G., & Pain, D. (1991). Signal peptide analogs derived from two chloroplast precursors interact with the signal recognition system of the chloroplast envelope. J Biol Chem. 266, 3335–3342.

Schnell, D.J., Blobel, G., & Pain, G. (1990). The chloroplast import receptor is an integral membrane protein of chloroplast envelope contact sites. J. Cell Biol. 111, 1825–1838.

Schwaiger, M., Herzog, V., & Neupert, W. (1987). Characterization of translocation sites involved in the import of mitochondrial proteins. J. Cell Biol. 105, 235–246.

Shackleton, J.B. & Robinson, C. (1991). Transport of proteins into chloroplasts: The thylakoidal processing peptidase is a signal-type peptidase with stringent substrate requirements at the -3 and -1 positions. J. Biol. Chem. 266, 12152–12156.

Sheffield, W.P., Shore, G.C., & Randall, S.K. (1990). Mitochondrial precursor protein. J. Biol. Chem. 265, 11069–11076.

Simbeni, R., Pon, L., Zinser, E., Paltauf, F., & Daum, G. (1991). Mitochondrial membrane contact sites of yeast. J. Biol. Chem. 266, 10047–10049.

Skerjanc, I.S., Sheffield, W.P., Randall, S.K., Slivius, J.R., & Shore, G.C. (1990). Import of precursor proteins into mitochondria: Site of polypeptide unfolding. J. Biol. Chem. 265, 9444–9451.

Smeekens, S., Bauerle, C., Hageman, J., Keegstra, K., & Weisbeek, P. (1986). The role of the transit peptide in the routing of precursors towards different chloroplast compartments. Cell 46, 365–376.

Smeekens, S., de Groot, M., van Binsbergen, J., & Weisbeek, P. (1985). Sequence of the precursor of the chloroplast thylakoid lumen protein plastocyanin. Nature 317, 456–458.

Smeekens, S., Geerts, D., Bauerle, C., & Weisbeek, P. (1989). Essential function in chloroplast recognition of the ferredoxin transit peptide processing region. Mol. Gen. Genet. 216, 178–182.

Smith, B.J. & Yaffe, M.P. (1991). A mutation in the yeast heat shock factor gene causes temperature sensitive defects in both mitochondrial protein import and the cell cycle. Mol. Cell. Biol. 11, 2647–2655.

Soll, J. & Alefsen, H. (1993). The protein import apparatus of chloroplasts. Physiol. Plant 87, 433–440.

Soll, J. & Waegeman, K. (1992). A functionally active protein import complex from chloroplasts. Plant J. 2, 253–256.

Söllner, T., Pfanner, N., & Neupert, W. (1988). Mitochondrial protein import: Differential recognition of various transport intermediates by antibodies. FEBS Lett. 229, 25–29.

Söllner, T., Rassow, J., Wiedmann, M., Schlossmann, J., Keil, P., Neupert, W., & Pfanner, N. (1992). Mapping of the import machinery in the mitochondrial outer membrane by crosslinking of translocation intermediates. Nature. 355, 84–87.

Steger, H.F., Söllner, T., Kiebler, M., Dietmeier, K.A., Pfaller, R., Trulzsch, K. S., Tropschug, M., Neupert, W., & Pfanner, N. (1990). Import of ADP/ATP carrier into mitochondria: Two receptors act in parallel. J. Cell Biol. 111, 2353–2363.

Tamm, L.K. (1991). Membrane insertion and lateral mobility of synthetic amphilic signal peptides in lipid model membranes. Biochim. Biophys. Acta Rev. Biomem. 1071, 123–148.

Theg, S.M., Bauerle, C., Olsen, L.J., Selman, B.R., & Keegstra, K. (1989). Internal ATP is the only energy requirement for the translocation of precursor proteins across chloroplastic membranes. J. Biol. Chem. 264, 6730–6736.

Tsugeki, R. & Nishimura, M. (1993). Interaction of homologues of hsp70 and cpn60 with ferredoxin-NADP+ reductase upon its import into chloroplasts. FEBS Lett. 320, 198–202.

van't Hof, R., Demel, R.A., Keegstra, K., & de Kruijff, B. (1991). Lipid-peptide interactions between fragments of the transit peptide of ribulose-1,5-bisphosphate carboxylase/oxygenase and chloroplast membrane lipids. FEBS Lett. 291, 350–354.

Verner, K. & Lemire, B.D. (1989). Tight folding of a passenger protein can interfere with the targeting function of a mitochondrial presequence. EMBO J. 8, 1491–1495.

Vestwebber, D., Brunner, J., Baker, A., & Schatz, G. (1989). A 42-kDa outer-membrane protein is a component of the yeast mitochondrial import site. Nature 341, 205–209.

von Heijne, G. (1988). Transcending the impenetrable: How proteins come to terms with membranes. Biochim. Biophys. Acta 947, 307–333.

von Heijne, G., Hirai, T., Klösgen, R.B., Steppuhn, J., Bruce, B.D., Keegstra, K., & Herrmann, R. (1991). CHLPEP: A database of chloroplast transit peptides. Plant Mol. Biol. Rep. 9, 104–126.

von Heijne, G. & Nishikawa, K. (1991). Chloroplast transit peptides: The perfect random coil? FEBS Lett. 278, 1–3.

von Heijne, G., Steppuhn, J., & Hermann, R.G. (1989). Eur. J. Biochem. 180, 535–545.

Waegemann, K., Paulsen, H., & Soll, J. (1990). Translocation of proteins into isolated chloroplasts requires cytosolic factors to obtain import competence. FEBS Lett. 261, 89–92.

Waegmann, K. & Soll, J. (1991). Characterization of the protein import apparatus in isolated outer envelopes of chloroplasts. Plant J. 1, 149–158.

Watanabe, M. & Blobel, G. (1989). SecB functions as a cytosolic recognition factor for protein export in *E. coli.* Cell 58, 695.

Weisbeek, P.H., J. Deboer, D. Pilon, R., & Smeekens, S. (1989). Import of proteins into the chloroplast lumen. J. Cell Sci. S11, 199–223.

Wiech, H., Sagstetter, M., Muller, G., & Zimmermann, R. (1987). The ATP requiring step in the assembly of M13 procoat protein into microsomes is related to preservation of transport competence of the precursor protein. EMBO J. 1011–1016.

Willey, D.L., Fischer, K., Wachter, E., Link, T.A., & Flügge, U.I. (1991). Molecular cloning and structural analysis of the phosphate translocator from pea chloroplasts and its comparison to the spinach phosphate translocator. Planta. 183, 451–461.

Yalovsky, S., Paulsen, H., Michaeli, Chitnis, P.R., & Nechushtai, R. (1992). Involvement of a chloroplast hsp70 heat shock protein in the integration of a protein (light-harvesting complex protein precursor) into the thylakoid membrane. Proc. Natl. Acad. Sci. USA 89, 5616–5619.

Yang, M., Jensen, R.E., Jaffe, M.P., Opplinger, W., & Scatz, G. (1988). Import of proteins into yeast mitochondria: The purified matrix processing protein contains two subunits which are encoded by the nuclear *MAS1* and *MAS2* genes. EMBO. 7, 3857–3862.

Yuan. J. & Cline, K. (1994). Plastocyanin and the 33-kDa subunit of the oxygen-evolving complex are transported into thylakoids with similar requirements as predicted from pathway specificity. J. Biol. Chem. 269, 18463–18467.

Yuan, J., Henry, R., & Cline, K. (1993). Stromal factor plays an essential role in the protein integration into thylakoids that cannot be replaced by unfolding or by heat shock protein hsp70. Proc. Natl. Acad. Sci. USA 90, 8552–8556.

INDEX

iated Protozoans, *Daniel E. Gottschling and Virginia Zakian, Fred Hutchinson Cancer Research Center.*

Volume 3, 1990, 275 pp. $97.50
ISBN 1-55938-013-6

CONTENTS: The Dictyostelium Discoideum Plasma Membrane: A Model System for the Study of Actin-Membrane Interactions, *Elizabeth J. Luna, Linda J. Wuestehube, Hilary M. Ingalls and Catherine P. Chia, Worcester Foundation for Experimental Biology.* Tektins and Microtubules, *R.W. Linck, University of Minnesota Medical School.* The Elusive Organization of the Spindle and the Kinetochore Fiber: A Conceptual Retrospective, *Andrew S. Bajer, University of Oregon.* Talin: Biochemistry and Cell Biology, *Keith Burridge and Leslie Molony, University of North Carolina, Medical School.* Digital Imaging Flourescence Microscopy: Statistical Analysis of Photobleaching and Passive Cellular Uptake Processing, *Zeljko Jericevic, B. Wiese, R. Homan, J. Bryan and L.C. Smith, Baylor College of Medicine.* Clathrin Assembly Proteins and the Organization of the Coated Membrane, *James H. Keen, Temple University of Medicine.* Control of the Mast Cell Secretion, *David Lagunoff, St. Louis University Medical School.* Pattern Formation: The Differentiation of Pigment Cells from Embryonic Neural Cells, *Sally K. Frost, University of Kansas.* Experimental Analysis of Centrosome Reproduction in Echinoderm Eggs, *Greenfield Sluder, Worcester Foundation for Experimental Biology.* Multiple Pathways of Protein Secretion in Exocrine Cells, *J. David Castle, University of Virginia Medical School.*

Please note: Volumes 1-3 published as *Advances in Cell Biology* and edited by *Kenneth R. Miller, Division of Biology and Medicine, Brown University*

Volume 4, 1992, 280 pp. $97.50
ISBN 1-55938-209-0

CONTENTS: Preface, *E. Edward Bittar.* The Centromere, *Jerome B. Rattner, University of Calgary, Calgary.* The Nuclear Matrix, *Ronald Berezney, SUNY at Buffalo.* The Relation between Scaffolds and Genome Function, *Susan M. Gasser, ISREC, Epalinges, Switzerland.* Signal Transduction in the Nucleus, *Eric A. Nigg, ISREC, Epalinges, Switzerland.* The Peroxisome, *Colin Masters and Dennis Crane, Griffith University, Nathan, Australia.* The Endoplasmic Reticulum, *Gordon Koch, University of Cambridge, UK.* The Golgi Complex, *Brian Storrie, VPI and State University, Blacksburg, Virginia.* The Lysosome, *Glenn Mortimore, Pennsylvania State University, Hershey, Pennsylvania.* Lysosomal Acidity, *Donald L. Schneider and Jean Chin, National Institutes of Health, Bethesda.* The Ribosome, *Richard Brimacombe, Max Planck Institut, Berlin.* Subject Index.

J A I P R E S S

Volume 5, Molecular Immunology
1992, 243 pp. $97.50
ISBN 1-55938-517-0

Edited by **Jacques F.A.P. Miller,** *The Walter and Eliza Hall Institute of Medical Research, Royal Melbourne Hospital, Victoria, Australia*

CONTENTS: Preface, *Jacques F.A.P. Miller.* Long-Term Human Hematopoiesis in Vitro Using Cloned Stromal Cell Lines and Highly Purified Progenitor Cells, *Flavia M. Cicuttiin, Michael Martin, Darryl Maher, and Andrew W. Boyd.* Multiple Routes for Late Intrathymic Precursor to Generate $CD4^{+}CD8^{+}$Thymocytes, Patrice *Hugo and Howard T. Petrie.* Immunity Versus Tolerance: The Cell Biology of Positive and Negative Signaling of B Lymphocytes, *G.J.V. Nossal.* Self-Tolerance in the T Cell Repertoire, *Jacques F.A.P. Miller and Grant Morahan.* Coordinate and Differential Regulation of GM-CSF and IL-3 Synthesis in Murine T Lymphocytes, *Antony B. Troutt, Nikki Tsoudis, and Anne Kelso.* Host-Parasite Interactions in Leishmaniasis, *Emanuela Handman.* The Non-obese Diabetic (NOD) Mouse: A Model for the Study of the Cell Biology of the Pathogenesis of an Organ-Specific Autoimmune Disease, *T.E. Mandal.* Epigenetic Regulation of the Early Development of the Nervous System, *Perry F. Bartlett and Mark Murphy.* Subject Index.

Volume 6, Extracellular Matrix
1993, 300 pp. $97.50
ISBN 1-55938-515-4

Edited by **Hynda K. Kleinman**, *National Institutes of Health, Bethesda*

CONTENTS: Preface, *Hynda K. Kleinman, National Institutes of Health, Bethesda.* Collagen: A Family of Proteins with Many Facets, *Michael van der Rest, Ecole Normale Superieure de Lyon, France, Robert Garrone, and Daniel Herbage, Institute for Bioliogy and Chemistry of Proteins, France.* Proteoglycan Gene Families, *John R. Hassell, Thomas C. Blochberger, Jody A. Rada, Shukti Chakravarti, University of Pittsburgh Medical School, and Douglas Noonan, Istituto Nazionale er la Ricerca Sul Cancro (Ist), Italy.* Structure-Function of Thrombospondins: Regulation of Fibrinolysis and Cell Adhesion, *Deane F. Mosher, Xi Sun, Jane Sottile, University of Wisconsin, and Philip J. Hogg, The Prince of Wales Hospital, Australia.* The Extracellular Matrix, *William C. Parks, Richard A. Pierce, Katherine A. Lee, and Robert P. Mecham, Washington University Medical Center, St. Louis.* Structure and Function of Basement Membrane Components: Laminin, Nidogen, Collagen IV, and BM-40, *Monique Aumailley, Universite Claude Bernard, France.* Extracellular Matrix and Bone Morphogenetic Proteins in Cartilage and Bone Development and Repair, *Slobodan Vukicevic, Vishwas M. Paralkar, National Institutes of Health, Bethesda and A. H. Reddi, Johns Hopkins Univer-*

JAI

sity School of Medicine, Baltimore. Integrins: Structure, Function, and Biological Properties, *David A. Cheresh, The Scripps Research Institute, La Jolla.* The Role of the Extracellular Matrix in Tumor Growth, *Rafael Fridman, Molecular Oncology Inc., Maryland.* Some Aspects of Inborn and Acquired Connective Tissue Diseases: A Special Emphasis on Renal Disease, *A. Noel, University of Liege, Belgium, J. A. Bruijn, E. C. Bergijk, University of Leiden, The Netherlands, and J. M. Foidart, University of Liege, Belgium.* Subject Index.

Volume 7, Biology of the Cancer Cell
1993, 268 pp. $97.50
ISBN 1-55938-624-X

Edited by **Gloria H. Heppner**, *Director, Breast Cancer Biology Program, Michigan Cancer Center*

CONTENTS: Introduction. Cellular Genetic Alterations; Models of Breast and Colon Cancer, *S.R. Wolman, D. Visscher, Wayne State University School of Medicine.* Altered Expression of Transforming Growth Factor α and Transforming Growth Factor–β Autocrine Loops in Cancer Cells, *M.G. Brattain, Medical College of Ohio, K.M. Mulder, Pennsylvania State University, School of Medicine, S.P. Wu, G. Howell, L. Sun, Baylor College of Medicine, J.K.V. Wilson, Case Western Reserve University, B.L. Ziober, Baylor College of Medicine.* Altered Signal Transduction in Carcinogenesis, *Catherine A. O'Brian, Nancy E. Ward, Constantin G. Ioannides, Universty of Texas, M.D. Anderson Cancer Center, Houston.* The Significance of the Extracellular Matrix in Mammary Epithelial Carcinogenesis, *Calvin D. Roskelley, Ole W. Petersen, Mina J. Bissell, Lawrence Berkeley Laboratory.* Epithelial-Stromal Cell Interactions and Breast Cancer, *Sandra Z. Haslam, Laura J. Counterman, Katherine A. Nummy, Michigan State University.* The Tissue Matrix and the Regulation of Gene Expression in Cancer Cells, *Kenneth J. Pienta, Wayne State University School of Medicine, Brian C. Murphy, Robert H. Getzenberg, Donald S. Coffey, Johns Hopkins School of Medicine.* Tumor Cell Interactions in Cancer Growth and Expression of the Malignant Phenotype, *Fred R. Miller, Bonnie E. Miller, Michigan Cancer Foundation.* Effect of Class I MHC Gene Products on the Immunobiological Properties of BL6 Melanoma, *Misoon Kim, Elieser Gorelik, Pittsburgh Cancer Institute.* The Role of Angiogenesis in Tumor Progression and Metastasis, *Janusz W. Rak, Erick J. Hegmann, Robert S. Kerbel, Reichmann Research Institute.* Subject Index.

Volume 8, Organelles In Vivo
1994, 191 pp. $97.50
ISBN 1-55938-636-3

Edited by **Lan Bo Chen,** *Dana-Farber Cancer Institute, Harvard Medical School*

CONTENTS: Confocal Redox Imaging of Cells, *Barry R. Masters, Walter Reed Army Medical Center.* Calcium Channels

J

A

I

P

R

E

S

S

and Vasodilation, *Alison M. Gurney and Lucie H. Clapp, UMDS, St. Thomas Hospital, London.* Changes in DNA Supercoiling Status of Cells Treated with Antineoplastic Drugs, *William D. Wright and J.L. Roti Roti, Washington University School of Medicine.* Functional Morphology of the Golgi Region: A Lectino-Electromicroscopic Exploration, *Margit Pavelka and Adolf Ellinger, Universität Innsbruk.* Receptor Mediated Endocytosis of Plasminogen Activators, *Gouujun Bu, Phillipa A. Morton, and Alan L. Schwartz, Washington University School of Medicine.* Lipid Trafficking in Hepatocytes: Relevance to Biliary Lipid Secretion, *Kristien J.M. Zaal, Jan Willem Kok, Folkert Kuipers, and Dick Hoekstra, University of Gro;ningen, Faculty of Medicine..* Identification and Characterization of Functional Secretory Cells: Advantages of Multiparameter Flow Cytometry Kinetics, *Elizabeth R. Simons and Theresa A. Davies, Boston University.* An Outline of Neurosecretion, *Jane Somsel Rodman, Tufts University School of Medicine.*

Volume 9. Homing Mechanisms and Cellular Targeting
ISBN 1-55938-686-X Approx. $97.50

Edited by **Bruce R. Zetter,** *Harvard Medical School and Childrens Hospital*

CONTENTS: Introduction, *Bruce R. Zetter.* Directed Cell Migration in Embryonic Blood Vessel Assembly, *Thomas J. Poole, SUNY Health Science Center at Syracuse.* Leukocyte Interaction with Endothelium and Extracellular Matrix: The Role of Selectins and CD44, *Ivan Stamenkovic, Massachusetts General Hospital and Harvard Medical School.* Oligosaccharide-dependent Mechanisms of Leukocyte Adhesion, *John B. Lowe, Univesity of Michigan Medical School.* Cell Adhesion Molecules: Novel Therapeutic Targets for Chronic Inflammatory Diseases of the Central Nervous System, *Gregory N. Dietsch, Gary M. Peterman and W. Michael Gallatin.* Molecular Mechanism of Targeting of Hemopoietic Stem Cells to the Bone Marrow After Intravenous Transplantation, *Mehdi Tavassoli, University of Mississippi School of Medicine.* Tumor Cell Adhesion and Growth in Organ Preference of Tumor Metastasis, *Gareth L. Nicolson, The University of Texas M.D. Anderson Cancer Center.* Cancer Cell Chemotaxis: Mechanisms and Influence on Site-Specific Tumor Metastasis, *F. William Orr, McMaster University.* Experimental Orthotopic Models of Organ-specific Metastasis by Human Neoplasms, *Isaiah J. Fidler, The University of Texas, M.D. Anderson Cancer Center.* Inter and Intracellular Targeting of Drugs, *Smadar Cohen, Ben Gurion University of the Negev and Robert Langer, Massachusetts Institute of Technology.* Chimeric Molecules Constructed with Endogenous Substances, *Gregory T. Lautenslager and Lance L. Simpson, Jefferson Medical College.* Organ-Specific Targeting of Synthetic and Natural Drug Carriers, *S. Moein Moghimi, Lisbeth Illum and Stanley S. Davis, University of Nottingham.*

www.ingramcontent.com/pod-product-compliance
Ingram Content Group UK Ltd.
Pitfield, Milton Keynes, MK11 3LW, UK
UKHW021439280726
14060UKWH00001BA/150